14.99

D1795098

SOUTH TYNESIDE LIBRARIES
WITHDRAWN FROM STOCK
Date 20/6/01   Price 80p

George + Son Ltd.

PHILIP'S

GEOGRAPHICAL DIGEST

1992-93

HEINEMANN-PHILIP
ATLASES

Published by
Heinemann Educational
a division of Heinemann Educational Books Limited
Halley Court
Jordan Hill
Oxford OX2 8EJ

OXFORD LONDON EDINBURGH MADRID
ATHENS BOLOGNA PARIS MELBOURNE
SYDNEY AUCKLAND SINGAPORE TOKYO
IBADAN NAIROBI HARARE GABORONE
PORTSMOUTH NH (USA)

in association with
George Philip Limited
59 Grosvenor Street
London W1X 9DA

© 1992 George Philip Limited

First published 1992

All rights reserved. Apart from any fair dealing for the purpose of private study,
research, criticism or review, as permitted under the Copyright Designs and Patents
Act, 1988, no part of this publication may be reproduced, stored in a retrieval
system, or transmitted in any form or by any means, electronic, electrical, chemical,
mechanical, optical, photocopying, recording or otherwise, without prior written
permission. All enquiries should be addressed to the Publishers.

British Library Cataloguing in Publication Data
A catalogue record for this book is available from the British Library

ISBN 0 435 34956 2

Compiled and edited by
Caroline Rayner, Ray Smith, Clara Willett, Patricia Willett and B.M. Willett

Designed and typeset by eMC Design, Bromham, Bedford

B20 534 217 9

SOUTH TYNESIDE LIBRARIES

584 31

910·7

OVERSIZE
SECTION

LAST
COPY

Contents

Introduction

The Geographical Digest is made of two parts: the statistical sections and the text part listing the more important geographical changes of the last two years. The statistical part consists of world lists for such topics as various aspects of demography and agriculture and over a hundred tables showing the production of many agricultural and metal products. Some population and trade data for the UK is at the end of the book; this includes totals from the 1991 Census and a brief summary of the trade and balance of payments for the UK. The world lists include as many countries as possible. The production lists contain countries that at some point in the latest or historic data produced over one thousandth part of the world total. As far as possible a world total is given, followed by the continental totals and then the countries in alphabetical order. The USSR is not included in the continental totals.

At the time of going to press many changes are coming about in the USSR and Yugoslavia. Little or no data exists for any of the parts of Russia which are becoming independent and no name has been selected to replace USSR. Therefore the data appears under the title of the USSR. A number of countries have adopted new names but for some of them the traditional English conventional name has been used: Brunei (Brunei Darussalam), Burma (Myanmar), Ivory Coast (Côte d'Ivoire), Surinam (Suriname). Wherever it has been possible or relevant, the statistics for the united Germany and the Yemen have been given.

Many statistics for Belgium include those of Luxembourg. In 1950 figures, Bangladesh is included within the totals for Pakistan.

The latest figures available have been given, often for 1989 and 1990. The 1990 figure is often provisional and all the world totals should be used cautiously and should be used to indicate a year-on-year trend.

We have used the measure and spelled it 'tonnes' to indicate that the statistics are recorded in metric tons.

In the production statistics the top-ten are indicated with the percentage of world total formed by each.

The contents list shows the first page for each section. The production statistics are arranged alphabetically within the broad groupings of agriculture and minerals.

We thank our readers who have written with comments and suggesting changes. We would welcome suggestions for future amendments to presentation and content.

The following are the sources that have been used in the compilation of the Digest.

United Nations: Demographic Yearbook, Energy Yearbook, Monthly Bulletin of Statistics, Population and Vital Statistics Report, Trade Yearbook, World Population Prospects.

Food and Agriculture Organization: Fertilizer Yearbook, Yearbook of Fishery Statistics, Yearbook of Forest Products, Production Yearbook.

World Bank: Development Report, World Tables.

Metallgesellschaft Aktiengesellschaft: Metal Statistics.

Office of Population, Censuses and Surveys: Census 1991, OPCS Monitor.

Department of Trade and Industry: Overseas Trade Statistics of the UK, December 1990.

Central Statistical Office: UK Balance of Payments, 1990.

Country by country summary

■ Introduction

It is the aim of these statistics to present for the largest countries a picture of their character and position in the world in such a way that comparisons between countries may be made and a wide variety of basic questions answered.

The chosen items of information are the most important within the general categories of area and demography, wealth and natural resources, agricultural and industrial production and trade. The arrangement of the columns corresponds to these categories, the first column being general.

In case some of the terms are unfamiliar, explanations are given below in the appropriate part of their notes.

■ Table arrangement

Country	Area and population	Production	Manufactures	Trade
COUNTRY NAME		1. Gross National Product (GNP) 1989, average annual rate of change 1980-89; GNP per capita 1989, average annual rate of change per capita, 1980-89. Origins of national wealth	1. Production/consumption of energy, million tonnes of coal-equivalent, 1989. Installed capacity for production of electricity, million kWh. (Percentage hydroelectric, nuclear and geothermal) 1989	Exports ·· Imports ·· Total $ million 1990 ·· Total $ million 1990 Leading commodities ·· Leading and countries ·· commodities and ·· countries
1. Form of government 2. Language 3. Currency 4. Exchange rates (Summer 1991)	1. Area 2. Population (1990) 3. Population density 4. Birth rate, Death rate, Average annual increase (estimates for the period 1990-95) 5. Infant mortality (estimates 1990-95), Median age 1990 6. Life expectancy (estimates 1990-95) 7. Urban population (1990) 8. Capital city 9. Land use (1989)	2. Agricultural production, thousand tonnes 3. Livestock, thousand head 4. Fish caught 1989, thousand tonnes 5. Timber cut 1988, thousand cubic metres 6. Minerals mined, thousand tonnes, except Coal - million tonnes Cobalt - tonnes Diamonds - thousand carats Gas - thousand terajoules Gold - tonnes Iron ore - million tonnes Manganese - tonnes Mercury - tonnes Molybdenum - tonnes Oil - million tonnes Silver - tonnes Tungsten - tonnes Uranium - tonnes Vanadium - tonnes	2. Manufactures - thousand tonnes 2a. Agricultural based (1989-90) 2b. Mineral based (1989-90) 3. Communications Telephones per 100 people, cars in use, thousands (1986-89) Railways, passenger-kilometres, tonne-kilometres (1988-90) Air, passenger-kilometres, tonne kilometres (1988-90) Sea cargo, loaded and unloaded, thousand tonnes (1988-90)	Export and import value of services Aid given or received, $ million (average aid received per person) 1989

■ General notes

As far as possible the figures refer to 1989-90.

Column one

The exchange rates for the sterling and the US dollar are shown. The CFA franc is the unit of currency used throughout the African countries associated with France. (CFA = Communauté financière africaine)

Column two

The area figure is for the total area of the country and includes the area of water bodies. The density figure is calculated using only the land area. The birth rate and the death rate are the number of births and deaths per thousand of the population. The annual change is expressed as an estimated average annual figure for the period 1990-95. This figure of course includes the natural net balance between births and deaths and migration. Infant mortality is the number of deaths under one year per thousand live births. The median age is the 'middle age' of the population, a figure which divides the population exactly in half, the one half older and the other younger than this age. Life expectancy is the age to which children born today could expect to live assuming the levels of mortality operating now. The land-use percentages are for the principal categories only.

Column three

The Gross National Product is a measure of the total production of goods and services and includes the balance of the import and export of services. For socialist countries the nearest figure to the GNP is the Net Material Product and this has been used. The second figure is for the GNP divided by the population. The average annual rate of change is given for both measures for the period 1980-89. The origin of the GDP is divided into three categories, agriculture, industry and services. Manufacturing is shown as a subdivision of industry. Services also includes items which cannot be included in the other two categories. Timber refers to the forest output of wood in its natural state as felled.

Column four

The production and consumption of energy has been converted to one measure - the heat energy obtained from burning one tonne of coal, the 'coal-equivalent'. The production figure is based on the home production of coal, lignite, crude oil, natural gas and hydro-, nuclear and geothermal electricity. Imported energy sources are included in the next figure, that of total energy consumption. Petrol is an abbreviation for all the refined oil products.

Column five

Exports are fob (free on board), that is a valuation of the cost of the goods plus insurance and freight to the border of the exporting country.

Imports are cif (cost, insurance, freight), a valuation of the cost of the goods plus insurance and freight to the border of the importing country.

The services import and export figures are the total for the country of the import and export of merchandise and services minus the goods part. The remainder, the services, are for such items as profit on investment, or the payment of the same, wages from workers working abroad, insurance and financial services, payment for transport, travel and tourism, diplomatic and military expenditure and aid given or received. Aid includes assistance from international, country and private donations.

Abbreviations

... data not available
– small or negligible quantity

c – carat
geo – geothermal
hl – hectolitre
hydr – hydroelectric
kg – kilogram
km – kilometre
kWh – kilowatt-hour
m – metre
nucl – nuclear
t – tonne (the metric ton)

Country	Area and population	Production	Manufactures	Trade

AFGHANISTAN

1. Republic
2. Pashto Persian
3. Afghani
4. $1 = 60.06
   £1 = 99.26

1. Area 652,090 sq km
2. Population 16,557,000
3. Density 25 per sq km
4. BR 52; DR 22; AI 0.3%
5. Inf. Mort. 162; Med.19
6. Life Exp. M 43, F 44
7. Urban population 22%
8. Kabul
9. Arable 12%; Past.46%; For.3%

1. GNP ...; $500
2. Cotton lint 20 Cotton seed 40 Maize 750 Wheat 1,925
3. Cattle 1,600 Goats 2,200 Pigs 4,400 Sheep 13,000
5. Timber 5,901
6. Natural gas 115 Phosphates 1,102

1. Energy 4.18/3.64; 489 kWh(59% hydr)
2a. Butter 10 Cheese 10 Meat 220
3. Telephones 0.2; Cars ...
   Air: 175 pass-km; 8 ton-km

Exports: $238
   Natural gas Fruit
   Exports to: USSR
   Pakistan India UK
   Germany

Imports: $798
   Wheat Veg. oil
   Sugar
   Imports from:
   USSR Japan

Aid received: $95

ALBANIA

1. Republic
2. Albanian
3. Lek
4. £1 = 6.22
   £1 = 10.28

1. Area 28,750 sq km
2. Population 3,245,000
3. Density 119 per sq km
4. BR 22; DR 6; AI 2.0%
5. Inf. Mort. 32; Med.24
6. Life Exp. M 70, F 75
7. Urban population 35%
8. Tirana
9. Arable 26%; Past.15%; For.38%

1. GNP $ 2,500mill (...); $800 (...)
2. Maize 280 Tobacco 20 Wheat 633
3. Goats 1,200 Sheep 1,560
4. Fish 15
5. Timber 2,330
6. Chromium 700 Coal 2.3 Cobalt 600 Copper 17 Natural gas 16 Oil 3 Nickel 8

1. Energy 6.31/4.07
2a. Cheese 14 Meat 79 Olive oil 4
2b. Cement 1,000 Copper 15 Nickel 5

Exports: $500

Imports: ...

Trade principally with socialist countries and India.

ALGERIA

1. Republic
2. Arabic
3. Dinar
4. $1 = 18.05
   £1 = 29.83

1. Area 2,381,740 sq km
2. Population 24,960,000
3. Density 10 per sq km
4. BR 35; DR 7; AI 2.9%
5. Inf. Mort. 61; Med.18
6. Life Exp. M 65, F 67
7. Urban population 45%
8. Algiers
9. Arable 3%; Past.13%; For.2%

1. GNP $53,116 mill (3.1%); $2,170 (0%)
   Agric. 16%, Ind. 44% (Man. 14%), Serv. 40%
2. Barley 700 Dates 212 Olives 179 Potatoes 1,107 Tomatoes 537 Wheat 750
3. Cattle 1,410 Goats 3,650 Sheep 12,550
4. Fish 107
5. Timber 2,065
6. Iron ore 2.9 Mercury 700 Natural gas 1,575 Oil 54.8 Phosphates 1,102 Salt 4,300

1. Energy 132.01/22.7; 4,136 kWh (7% hydr)
2a. Meat 239 Olive oil 10 Wine 100
2b. Cement 8,000 Iron 1,500 Petrol 37.9 Radios 160 TV 318
3. Telephones 3.8; Cars 629
   Rail: 1,972 pass-km; 2,937 ton-km
   Air: 3,303 pass-km; 13 ton-km

Exports: $8,164
   Exports to: France
   USA Italy Netherlands
   Oil Petroleum Natural
   gas Wine

Imports: $7,396
   Machinery
   Transport equip.
   Food
   Imports from:
   France USA Italy
   Netherlands

Services: Exports $645, Imports $3,706
Aid received: $153 (6)

ANGOLA

1. Republic
2. Portuguese
3. Kwanza
4. $1 = 59.09
   £1 = 97.65

1. Area 1,246,700 sq km
2. Population 10,020,000
3. Density 8 per sq km
4. BR 47; DR 19; AI 2.6%
5. Inf. Mort. 127; Med.18
6. Life Exp. M 45, F 48
7. Urban population 28%
8. Luanda
9. Arable 3%; Past.23%; For.43%

1. GNP $6,010; $620
2. Cassava 1,920 Coffee 5 Maize 180 Millet 63 Palm oil 40 Sugar cane 330
3. Cattle 3,100 Goats 985
4. Fish 102
5. Timber 5,258
6. Diamonds 1,300 Oil 23.6

1. Energy 32.73/0.87; 617 kWh(67% hydr)
2a. Meat 90 Sugar 29
3. Telephones 0.4; Cars ...
   Rail: 326 pass-km; 1,720 ton-km

Exports: $2,190
   Exports to: USA Neths
   Antilles Bahamas UK
   Minerals Oil Precious
   stones

Imports: $451
   Imports from:
   Portugal France
   USA Neths Brazil

Aid received: $140 (14)

ARGENTINA

1. Republic
2. Spanish
3. Austral
4. $1 = 9906
   £1 = 16,370.66

1. Area 2,766,890 sq km
2. Population 32,322,000
3. Density 12 per sq km
4. BR 20; DR 9; AI 1.4%
5. Inf. Mort. 29; Med.28
6. Life Exp. M 68, F 75
7. Urban population 86%
8. Buenos Aires
9. Arable 13%; Past.52%; For.22%

1. GNP $68,780 mill (-0.3%); $2,160 (-1.6%)
   Agric. 14%, Ind. 33% (Man. 35%), Serv. 53%
2. Grapes 2,950 Linseed 509 Maize 5,049 Oranges 750 Soybeans 11,000 Wheat 11,000 Wool 93
3. Cattle 50,581 Goats 3,300 Sheep 28,571
4. Fish 491
5. Timber 10,819
6. Lead 27 Natural gas 785 Oil 23.9 Silver 83 Uranium 100 Zinc 43

1. Energy 64.99/61.86; 16,600 kWh (40% hydr)
2a. Beer 5.9 Butter 37 Cheese 296 Meat 3,325 Olive oil 10 Paper 1 Sugar 915 Wine 2,000
2b. Aluminium 169 Cars 114 Cement 6,000 Commercial veh. 19 Iron 2,100 Lead 24.5 Petrol 20 Ships 10 Steel 4,000 TV 695
3. Telephones 11.6; Cars 3,856
   Rail: 12,480 pass-km; 7,920 ton-km
   Air: 8,652 pass-km; 198 ton-km
   Sea: goods loaded 25,596; unloaded 7,440

Exports: $9,579
   Exports to: USA Neths
   USSR Brazil Germany
   Agricultural prods
   Textiles

Imports: $4,204
   Machinery
   Chemicals
   Minerals
   Imports from: USA
   Brazil Germany
   Japan Italy

Services: Exports $2,483, Imports $9,492
Aid received: $215 (7)

AUSTRALIA

1. Commonwealth
2. English
3. Dollar
4. $1 = 1.29
   £1 = 2.13

1. Area 7,686,850 sq km
2. Population 16,873,000
3. Density 2.2 per sq km
4. BR 14; DR 8; AI 1.4%
5. Inf. Mort. 7; Med.32
6. Life Exp. M 74, F 80
7. Urban population 86%
8. Canberra
9. Arable 6%; Past.56%; For.14%

1. GNP $242,131 mill (3.2%); $14,440 (1.8)
   Agric. 4%, Ind. 32%(Man. 15%), Serv. 64%
2. Barley 3,550 Cotton 295 Grapes 782 Oats 1,615 Oranges 460 Potatoes 1,107 Sugar cane 26,226 Wheat 15,700 Wool 711
3. Cattle 22,602 Pigs 2,610 Sheep 167,781
4. Fish 202
5. Timber 20,677
6. Bauxite 38,583 Coal 210 Cobalt 2,000 Copper 296 Diamonds 36,000 Gold 241 Iron ore 70.3 Lead 495 Natural gas 702 Nickel 65 Oil 26.8 Salt 74,000 Silver 1,075 Tungsten 1,300 Uranium 3,520 Zinc 803

1. Energy 192.99/120.6; 36,782 kWh (20% hydr)
2a. Beer 18.6 Butter 116 Cheese 171 Meat 2,770 Paper 1.7 Sugar 3,679 Wine 494
2b. Aluminium 1,318 Cars 361 Cement 7,068 Commercial veh. 23 Copper 296 Iron 6,192 Lead 204 Manganese 2,300 Nickel 42.9 Petrol 29.2 Radios 238 Ships 25 Steel 6,684 TV 211 Zinc 296
3. Telephones 63; Cars 7,073
   Rail: ... pass-km; 40,000 ton-km
   Air: 35,596 pass-km;995 ton-km
   Sea: goods loaded 276,156; unloaded 34,956

Exports: $39,539
   Agricultural prods Oil
   and gas Metallic ores
   Textile fibres
   Machinery Transport
   equip. Timber Dairy
   prod.
   Exports to: Japan USA
   New Zealand S. Korea

Imports: $38,843
   Machinery
   Transport equip.
   Chemicals Petrol
   Agricultural prods.
   Imports from: USA
   Japan UK
   Germany

Services: Exports $12,809, Imports $26,958
Aid given: $1,020

AUSTRIA

1. Federal Republic
2. German
3. Schilling
4. $1 = 12.59
   £1 = 20.81

1. Area 83,850 sq km
2. Population 7,583,000
3. Density 93 per sq km
4. BR 12; DR 12; AI 0.1%
5. Inf. Mort. 9; Med.36
6. Life Exp. M 72, F 79
7. Urban population 58%
8. Vienna
9. Arable 18%; Past.24%; For.39%

1. GNP $131,899 mill (1.9%); $17,360 (1.9%)
   Agric. 3%, Ind. 37%(Man. 27%), Serv. 50%
2. Barley 1,353 Maize 1,400 Oats 262 Potatoes 850 Rye 375 Wheat 1,370
3. Cattle 2,562 Goats 36 Pigs 3,766
4. Fish 5
5. Timber 14,830
6. Coal 2.5 Copper 46.3 Iron ore 0.7 Magnesite 1,100 Natural gas 46 Oil 1.2 Salt 650 Tungsten 1,400

1. Energy 8.52/30.45; 16,807 kWh (64% hydr)
2a. Beer 8.6 Butter 35 Cheese 109 Meat 779 Paper 2.7 Sugar 446 Wine 350
2b. Aluminium 127 Cement 4,908 Iron 3,444 Lead 22 Petrol 7.2 Steel 4,548 TV 569 Zinc 27
3. Telephones 53; Cars 2,649
   Rail: 8,460 pass-km; 12,684 ton-km
   Air: 1,691 pass-km; 20 ton-km

Exports: $41,881
   Machinery and
   transport equip. Iron
   and steel
   Paper Timber
   Foodstuffs
   Exports to: Germany
   Italy Switzerland UK
   France

Imports: $50,017
   Machinery and
   transport equip.
   Chemicals Mineral
   Fuels Foodstuffs
   Imports from:
   Germany Italy UK
   Switzerland

Services: Exports $24,795, Imports $17,547
Aid given: $283

For detailed table headings and notes see the first page of this section

Country	Area and population	Production	Manufactures	Trade	

BANGLADESH

1. Republic
2. Bengali
3. Taka
4. $1 = 34.49
   £1 = 57.00

1. Area 144,000 sq km
2. Population 115,593,000
3. Density 888 per sq km
4. BR 41; DR 14; AI 2.7%
5. Inf. Mort. 108; Med.18
6. Life Exp. M 53, F 53
7. Urban population 14%
8. Dacca
9. Arable 69%; Past. 4%;For. 15%

1. GNP $19,913 mill (3.5%); $180 (0.7%)
   Agric. 44%, Ind. 14%(Man. 7%), Serv. 41%
2. Bananas 600 Jute 849 Linseed 44
   Potatoes 1,100 Rice 29,400 Sugar cane 6,700
3. Cattle 23,113 Goats 11,071 Sheep 1,160
4. Fish 829
5. Timber 29,368
6. Natural gas 165 Salt 400

1. Energy 5.41/7.74; 2,348 kWh(10% hydr)
2a. Butter 13 Meat 366 Sugar 119
3. Rail: 5,244 pass-km; 660 ton-km
   Air: 1,876 pass-km; 83 ton-km
   Sea: goods loaded 1,344; unloaded 7,788

Exports: $1,305
Jute and jute prods
Fish Hides and skins
Exports to: USA Italy
Japan UK

Imports: $3,524
Machinery Food
Petroleum
Imports from:
Japan USA UAE
Singapore Canada
S. Korea

Services: Exports $391, Imports $540
Aid received: $1,791 (16)

BELGIUM

1. Kingdom
2. Flemish French German
3. Franc
4. $1 = 36.88
   £1 = 60.95

1. Area 33,100 sq km
2. Population 9,845,000
3. Density 311 per sq km
4. BR 12; DR 12; AI 0%
5. Inf. Mort. 8; Med.37
6. Life Exp. M 72, F 79
7. Urban population 97%
8. Brussels
9. Arable 25%; Past. 21%; For. 21%

1. GNP $162,026 mill (1.7%); $16,390 (1.7%)
   Agric. 2%, Ind. 31%(Man. 22%), Serv. 67%
2. Barley 625 Potatoes 1,750 Sugar beet 6,200
   Wheat 1,527
3. Cattle 3,069 Pigs 6,350
4. Fish 42
5. Timber 4,018
6. Coal 2.3

1. Energy 8.45/57.22; 14,082 kWh (10% hydr; 39% nucl)
2a. Beer 14 Butter 90 Cheese 60 Meat 1,354
   Paper 0.8 Sugar 1,038
2b. Cars 1,170 Cement 6,924 Commercial veh. 76
   Copper 334 Iron 9,420 Lead 93
   Petrol 23.9 Radios 1,165 Rubber-synth. 120
   Ships 57 Steel 11,424 TV 917 Tin 6 Zinc 307
3. Telephones 48; Cars 3,453
   Rail: 6,552 pass-km; 8,352 ton-km
   Air: 5,973 pass-km; 536 ton-km
   Sea: goods loaded 48,732; unloaded 78,012

Exports: $118,295
Vehicles Chemicals
Foodstuffs
Iron and steel
Precious stones Petrol
Exports to: France
Germany Neths UK
Italy

Imports: $120,067
Vehicles and parts
Chemicals
Foodstuffs
Petroleum
Diamonds
Imports from:
Germany Neths
France UK USA

Trade includes Luxembourg
Services: Exports $71,850, Imports $67,902
Aid given: $703

BENIN

1. Republic
2. French Fon
3. CFA franc
4. $1 = 303.4
   £1 = 501.40

1. Area 112,620 sq km
2. Population 4,630,000
3. Density 43 per sq km
4. BR 49; DR 18; AI 3.0%
5. Inf. Mort. 85; Med.17
6. Life Exp. M 46, F 50
7. Urban population 42%
8. Lomé
9. Arable 17%; Past. 4%;For. 32%

1. GNP $1,753 mill (1.3%); $380 (-1.8%)
   Agric. 46%, Ind. 12% (Man. 5%), Serv. 42%
2. Cassava 1,002 Maize 460 Palm oil 40
   Yams 1,049
3. Cattle 914 Goats 1,000
4. Fish 38
5. Timber 4,845

1. Energy 0.42/0.22; 15 kWh(0% hydr)
2a. Meat 62
3. Telephones 0.5; Cars 25
   Rail: 137 pass-km; 164 ton-km
   Air: 228 pass-km; 16 ton-km

Exports: $124
Fuels Cotton Palm
products
Exports to: USA Neths
Japan France

Imports: $216
Manufactures
Textiles
Imports from:
France UK Neths
China

Services: Exports $97, Imports $152
Aid received: $247 (54)

BOLIVIA

1. Republic
2. Spanish
3. Boliviano
4. $1 = 3.57
   £1 = 5.90

1. Area 1,098,580 sq km
2. Population 7,314,000
3. Density 6.8 per sq km
4. BR 41; DR 12; AI 2.8%
5. Inf. Mort. 93; Med.18
6. Life Exp. M 54, F 58
7. Urban population 51%
8. La Paz Sucre
9. Arable 3%; Past. 25%; For. 51%

1. GNP $4,301 mill (-0.8%); $600 (-3.5%)
   Agric. 32%, Ind. 30% (Man. 13%), Serv. 38%
2. Maize 370 Potatoes 534 Soybeans 225
   Sugar cane 1,900
3. Cattle 8,073 Goats 2,450 Pigs 2,200
   Sheep 12,400
4. Fish 4
5. Timber 1,417
6. Antimony 8.5 Lead 16 Natural gas 119
   Silver 295 Tin 15.8 Tungsten 950 Zinc 75

1. Energy 5.61/2.7; 735 kWh(47% hydr)
2a. Beer 0.8 Meat 251 Sugar 175
2b. Tin 9.7
3. Telephones 3; Cars 33
   Rail: 500 pass-km; 505 ton-km
   Air: 912 pass-km; 26 ton-km

Exports: $900
Tin Zinc Antimony
Natural gas
Exports to: Argentina
USA UK Germany

Imports: $715
Consumer goods
Transport
equipment
Imports from: USA
Brazil Argentina
Japan

Services: Exports $167, Imports $581
Aid received: $232 (61)

BRAZIL

1. Federal Republic
2. Portuguese
3. Cruzeiro
4. $1 = 323.66
   £1 = 534.88

1. Area 8,511,970 sq km
2. Population 150,638,000
3. Density 18 per sq km
4. BR 26; DR 8; AI 2.2%
5. Inf. Mort. 57; Med.23
6. Life Exp. M 64, F 69
7. Urban population 77%
8. Brasilia
9. Arable 9%; Past. 20%; For. 66%

1. GNP $375,146 mill (3.1%); $2,550 (0.9%)
   Agric. 9%, Ind. 43% (Man. 31%), Serv. 48%
2. Bananas 5,552 Cassava 23,247 Coffee 1,440
   Cotton 660 Maize 21,405 Oranges 17,781
   Pineapples 720 Potatoes 2,239 Rice 7,457
   Silk 1,900 Soybeans 19,981
   Sugar cane 272,540 Tobacco 468
3. Cattle 140,000 Goats 11,500
   Pigs 33,200 Sheep 21,000
4. Fish 750
5. Timber 245,751
6. Bauxite 7,894 Coal 6.5 Copper 47
   Diamonds 500 Gold 75 Iron ore 104.5
   Magnesite 700 Natural gas 136 Nickel 13.7
   Oil 31.6 Silver 64 Tin 50.2 Tungsten 500
   Uranium 50 Zinc 106

1. Energy 79.1/117.61; 52,071 kWh (86% hydr; 1% nucl)
2a. Beer 33.9 Butter 80 Cheese 60 Meat 5,693
   Paper 4.6 Sugar 8,352
2b. Aluminium 946 Cars 268 Cement 25,848
   Commercial veh. 672 Copper 173 Iron 21,084
   Lead 86 Manganese 1,850 Radios 8,676
   Rubber-synth. 257 Ships 105 Steel 20,580
   Tin 44.4 TV 2,902 Zinc 162
3. Telephones 9; Cars 9,293
   Rail: 15,782 pass-km; 103,677 ton-km
   Air: 22,613 pass-km; 1,014 ton-km
   Sea: goods loaded 160,044; unloaded 57,876

Exports: $34,392
Transport equip.
Soybeans Coffee
Iron ore Machinery
Exports to: USA
Neths Japan Germany
Italy

Imports: $18,281
Primary prods
Capital goods Oil
and prods
Imports from: USA
Iraq Germany
Saudi Arabia
Japan

Services: Exports $2,900, Imports $18,900
Aid received: $189 (1)

BULGARIA

1. Republic
2. Bulgarian
3. Lev
4. $1 = 18.77
   £1 = 31.02

1. Area 110,910 sq km
2. Population 9,010,000
3. Density 82 per sq km
4. BR 12; DR 12; AI 0.2%
5. Inf. Mort. 14; Med.37
6. Life Exp. M 70, F 76
7. Urban population 70%
8. Sofia
9. Arable 37%; Past. 18%; For. 35%

1. GNP $20,860 mill (...); $2,320 (...)
   Agric. 11%, Ind. 59%, Serv. 29%
2. Grapes 800 Maize 1,200 Potatoes 475
   Tobacco 110 Wheat 5,225
3. Cattle 1,577 Goats 409
   Pigs 4,352 Sheep 7,988
4. Fish 117
5. Timber 4,471
6. Coal 31.5 Copper 50 Iron ore 0.3
   Lead 85 Molybdenum 200 Silver 20 Zinc 87

1. Energy 18.02/44.03; 11,103 kWh (18% hydr; 25% nucl)
2a. Beer 6.2 Butter 21 Cheese 192 Meat 814
   Paper 0.5 Sugar 82 Wine 340
2b. Cement 4,680 Copper 55.8 Iron 1,140
   Petrol 10.2 Ships 81 Steel 2,184
3. Telephones 14; Cars ...
   Rail: 7,788 pass-km; 14,124 ton-km

Exports: $13,347
Machinery Food Wine
Tobacco

Imports: $12,893
Machinery Fuels
Minerals and
Metals Chemicals
Rubber

Principal trade is with USSR and eastern Europe

BURKINA FASO

1. Republic
2. French
3. CFA franc
4. $1 = 303.4
   £1 = 501.40

1. Area 274,200 sq km
2. Population 8,996,000
3. Density 33 per sq km
4. BR 47; DR 17; AI 2.6%
5. Inf. Mort. 127; Med.18
6. Life Exp. M 48, F 51
7. Urban population 9%
8. Ouagadougou
9. Arable 13%; Past. 37%; For. 25%

1. GNP $2,716 mill (5%); $310 (2.3%)
   Agric. 35%, Ind. 26%(Man. 15%), Serv. 39%
2. Cotton lint 68 Cotton seed 110
   Groundnuts 150 Millet 597 Sorghum 917
3. Cattle 2,850 Goats 5,400 Sheep 3,050
4. Fish 7
5. Timber 8,298

1. Energy .../0.25; 59 kWh
2a. Meat 91
3. Telephones 0.2; Cars 11
   Air: 243 pass-km; 16 ton-km

Exports: $489
Cotton Livestock and
prods
Exports to: Ivory Coast
France UK Germany
Japan

Imports: $142
Transport equip.
Machinery
Petroleum
Imports from:
France
Ivory Coast USA

Services: Exports $53, Imports $237
Aid received: $284 (32)

Country	Area and population	Production	Manufactures	Trade	
BURMA (MYANMAR)		1. GNP $ 20,000mill (...); $500 (...) Agric. 51%, Ind. 10%, Serv. 39%	1. Energy .../0.53; 1,086 kWh (24%hydr)	Exports: $215 Rice Timber Metals	Imports: $191 Machinery Base
1. Republic 2. Burmese 3. Kyat 4. $1 = 6.53 £1 = 10.79	1. Area 676,550 sq km 2. Population 41,675,000 3. Density 63 per sq km 4. BR 30; DR 9; AI 1.8% 5. Inf. Mort. 59; Med.21 6. Life Exp. M 61, F 64 7. Urban population 25% 8. Rangoon 9. Arable 15%; Past .1%; For. 49%	2. Groundnuts 524 Jute 80 Rice 13,623 Tobacco 47 3. Buffaloes 2,020 Cattle 9,150 Goats 1,050 Pigs 3,000 4. Fish 705 5. Timber 21,033 6. Lead 5 Natural gas 42 Salt 200 Silver 7 Tin 0.6 Tungsten 450	2a. Butter 13 Cheese 36 Meat 323 2b. Lead 3.8 3. Telephones 0.09; Cars 60 Rail: 4,486 pass-km; 545 ton-km Sea: goods loaded 576; unloaded612	Exports to: Japan Indonesia Vietnam Aid received: $220 (5)	metals Transport equip. Imports from: Japan China Germany UK
BURUNDI		1. GNP $1,149 mill (4.5%); $220 (1.6%) Agric. 56%, Ind. 15% (Man. 10%), Serv. 29%	1. Energy 0.02/0.1;... kWh 3. Cars 11	Exports: $129 Coffee Tea	Imports: $204 Petrol Food
1. Republic 2. French Kirundi 3. Franc 4. $1 = 6.53 £1 = 10.79	1. Area 27,830 sq km 2. Population 5,472,000 3. Density 213 per sq km 4. BR 47; DR 16; AI 3.0% 5. Inf. Mort. 110; Med.17 6. Life Exp. M 48, F 51 7. Urban population 7% 8. Bujumbura 9. Arable 52%; Past .36%; For. 3%	2. Bananas 1,600 Cassava 600 Coffee 40 Groundnuts 87 Sorghum 87 Sweet potatoes 426 3. Cattle 345 Goats 770 Sheep 370 4. Fish 12 5. Timber 3,966		Exports to: Germany Finland Services: Exports $22, Imports $132 Aid received: $198 (37)	Vehicles Textiles Imports from: Belgium Iran France
CAMBODIA		1. GNP $4,000 mill (...); $450 (...) Agric. 41%, Ind. 17%, Serv. 42%	1. Energy ...; 35 kWh(29% hydr) 2a. Meat 76	Exports: ... Iron Steel Rubber	Imports: ... Machinery
1. Republic 2. Khmer 3. Riel 4. $1 = 795.64 £1 = 1,314.87	1. Area 181,040 sq km 2. Population 8,246,000 3. Density 47 per sq km 4. BR 37; DR 15; AI 2.6% 5. Inf. Mort. 116; Med.23 6. Life Exp. M 50, F 52 7. Urban population 12% 8. Phnom Penh 9. Arable 17%; Past .3%; For. 76%	2. Bananas 115 Cassava 110 Coconut 48 Rice 2,400 3. Cattle 2,100 Pigs.1,585 4. Fish 70 5. Timber 5,677	3. Telephones 0.9; Cars ...	 Aid received: $25	Transport equip. Chemicals
CAMEROON		1. GNP $11,661 mill (3.9%); $1,010 (0.7%) Agric. 27%, Ind. 27% (Man. 15%), Serv. 46%	1. Energy 11.77/2.9; 605 kWh (87% hydr) 2a. Beer 5.9 Meat 133	Exports: $924 Aluminium prods	Imports: $1,271 Transport equip.
1. Republic 2. French English 3. CFA franc 4. $1 = 303.4 £1 = 501.40	1. Area 475,440 sq km 2. Population 11,833,000 3. Density 25 per sq km 4. BR 47; DR 13; AI 3.2% 5. Inf. Mort. 86; Med.17 6. Life Exp. M 54, F 57 7. Urban population 49% 8. Yaoundé 9. Arable 15%; Past .18%; For. 53%	2. Cassava 1,530 Cocoa 109 Coffee 100 Groundnuts 143 Palm oil 108 Plantains 1,150 Rubber 38 Sugar cane 1,300 Yams 230 3. Cattle 4,600 Goats 3,300 Pigs 1,300 Sheep 3,200 4. Fish 83 5. Timber 12,574 6. Oil 8.3	2b. Aluminium 87 Petrol 1.8 3. Telephones 0.6; Cars 388 Rail: 492 pass-km; 684 ton-km Air: 610 pass-km; 37 ton-km Sea: goods loaded 1,164;unloaded 2,412	Coffee Cocoa Cotton Timber Oil Exports to: Netherlands France Italy USA Services: Exports $378, Imports $1,166 Aid received: $470 (41)	Iron and steel Medicines Textiles Imports from: France Germany Japan
CANADA		1. GNP $500,337 mill (3.6%); $19,020 (2.6%) Agric. 3%, Ind. 21% (Man 17%), Serv. 76%	1. Energy 369.51/287.06; 98,890 kWh(59% hydr; 12% nucl)	Exports: $126,995 Cars and trucks	Imports: $116,461 Machinery and
1. Commonwealth 2. English French 3. Dollar 4. $1 = 1.14 £1 = 1.88	1. Area 9,916140 sq km 2. Population 26,521,000 3. Density 2.9 per sq km 4. BR 13; DR 8; AI 1.0% 5. Inf. Mort. 7; Med.33 6. Life Exp. M 74, F 81 7. Urban population 76% 8. Ottawa 9. Arable 5%; Past .4%; For. 39%	2. Barley 13,232 Linseed 531 Maize 7,033 Oats 3,507 Potatoes 2,870 Rapeseed 3,325 Rye 939 Soybeans 1,327 Wheat 31,798 3. Cattle 4,600 Pigs 10,532 Sheep 3,200 4. Fish 1597 5. Timber 179,957 6. Asbestos 650 Coal 68.4 Cobalt 2,500 Copper 732 Gold 165 Iron ore 21.7 Lead 275 Molybdenum 12,400 Natural gas 4,088 Nickel 203 Oil 76.3 Potash 7,000 Salt 11,100 Silver 1,306 Tin 2.8 Uranium 8,700 Zinc 1,215	2a. Beer 23.5 Butter 110 Cheese 274 Meat 2,871 Paper 16.6 Sugar 135 Wine 57 2b. Aluminium 1,654 Cars 941 Cement 11,808 Commercial veh. 808 Copper 547 Iron 11,000 Lead 245 Nickel 129 Petrol 85.2 Steel 15,500 Zinc 670 3. Telephones 78; Cars 11,477 Rail: 2,220 pass-km; 263,436 ton-km Air: 40,241 pass-km; 1,007 ton-km	Machinery Chemicals Food Beverages Tobacco Metals Exports to: USA Japan UK Services: Exports $21,500, Imports $46,790 Aid given: $2,320	transport equip. Clothing Paper Textiles Iron and steel Imports from: USA Japan UK Germany
CENTRAL AFRICAN REPUBLIC		1. GNP $1,144 mill (1.2%); $390 (-1.5%) Agric. 42%, Ind. 15% (Man. 8%), Serv. 43%	1. Energy 0.009/0.14; 43 kWh(51% hydr) 2a. Meat 72	Exports: $134 Coffee Cotton Cork	Imports: $150 Cereals Machinery
1. Republic 2. French 3. CFA franc 4. $1 = 303.4 £1 = 501.40	1. Area 622,980 sq km 2. Population 3,039,000 3. Density 4.9 per sq km 4. BR 45; DR 16; AI 2.7% 5. Inf. Mort. 95; Med.18 6. Life Exp. M 48, F 53 7. Urban population 47% 8. Bangui 9. Arable 3%; Past .5%; For. 57%	2. Bananas 88 Cassava 540 Coffee 27 Cotton seed 15 Groundnuts 100 3. Cattle 2,500 Goats 11,210 4. Fish 13 5. Timber 3,449 6. Diamonds 500	3. Telephones 0.3; Cars 41 Air: 226 pass-km; 16 ton-km	Timber Diamonds Exports to: France Belgium Israel USA Services: Exports $58, Imports $168 Aid received: $189 (64)	Transport equip. Medicines Imports from: France Zaïre Japan USA
CHAD		1. GNP $1,138 mill (6.3%); $190 (3.9%) Agric. 36%, Ind. 20% (Man. 16%), Serv. 44%	1. Energy .../0.1; 31 kWh (0% hydr) 2a. Meat 67	Exports: $141 Cotton Cattle	Imports: $419 Cereals Petroleum
1. Republic 2. French 3. CFA franc 4. $1 = 303.4 £1 = 501.40	1. Area 1,284,000 sq km 2. Population 5,678,000 3. Density 4.5 per sq km 4. BR 43; DR 18; AI 2.4% 5. Inf. Mort. 122; Med.19 6. Life Exp. M 46, F 49 7. Urban population 33% 8. Ndjamena 9. Arable 3%; Past .36%; For. 10%	2. Cotton lint 55 Cotton seed 87 Groundnuts 80 Millet 172 Yams 240 3. Cattle 4,200 Goats 2,350 Sheep 2,350 4. Fish 110 5. Timber 3,834	3. Telephones 0.1; Cars 11 Air: 221 pass-km; 16 ton-km	Exports to: France Nigeria Cameroon Services: Exports $79, Imports $230 Aid received: $239 (43)	prods Chemicals Imports from: France Nigeria Netherlands

For detailed table headings and notes see the first page of this section

Country	Area and population	Production	Manufactures	Trade	

CHILE
1. Republic
2. Spanish
3. Peso
4. $1 = 350.27
£1 = 578.86
	1. Area 756,950 sq km
	2. Population 13,173,000
	3. Density 18 per sq km
	4. BR 23; DR 6; AI 1.7%
	5. Inf. Mort. 19; Med.25
	6. Life Exp. M 69, F 76
	7. Urban population 86%
	8. Santiago
	9. Arable 6%; Past .18%; For. 12%
		1. GNP $22,910 mill (2.7%); $1,790 (1.0%)
		Agric. 8%, Ind. 38% (Man. 21%), Serv. 54%
		2. Grapes 1,050 Maize 823 Oats 205
		Potatoes 829 Tobacco 14 Wheat 1,718
		Wool 20
		3. Cattle 3,700 Goats 600
		Pigs 1,450 Sheep 6,700
		4. Fish 5210
		5. Timber 16,761
		6. Coal 1.9 Copper 1,609 Gold 31.6 Iron 5.3
		Lead 1 Manganese 30 Molybdenum 17,000
		Natural gas 593 Potash 10 Silver 536 Zinc 18
			1. Energy 7.35/15.63; 4,079 kWh(56% hydr)
			2a. Beer 2.5 Butter 5 Cheese 25 Meat 449
			Olive oil 1 Paper 0.4 Sugar 448
			Wine 390
			2b. Cars 4 Cement 1,800 Commercial veh. 6
			Copper 1,071 Fertilizers 130 Iron 680
			Petrol 5.9 Steel Tyres 1,200
			3. Telephones 7; Cars 619
			Rail: 1,056 pass-km; 2,952 ton-km
			Air: 2,117 pass-km; 186 ton-km
			Sea: goods loaded 19,968; unloaded 9,444
				Exports: $8,580
				  Copper Fruit Fish
				  meal Iron ore
				Exports to: USA Japan
				  Germany Brazil UK
				  Italy
				
				Services: Exports $1,638, Imports $4,361
				Aid received: $61 (5)
					Imports: $7,272
					  Machinery
					  Minerals
					  Chemicals Food
					Imports from: USA
					  Japan Brazil
					  Germany

CHINA
1. Republic
2. Chinese Mandarin
  and others
3. Renminbi yuan
4. $1 = 5.33
£1 = 8.81
	1. Area 9,596,960 sq km
	2. Population 1,139,060,000
	3. Density 122 per sq km
	4. BR 23; DR 6; AI 1.4%
	5. Inf. Mort. 27; Med.26
	6. Life Exp. M 69, F 73
	7. Urban population 21%
	8. Beijing
	9. Arable 10%; Past .34%; For. 13%
		1. GNP $393,006 mill (9.6%); $360 (8.2%)
		Agric. 32%, Ind. 48% (Man. 34%), Serv. 20%
		2. Cotton lint 4,421 Grapes 1,050
		Groundnuts 6,100 Jute 590 Maize 82,345
		Millet 4,501 Oranges 3,332 Potatoes 28,050
		Rapeseed 6,600 Rice 188,403 Rye 1,000
		Seed cotton 11,757 Silk 42,024 Sorghum 5,770
		Soybeans 11,508 Sugar beet 10,000
		Sugar cane 61,767 Tea 521 Tobacco 2,711
		Wheat 96,004
		3. Cattle 76,980 Goats 98,332 Pigs 360,594
		Sheep 113,508
		4. Fish 10359
		5. Timber 276,061
		6. Aluminium 3,650 Antimony 29 Asbestos 150
		Coal 1,132 Cobalt 250 Copper 380
		Diamonds 1,000 Gold 80 Iron 84.7 Lead 341
		Magnesite 3,500 Manganese 2,700
		Mercury 1,042 Molybdenum 2,000
		Natural gas 600 Nickel 27.5 Oil 138
		Salt 28,600 Silver 165 Tin 33 Tungsten 20,000
		Vanadium 4,500 Zinc 620
			1. Energy 971.91/892.48; 98,00 kWh(31% hydr)
			2a. Beer 54 Butter 63 Cheese 140 Meat 27,178
			Paper 14.1 Sugar 5,634 Wine 85
			2b. Aluminium 744 Cement 204,000 Copper 470
			Fertilizers 17,774 Iron 64,548 Lead 302
			Nickel 26.3 Petrol 84.7 Radios 17,638
			Ships 453 Steel 64,656 TV 19,344
			Tin 28.3 Zinc 451
			3. Telephones 0.7; Cars ...
			Rail: 261,012 pass-km; 1,060,116 ton-km
			Air: 18,623 pass-km; 648 ton-km
			Sea: goods loaded 69,684; unloaded 67,536
				Exports: $62,089
				  Manufactures Food
				  Mineral fuels
				  Chemicals
				Exports to: Hong Kong
				  Japan USA
				
				Services: Exports $5,373, Imports $8,116
				Aid received: $2,227 (2)
					Imports: $53,369
					  Machinery and
					  transport equip.
					  Chemicals Raw
					  materials Food
					Imports from:
					  Japan Hong Kong
					  USA Germany

COLOMBIA
1. Republic
2. Spanish
3. Peso
4. $1 = 621.24
£1 = 1.026.66
	1. Area 1,138,910 sq km
	2. Population 32,978,000
	3. Density 32 per sq km
	4. BR 26; DR 6; AI 2.1%
	5. Inf. Mort. 37; Med.22
	6. Life Exp. M 66, F 72
	7. Urban population 70%
	8. Bogotá
	9. Arable 5%; Past .39%; For. 49%
		1. GNP $38,607 mill (3%); $1,190 (0.9%)
		Agric. 17%, Ind. 36% (Man. 21%), Serv. 47%
		2. Bananas 1,340 Cassava 1,396 Coffee 780
		Cotton lint 122 Maize 1,215 Palm oil 260
		Plantains 2,463 Potatoes 2,649 Rice 2,050
		Sorghum 809 Sugar cane 28,000
		3. Cattle 25,000 Goats 1,020
		Pigs 2,630 Sheep 2,650
		4. Fish 85
		5. Timber 18,163
		6. Coal 20 Gold 32.5 Magnesite 15
		Natural gas 170 Nickel 16.9 Oil 21.9 Salt 700
			1. Energy 56.54/26.45; 8,589 kWh (74% hydr)
			2a. Beer 15.4 Butter 15 Cheese 51 Meat 1,008
			Paper 0.5 Sugar 1,425
			2b. Cars 40 Cement 6,360 Commercial veh. 14
			Fertilizers 119 Nickel 16.9 TV 112 Tyres 1,899
			3. Telephones 8; Cars 842
			Rail: 156 pass-km; 360 ton-km
			Air: 4,230 pass-km; 398 ton-km
			Sea: goods loaded 17,892; unloaded 5,580
				Exports: $5,739
				  Coffee Oil Bananas
				  Cotton
				Exports to: USA
				  Germany Netherlands
				  Japan
				
				Services: Exports $1,619, Imports $4,086
				Aid received: $62 (2)
					Imports: $5,010
					  Machinery
					  Chemicals Vehicles
					Imports from: USA
					  Japan Germany
					  Brazil

CONGO
1. Republic
2. French
3. CFA franc
4. $1 = 303.4
£1 = 501.40
	1. Area 342,000 sq km
	2. Population 2,271,000
	3. Density 6.6 per sq km
	4. BR 46; DR 13; AI 3.1%
	5. Inf. Mort. 65; Med.17
	6. Life Exp. M 52, F 57
	7. Urban population 42%
	8. Brazzaville
	9. Past .29%; For. 62%
		1. GNP $2,045 mill (3.6%); $930 (0.1%)
		Agric. 14%, Ind. 35% (Man. 9%), Serv. 51%
		2. Cassava 760 Palm oil 16
		3. Cattle 70 Goats 245 Sheep 102
		4. Fish 42
		5. Timber 3,315
		6. Oil 7.6
			1. Energy 10.58/0.78; 149 kWh (81% hydr)
			2a. Meat 16 Sugar 36
			2b. Cement 50
			3. Telephones 1; Cars 26
			Rail: 400 pass-km; 449 ton-km
			Air: 259 pass-km; 17 ton-km
				Exports: $751
				  Oil Timber Diamonds
				Exports to: USA Spain
				  France Neths
				
				Services: Exports $, Imports $
				Aid received: $91 (41)
					Imports: $544
					  Machinery Iron
					  and steel
					  Transport equip.
					  Food
					Imports from:
					  France Italy USA
					  Germany

COSTA RICA
1. Republic
2. Spanish
3. Colon
4. $1 = 128
£1 = 212
	1. Area 51,100 sq km
	2. Population 3,015,000
	3. Density 59 per sq km
	4. BR 26; DR 4; AI 2.3%
	5. Inf. Mort. 17; Med.22
	6. Life Exp. M 73, F 78
	7. Urban population 54%
	8. San José
	9. Arable 10%; Past .45%; For. 32%
		1. GNP $4,898 mill (2.7%); $1,790 (0.4%)
		Agric. 18%, Ind. 27%, Serv. 55%
		2. Bananas 1,530 Coffee 172 Oranges 86
		Plantains 110 Sugar cane 2,600
		3. Cattle 1,735 Pigs 223
		4. Fish 20
		5. Timber 3,961
		6. Gold -
			1. Energy 0.41/1.46;915 kWh(80% hydr)
			3. Telephones 15; Cars 101
			Rail: 604 pass-km; 98 ton-km
				Exports: $1,394
				  Manufactures
				  Agricultural prods
				Exports to: USA
				  Germany
				
				Services: Exports $570,Imports $857
				Aid received: $224 (82)
					Imports: $2,044
					  Machinery and
					  transport equip.
					  Chemicals
					Imports from: USA
					  Venezuela

CUBA
1. Republic
2. Spanish
3. Peso
4. $1 = 0.79
£1 = 1.31
	1. Area 110,860 sq km
	2. Population 10,608,000
	3. Density 96 per sq km
	4. BR 17; DR 7; AI 0.9%
	5. Inf. Mort. 13; Med.28
	6. Life Exp. M 74, F 78
	7. Urban population 75%
	8. Havana
	9. Arable 30%; Past .25%; For. 25%
		1. GNP $16,200 mill (...); $1,500 (...%)
		Agric. 12%, Ind. 36%, Serv. 52%
		2. Cassava 305 Coffee 24 Grapefruit 280
		Oranges 520 Rice 500 Sugar cane 85,900
		Tobacco 44
		3. Cattle 4,920 Pigs 1,850
		4. Fish 231
		5. Timber 3,283
		6. Chromium 55 Cobalt 2,000 Nickel 46.5
		Salt 230
			1. Energy 1.08/16.09; 3,988 kWh(1% hydr)
			2a. Beer 3.3 Butter 10 Cheese 17 Meat 337
			Sugar 8,188
			2b. Cement 3,600 Fertilizers 142 Nickel 26.5
			Petrol 7.3 Radios 227 TV 56 Tyres 228
			3. Telephones 5; Cars 230
			Rail: 2,892 pass-km; 2,052 ton-km
			Air: 1,997 pass-km; 26 ton-km
			Sea: goods loaded 2,628; unloaded 2,760
				Exports: $5,518
				  Sugar Copper Nickel
				  Tobacco
				Exports to: USSR
				  Germany
					Imports: $7,579
					  Mineral fuels
					  Machinery
					  Transport equip.
					  Foods
					Imports from:
					  USSR Germany

For detailed table headings and notes see the first page of this section

Country	Area and population	Production	Manufactures	Trade

CYPRUS

1. Republic (split into Greek and Turkish parts)
2. Greek Turkish English
3. Pound
4. $1 = 0.48
   £1 = 0.79

1. Area 9,250 sq km
2. Population 701,000
3. Density 76 per sq km
4. BR 17; DR 8; AI 1.1%
5. Inf. Mort. 10; Med.30
6. Life Exp. M 74, F 79
7. Urban population 53%
8. Nicosia
9. Arable 17%; Past. 1%; For. 13%

1. GNP $4,893 mill (5.8%); $7,040 (4.6%)
   Agric. 7%, Ind. 27% (Man. 15%), Serv. 66%
2. Grapes 212 Lemons 34 Olives 18 Oranges 50
3. Cattle 46 Pigs 284 Sheep 300
4. Fish 3
5. Timber 76
6. Asbestos ... Copper 59

1. Energy 0/1.75; 471 kWh
2a. Cheese 8 Meat 64 Olive oil 2 Wine 61
2b. Cement 960
3. Telephones 40; Cars 143
   Air: 1,586 pass-km; 30 ton-km
   Sea: goods loaded 2,628; unloaded 4,584

Exports: $793
  Vegetables Fruit
  Clothing Shoes
  Exports to: UK Libya
  Lebanon Greece
  Saudi Arabia

Imports: $2,281
  Food Oil
  Chemicals
  Imports from: UK
  Italy Japan
  Germany Greece

Services: Exports $1,874, Imports $704

CZECHOSLOVAKIA

1. Republic
2. Czech Slovak Hungarian
3. Koruna
4. $1 = 29.8
   £1 = 49.25

1. Area 127,870 sq km
2. Population 15,667,000
3. Density 125 per sq km
4. BR 14; DR 11; AI 0.2%
5. Inf. Mort. 13; Med.34
6. Life Exp. M 69, F 76
7. Urban population 69%
8. Prague
9. Arable 41%; Past. 13%; For. 37%

1. GNP $63,000 mill (...); $4,000 (...)
   Agric. 6%, Ind. 57%, Serv. 36%
2. Barley 3,650 Hops 12 Maize 508
   Oats 414 Potatoes 2,800 Rye 530
   Tomatoes 125 Wheat 6,712
3. Cattle 5,119 Goats 50 Pigs 7,400
4. Fish 21
5. Timber 18,435
6. Antimony 0.3 Coal 108 Iron 500
   Lead 3 Magnesite 2,800 Mercury 49.9
   Natural gas 23 Salt 350 Silver 40
   Tin 0.3 Tungsten 50

1. Energy 63.84/93.13; 17,392 kWh (17% hydr; 19% nucl)
2a. Beer 22.2 Butter 150 Cheese 234 Meat 1,635
   Paper 1.3 Sugar 755 Wine 139
2b. Aluminium 69 Cars 191 Cement 10,368
   Commercial veh. 58 Fertilizers 1,483 Iron 9,840
   Lead 26 Nickel 3.5 Petrol 13 Radios 200
   Steel 14,868 TV 507 Tyres 5,316
3. Telephones 25; Cars 2,904
   Air: 2,063 pass.-km; 16 ton-km

Exports: $11,882
  Machinery Road
  vehicles Iron and steel
  Chemicals
  Exports to: USSR
  Poland Germany
  Hungary

Imports: $13,106
  Machinery and
  Transport equip.
  Fuel and electricity
  Consumer goods
  Foods
  Imports from:
  USSR Germany
  Poland Hungary

DENMARK

1. Kingdom
2. Danish
3. Krone
4. $1 = 6.93
   £1 = 11.45

1. Area 43,070 sq km
2. Population 5,143,000
3. Density 121 per sq km
4. BR 11; DR 11; AI 0.0%
5. Inf. Mort. 6; Med.37
6. Life Exp. M 73, F 79
7. Urban population 86%
8. Copenhagen
9. Arable 61%; Past. 5%; For. 12%

1. GNP $105,263 mill (2.2%); $20,510 (2.2%)
   Agric. 4%, Ind. 29% (Man. 20%), Serv. 67%
2. Barley 4,923 Oats 109 Potatoes 1,614
   Rapeseed 904 Rye 565 Wheat 4,101
3. Cattle 2,190 Pigs 9,300
4. Fish 1,972
5. Timber 2,082
6. Oil 5.9 Salt 550

1. Energy 11.54/22.68; 8,817 kWh
2a. Beer 8.8 Butter 93 Cheese 302 Meat 1,497
   Paper 0.3 Sugar 530
2b. Cement 1,656 Fertilizers 130 Petrol 8
   Radios 47 Ships 405 TV 110
3. Telephones 86; Cars 1,602
   Rail: 4,977 pass-km; 1,693 ton-km
   Air: 3,748 pass-km; 108 ton-km
   Sea: goods loaded 14,724; unloaded 30,696

Exports: $35,087
  Meat Dairy prods. Fish
  Machinery Chemicals
  Exports to: Germany
  UK Sweden Norway
  USA

Imports: $31,743
  Machinery and
  transport equip.
  Chemicals
  Foodstuffs Mineral
  oils
  Imports from:
  Germany Sweden
  UK Neths USA

Services: Exports $14,050, Imports $17,754
Aid given: $937

DOMINICAN REPUBLIC

1. Republic
2. Spanish
3. Peso
4. $1 = 12.73
   £1 = 21.04

1. Area 48,730 sq km
2. Population 7,170,000
3. Density 148 per sq km
4. BR 28; DR 6; AI 1.9%
5. Inf. Mort. 57; Med.21
6. Life Exp. M 65, F 70
7. Urban population 60%
8. Santo Domingo
9. Arable 30%; Past. 43%; For. 13%

1. GNP $5,513 mill (2.2%); $790 (-0.1)
   Agric. 15%, Ind. 26% (Man 11%), Serv. 59%
2. Bananas 396 Cassava 156 Cocoa 50
   Coffee 49 Copra 25 Groundnuts 34
   Mangoes 193 Plantains 809
   Rice 450 Sugar cane 9,000 Tobacco 19
3. Cattle 2,245 Goats 543 Pigs 429 Sheep 110
4. Fish 22
5. Timber 982
6. Aluminium 165 Gold 6.5 Nickel 31.3 Silver 38

1. Energy 0.12/1.46; 1,065 kWh (15% hydr)
2a. Beer 1.3
3. Cars 152
   Air: 214 pass-km; 22 ton-km

Exports: $924
  Sugar Coffee Gold
  and silver
  Exports to: USA Neths

Imports: $1,964
  Food Petrol
  Machinery Vehicles
  Imports from: USA
  Venezuela Mexico

Services: Exports $1,224, Imports $712
Aid received: $488 (20)

ECUADOR

1. Republic
2. Spanish
3. Sucre
4. $1 = 1,022
   £1 = 1,688.96

1. Area 283,560 sq km
2. Population 10,587,000
3. Density 39 per sq km
4. BR 31; DR 7; AI 2.7%
5. Inf. Mort. 57; Med.20
6. Life Exp. M 65, F 69
7. Urban population 57%
8. Quito
9. Arable 10%; Past. 18%; For. 42%

1. GNP $10,774 mill (2.2%); $1,040 (-0.5%)
   Agric. 15%, Ind. 39% (Man. 21%), Serv. 47%
2. Bananas 2,817 Cocoa beans 95 Coffee 134
   Oranges 255 Palm oil 138 Plantains 959
   Potatoes 398 Rice 760
3. Cattle 4,100 Goats 310
   Pigs 4,160 Sheep 1,900
4. Fish 769
5. Timber 9,336
6. Gold 9.5 Oil 14.7

1. Energy 21.6/6.84; 1,914 kWh (47% hydr)
2a. Cheese 13 Meat 297
   Sugar 325
2b. Cement 3,000 Petrol 5.3
3. Telephones 4; Cars 141
   Rail: 63 pass-km; 8 ton-km
   Air: 1,073 pass-km; 70 ton-km

Exports: $2,722
  Oil Seafoods Bananas
  Cocoa Coffee
  Exports to: USA Peru
  Germany

Imports: $1,862
  Fuels Transport
  equip.
  Imports from: USA
  Japan Brazil
  Germany

Services: Exports $529, Imports $1,819
Aid received: $162 (16)

EGYPT

1. Republic
2. Arabic
3. Pound
4. $1 = 3.28
   £1 = 15.42

1. Area 1,001,450 sq km
2. Population 52,426,000
3. Density 53 per sq km
4. BR 31; DR 9; AI 2.5%
5. Inf. Mort. 57; Med.21
6. Life Exp. M 60, F 63
7. Urban population 49%
8. Cairo
9. Arable 3%

1. GNP $32,501 mill (...); $630 (...)
   Agric. 19%, Ind. 30% (Man. 14%), Serv. 52%
2. Cotton lint 341 Cotton seed 530 Dates 580
   Maize 3,750 Millet 215 Oranges 1,420
   Potatoes 921 Rice 2,700 Seed cotton 821
   Sorghum 600 Sugar cane 10,850
   Tomatoes 5,000
   Wheat 4,000
3. Cattle 2,000 Goats 3,880 Sheep 3,800
4. Fish 250
5. Timber 2,211
6. Iron 1,200 Natural gas 250 Oil 45.8
   Phosphates 1,280 Salt 900

1. Energy 73.09/38.27; 11,845 kWh (0% hydr)
2a. Butter 82 Cheese 320 Meat 581
   Sugar 977
2b. Aluminium 180 Cars 8 Cement 9,900
   Commercial veh. 4 Fertilizers 879
   Iron 140 Petrol 22.4 Radios 19
   Ships 20 Steel 750 TV 333 Tyres 1,008
3. Telephones 3; Cars 999
   Rail: 23,796 pass-km; 3,021 ton-km
   Air: 4,906 pass-km; 112 ton-km
   Sea: goods loaded 12,024; unloaded 12,024

Exports: $2,565
  Minerals Oil Textiles
  Exports to: USSR Italy
  Israel Neths. USA
  France

Imports: $7,434
  Foodstuffs
  Machinery and
  electrical equip.
  Imports from: USA
  Germany France
  Italy Japan

Services: Exports $6,618, Imports $6,035
Aid received: $1,578 (31)

EL SALVADOR

1. Republic
2. Spanish
3. Colon
4. $1 = 7.99
   £1 = 13.21

1. Area 21,040 sq km
2. Population 5,252,000
3. Density 253 per sq km
4. BR 36; DR 7; AI 2.5%
5. Inf. Mort. 53; Med.17
6. Life Exp. M 64, F 69
7. Urban population 44%
8. San Salvador
9. Arable 35%; Past. 29%; For. 5%

1. GNP $5,356 mill (0.3%); $1,040 (-1.1%)
   Agric. 21%, Ind. 21% (Man. 16%), Serv. 58%
2. Bananas 36 Coconuts 75 Coffee 165
   Maize 196 Oranges 90 Plantains 12
   Sorghum 166 Sugar cane 2,500
3. Cattle 1,162 Goats 543 Pigs 450 Sheep 110
4. Fish 12
5. Timber 4,315

1. Energy 0.23/1.16; 703 kWh (58% hydr, 14% geo)
3. Telephones 3; Cars 85
   Rail: 5 pass-km; 24 ton-km
   Air: 624 pass-km; 64 ton-km

Exports: $412
  Coffee
  Exports to: USA
  Germany Guatemala

Imports: $902
  Petrol Chemicals
  Vehicles
  Manufactures
  Imports from: USA
  Guatemala Mexico

Services: Exports $306, Imports $343
Aid received: $446 (87)

For detailed table headings and notes see the first page of this section

Country	Area and population	Production	Manufactures	Trade	
ETHIOPIA		1. GNP $5,953 mill (1.8%); $120 (-1.1%) Agric. 42%, Ind. 16% (Man. 11%), Serv. 42%	1. Energy 0.08/1.15; 343 kWh (67% hydr) 2a. Beer 0.8 Butter 9 Meat 588 Sugar 190	Exports: $446 Coffee Hides and skins	Imports: $1,075 Machinery and transport equip. Oil
1. Republic 2. Amharic 3. Birr 4. $1 = 2.04 £1 = 3.37	1. Area 1,221,900 sq km 2. Population 49,240,000 3. Density 45 per sq km 4. BR 48; DR 18; AI 2.4% 5. Inf. Mort. 122; Med.17 6. Life Exp. M 45, F 49 7. Urban population 13% 8. Addis Ababa 9. Arable 13%; Past. 41% For. 25%	2. Barley 1,188 Coffee 195 Linseed 37 Maize 2,000 Millet 133 Sorghum 984 3. Cattle 30,000 Goats 17,200 Sheep 22,960 4. Fish 4 5. Timber 38,896	3. Telephones 0.3; Cars 48 Rail: 315 pass-km; 150 ton-km Air: 1,244 pass-km; 86 ton-km	Exports to: Germany USA Japan Neths	Textile yarns Chemicals Imports from: USSR USA Germany Italy UK Japan
				Services: Exports $287, Imports $254 Aid received: $702 (14)	
FINLAND		1. GNP $109,705 mill (3.4%); $22,060 (2.9%) Agric. 6%, Ind. 36% (Man. 22%), Serv. 58%	1. Energy 5.85/28.68; 12,706 kWh (20% hydr; 18% nucl) 2a. Beer 3.4 Butter 60 Cheese 93 Meat 321	Exports: $26,743 Machinery Paper and board Wood and	Imports: $27,108 Machinery and transport equip.
1. Republic 2. Finnish Swedish 3. Markka 4. $1 = 4.29 £1 = 7.09	1. Area 338,130 sq km 2. Population 4,975,000 3. Density 16 per sq km 4. BR 12; DR 10; AI 0.4% 5. Inf. Mort. 5; Med.36 6. Life Exp. M 72, F 80 7. Urban population 68% 8. Helsinki 9. Arable 8%; For. 76%	2. Barley 1,641 Oats 1,661 Potatoes 820 Rye 244 Wheat 627 3. Cattle 1,400 Goats Pigs 1,340 Sheep 59 4. Fish 121 5. Timber 48,620 6. Chromium 450 Copper 15 Lead ... Mercury 100 Nickel 10.5 Phosphates 580 Silver 31 Zinc 58	Paper 8.7 Sugar 160 2b. Cement 1,668 Copper 55.7 Fertilizers 1,233 Iron 2,280 Nickel 13.4 Petrol 7.8 Ships 161 Steel 2,856 TV 627 Tyres 20,000 Zinc 163 3. Telephones 70; Cars 1,699 Rail: 2,736 pass-km; 8,352 ton-km Air: 3,589 pass-km; 90 ton-km Sea: goods loaded 24,048; unloaded 34,824	timber Exports to: USSR Sweden UK Germany USA France Services: Exports $6,848, Imports $11,098 Aid given: $706	Textiles Metals Petrol Chemicals Foodstuffs Imports from: Germany USSR Sweden Japan UK USA
FRANCE		1. GNP $1,000,866 mill (2.0%); $17,830 (1.6%) Agric. 3%, Ind. 29% (Man. 21%), Serv. 67%	1. Energy 66.58/220.9; 100,140 kWh (25% hydr. 52% nucl) 2a. Beer 19 Butter 539 Cheese 1,400 Meat 5,449	Exports: $209,958 Machinery and transport equip.	Imports: $233,140 Machinery and transport equip.
1. Republic 2. French 3. Franc 4. $1 = 6.07 £1 = 10.03	1. Area 551,500 sq km 2. Population 56,138,000 3. Density 103 per sq km 4. BR 13; DR 10; AI 0.4% 5. Inf. Mort. 7; Med.35 6. Life Exp. M 73, F 81 7. Urban population 74% 8. Paris 9. Arable 36%; Past. 21%; For. 27%	2. Apples 2,400 Barley 10,077 Grapes 7,800 Maize 8,996 Oats 875 Peaches 551 Pears 344 Potatoes 6,000 Rapeseed 2,011 Rye 248 Sugar beet 29,925 Wheat 33,363 3. Cattle 21,200 Goats 1,220 Pigs 12,200 Sheep 11,900 4. Fish 898 5. Timber 42,643 6. Aluminium 720 Coal 13.4 Iron 2.6 Lead 1 Natural gas 165 Oil 3.1 Potash 1,300 Salt 8,000 Silver 19 Uranium 2,800 Zinc 27	Olive oil 2 Paper 6.3 Sugar 4,130 Wine 5,891 2b. Aluminium 568 Cars 3,215 Cement 26,508 Commercial veh. 553 Copper 193 Fertilizers 3,801 Iron 14,412 Lead 267 Nickel 9.9 Petrol 67.5 Radios 2,080 Ships 108 Steel 16,800 TV 2,184 Tin 0.3 Tyres 61,368 Zinc 294 3. Telephones 66; Cars 21,970 Rail: 63,588 pass-km; 51,528 ton-km Air: 4,443 pass-km; 3,386 ton-km Sea: goods loaded 41,812; unloaded 177,036	Chemicals Foods Exports to: Germany Italy UK Belgium USA Neths Services: Exports $102,050, Imports $87,580 Aid given: $7,450	Chemicals Food Petrol Textiles Iron and steel Imports from: Germany Italy Belgium USA UK Neths
GABON		1. GNP $3,060 mill (1.0%); $1,105 (3.7%) Agric. 10%, Ind. 47% (Man. 10%), Serv. 43%	1. Energy 15.77/1.42; 200 kWh(63% hydr) 2a. Beer 0.9 Meat 22	Exports: $1,288 Oil Manganese	Imports: $732 Imports from:
1. Republic 2. French 3. CFA franc 4. $1 = 303.4 £1 = 501.40	1. Area 267,670 sq km 2. Population 1,172,000 3. Density 4.5 per sq km 4. BR 43; DR 16; AI 3.8% 5. Inf. Mort. 94; Med.23 6. Life Exp. M 52, F 47 7. Urban population 46% 8. Libreville 9. Arable 2%; Past. 18%; For. 78%	2. Cassava 265 Plantains 190 Sugar cane 160 3. Cattle 9 Goats 63 Pigs 150 Sheep 84 4. Fish 21 5. Timber 3,618 6. Manganese 2,200 Oil 13.8 Uranium 710	3. Telephones ...; Cars 16 Air: 404 pass-km; 29 ton-km	Timber Uranium Exports to: France USA Spain Services: Exports $257, Imports $1,181 Aid received: $134 (121)	France USA Japan Germany
GERMANY		1. GNP $1,272,959 mill (2.0%); $20,750 (2.1%) Agric. 2%, Ind. 37% (Man. 32%), Serv. 62%	1. Energy 243.07/453.95; 121,868 kWh (15% hydr; 31% nucl) 2a. Beer 111 Butter 695 Cheese 1,087 Meat 7,366	Exports: $398,446 Machinery Transport equip. Chemicals	Imports: $342,586 Machinery Transport equip.
1. Republic 2. German 3. Mark 4. $1 = 1.79 £1 = 2.96	1. Area 356,910 sq km 2. Population 77,573,000 3. Density 229 per sq km 4. BR 11; DR 12; AI -0.1% 5. Inf. Mort. 8; Med.37 6. Life Exp. M 72, F 78 7. Urban population 82% 8. Berlin 9. Arable 35%; Past. 16%; For. 30%	2. Apples 2,658 Barley 13,309 Grapes 1,902 Hops 3,940 Maize 1,440 Oats 2,123 Potatoes 17,520 Rapeseed 2,153 Rye 4,025 Sugar beet 28,900 Wheat 15,793 3. Cattle 20,287 Pigs 34,178 Sheep 4,136 4. Fish 388 5. Timber 45,828 6. Coal 433 Copper 5 Lead 9 Natural gas 530 Oil 3.7 Potash 4,850 Salt 16,300 Silver 69 Tin 2 Uranium 3,000 Zinc 64	Paper 11.6 Sugar 160 2b. Aluminium 1,333 Cars 4,618 Cement 40,000 Copper 749 Commercial veh. 349 Fertilizers 8,718 Iron 32,724 Lead 500 Nickel 2.7 Petrol 96 Radios 6,381 Ships 863 Steel 45,600 TV 4,206 Tin 6.1 Tyres 58,380 Zinc 481 3. Telephones 55; Cars 31,500 Rail: 81,180 pass-km; 115,980 ton-km Air: 31,810 pass-km; 3,248 ton-km Sea: goods loaded 43,704; unloaded 96,096	Manufactures Textiles Exports to: France UK Italy Neths USA Belg Switz Services: Exports $98,310, Imports $101,130 Aid given: $4,949	Food Chemicals Oil Clothing Textiles Iron and steel Imports from: France Neths, Italy Belgium USA
GHANA		1. GNP $5,503 mill (2.6%); $14,425 (3.4%) Agric. 49%, Ind. 17% (Man. 10%), Serv. 34%	1. Energy 0.59/1.58; 1,186 kWh (90% hydr) 2a. Meat 139	Exports: $1,014 Cocoa Gold Timber	Imports: $907 Fuels Machinery
1. Republic 2. English Akan 3. Cedi 4. $1 = 366.44 £1 = 605.58	1. Area 238,540 sq km 2. Population 15,028,000 3. Density 65 per sq km 4. BR 44; DR 12; AI 3.4% 5. Inf. Mort. 81; Med.17 6. Life Exp. M 54, F 58 7. Urban population 33% 8. Accra 9. Arable 12%; Past. 15%; For. 36%	2. Cassava 3,327 Cocoa beans 299 Groundnuts 180 Maize 493 Millet 124 Palm oil 85 Plantains 1,036 Sorghum 156 Yams 782 3. Goats 2,000 Sheep 2,250 4. Fish 360 5. Timber 17,025 6. Aluminium 348 Diamonds 200 Gold 17.3 Manganese 260	2b. Aluminium 169 3. Telephones 0.6; Cars 34 Rail: 130 pass-km; 77 ton-km Air: 286 pass-km; 12 ton-km	Manganese ore Diamonds Exports to: Switzerland UK USSR Japan Germany Services: Exports $82, Imports $403 Aid received: $543 (38)	and transport equip. Chemicals Foods Imports from: Nigeria UK Germany USA

For detailed table headings and notes see the first page of this section

Country	Area and population	Production	Manufactures	Trade	

GREECE

1. Republic
2. Greek
3. Drachma
4. $1 = 194.73
   £1 = 321.81

1. Area 131,990 sq km
2. Population 10,047,000
3. Density 77 per sq km
4. BR 12; DR 10; AI 0.4%
5. Inf. Mort. 13; Med.36
6. Life Exp. M 74, F 79
7. Urban population 63%
8. Athens
9. Arable 30%; Past. 40%; For. 20%

1. GNP $53,626 mill (1.0%); $5,340 (0.6%)
   Agric. 16%, Ind. 29% (Man. 18%), Serv. 55%
2. Cotton seed 515 Cotton lint 265 Grapes 1,600
   Maize 1,500 Olives 1,100 Oranges 831
   Peaches 582 Potatoes 1,100 Seed cotton 752
   Tobacco 132 Wheat 1,656
3. Cattle 731 Goats 3,488
   Pigs 1,300 Sheep 10,400
4. Fish 129
5. Timber 3,303
6. Aluminium 2,576 Chromium 65 Coal 51.7
   Lead 23 Magnesite 800 Manganese 260
   Nickel 16.1 Silver 62 Zinc 25

1. Energy 11.41/31.31; 8,346 kWh (28% hydr)
2a. Beer 3.3 Butter 4 Cheese 221 Meat 541
   Olive oil 210 Paper 0.3 Sugar 421
   Wine 497
2b. Aluminium 154 Cement 13,944
   Fertilizers 636 Lead 7 Nickel 16.1 Petrol 15.1
   Ships 3 Steel 960
3. Telephones 41.3; Cars 1,433
   Rail: 1,536 pass-km; 660 ton-km
   Air: 7,122 pass-km; 104 ton-km
   Sea: goods loaded 24,444; unloaded 31,644

Exports: $7,543
  Foods Olive oil
  Tobacco Textiles
  Petrol
  Exports to: Germany
  Italy France UK USA

Imports: $16,126
  Machinery
  Transport equip.
  Foodstuffs
  Beverages Oil
  Chemicals
  Imports from:
  Germany USA
  France Neths
  Japan

Services: Exports $5,179, Imports $4,352
Aid received: $33 (3)

GUATEMALA

1. Republic
2. Spanish
3. Quetzal
4. $1 = 4.87
   £1 = 8.05

1. Area 108,890 sq km
2. Population 9,197,000
3. Density 85 per sq km
4. BR 39; DR 8; AI 2.9%
5. Inf. Mort. 48; Med.17
6. Life Exp. M 62, F 67
7. Urban population 42%
8. Guatemala City
9. Arable 17%; Past .13%; For .36%

1. GNP $8,205 mill (0.2%); $920 (-2.6%)
   Agric. 18%, Ind. 26%, Serv. 56%
2. Bananas 478 Coffee 240 Cotton lint 48
   Maize 1,277 Plantains 55 Rubber 16
   Sugar cane 7,400 Tobacco 12
3. Cattle 2,023 Pigs 800 Sheep 660
4. Fish 3.2
5. Timber 7,390
6. Antimony 1.2 Lead - Tungsten - Zinc -

1. Energy 0.51/1.76;696 kWh (63% hydr)
3. Telephones 2
   Air: 160 pass-km; 26 ton-km
   Rail: 612 pass-km; 512 ton-km

Exports: $1,108
  Coffee Bananas
  Sugar
  Exports to: USA
  El Salvador

Imports: $1,654
  Fuels Chemicals
  Machinery Vehicles
  Imports from: USA
  Venezuela Mexico

Services: Exports $330,Imports $595
Aid received: $256 (29)

GUINEA

1. Republic
2. French
3. Franc
4. $1 = 616.61
   £1 = 1,019.01

1. Area 245,860 sq km
2. Population 5,755,000
3. Density 23 per sq km
4. BR 51; DR 20; AI 2.6%
5. Inf. Mort. 134; Med.17
6. Life Exp. M 44, F 45
7. Urban population 26%
8. Conakry
9. Arable 3%; Past .12%; For .40%

1. GNP $2,372 mill (...); $430 (...)
   Agric. 30%, Ind. 33% (Man. 3%), Serv. 38%
2. Bananas 111 Cassava 358 Coffee 26
   Plantains 350 Palm oil 45 Rice 500
3. Cattle 1,800
4. Fish 34
5. Timber 4,559
6. Aluminium 17,500 Diamonds 100 Iron...

1. Energy 0.02/0.49; 176 kWh (27% hydr)
2a. Meat 43
3. Telephones ...; Cars 31

Exports: $584
  Bauxite Iron ore
  Coffee

Imports: $468
  Machinery Food

Main trade is with France and USA
Aid received: $346 (62)

HAITI

1. Republic
2. French and Creole
3. Gourde
4. $1 = 5
   £1 = 18.26

1. Area 27,750 sq km
2. Population 6,513,000
3. Density 235 per sq km
4. BR 35; DR 12; AI 2.0%
5. Inf. Mort. 86; Med.20
6. Life Exp. M 55, F 58
7. Urban population 30%
8. Port-au-Prince
9. Arable 33%; Past .18%;For .2%

1. GNP $2,556 mill (1.1%); $400 (-0.7%)
   Agric. 31%, Ind. 38% (Man. 15%), Serv. 31%
2. Bananas 230 Cocoa 5 Coffee 33
   Mangoes 353 Sisal 10
3. Cattle 1,500 Goats 1,300
   Pigs 960
4. Fish 8
5. Timber 5,629

1. Energy 0.04/0.33; 153 kWh (46% hydr)
2a. Meat 85 Sugar 45
3. Telephones ...; Cars 21
   Air: 0 pass-km; 5 ton-km

Exports: $165
  Coffee Cocoa
  Essential oils
  Handicrafts
  Exports to: USA
  France Italy

Imports: $344
  Food Machinery
  Transport equip.
  Fuels
  Imports from: USA
  Neths. Antilles
  Canada Japan

Aid received: $198 (31)

HONDURAS

1. Republic
2. Spanish and others
3. Lempira
4. $1 = 5.59
   £1 = 9.23

1. Area 112,090 sq km
2. Population 5,138,000
3. Density 46 per sq km
4. BR 37; DR 7; AI 3.0%
5. Inf. Mort. 57; Med.17
6. Life Exp. M 64, F 68
7. Urban population 44%
8. Tegucigalpa
9. Arable 16%; Past .23%; For .31%

1. GNP $4,495 mill (2.3%); $900 (-1.2)
   Agric. 21%, Ind. 25% (Man. 14%), Serv. 54%
2. Bananas 1,050 Coffee 104 Grapefruit 36
   Maize 580 Oranges 49 Plantains 180
   Sugar cane 2,700
3. Cattle 2,601 Pigs 600
4. Fish 18
5. Timber 5,957
6. Lead 6 Silver 37 Zinc 33

1. Energy 0.11/0.89;290 kWh (45% hydr)
3. Telephones 1; Cars 34
   Air: 471 pass-km; 62 ton-km

Exports: $869
  Coffee Bananas
  Seafood Timber
  Exports to: USA
  Germany Japan

Imports: $933
  Machinery Vehicles
  Chemicals
  Imports from: USA
  Venezuela

Services: Exports $138, Imports $488
Aid received: $256 (52)

HONG KONG

1. UK colony
2. English Chinese
3. Dollar
4. $1 = 7.73
   £1 = 12.77

1. Area 1,040 sq km
2. Population 5,851,000
3. Density 5,859 per sq km
4. BR 12; DR 6; AI 1.5%
5. Inf. Mort. 6; Med.32
6. Life Exp. M 75, F 80
7. Urban population 93%
8. Hong Kong
9. Arable 7%; Past .1%; For .12%

1. GNP $59,202 mill (7.2%); $10,320 (5.7%)
   Agric. 0%, Ind. 28% (Man. 21%), Serv. 72%
4. Fish 238
5. Timber 186

1. Energy .../11.54; 7,456 kWh (0% hydr)
2a. Meat 290
2b. Cement 2,160 Radios 29,618 TV ...
3. Telephones 47.0; Cars 162
   Rail: 2,460 pass-km; 72 ton-km
   Sea: goods loaded 22,428; unloaded 52,272

Exports: $82,160
  Textiles Clothing
  Plastic and light metal
  prods Electronic
  goods Printing
  Exports to: China USA
  Japan Germany

Imports: $82,496
  Machinery
  Manufactures
  Chemicals Fuels
  Imports from:
  China Japan USA
  Korea

Services: Exports $7,333, Imports $5,410
Aid received: $23 (4)

HUNGARY

1. Republic
2. Hungarian
3. Forint
4. $1 = 76.24
   £1 = 125.99

1. Area 93,030 sq km
2. Population 10,552,000
3. Density 114 per sq km
4. BR 12; DR 13; AI 0.2%
5. Inf. Mort. 17; Med.37
6. Life Exp. M 68, F 75
7. Urban population 60%
8. Budapest
9. Arable 57%; Past .13%; For .18%

1. GNP $27,078 mill (1.3%); $2,560 (1.4%)
   Agric. 14%, Ind. 36%, Serv. 50%
2. Apples 900 Hemp 8 Maize 4,500 Oats 150
   Potatoes 1,200 Rye 240 Sugar beet 5,000
   Tobacco 15 Wheat 6,159
3. Cattle 1,599 Pigs 7,660 Sheep 2,069
4. Fish 38
5. Timber 6,589
6. Aluminium 2,352 Coal 17.5 Manganese 80
   Natural gas 183 Oil 2

1. Energy 20.76/38.76; 6,793 kWh(1% hydr; 24% nucl)
2a. Beer 9 Butter 38 Cheese 88 Meat 1,588
   Paper 0.5 Sugar 576 Wine 450
2b. Aluminium 75 Cement 3,936 Copper 13.1
   Fertilizers 811 Iron 1,692 Petrol 7.5 Radios 100
   Steel 2,808 TV 446 Tyres 852
3. Telephones 15.2; Cars 1,660
   Rail: 11,400 pass-km; 16,776 ton-km
   Air: 1,181 pass-km; 7 ton-km

Exports: $9,707
  Machinery Transport
  equip. Foodstuffs
  Manufactures
  Exports to: USSR
  Germany Austria
  Czech.

Imports: $8,764
  Machinery
  Transport equip.
  Manufactures Fuel
  Electricity
  Consumer goods
  Foodstuffs
  Imports from:
  USSR Germany
  Austria Czech.

Services: Exports $2,305, Imports $4,050

For detailed table headings and notes see the first page of this section

Country	Area and population	Production	Manufactures	Trade	
ICELAND 1. Republic 2. Icelandic 3. Krona 4. $1 = 62.62 £1 = 103.49	1. Area 103,000 sq km 2. Population 253,000 3. Density 2.5 per sq km 4. BR 15; DR 7; AI 1.1% 5. Inf. Mort. 5; Med.30 6. Life Exp. M 75, F 81 7. Urban population 90% 8. Reykjavik 9. Arable 0%; Past. 23%; For. 1%	1. GNP $5,351 mill (2.8%); $21,240 (1.7%) 2. Potatoes 8 3. Cattle 72 Sheep 560 4. Fish 1759	1. Energy 0.56/1.43; 960 kWh (79% hydr; 5% geo) 2a. Meat 20 2b. Aluminium 89 3. Telephones ...; Cars 122 Air: 2,586 pass-km; 86 ton-km	Exports: $1,401 Fish and fish products Aluminium Exports to: UK USA Germany Portugal Japan Services: Exports $540, Imports $765	Imports: $1,395 Foodstuffs Live animals Crude materials Oil Imports from: Germany Denmark Sweden Norway
INDIA 1. Republic 2. Hindi English and others 3. Rupee 4. $1 = 25.42 £1 = 42.01	1. Area 3,287,590 sq km 2. Population 853,094,000 3. Density 278 per sq km 4. BR 31; DR 10; AI 2.2% 5. Inf. Mort. 88; Med.22 6. Life Exp. M 60, F 61 7. Urban population 28% 8. New Delhi 9. Arable 57%; Past. 4%; For. 22%	1. GNP $287,383 mill (5.4%); $350 (3.2%) Agric. 30%, Ind. 29% (Man. 18%), Serv. 41% 2. Butter 840 Coconuts 4,739 Cotton lint 1,497 Groundnuts 7,500 Jute 1,920 Linseed 349 Maize 9,000 Mangoes 9,500 Millet 10,500 Oranges 1,854 Potatoes 15,000 Rapeseed 4,220 Rice 112,500 Seed cotton 4,430 Silk 10,500 Sorghum 13,000 Sugar cane 210,000 Tea 735 Wheat 49,652 3. Cattle 197,300 Goats 110,000 Pigs 10,400 Sheep 54,588 4. Fish 3146 5. Timber 264,412 6. Aluminium 4,345 Asbestos 35 Chromium 750 Coal 209 Copper 53 Diamonds 15 Gold 1.9 Iron 34 Lead 25 Magnesite 500 Manganese 1,300 Natural gas 400 Oil 33.3 Phosphates 800 Salt 9,000 Silver 36 Uranium 100 Zinc 65	1. Energy 236.1/256.88; 69,873 kWh (26% hydr; 2% nucl) 2a. Beer 1.5 Butter 840 Meat 1,584 Paper 1.9 Sugar 10,200 2b. Aluminium 423 Cars 180 Cement 43,200 Commercial veh. 115 Copper 41.8 Fertilizers 9,002 Iron 12,000 Lead 37.3 Petrol 39.6 Radios 1,178 Steel 13,000 TV 972 Tyres 8,064 Zinc 71 3. Telephones 0.6; Cars 1,414 Rail: 277,272 pass-km; 233,292 ton-km Air: 17,280 pass-km; 645 ton-km	Exports: $17,786 Gems and jewellery Clothing Leather and leather goods Machinery and transport equip. Textiles Tea Exports to: USSR USA Japan UK Germany Services: Exports $4,591, Imports $7,177 Aid received: $1,874 (2)	Imports: $23,382 Machinery Fuels Iron and steel Precious and semi- precious stones Transport equip. Imports from: USA Japan USSR Germany UK
INDONESIA 1. Republic 2. Bahasa Indonesia 3. Rupiah 4. $1 = 1950 £1 = 3,222.57	1. Area 1,904,570 sq km 2. Population 184,283,000 3. Density 99 per sq km 4. BR 27; DR 9; AI 2.0% 5. Inf. Mort. 65; Med.22 6. Life Exp. M 61, F 65 7. Urban population 29% 8. Jakarta 9. Arable 12%; Past. 7%; For. 63%	1. GNP $87,936 mill (5.7%); $490 (3.6%) Agric. 23%, Ind. 37% (Man. 17%), Serv. 39% 2. Cassava 16,581 Coconuts 12,300 Coffee 390 Copra 1,260 Groundnuts 919 Maize 6,741 Mangoes 500 Oranges 412 Palm oil 2,200 Rice 44,490 Rubber 1,600 Soybeans 1,412 Sugar cane 26,800 3. Cattle 10,300 Goats 10,950 Pigs 6,600 Sheep 5,700 4. Fish 2703 5. Timber 173,598 6. Aluminium 862 Copper 149 Diamonds 30 Natural gas 1,239 Nickel 59.6 Oil 74.5 Salt 600 Silver 68 Tin 31.3	1. Energy 138.7/49.58; 11,030 kWh (17% hydr) 2a. Beer 0.8 Meat 1,003 Sugar 1,917 2b. Aluminium 197 Cars 57 Cement 15,972 Commercial veh. 207 Fertilizers 2,585 Nickel 5 Radios 997 Ships 12 TV 575 Tin 29.9 Tyres 8028 3. Telephones 0.5; Cars 1,170 Rail: 7,860 pass-km; 2,364 ton-km Air: 11,728 pass-km; 340 ton-km Sea: goods loaded 109,560; unloaded 30,840	Exports: $25,675 Oil Natural gas Wood and cork manfactures Coffee Tea Cocoa Spices Rubber Timber Exports to: Japan USA Singapore S. Korea Services: Exports $2,397, Imports $10,475 Aid received: $1,830 (10)	Imports: $21,837 Machinery Chemicals Iron and steel Petrol Transport equip. Crude materials Food Imports from: Japan USA Singapore Germany Saudi A.
IRAN 1. Islamic Republic 2. Farsi 3. Rial 4. $1 = 67.78 £1 = 112.01	1. Area 1,648,000 sq km 2. Population 54,607,000 3. Density 33 per sq km 4. BR 33; DR 7; AI 3.5% 5. Inf. Mort. 40; Med.18 6. Life Exp. M 67, F 68 7. Urban population 55% 8. Tehran 9. Arable 9%; Past. 27%; For. 11%	1. GNP $160,000 mill (...); $3,190 (...) Agric. 23%, Ind. 15% (Man. 7%), Serv. 62% 2. Cotton lint 144 Dates 540 Grapes 1,050 Oranges 1,262 Potatoes 1,450 Rice 1,800 Seed cotton 394 Silk 850 Tobacco 21 Wheat 7,000 3. Cattle 8,000 Goats 13,500 Sheep 34,000 4. Fish 156 5. Timber 6,817 6. Aluminium 100 Chromium 60 Copper 68 Lead 11 Magnesite 5 Natural gas 955 Oil 157 Salt 700 Silver 25 Zinc 25	1. Energy 235.27/81.84; 15,504 kWh (12% hydr) 2a. Butter 71 Cheese 113 Meat 755 Olive oil 1 Sugar 603 2b. Aluminium 65 Cement 15,000 Copper 40 Fertilizers 114 Lead 9 Petrol 33.1 Radio 245 TV 411 3. Telephones 3.5; Cars 72 Rail: 5,784 pass-km; 6,762 ton-km Air: 4,874 pass-km; 138 ton-km	Exports: $21,378 Oil and oil products Exports to: USA Japan Italy Neths. India Turkey Services: Exports $437, Imports $2,355 Aid received: $89 (2)	Imports: $9,738 Machinery Transport equip. Iron and steel Foodstuffs Chemicals Imports from: Germany Japan UK Italy Turkey USSR Netherlands Australia
IRAQ 1. Republic 2. Arabic 3. Dinar 4. $1 = 0.36 £1 = 0.59	1. Area 438,320 sq km 2. Population 18,920,000 3. Density 43 per sq km 4. BR 41; DR 7; AI 3.6% 5. Inf. Mort. 56; Med.17 6. Life Exp. M 65, F 67 7. Urban population 74% 8. Baghdad 9. Arable 12%; Past. 9%; For. 4%	1. GNP $38,000 mill (...); $2,000 (...) 2. Barley 1,000 Dates 490 Oranges 180 Wheat 805 3. Cattle 1,700 Goats 1,650 Sheep 9,500 4. Fish 18 5. Timber 149 6. Natural gas 190 Oil 101	1. Energy 204.48/19.36; 3,920 kWh (3% hydr) 2a. Butter 8 Cheese 34 Meat 296 2b. Cement 10,000 Fertilizers 744 Petrol 19.9 3. Telephones 3.5; Cars 251 Rail: 1,150 pass-km; 1,534 ton-km	Exports: $21,434 Fuels, mainly oil Exports to: Brazil Italy Aid received: $5 (0.3)	Imports: $3,854 Machinery and Transport equip. Manufactured Foodstuffs Chemicals Imports from: Germany
IRELAND 1. Republic 2. Irish English 3. Punt 4. $1 = 0.67 £1 = 1.11	1. Area 70,280 sq km 2. Population 3,720,000 3. Density 51 per sq km 4. BR 18; DR 8; AI 0.9% 5. Inf. Mort. 8; Med.28 6. Life Exp. M 73, F 78 7. Urban population 59% 8. Dublin 9. Arable 14%; Past. 68%; For. 5%	1. GNP $30,054 mill (1.2%); $8,500 (0.8%) Agric. 11%, Ind. 10% (Man. 3%), Serv. 79% 2. Barley 1,337 Oats 136 Potatoes 687 Wheat 603 3. Cattle 5,899 Pigs 995 Sheep 5,782 4. Fish 253 5. Timber 1,282 6. Lead 32 Natural gas 92 Zinc 169	1. Energy 4.74/13.39; 3,805 kWh (13% hydr.) 2a. Beer 4.4 Butter 132 Cheese 78 Meat 706 Sugar 228 2b. Cement 10,000 Fertilizers 268 Lead 12 Petrol 1.5 3. Telephones 28; Cars 743 Rail: 1,140 pass-km; 516 ton-km Air: 2,737 pass-km; 81 ton-km	Exports: $23,788 Machinery and Transport equip. Food and live animals Chemicals Exports to: UK Germany France USA Neths. Aid given: $49	Imports: $20,716 Machinery and Transport equip. Chemicals Foodtuffs Petroleum Clothing Imports from: UK USA Germany France
ISRAEL 1. Republic 2. Hebrew Arabic 3. Shekel 4. $1 = 2.39 £1 = 3.95	1. Area 21,060 sq km 2. Population 4,600,000 3. Density 226 per sq km 4. BR 21; DR 7; AI 1.7% 5. Inf. Mort. 10; Med.26 6. Life Exp. M 74, F 78 7. Urban population 92% 8. Jerusalem 9. Arable 21%; Past. 7%; For. 5%	1. GNP $44,131 mill (3.2%); $9,750 (1.4%) 2. Cotton lint 45 Cotton seed 74 Grapefruit 390 Oranges 630 Tomatoes 425 Wheat 250 3. Cattle 357 Goats 125 Pigs 130 Sheep 375 4. Fish 28 5. Timber 118 6. Natural gas 2 Phosphates 3,516 Potash 1,260 Salt 400	1. Energy 0.08/13.72; 4,137 kWh 2a. Cheese 74 Meat 214 Olive oil 2 Wine 15 2b. Cement 2,868 Fertilizers 1,291 Petrol 6.7 TV 35 Tyres 756 3. Telephones 44.5; Cars 614 Rail: 144 pass-km; 1,020 ton-km Air: 7,360 pass-km; 643 ton-km Sea: goods loaded 7,896; unloaded 8,604	Exports: $11,576 Machinery Diamonds Chemicals Textiles Foodstuffs Exports to: USA UK Japan Germany Services: Exports $5,750, Imports $7,659 Aid received: $1,192 (264)	Imports: $15,104 Diamonds Capital goods Consumer goods Fuels Imports from: USA Belgium Germany UK Switz.

For detailed table headings and notes see the first page of this section

Country	Area and population	Production	Manufactures	Trade

ITALY

1. Republic
2. Italian
3. Lira
4. $1 = 1331
   £1 = 2,199.61

1. Area 301,270 sq km
2. Population 57,061,000
3. Density 196 per sq km
4. BR 11; DR 11; AI 0.1%
5. Inf. Mort. 9; Med.37
6. Life Exp. M 73, F 80
7. Urban population 69%
8. Rome
9. Arable 41%; Past . 17%; For. 23%

1. GNP $871,955 mill (2.4%); $15,150 (2.1%)
   Agric. 4%, Ind. 34% (Man. 22%), Serv. 63%
2. Apples 1,950 Cheese 722 Grapes 10,000
   Lemons 616 Maize 5,788 Oats 313
   Olives 2,018 Oranges 1,824 Peaches 1,650
   Pears 797 Potatoes 2,454 Soybeans 1,491
   Sugar beet 13,855 Tobacco 205
   Tomatoes 5,796 Wheat 8,245
3. Cattle 8,746 Goats 1,224
   Pigs 9,254 Sheep 11,569
4. Fish 559
5. Timber 9,733
6. Aluminium 12 Asbestos 95 Coal 1.5 Lead 15
   Manganese 9 Natural gas 671 Oil 4.6
   Potash 100 Salt 4,400 Silver 14 Zinc 44

1. Energy 32.01/217.76; 57,448 kWh (32% hydr; 2% nucl; 1% geo)
2a. Beer 11.5 Butter 80 Cheese 722 Meat 3,867
    Olive oil 382 Paper 4.5 Sugar 1,880
    Wine 5,980
2b. Aluminium 610 Cars 1,970 Cement 39,600
    Commercial veh. 246 Copper 291
    Fertilizers 1,869 Iron 12,000 Lead 181
    Petrol 77.2 Ships 352 Steel 25,000
    TV 2,233 Tyres 44,928 Zinc 252
3. Telephones 48.8; Cars 23,496
   Rail: 44,328 pass-km; 20,856 ton-km
   Air: 18,647 pass-km; 912 ton-km
   Sea: goods loaded 38,160; unloaded 213,048

Exports: $168,680   Imports: $180,105
Machinery Textiles   Machinery Textiles
Clothing Vehicles   Clothes Vehicles
Chemicals Footwear   Chemicals Oil
Petrol Iron and steel Imports from:
Exports to: Germany   Germany France
France USA UK Switz. USA Neths. UK

Services: Exports $42,510, Imports $45,950
Aid given: $3,613

IVORY COAST

1. Republic
2. French
3. CFA franc
4. $1 = 303.4
   £1 = 501.40

1. Area 322,460 sq km
2. Population 11,997,000
3. Density 38 per sq km
4. BR 50; DR 13; AI 3.9%
5. Inf. Mort. 87; Med.16
6. Life Exp. M 53, F 56
7. Urban population 47%
8. Abidjan
9. Arable 12%; Past . 9%; For. 18%

1. GNP $9,305 mill (0.9%); $790 (-3%)
   Agric. 46%, Ind. 24% (Man. 17%), Serv. 30%
2. Cassava 1,300 Cocoa beans 740 Coffee 220
   Maize 670 Palm oil 220 Pineapples 189
   Plantains 1,030 Rice 650 Rubber 68
   Yams 2,370
3. Cattle 991 Goats 1,550
   Pigs 450 Sheep 1,500
4. Fish 100
5. Timber 12,813
6. Diamonds 700

1. Energy 0.43/2.04; 1,173 kWh (76% hydr)
2a. Beer 1.4 Sugar 145
3. Telephones...; Cars 164
   Rail: 858 pass-km; 530 ton-km

Exports: $2,792   Imports: $2,100
Coffee Cocoa Fuels Oil Machinery and
Timber Canned fish transport equip.
and fruit   Chemicals Vehicles
Exports to: Neths. Imports from:
France USA Italy   France Nigeria
Germany   Japan Germany

Services: Exports $524, Imports $2,306
Aid received: $409 (35)

JAMAICA

1. Commonwealth
2. English Creole
3. Dollar
4. $1 = 9.96
   £1 = 16.46

1. Area 10,990 sq km
2. Population 2,456,000
3. Density 223 per sq km
4. BR 22; DR 6; AI 1.4%
5. Inf. Mort. 14; Med.22
6. Life Exp. M 71, F 76
7. Urban population 52%
8. Kingston
9. Arable 25%; Past . 18%; For. 17%

1. GNP $3,011 mill (-0.4%); $1,260 (-1.7%)
   Agric. 5%, Ind. 45% (Man. 18%), Serv. 50%
2. Bananas 130 Oranges 2,200 Plantains 26
   Sugar cane 1,750
3. Cattle 290 Goats 440 Pigs 250
4. Fish 10
5. Timber 220
6. Aluminium 9,395

1. Energy 0.01/1.07; 732 kWh (3% hydr)
2a. Meat 45 Sugar 192
3. Air: 2,125 pass-km; 24 ton-km

Exports: $1,029   Imports: $1,809
Alumina Bauxite   Mineral fuels Food
Clothing Sugar Fruit Construction
Exports to: USA UK materials
Canada Neths.   Machinery and
                    transport equip.
                    Imports from: USA
                    Canada UK

Services: Exports $1,209, Imports $1,167
Aid received: $258 (108)

JAPAN

1. Monarchy
2. Japanese
3. Yen
4. $1 = 136.61
   £1 = 225.76

1. Area 377,800 sq km
2. Population 123,460,000
3. Density 328 per sq km
4. BR 12; DR 8; AI 0.6%
5. Inf. Mort. 5; Med.37
6. Life Exp. M 76, F 82
7. Urban population 77%
8. Tokyo
9. Arable 12%; Past . 2%; For. 67%

1. GNP $2,920,310 mill (4.1%); $23,730 (3.5%)
   Agric. 3%, Ind. 41% (Man. 30%), Serv. 56%
2. Apples 1,069 Barley 365 Oranges 310
   Pears 472 Potatoes 3,500 Rice 12,500
   Silk 7,000 Soybeans 270 Tangerines 2,100
   Tea 90 Tobacco 74 Tomatoes 770
   Wheat 1,000
3. Cattle 4,770 Goats 40 Pigs 11,880
4. Fish 11,897
5. Timber 28,371
6. Chromium 10 Coal 8.6 Copper 13 Gold 7.5
   Lead 19 Molybdenum 100 Natural gas 84
   Salt 1,400 Silver 156 Tungsten 250 Zinc 132

1. Energy 47.12/495.9; 185,133 kWh (20% hydr; 16% nucl; 0.2% geo)
2a. Beer 54.9 Butter 85 Cheese 85 Meat 3,605
    Paper 24.6 Sugar 945 Wine 61
2b. Aluminium 1,135 Cars 9,948 Cement 84,444
    Commercial veh. 3,550 Copper 1,624
    Fertilizers 1,467 Iron 81,360 Lead 332 Nickel 107
    Petrol 144 Radios 10,496 Ships 6,531
    Steel 110,328 TV 14,777 Tin 4.7 Tyres 155,724
    Zinc 843
3. Telephones 62; Cars 29,478
   Rail: 38,352 pass-km; 26,652 ton-km
   Air: 76,506 pass-km; 4,326 ton-km
   Sea: goods loaded 84,312; unloaded 711,612

Exports: $286,948   Imports: $234,806
Machinery and   Oil Food and live
transport equip.   animals Machinery
Electric and electronic and transport
machinery Iron and   equip. Metals
steel Chemicals   Chemicals Timber
Textiles Ships   Textile fibres
Exports to: USA   Imports from: USA
Germany S. Korea   S. Korea China
Hong Kong   Austral Indon

Services: Exports $111,780, Imports $123,050
Aid given: $8,949

JORDAN

1. Kingdom
2. Arabic
3. Dinar
4. $1 = 0.68
   £1 = 1.12

1. Area 89,210 sq km
2. Population 4,009,000
3. Density 45 per sq km
4. BR 39; DR 5; AI 3.3%
5. Inf. Mort. 36; Med.17
6. Life Exp. M 66, F 70
7. Urban population 68%
8. Amman
9. Arable 4%; Past . 9%; For. 1%

1. GNP $5,291 mill (0.6); $1,730 (-3)
   Agric. 6%, Ind. 29% (Man. 16%), Serv. 65%
2. Cucumbers 68 Lemons 35 Olives 50
3. Goats 500 Sheep 1,225
4. Fish -
5. Timber 8
6. Phosphates 5,925 Potash 842

1. Energy 0.03/3.91;991 kWh (1%hydr)
2a. Olive oil 10
2b. Cement 1,800 Fertilizers 1,174
3. Telephones 3; Cars 166
   Air: 1,120 pass-km; 503 ton-km
   Rail: 20 pass-km; 3 ton-km

Exports: $922   Imports: $2,603
Phosphates   Oil Chemicals
Chemicals Food Live Manufactures
animals   Machinery
Exports to: Iraq India Imports from: USA
Saudi Arabia   Iraq Saudi Arabia

Services: Exports $1,344, Imports $1,270
Aid received: $280 (72)

KENYA

1. Republic
2. Swahili English
3. Shilling
4. $1 = 28.71
   £1 = 47.45

1. Area 580,370 sq km
2. Population 24,031,000
3. Density 42 per sq km
4. BR 47; DR 10; AI 3.8%
5. Inf. Mort. 64; Med.15
6. Life Exp. M 59, F 63
7. Urban population 24%
8. Nairobi
9. Arable 4%; Past . 7%; For. 6%

1. GNP $8,785 mill (4.2%); $380 (0.4%)
   Agric. 31%, Ind. 20%(Man. 12%), Serv. 49%
2. Bananas 150 Cassava 620 Coffee 96
   Maize 2,700 Pineapples 226 Plantains 272
   Sisal 43 Sugar cane 4,500 Tea 193
3. Cattle 13,793 Goats 7,600 Sheep 6,350
4. Fish 137
5. Timber 36,214

1. Energy 0.34/2.45; 719 kWh(69% hydr; 6% geo)
2a. Beer 3.1 Meat 363 Sugar 470
2b. Cement 1,500 Petrol 2
3. Telephones 1.4; Cars 122
   Rail: 484 pass-km; 2,034 ton-km
   Air: 1,239 pass-km; 54 ton-km

Exports: $1,054   Imports: $2,227
Coffee Tea Petrol Machinery and
Vegetables and fruit transport equip. Oil
Exports to: UK   Chemicals
Germany Neths. USA Imports from: UK
Uganda   France Germany
                   Japan

Services: Exports $1,009, Imports $941
Aid received: $967 (41)

KOREA (NORTH)

1. Republic
2. Korean
3. Won
4. $1 = 0.96
   £1 = 1.59

1. Area 120,540 sq km
2. Population 21,773,000
3. Density 181 per sq km
4. BR 5; DR 5; AI 1.8%
5. Inf. Mort. 24; Med.24
6. Life Exp. M 68, F 74
7. Urban population 67%
8. Pyongyang
9. Arable 20%; Past . 0%; For. 74%

1. GNP ...
2. Barley 630 Hemp 9 Maize 4,400
   Potatoes 2,100 Rice 5,500 Silk 3,000
   Soybeans 455 Tobacco 65
   Wheat 900
3. Cattle 290 Goats 1,280 Pigs 3,200 Sheep 380
4. Fish 1700
5. Timber 4,705
6. Coal 55 Copper 12 Gold 5.0 Lead 80
   Magnesite 3,000 Phosphates 550 Salt 550
   Silver 300 Tungsten 500 Zinc 200

1. Energy 52.2/59.93; 9,500 kWh (53% hydr)
2a. Meat 262
2b. Cement 10,000 Copper 40 Fertilizers 797
    Iron 2,500 Lead 75 Steel 5,000 Zinc 240

Exports: ...   Imports: ...
Iron and other metals Oil Coal Machinery
Agricultural prods and transport
Textiles   equip. Chemicals
Exports to: USSR   Imports from:
Japan China   USSR China Japan

For detailed table headings and notes see the first page of this section

Country	Area and population	Production	Manufactures	Trade
KOREA (SOUTH) 1. Republic 2. Korean 3. Won 4. $1 = 729.73 £1 = 1,205.95	1. Area 99,020 sq km 2. Population 42,793,000 3. Density 433 per sq km 4. BR 15; DR 6; AI 1.5% 5. Inf. Mort. 21; Med.27 6. Life Exp. M 68, F 74 7. Urban population 72% 8. Seoul 9. Arable 22%; Past .1%;For. 66%	1. GNP $186,467 mill (10.1%); $4,400 (8.8%) Agric. 10%, Ind. 44% (Man. 26%), Serv. 46% 2. Barley 416 Potatoes 500 Rice 7,786 Silk 1,400 Soybeans 260 3. Cattle 2,050 Goats 139 Pigs 4,900 4. Fish 2,727 5. Timber 6,803 6. Aluminium 72 Coal 20.8 Gold 12.0 Lead 17 Molybdenum 270 Salt 1,000 Silver 70 Tungsten 1,250 Zinc 32	1. Energy 19.74/93.11; 23,522 kWh (10% hydr; 32% nucl) 2a. Beer 8.8 Butter 33 Meat 764 Paper 3.7 2b. Cars 968 Cement 33,912 Commercial veh. 322 Copper 223 Fertilizers 1,167 Iron 15,528 Lead 106 Petrol 34.4 Radios 1,425 Ships 3,295 Steel 24,468 TV 14,922 Tin 2.4 Tyres 27,900 Zinc 215 3. Telephones 25.5; Cars 844 Rail: 29,868 pass-km; 13,476 ton-km Air: 14,491 pass-km; 1,621 ton-km Sea: goods loaded 47,508; unloaded172,272	Exports: $64,933 Machinery and transport equip. Electrical machinery Footwear Textiles Exports to: USA Japan Hong Kong Imports: $68,771 Oil Electronic components Chemicals Imports from: Japan USA Germany Services: Exports $12,643, Imports $12,432
LAOS 1. Republic 2. Laotian French 3. Kip 4. $1 = 696.19 £1 = 1,150.52	1. Area 236,800 sq km 2. Population 4,139,000 3. Density 18 per sq km 4. BR 44; DR 15; AI 2.6% 5. Inf. Mort. 40; Med.18 6. Life Exp. M 50, F 53 7. Urban population 19% 8. Vientiane 9. Arable 4%; Past.3%; For. 66%	1. GNP $693 mill (3.0%); $170 (0.0%) 2. Rice 1,500 3. Cattle 840 Pigs 1,350 4. Fish 20 5. Timber 3,878 6. Tin 0.3	1. Energy 0.14/0.16; 225 kWh (89% hydr) 2a. Meat 117 3. Telephones 0.3; Cars ...	Exports: $81 Coffee Tea Cocoa Spices Timber and cork Exports to: Thailand Malaysia Hong Kong Imports: $162 Machinery and transport equip. Oil Cotton goods Imports from: Japan Thailand France Germany Aid received: $141 (35)
LEBANON 1. Republic 2. Arabic French 3. Pound 4. $1 = 892.59 £1 = 1,475.09	1. Area 10,400 sq km 2. Population 2,701,000 3. Density 264 per sq km 4. BR 30; DR 8; AI 0.1% 5. Inf. Mort. 40; Med.21 6. Life Exp. M 65, F 69 7. Urban population 84% 8. Beirut 9. Arable 29%; Past.1%; For. 8%	1. GNP $5,000 mill (...); $2,000 (...) 2. Apples 205 Grapes 215 Olives 72 Oranges 280 Tomatoes 120 3. Goats 480 4. Fish 2 5. Timber 488	1. Energy 0.06/3.88; 819 kWh (30% hydr) 2a. Cheese 10 Meat 92 Olive oil 5 3. Air: 642 pass-km; 195 ton-km	Exports: ... Clothing Jewellery Aluminium Pharmaceuticals Exports to: Saudi Arabia Syria Libya Kuwait Imports: $ ... Food Petrol Machinery Imports from: USA Germany France Italy UK Aid received: $132
LIBYA 1. Republic 2. Arabic 3. Dinar 4. $1 = 0.3 £1 = 0.50	1. Area 1,759,540 sq km 2. Population 4,545,000 3. Density 2.6 per sq km 4. BR 43; DR 8; AI 4.1% 5. Inf. Mort. 68; Med.17 6. Life Exp. M 62, F 65 7. Urban population 70% 8. Tripoli 9. Arable 1%; Past.8%	1. GNP $23,000 mill (-6.0%); $6,000 (-9.9%) Agric. 5%, Ind. 50% (Man. 7%), Serv. 45% 2. Barley 130 Dates 108 Olives 118 Tomatoes 218 3. Goats 970 Sheep 5,850 4. Fish 9 5. Timber 640 6. Natural gas 201 Oil 65.2	1. Energy 88.62/17.6; 3,000 kWh (0% hydr) 2a. Meat 176 Olive oil 24 2b. Cement 3,000 Fertilizers 144 3. Telephones ...; Cars 485 Air: 1,447 pass-km; 9 ton-km	Exports: $6,683 Oil Natural gas Exports to: Italy France Neths. Spain Greece Imports: $5,879 Machinery and transport equip. Foodstuffs Consumer goods Imports from: Italy Germany UK Japan France Services: Exports $889, Imports $2,071 Aid received: $11 (3)
LUXEMBOURG 1. Grand Duchy 2. Letzeburgish French German 3. Franc 4. $1 = 36.88 £1 = 60.95	1. Area 2,590 sq km 2. Population 373,000 3. Density 144 per sq km 4. BR 12; DR 11; AI 0.2% 5. Inf. Mort. 9; Med.37 6. Life Exp. M 72, F 79 7. Urban population 83% 8. Luxembourg 9. included in Belgium	1. GNP $9,408 mill (4.0%); $24,860 (3.6%) Agric. 3%, Ind. 40% (Man. 30%), Serv. 72% 2. Maize 89 3. Cattle 214 Pigs 76	1. Energy 0.1/4.64; 1,238 kWh (91% hydr) 2b. Iron 2,724 Steel 3,480 3. Telephones ...; Cars 169 Rail: 264 pass-km; 720 ton-km	Trade - see Belgium
MADAGASCAR 1. Republic 2. Malagasy French 3. Franc 4. $1 = 1763 £1 = 2,913.53	1. Area 587,040 sq km 2. Population 12,004,000 3. Density 19 per sq km 4. BR 45; DR 13; AI 3.2% 5. Inf. Mort. 110; Med.16 6. Life Exp. M 54, F 57 7. Urban population 25% 8. Antananarivo 9. Arable 5%; Past.58%; For. 25%	1. GNP $2,543 mill (0.1%); $230 (-2.6%) Agric. 31%, Ind. 14% (Man. 12%), Serv. 54% 2. Bananas 220 Cassava 2,250 Coffee 89 Groundnuts 32 Mangoes 196 Rice 2,400 Sisal 20 3. Cattle 10,250 Goats 1,250 Pigs 1,425 4. Fish 80 5. Timber 7,634 6. Chromium 70	1. Energy 0.04/0.45; 220 kWh (48% hydr) 2a. Meat 271 Sugar 120 3. Telephones 0.4; Cars 44 Air: 423 pass-km; 21 ton-km Sea: goods loaded 540; unloaded 792	Exports: $312 Coffee Vanilla Sugar Cloves Exports to: France USA Japan Germany Neths. Imports: $340 Oil Chemicals Machinery and transport equip. Imports from: France USA USSR Germany Japan Services: Exports $146, Imports $475 Aid received: $320 (28)
MALAWI 1. Republic 2. Chichewa English 3. Kwacha 4. $1 = 2.91 £1 = 4.81	1. Area 118,480 sq km 2. Population 8,754,000 3. Density 88 per sq km 4. BR 55; DR 19; AI 3.5% 5. Inf. Mort. 138; Med.21 6. Life Exp. M 48, F 50 7. Urban population 15% 8. Lilongwe 9. Arable 25%; Past.20%; For. 45%	1. GNP $1,475 mill (3.3%); $180 (-0.1%) Agric. 35%, Ind. 19% (Man. 11%), Serv. 45% 2. Groundnuts 193 Maize 1,843 Tea 42 Tobacco 91 3. Cattle 1,100 Goats 1,200 4. Fish 89 5. Timber 7,407	1. Energy 0.07/0.34; 185 kWh (79% hydr) 2a. Meat 39 Sugar 175 3. Telephones 0.6; Cars 15 Rail: 108 pass-km; 720 ton-km Air: 64 pass-km; 1 ton-km	Exports: $418 Tobacoo Tea Sugar Exports to: UK USA S. Africa Germany Japan Imports: $573 Fuel Vehicles Clothing Imports from: S. Africa UK Japan Zimbabwe Services: Exports $38, Imports $354 Aid received: $393 (48)
MALAYSIA 1. Republic 2. Malay Chinese English 3. Ringgit 4. $1 = 2.77 £1 = 4.58	1. Area 329,750 sq km 2. Population 17,891,000 3. Density 54 per sq km 4. BR 28; DR 5; AI 2.7% 5. Inf. Mort. 20; Med.21 6. Life Exp. M 69, F 73 7. Urban population 42% 8. Kuala Lumpur 9. Arable 15%; For.59%	1. GNP $37,005 mill (4.6%); $180 (-0.1%) Agric. 22%, Ind. 49% (Man. 30%), Serv. 44% 2. Bananas 505 Cocoa beans 250 Copra 93 Palm oil 6,200 Pineapples 207 Rice 1,800 Rubber 1,420 3. Pigs 2,400 4. Fish 604 5. Timber 44,431 6. Aluminium 355 Copper 24 Iron 0.3 Natural gas 575 Oil 29.7 Silver 13 Tin 32	1. Energy 60.35/24.07; 4,967 kWh (29% hydr) 2a. Meat 463 2b. Cars 103 Cement 5,880 Commercial veh. 50 Fertilizers 247 Lead 16 Petrol 8 TV 1,240 Tin 51.9 Tyres 6,744 3. Telephones 9.1; Cars 1,504 Rail: 1,668 pass-km; 1,404 ton-km Air: 7,605 pass-km; 343 ton-km Sea: goods loaded 18,708; unloaded 29,340	Exports: $25,113 Electronic components Oil Timber Rubber Palm oil Natural gas Exports to: Singapore USA Japan S. Korea Imports: $22,541 Electronic components Petrol Steel Grain Oil Imports from: Japan USA Singapore Services: Exports $3,985, Imports $7,979 Aid received: $139 (8)

For detailed table headings and notes see the first page of this section

Country	Area and population	Production	Manufactures	Trade

MALI

Country
1. Republic
2. French
3. CFA franc
4. $1 = 303.4
   £1 = 501.40

Area and population
1. Area 1,240,190 sq km
2. Population 9,214,000
3. Density 6.7 per sq km
4. BR 51; DR 19; AI 3.0%
5. Inf. Mort. 159; Med.17
6. Life Exp. M 44, F 48
7. Urban population 19%
8. Bamako
9. Arable 2%; Past .25%; For. 7%

Production
1. GNP $2,109 mill (3.5%); $260 (1.0%)
   Agric. 50%, Ind. 12% (Man. 6%), Serv. 38%
2. Cotton lint 97 Cotton seed 135
   Groundnuts 162 Millet 695
3. Cattle 4,900 Goats 5,750 Sheep 5,700
4. Fish 56
5. Timber 5,358

Manufactures
1. Energy 0.02/0.22; 87 kWh (52% hydr)
2a. Meat 160
3. Telephones ...; Cars 19
   Rail: 173 pass-km; 241 ton-km

Trade
Exports: $271
  Cotton Live animals
  Exports to: Ivory Coast
  France Senegal
  Germany
Imports: $500
  Industrial prods
  Petrol
  Imports from:
  France Ivory Coast
  Senegal Germany

Services: Exports $82, Imports $364
Aid received: $470 (57)

MALTA

Country
1. Commonwealth
2. Maltese English
3. Pound
4. $1 = 0.34
   £1 = 0.56

Area and population
1. Area 320 sq km
2. Population 353,000
3. Density 1,094 per sq km
4. BR 13; DR 19; AI -0.3%
5. Inf. Mort. 9; Med.33
6. Life Exp. M 72, F 76
7. Urban population 85%
8. Valletta
9. Arable 41%;

Production
1. GNP $2,041 mill (2.5%); $5,820 (3.1%)
   Agric. 4%, Ind. 41% (Man. 27%), Serv. 55%
2. Potatoes 14 Tomatoes 17 Wheat 6
3. Cattle 14 Pigs 95
4. Fish 1

Manufactures
1. Energy 0/0.72; 250 kWh (0% hydr)
2a. Meat 13
3. Telephones 37; Cars 90
   Air: 633 pass-km; 5 ton-km
   Sea: goods loaded 96; unloaded 2,472

Trade
Exports: $858
  Manufactures
  Clothing Machinery
  and transport equip.
  Exports to: Germany
  Italy UK Libya USA
Imports: $1,505
  Food and live
  animals Fuels
  Textile yarns
  Imports from: Italy
  UK Germany USA

Services: Exports $823, Imports $462

MAURITANIA

Country
1. Republic
2. Arabic French
3. Ouguiya
4. $1 = 82.6
   £1 = 136.50

Area and population
1. Area 1,025,220 sq km
2. Population 2,024,000
3. Density 2 per sq km
4. BR 46; DR 18; AI 2.7%
5. Inf. Mort. 117; Med.18
6. Life Exp. M 46, F 50
7. Urban population 40%
8. Nouakchott
9. Past .38%;For. 15%

Production
1. GNP $953 mill (0.4%); $490 (-2.2%)
   Agric. 37%, Ind. 24%, Serv. 38%
2. Dates 13 Millet 7
3. Cattle 1,270 Goats 3,350 Sheep 4,300
4. Fish 98
5. Timber 12
6. Iron 6.5

Manufactures
1. Energy 0.003/1.41; 114 kWh (0% hydr)
2a. Meat 40
3. Telephones ...; Cars 11
   Air: 288 pass-km; 17 ton-km

Trade
Exports: $437
  Iron Fish Fish
  products
  Exports to: Japan
  France Spain
Imports: $222
  Food Vehicles
  Fuels
  Imports from:
  France Spain
  Senegal

Services: Exports $57, Imports $280
Aid received: $195 (102)

MEXICO

Country
1. Republic
2. Spanish
3. Peso
4. $1 = 3008
   £1 = 4,971.02

Area and population
1. Area 1,958,200 sq km
2. Population 88,598,000
3. Density 45 per sq km
4. BR 27; DR 5; AI 2.3%
5. Inf. Mort. 36; Med.20
6. Life Exp. M 67, F 74
7. Urban population 73%
8. Mexico City
9. Arable 13%; Past .39%; For. 23%

Production
1. GNP $170,053 mill (0.6%); $1,990 (-1.5%)
   Agric. 9%, Ind. 32% (Man. 23%), Serv. 59%
2. Bananas 1,065 Cotton lint 168 Cotton
   seed 263 Coffee 309 Copra 183 Lemons 612
   Maize 12,019 Mangoes 800 Oranges 2,200
   Pineapples 324 Potatoes 810 Seed cotton 600
   Sisal 55 Sorghum 5,900 Sugar cane 34,893
   Tobacco 52 Tomatoes 1,746 Wheat 3,630
3. Cattle 28,200 Goats 9,086
   Pigs 17,300 Sheep 5,500
4. Fish 1363
5. Timber 22,302
6. Antimony 1.9 Coal 10 Copper 291 Gold 9.2
   Lead 163 Manganese 450 Mercury 195
   Molybdenum 4,300 Natural gas 1,005 Oil 133
   Phosphates 650 Salt 7,900 Silver 2,306
   Tungsten 100 Zinc 284

Manufactures
1. Energy 25,089/146.49; 27,338 kWh (29% hydr; 3% geo)
2a. Beer 31.5 Butter 33 Cheese 141 Meat 3,724
   Paper 3.4 Sugar 3,678 Wine 216
2b. Cars 614 Cement 24,504 Commercial veh. 124
   Copper 144 Fertilizers 375 Iron 5,280
   Lead 174 Petrol 68.6 Radios 207 Steel 8,220
   TV 609 Tin 3 Tyres 11,772 Zinc 195
3. Telephones 10; Cars 5,312
   Rail: 5,952 pass-km; 38,112 ton-km
   Air: 17,649 pass-km; 169 ton-km
   Sea: goods loaded 112,800; unloaded 44,736

Trade
Exports: $22,818
  Oil Engines Vehicle
  parts Machinery
  Coffee
  Exports to: USA
  France Spain
Imports: $23,633
  Car parts for
  assembly
  Machinery
  Foodstuffs
  Imports from: USA
  Germany Japan

Services: Exports $13,135, Imports $18,609
Aid received: $97 (1)

MONGOLIA

Country
1. Republic
2. Mongol
3. Tugrik
4. $1 = 3.34
   £1 = 5.52

Area and population
1. Area 1,566,500 sq km
2. Population 2,190,000
3. Density 1.4 per sq km
4. BR 34; DR 8; AI 2.8%
5. Inf. Mort. 60; Med.19
6. Life Exp. M 62, F 65
7. Urban population 55%
8. Ulan Bator
9. Arable 1%; Past .79%; For. 9%

Production
1. GNP $600 mill (...); $400 (...)
2. Oats 20 Wheat 500
3. Cattle 2,600 Goats 4,400 Sheep 13,500
5. Timber 2,390
6. Molybdenum 1,100 Tungsten 1,500

Manufactures
1. Energy 3.12/3.99; 901 kWh (0% hydr)
2a. Butter 10 Meat 235
3. Rail: 486 pass-km; 6,180 ton-km

Trade
Exports: $436
Main trade is with USSR
Aid received: $4 (2)
Imports: $655

MOROCCO

Country
1. Kingdom
2. Arabic Berber
3. Dirham
4. $1 = 9.08
   £1 = 15.01

Area and population
1. Area 446,550 sq km
2. Population 25,061,000
3. Density 56 per sq km
4. BR 33; DR 8; AI 2.6%
5. Inf. Mort. 68; Med.19
6. Life Exp. M 62, F 65
7. Urban population 49%
8. Rabat
9. Arable 20%; Past .47%; For. 12%

Production
1. GNP $22,069 mill (4.1%); $900 (1.3%)
   Agric. 16%, Ind. 34% (Man. 17%), Serv. 50%
2. Barley 2,224 Dates 46 Grapes 189
   Maize 436 Olives 350 Oranges 844
   Potatoes 851 Wheat 3,614
3. Cattle 3,600 Goats 6,071 Sheep 17,500
4. Fish 551
5. Timber 2,074
6. Antimony 0.1 Cobalt 200 Copper 135
   Iron 0.1 Lead 65 Manganese 30
   Phosphates 21,000 Silver 195 Zinc 17

Manufactures
1. Energy 0.79/8.99; 2,332 kWh (27% hydr)
2a. Butter 14 Meat 314 Olive oil 40 Sugar 610
   Wine 36
2b. Cars 48 Cement 4,500 Commercial veh. 10
   Fertilizers 1,203 Lead 66.2 Petrol 4.7
3. Telephones 1.5; Cars 554
   Rail: 2,160 pass-km; 4,680 ton-km
   Air: 2,218 pass-km; 51 ton-km
   Sea: good loaded 19,476; unloaded 16,068

Trade
Exports: $3,308
  Food and beverages
  Phosphoric acid
  Phosphates Clothing
  Exports to: France
  India Spain Germany
  Italy
Imports: $5,492
  Machinery Oil
  Consumer goods
  Food Sulphur
  Imports from:
  France Spain USA
  Italy Canada

Services: Exports $1,700, Imports $2,431
Aid received: $443 (18)

MOZAMBIQUE

Country
1. Republic
2. Potugese Bantu
3. Metical
4. $1 = 1519
   £1 = 2,510.30

Area and population
1. Area 799,380 sq km
2. Population 15,656,000
3. Density 20 per sq km
4. BR 44; DR 17; AI 2.6%
5. Inf. Mort. 130; Med.18
6. Life Exp. M 47, F 50
7. Urban population 27%
8. Maputo
9. Arable 4%; Past .56%; For. 19%

Production
1. GNP $1,193 mill (-3.5%); $80 (-6.0%)
   Agric. 64%, Ind. 22%, Serv. 14%
2. Bananas 24 Cassava 3,400 Copra 70
   Cotton seed 64 Groundnuts 70 Maize 453
   Sorghum 135
3. Cattle 1,380 Goats 385
4. Fish 34
5. Timber 16,002
6. Coal 0.4

Manufactures
1. Energy 0.05/0.5; 2,358 kWh (88% hydr)
2a. Meat 74 Sugar 24
3. Telephones 0.4; Cars 33
   Rail: 105 pass-km; 307 ton-km
   Air: 426 pass-km; 9 ton-km

Trade
Exports: $103
  Shrimps Cashew nuts
  Cotton Sugar Copra
  Exports to: USA
  Germany Japan Spain
Imports: $715
  Foodstuffs
  Machinery Petrol
  Chemicals
  Imports from:
  USSR S. Africa
  France USA

Services: Exports $167, Imports $388
Aid received: $759 (49)

Country	Area and population	Production	Manufactures	Trade	

NEPAL
1. Kingdom
2. Nepalese
3. Rupee
4. $1 = 42.47
  £1 = 70.19
	1. Area 140,800 sq km
	2. Population 19,143,000
	3. Density 138 per sq km
	4. BR 36; DR 13; AI 2.6%
	5. Inf. Mort. 118 ; Med.19
	6. Life Exp. M 54, F 53
	7. Urban Population 10%
	8. Katmandu
	9. Arable 17%; Past .15%; For. 17%.
		1. GNP $3,206 mill (3.7%); $170 (2.1%)
		  Agric. 58%, Ind. 14% (Man. 6%), Serv. 28%
		2. Jute 19 Maize 950 Millet 240 Potatoes 658
		  Rice 3,300 Wheat 850
		3. Cattle 6,350 Goats 5,210 Sheep 880
		4. Fish 12
		5. Timber 17,388
			1. Energy 0.07/0.44; 202 kWh (79% hydr)
			2a. Butter 12 Meat 144
			3. Telephones ...; Cars ...
			  Air: 338 pass-km; 6 ton-km.
				Exports: $156
				  Manufactures
				  Foodstuffs Machinery
				  and transport equip.
				  Exports to: India USA
				  Germany
					Imports: $580
					  Manufactures
					  Machinery and
					  transport equip.
					  Chemicals
					  Foodstuffs Fuels
					  Imports from: India
					  Japan China S.
					  Korea
					
					Services: Exports $241, Imports $151
					Aid received: $488 (27)

NETHERLANDS
1. Monarchy
2. Dutch
3. Guilder
4. $1 = 2.01
  £1 = 3.32
	1. Area 37,330 sq km
	2. Population 14,951,000
	3. Density 440 per sq km
	4. BR 13; DR 9; AI 0.6%
	5. Inf. Mort. 7 ; Med.35
	6. Life Exp. M 74, F 81
	7. Urban Population 89%
	8. Amsterdam
	9. Arable 27%; Past .32%; For.9%
		1. GNP $237,415 mill (1.8%); $16,010 (1.3%)
		  Agric. 4%, Ind. 31% (Man. 20%), Serv. 65%
		2. Apples 333 Barley 213 Cheese 596
		  Potatoes 6,500 Sugar beet 7,650
		  Tomatoes 595 Wheat 1,076
		3. Cattle 4,731 Pigs 14,200
		4. Fish 399
		5. Timber 1,141
		6. Natural gas 2,296 Oil 3.5 Salt 3,700
			1. Energy 83.24/98.62; 17,291 kWh (3% nucl)
			2a. Beer 17.6 Butter 177 Cheese 596 Meat 2,570
			  Paper 2.5 Sugar 1,240
			2b. Aluminium 404 Cement 3,706 Commercial veh. 28
			  Fertilizers 2,124 Iron 4,956 Lead 41.5 Petrol 54.6
			  Ships 163 Steel 5,412 Tin 4.8 Zinc 203
			3. Telephones 63.9; Cars 5,118
			  Rail: 10,836 pass-km; 3,072 ton-km.
			  Air: 22,605 pass-km.; 1,734 ton-km.
			  Sea: goods loaded 92,676; unloaded 280,488
				Exports: $131,839
				  Machinery and
				  transport equip.
				  Foodstuffs Chemicals
				  Exports to: Germany
				  Belgium France UK
				  Italy
					Imports: $126,195
					  Machinery and
					  transport equip.
					  Foodstuffs
					  Chemicals Mineral
					  fuels
					  Imports from:
					  Germany Belgium
					  UK France USA
					
					Services: Exports $44,210, Imports $43,300
					Aid given: $2,094

NEW ZEALAND
1. Commonwealth
2. English
3. Dollar
4. $1 = 1.77
  £1 = 2.93
	1. Area 270,990 sq km
	2. Population 3,392,000
	3. Density 13 per sq km
	4. BR 16; DR 8; AI 0.9%
	5. Inf. Mort. 9 ; Med.31
	6. Life Exp. M 73, F 79
	7. Urban Population 84%
	8. Wellington
	9. Arable 2%; Past .51%; For. 27%
		1. GNP $39,437 mill (1.7%); $11,800 (0.9%)
		  Agric. 8%, Ind. 28% (Man. 17%), Serv. 64%
		2. Apples 380 Barley 462 Pears 11 Tomatoes 44
		  Wheat 220 Wool 272
		3. Cattle 7,999 Pigs 410 Sheep 58,334
		4. Fish 503
		5. Timber 10,153
		6. Coal 0.2 Gold 2.4 Natural gas 192 Oil 1.7
			1. Energy 13.96/17.02; 6,964 kWh (62% hydr; 2% geo)
			2a. Beer 4.1 Butter 238 Cheese 129
			  Meat 1,200 Paper 0.7 Wine 49
			2b. Aluminium 263 Cars 60 Cement 804 Commercial
			  veh. 13 Fertilizers 220 Petrol 4.2 TV 90
			  Tyres 1,200
			3. Telephones 69.7; Cars 1,531
			  Rail: 5,000 pass-km; 37,980 ton-km
			  Air: 22,605 pass-km.; 1,734 ton-km
			  Sea: goods loaded 12,204; unloaded 8,316
				Exports: $9,435
				  Meat and meat prods
				  Wool Fruit Vegetables
				  Timber Dairy produce
				  Leather
				  Exports to: Australia
				  Japan USA UK
					Imports: $9,489
					  Machinery and
					  transport equip.
					  Chemicals Textiles
					  Electrical goods
					  Clothing Footwear
					  Imports from:
					  Australia Japan
					  USA UK
					
					Services: Exports $3,438, Imports $6,570
					Aid given: $87

NICARAGUA
1. Republic
2. Spanish
3. Cordoba
4. $1 = 4.97
  £1 = 8.21
	1. Area 130,000 sq km
	2. Population 3,871,000
	3. Density 33 per sq km
	4. BR 39; DR 7; AI 3.4%
	5. Inf. Mort. 50; Med.17
	6. Life Exp. M 65, F 68
	7. Urban Population 60%
	8. Managua
	9. Arable 11%; Past .45%;For.30%
		1. GNP $2,800 mill (-1.4%); $1,000 (...)
		  Agric. 29%, Ind. 23% (Man. 19%), Serv. 48%
		2. Bananas 108 Coffee 51 Cotton lint 30
		  Cotton seed 42
		3. Cattle 1,650 Pigs 680
		4. Fish 5
		5. Timber 3,870
		6. Gold 0.8
			1. Energy 0.07/1.05; 395 kWh (26% hydr; 9% geo)
			2a. Cheese 9 Meat 54 Sugar 205
			3. Telephones 1.9; Cars ...
			  Rail: 66 pass-km; 4 ton-km.
				Exports: $300
				  Coffee Cotton Sugar
				  Bananas
				  Exports to: Japan
				  Germany USA Spain
					Imports: $923
					  Petrol Machinery
					  and transport
					  equip.
					  Imports from:
					  USSR Mexico USA
					
					Services: Exports $58, Imports $331
					Aid received: $227 (61)

NIGER
1. Republic
2. French
3. CFA franc
4. $1 = 303.4
  £1 = 501.40
	1. Area 1,267,000 sq km
	2. Population 7,731,000
	3. Density 6.1 per sq km
	4. BR 51; DR 19; AI 3.3%
	5. Inf. Mort. 124 ; Med.16
	6. Life Exp. M 45, F 48
	7. Urban Population 20%
	8. Niamey
	9. Arable 3%; Past .7%; For.2%
		1. GNP $2,195 mill (-1.7%); $290 (-5.0%)
		  Agric. 36%, Ind. 13% (Man. 8%), Serv. 51%
		2. Groundnuts 60 Millet 1,133 Sorghum 415
		3. Cattle 3,700 Goats 7,800 Sheep 3,500
		4. Fish 2
		5. Timber 4,285
		6. Molybdenum 15 Uranium 2,800
			1. Energy 0.16/0.45; 63 kWh (0% hydr)
			2a. Meat 114
			3. Telephones 0.3; Cars 23
			  Air: 237 pass-km.; 16 ton-km.
				Exports: $209
				  Uranium Livestock
				  Exports to: France
				  Nigeria Japan Spain
					Imports: $345
					  Manufactured
					  goods
					  Imports from:
					  France USA Ivory
					  Coast Nigeria
					
					Services: Exports $58, Imports $203
					Aid received: $296 (40)

NIGERIA
1. Republic
2. English
3. Naira
4. $1 = 10.69
  £1 = 17.67
	1. Area 923,770 sq km
	2. Population 108,542,000
	3. Density 119 per sq km
	4. BR 47; DR 14; AI 3.3%
	5. Inf. Mort. 96 ; Med.16
	6. Life Exp. M 51, F 54
	7. Urban Population 35%
	8. Lagos (Abuja)
	9. Arable 34%; Past .23% ;For. 15%
		1. GNP $28,314 mill (-0.3%); $250 (-3.6%)
		  Agric. 31%, Ind. 44% (Man. 10%), Serv. 25%
		2. Cassava 16,500 Cocoa beans 170
		  Cotton lint 42 Cotton seed 108
		  Groundnuts 680 Maize 1,832 Millet 4,000
		  Palm oil 790 Plantains 1,800 Rice 2,000
		  Rubber 80 Sorghum 4,000 Yams 16,000
		3. Cattle 12,200 Goats 26,000
		  Pigs 1,100 Sheep 13,200
		4. Fish 261
		5. Timber 104,881
		6. Natural gas 164 Oil 86.5 Tin 1.2
			1. Energy 127.83/20.19; 4,040 kWh (47% hydr)
			2a. Beer 6.7 Butter 8 Cheese Meat 935
			2b. Cement 3,000 Fertilizers 243 Petrol 7.5
			  TV 34 Tin 0.3
			3. Telephones 0.4; Cars 377
			  Air: 1,632 pass-km.; 37 ton-km.
				Exports: $8,138
				  Mineral fuels and
				  lubricants Foodstuffs
				  Exports to: USA
				  France Neths. Italy UK
					Imports: $3,419
					  Machinery and
					  transport equip.
					  Manufactures
					  Food and live
					  animals
					  Imports from: UK
					  Germany USA
					  France Japan
					
					Services: Exports $281, Imports $3,361
					Aid received: $339 (3)

NORWAY
1. Kingdom
2. Norwegian
3. Krone
4. $1 = 6.98
  £1 = 11.54
	1. Area 323,900 sq km
	2. Population 4,212,000
	3. Density 14 per sq km
	4. BR 13; DR 11; AI 0.3%
	5. Inf. Mort. 6 ; Med.36
	6. Life Exp. M 74, F 81
	7. Urban Population 74%
	8. Oslo
	9. Arable 3%; Past .0%; For. 27%
		1. GNP $92,097 mill (3.9%); $21,850 (3.5%)
		  Agric. 3%, Ind. 34% (Man. 15%), Serv. 63%
		2. Barley 600 Oats 596 Potatoes 499
		3. Cattle 940 Sheep 2,248
		4. Fish 1826
		5. Timber 10,984
		6. Copper 20 Iron 1.3 Lead 3 Natural gas 1,121
		  Nickel 1.3 Oil 81.9 Zinc 15
			1. Energy 166.29/30.18; 26,715 kWh (99% hydr)
			2a. Beer 2.1 Butter 25 Cheese 77 Meat 224
			  Paper 1.7
			2b. Aluminium 867 Cement 1,260 Copper 45.2
			  Fertilizers 622 Iron 900 Nickel 54.9
			  Petrol 9.8 Ships 361 Steel 396 Zinc 121
			3. Telephones 67; Cars 1,622
			  Rail: 2,136 pass-km; 2,352 ton-km
			  Air: 5,389 pass-km.; 116 ton-km
			  Sea: goods loaded 89,100; unloaded 18,900
				Exports: $34,072
				  Oil Natural gas Metal
				  prods Machinery and
				  transport equip.
				  Foodstuffs
				  Exports to: UK
				  Germany Sweden
				  France Neths.
					Imports: $26,905
					  Machinery and
					  transport equip.
					  Metals and metal
					  prods Foodstuffs
					  Imports from:
					  Sweden Germany
					  UK Denmark USA
					
					Services: Exports $14,233, Imports $16,817
					Aid given: $917

For detailed table headings and notes see the first page of this section

Country	Area and population	Production	Manufactures	Trade	

PAKISTAN

1. Republic
2. Urdu Punjabi
3. Rupee
4. $1 = 23.6
   £1 = 39.00

1. Area 796,100 sq km
2. Population 122,626,000
3. Density 145 per sq km
4. BR 42; DR 11; AI 3.7%
5. Inf. Mort. 98; Med.17
6. Life Exp. M 59, F 59
7. Urban Population 32%
8. Islamabad
9. Arable 27%; Past .6%;For. 4%

1. GNP $40,134 mill (6.3%); $370 (2.9%)
   Agric. 27%, Ind. 24% (Man. 16%), Serv. 49%
2. Cotton lint 1,541 Cotton seed 3,082 Dates 302
   Maize 1,279 Mangoes 761 Oranges 1,139
   Potatoes 670 Rapeseed 233 Rice 5,250
   Seed cotton 4,331 Sugar cane 38,000
   Tobacco 63 Wheat 14,300
3. Cattle 17,573 Goats 35,000 Sheep 29,239
4. Fish 445
5. Timber 23,928
6. Chromium 5 Magnesite 4 Natural gas 490
   Salt 700

1. Energy 20.47/31.74; 8,467 kWh (34% hydr;
   2% nucl)
2a. Butter 332 Meat 1,233 Sugar 2,011
2b. Cars 26 Cement 7,488 Commercial veh. 16
   Fertilizers 1,214 Petrol 5.3 TV 199 Tyres 912
3. Telephones 0.7; Cars 540
   Rail: 20,052 pass-km; 6,612 ton-km
   Air: 7,761 pass-km; 328 ton-km
   Sea: goods loaded 4,740; unloaded19,716

Exports: $5,522
Raw cotton Cotton
yarn and fabrics Rice
Leather Carpets
Exports to: Japan USA
Germany USA UK Italy
Saudi Arabia

Imports: $7,356
Machinery and
transport equip.
Mineral oils
Chemicals
Manufactures
Foodstuffs
Imports from:
Japan USA Kuwait
Germany UK Saudi

Services: Exports $1,088, Imports $2,556
Aid received: $1,119 (10)

PANAMA

1. Republic
2. Spanish
3. Balboa
4. $1 = 1
   £1 = 1.65

1. Area 77,080 sq km
2. Population 2,418,000
3. Density 32 per sq km
4. BR 25; DR 5; AI 2.1%
5. Inf. Mort. 21; Med.22
6. Life Exp. M 71, F 75
7. Urban Population 55%
8. Panama
9. Arable 8%; Past .18%; For. 52%

1. GNP $4,211 mill (0.1%); $1,780 (-2.1%)
   Agric. 11%, Ind. 15% (Man. 7%), Serv. 75%
2. Bananas 1,250 Coffee 13 Oranges 37
   Sugar cane 1,600
3. Cattle 1,500 Pigs 240
4. Fish 112
5. Timber 2,047

1. Energy 0.27/1.4; 898 kWh (61% hydr)
2a. Beer 1 Meat 95 Sugar 110
2b. Petrol 1.1
3. Telephones 10.7; Cars 145
   Air: 491 pass-km; 15 ton-km
   Sea: goods loaded 94,656; unloaded 67,068

Exports: $321
Bananas Coffee
Shrimps Sugar
Exports to: USA
Germany Costa Rica

Imports: $1,489
Mineral products
Electrical goods
Machinery and
transport equip.
Imports from: USA
Japan Mexico
Venezuela

Services: Exports $1,880, Imports $1,311
Aid received: $17 (7)

PAPUA NEW GUINEA

1. Commonwealth
2. Motu English
3. Kina
4. $1 = 0.95
   £1 = 1.57

1. Area 462,840 sq km
2. Population 3,874,000
3. Density 8.2 per sq km
4. BR 33; DR 11; AI 2.3%
5. Inf. Mort. 53; Med.19
6. Life Exp. M 52, F 57
7. Urban Population 16%
8. Port Moresby
9. Arable 1%; Past .0%; For. 84%

1. GNP $3,444 mill (1.8%); $900 (-0.7%)
   Agric. 28%, Ind. 30% (Man. 10%), Serv. 42%
2. Cocoa beans 40 Coffee 65 Copra 116
   Palm oil 143
3. Cattle 101 Pigs 1,800
4. Fish 26
5. Timber 8,231
6. Copper 205 Gold 33.6 Silver 92

1. Energy 0.06/1.12; 490 kWh (32% hydr)
2a. Meat 50
3. Telephones 2.1; Cars 17
   Air: 542 pass-km; 9 ton-km.

Exports: $1,281
Copper ore Coffee
Timber Cocoa beans
Exports to: Japan
Germany Australia
Philippines

Imports: $1,335
Machinery and
transport equip.
Manufactures
Foodstuffs
Imports from:
Australia Japan
USA Singapore

Services: Exports $200, Imports $691
Aid received: $334 (88)

PARAGUAY

1. Republic
2. Spanish
3. Guarani
4. $1 = 1317
   £1 = 2,176.47

1. Area 406,750 sq km
2. Population 4,277,000
3. Density 11 per sq km
4. BR 33; DR 6; AI 3.1%
5. Inf. Mort. 39; Med.20
6. Life Exp. M 65, F 70
7. Urban Population 48%
8. Asuncion
9. Arable 6%; Past .52%;For. 38%

1. GNP $4,299 mill (1.6%); $1,030 (-1.5%)
   Agric. 29%, Ind. 22% (Man. 16%), Serv. 48%
2. Cassava 4,000 Cotton lint 230 Cotton
   seed 375 Maize 1,139 Oranges 366
   Seed cotton 515 Soybeans 1,826
3. Cattle 8,100 Pigs 2,305
4. Fish 10
5. Timber 8,358

1. Energy 0.34/0.91; 5,450 kWh (100% hydr)
2a. Beer 0.9 Meat 285 Sugar 105
3. Telephones 2.1; Cars 80
   Rail: 3 pass-km; 30 ton-km

Exports: $1,163
Cotton Oil seeds
Timber
Exports to:
Netherlands Brazil
Argentina Switz.

Imports: $695
Petrol Chemicals
Machinery and
transport equip.
Imports from: Brazil
USA Algeria
Argentina

Services: Exports $341, Imports $587
Aid received: $91 (22)

PERU

1. Republic
2. Spanish Quechua
3. Sol
4. $1 = 0.8
   £1 = 1.32

1. Area 1,285,220 sq km
2. Population 21,550,000
3. Density 17 per sq km
4. BR 29; DR 8; AI 2.2%
5. Inf. Mort. 76; Med.21
6. Life Exp. M 63, F 67
7. Urban Population 70%
8. Lima
9. Arable 3%; Past .21%; For. 54%

1. GNP $23,009 mill (0.6%); $1,090 (-1.6%)
   Agric. 8%, Ind. 30% (Man. 21%), Serv. 62%
2. Coffee 101 Cotton lint 93 Cotton seed 160
   Maize 705 Oranges 165 Plantains 420
   Potatoes 1,300 Rice 1,037 Seed cotton 321
3. Cattle 3,800 Goats 1,700 Pigs 2,300
   Sheep 12,750
4. Fish 6637
5. Timber 8,780
6. Antimony 0.3 Copper 364 Gold 5.2 Iron 1.9
   Lead 192 Molybdenum 2,400 Natural gas 20
   Oil 6.9 Salt 350 Silver 1,840 Tin 5.1
   Tungsten 1,200 Zinc 597

1. Energy 12.11/10.64; 3,986 kWh (57% hydr)
2a. Beer 8.6 Cheese 17 Meat 476
   Paper 0.3 Sugar 580
2b. Cars 1 Cement 2,184 Commercial Veh. 3
   Copper 224 Iron 250 Lead 74.4 Petrol 7.5 TV 116
   Zinc 138
3. Telephones 3.2; Cars 377
   Rail: 594 pass-km; 594 ton-km
   Air: 2,670 pass-km; 79 ton-km.

Exports: $3,562
Copper Petrol Lead
Zinc Fish Coffee
Exports to: USA Japan
Germany

Imports: $1,839
Fuels Machinery
Chemicals Food
Tobacco
Imports from: USA
Germany Brazil
Argentina

Services: Exports $1,057, Imports $2,106
Aid received: $300 (14)

PHILIPPINES

1. Republic
2. Philipino Spanish
   English
3. Peso
4. $1 = 25.72
   £1 = 42.50

1. Area 300,000 sq km
2. Population 62,413,000
3. Density 206 per sq km
4. BR 30; DR 7; AI 2.6%
5. Inf. Mort. 40; Med.20
6. Life Exp. M 63, F 67
7. Urban Population 42%
8. Manila
9. Arable 27%; Past .4%; For. 36%

1. GNP $42,754 mill (0.6%); $700 (-1.8%)
   Agric. 24%, Ind. 33% (Man. 22%), Serv. 43%
2. Bananas 3,250 Coconuts 8,300 Coffee 105
   Copra 2,072 Maize 4,200 Mangoes 375
   Palm oil 65 Pineapples 1,200 Rice 9,600
   Rubber 180 Sugar cane 19,600 Tobacco 71
3. Cattle 3,800 Goats 2,187
   Pigs 2,300 Sheep 12,750
4. Fish 2042
5. Timber 38,214
6. Chromium 170 Copper 193 Gold 37.5
   Manganese 1 Nickel 15.4 Salt 500
   Silver 51

1. Energy 2.76/17.98; 7,038 kWh (31% hydr; 13% geo)
2a. Beer 11.1 Meat 980 Paper 0.3 Sugar 1,580
2b. Cars 27 Cement 6,348 Copper 132 Fertilizers 327
   Petrol 8.5 TV 300 Tyres 2,016
3. Telephones 1.7; Cars 353
   Rail: 240 pass-km; 60 ton-km
   Air: 9,259 pass-km.; 239 ton-km
   Sea: goods loaded 13,032; unloaded 28,668

Exports: $7,747
Electric and electronic
goods Clothing
Coconut oil Copper
Fish Timber
Exports to: USA Japan

Imports: $10,732
Machinery and
transport equip.
Mineral fuels
Electrical
machinery Base
metals Chemicals
Parts for electrical
equip.
Imports from: USA
Japan

Services: Exports $4,586, Imports $4,283
Aid received: $831 (14)

POLAND

1. Republic
2. Polish
3. Zloty
4. $1 = 11317
   £1 = 18,702.47

1. Area 312,680 sq km
2. Population 38,423,000
3. Density 125 per sq km
4. BR 15; DR 10; AI 0.8%
5. Inf. Mort. 17; Med.32
6. Life Exp. M 68, F 76
7. Urban Population 63%
8. Warsaw
9. Arable 49%; Past .13%; For. 29%

1. GNP $66,974 mill (2.6%); $1,760 (1.8%)
   Agric. 24%, Ind. 33% (Man. 22%), Serv. 43%
2. Apples 740 Barley 4,000 Oats 2,078
   Potatoes 36,300 Rapeseed 1,200 Rye 5,800
   Sugar beet 15,200 Tobacco 66 Wheat 8,635
3. Cattle 10,600 Goats 10 Pigs 19,800
   Sheep 4,300
4. Fish 655
5. Timber 22,848
6. Coal 216 Copper 385 Lead 47
   Magnesite 25 Natural gas 121 Salt 5,600
   Silver 1,083 Zinc 170

1. Energy 167.63/173.11; 30,750 kWh (6% hydr)
2a. Beer 11.9 Butter 290 Cheese 440 Meat 2,801
   Paper 1.5 Sugar 1,850
2b. Aluminium 48 Cars 264 Cement 12,564
   Commercial veh. 43 Copper 390 Fertilizer 2,584
   Iron 8,640 Lead 78.2 Petrol 12.1 Radios 2,833
   Ships 136 Steel 13,644 TV 647 Tyres 6,024
   Zinc 164
3. Telephones 12.2; Cars 4,232
   Rail: 50,376 pass-km; 83,532 ton-km.
   Air: 2,301 pass-km; 10 ton-km
   Sea: goods loaded 30,732; unloaded12,504

Exports: $13,627
Machinery and
transport equip.
Chemicals Fuel and
power Metals Textiles
and clothing
Exports to: USSR
Germany Czech. UK

Imports: $8,160
Machinery and
transport equip.
Chemicals Fuel
and power Iron
and steel
Consumer
products
Imports from:
USSR Germany
Czech.

Services: Exports $3,611, Imports $6,676

For detailed table headings and notes see the first page of this section

Country	Area and population	Production	Manufactures	Trade	
PORTUGAL		1. GNP $44,058 mill (2.7%); $4,260 (2.1%) Agric. 9%, Ind. 37%, Serv. 54%	1. Energy 0.87/18.59; 7,342 KWh (46% hydr)	Exports: $16,348	Imports: $25,072
1. Republic 2. Portuguese 3. Escudo 4. $1 = 153.4 £1 = 253.51	1. Area 92,390 sq km 2. Population 10,285,000 3. Density 115 per sq km 4. BR 13; DR 10; AI 0.5% 5. Inf. Mort. 13 ; Med.33 6. Life Exp. M 71, F 78 7. Urban Population 33% 8. Lisbon 9. Arable 30%; Past .6%; For. 40%	2. Grapes 1,050 Maize 776 Olives 250 Tomatoes 1,005 Wheat 268 3. Cattle 1,359 Goats 746 Pigs 2,300 Sheep 5,380 4. Fish 347 5. Timber 10,151 6. Copper 103 Salt 700 Silver 19 Tin 0.1 Tungsten 1,400 Uranium 100	2a. Beer 5.1 Butter 11 Cheese 51 Meat 499 Olive oil 41 Paper 0.6 Wine 845 2b. Cars 75 Cement 6,000 Commercial veh. 57 Iron 360 Lead 7 Petrol 9.4 Radios 1,188 Ships 24 TV 485 Tin 0.1 Tyres 3,120 3. Telephones 20.2; Cars 1,947 Rail: 5,568 pass-km; 1,584 ton-km Air: 5,014 pass-km; 125 ton-km	Machinery and transport equip. Chemicals Fuel Metals Textiles Clothing Exports to: France Germany UK Spain USA Neths.	Machinery and transport equip. Chemicals Fuel and power Consumer goods Iron and steel goods Imports from: Germany Spain France Italy UK
				Services: Exports $3,220, Imports $3,033 Aid received: $79 (8)	
ROMANIA		1. GNP $45,000 mill (...); $2,000 (...)	1. Energy 78.74/103.9; 22,904 kWh (24% hydr)	Exports: $5,962	Imports: $9,156
1. Republic 2. Romanian 3. Leu 4. $1 = 61.29 £1 = 101.29	1. Area 237,500 sq km 2. Population 23,272,000 3. Density 101 per sq km 4. BR 15; DR 11; AI 0.5% 5. Inf. Mort. 19 ; Med.33 6. Life Exp. M 69, F 74 7. Urban Population 50% 8. Bucharest 9. Arable 46%; Past .19%; For. 28%	2. Grapes 2,200 Hemp 35 Linseed 40 Maize 9,200 Oats 135 Potatoes 7,600 Soybeans 420 Sunflower seed 770 Tobacco 32 Sugar beet 7,000 Tomatoes 2,350 Wheat 8,000 Wool 29 3. Cattle 7,167 Goats 1,050 Pigs 15,510 Sheep 19,000 4. Fish 268 5. Timber 20,369 6. Coal 37 Copper 43 Iron 0.1 Lead 38 Manganese 65 Natural gas 1,157 Oil 7.9 Salt 5,400 Silver 20 Zinc 45	2a. Beer 10.6 Butter 41 Cheese 101 Meat 1,628 Paper 0.8 Sugar 400 Wine 1,000 2b. Cars 120 Cement 13,200 Commercial veh. 5 Copper 48 Fertilizers 2,620 Iron 9,200 Lead 45 Petrol 28.1 Radios 618 Ships 5 Steel 13,000 TV 484 Zinc 50 3. Rail: 31,082 pass-km; 74,215 ton-km Air: 1,544 pass-km; 12 ton-km	Machinery and transport equip. Fuels Chemicals Exports to: USSR Germany	Mineral fuels Machinery Chemicals Imports from: USSR Germany
RWANDA		1. GNP $2,157 mill (1.3%); $310 (-1.9%) Agric. 37%, Ind. 23% (Man. 15%), Serv. 41%	1. Energy 0.02/0.21;60 kWh (93%hydr)	Exports: $101	Imports: $369
1. Republic 2. French Kinyarwanda 3. Franc 4. $1 = 127 £1 = 210	1. Area 26,340 sq km 2. Population 7,237,000 3. Density 288 per sq km 4. BR 50; DR 16; AI 3.5% 5. Inf. Mort. 112; Med.16 6. Life Exp. M 49, F 52 7. Urban population 8% 8. Kigali 9. Arable 45%; Past .16%; For. 20%	2. Cassava 360 Coffee 37 Groundnuts 18 Plantains 2,150 Sorghum 155 Tea 12 3. Cattle 630 Goats 1,100 Pigs 303 Sheep 367 4. Fish 1.5 5. Timber 5,842 6. Cassiterite ... Natural gas ...	3. Telephones 0.2	Coffee Tea Tin Exports to: Belgium Italy Kenya	Fuels Manufactures Vehicles Imports from: Belgium Kenya Japan
				Services: Exports $52, Imports $143 Aid received: $238 (35)	
SAUDI ARABIA		1. GNP $89,986 mill (-1.1%); $6,230 (-5.9%) Agric. 8%, Ind. 45% (Man. 8%), Serv. 48%	1. Energy 417.1/86.49; 17,150 kWh (0% hydr)	Exports: $28,369	Imports: $21,153
1. Kingdom 2. Arabic 3. Riyal 4. $1 = 3.75 £1 = 6.20	1. Area 2,149,690 sq km 2. Population 14,134,000 3. Density 6.9 per sq km 4. BR 42; DR 7; AI 4.2% 5. Inf. Mort. 58 ; Med.18 6. Life Exp. M 64, F 68 7. Urban Population 77% 8. Riyadh 9. Arable 1%; Past .40%;For. 1%	2. Dates 500 Grapes 100 Tomatoes 390 Wheat 3,100 3. Goats 3,800 Sheep 7,700 4. Fish 47 6. Natural gas 1,069 Oil 323	2a. Meat 343 2b. Cement 13,000 Fertilizers 417 Petrol 77.2 3. Telephones 18; Cars ... Rail: 87 pass-km; 194 ton-km Air: 15,640 pass-km; 453 ton-km	Oil Exports to: Japan UK Germany	Machinery and transport equip. Foodstuffs Textiles Clothing Chemicals Metals and metal products Imports from: Japan USA UK Germany Italy
				Services: Exports $12,745, Imports $15,434 Aid given: $1,171	
SENEGAL		1. GNP $4,716 mill (3.0%); $650 (0.0%) Agric. 22%, Ind. 31% (Man. 20%), Serv. 47%	1. Energy 0.1/0.39; 231 kWh	Exports: $606	Imports: $1,023
1. Republic 2. French 3. CFA franc 4. $1 = 303.4 £1 = 501.40	1. Area 196,720 sq km 2. Population 7,327,000 3. Density 38 per sq km 4. BR 44; DR 16; AI 2.8% 5. Inf. Mort. 80 ; Med.17 6. Life Exp. M 48, F 50 7. Urban Population 38% 8. Dakar 9. Arable 27%; Past .30%; For. 31%	2. Groundnuts 800 Millet 508 3. Cattle 2,700 Goats 1,300 Sheep 3,900 4. Fish 257 5. Timber 4,283 6. Phosphates 2,289 Salt 25	2a. Meat 109 3. Telephones ... Cars 39 Rail: 143 pass-km; 492 ton-km Air: 233 pass-km; 16 ton-km	Groundnuts Phosphates Textiles Exports to: France Ivory Coast Mauritania	Oil Cereals Vehicles Chemicals Imports from: France Italy Germany Ivory Coast
				Services: Exports $505, Imports $737 Aid received: $652 (91)	
SIERRA LEONE		1. GNP $813 mill (-0.8); $200 (-3.2) Agric. 46%, Ind. 11% (Man. 6%), Serv. 43%	1. Energy 0/0.31;110 kWh (2% hydr)	Exports: $138	Imports: $189
1. Republic 2. English, Creole, Mende, Limba 3. Leone 4. $1 = 273 £1 = 451	1. Area 71,740 sq km 2. Population 4,151,000 3. Density 58 per sq km 4. BR 48; DR 22; AI 2.7% 5. Inf. Mort. 143; Med.18 6. Life Exp. M 41, F 45 7. Urban population 32% 8. Freetown 9. Arable 25%; Past .31%; For. 29%	2. Cassava 116 Cocoa 9 Coffee 9 Groundnuts 19 Palm oil 44 Plantains 28 Rice 450 3. Cattle 330 Goats 180 Sheep 330 4. Fish 53 5. Timber 2,938 6. Bauxite 1,548 Diamonds 700 Rutile 114	3. Cars 24	Diamonds Rutile Coffee Bauxite Cocoa Exports to: Belgium Neths. UK	Food Mineral fuels Machinery Vehicles Imports from: UK USA Germany Nigeria
				Services: Exports $52, Imports $55 Aid received: $99 (25)	
SINGAPORE		1. GNP $28,058 mill (6.9%); $10,450 (5.7%) Agric. 0%, Ind. 37% (Man. 26%), Serv. 63%	1. Energy 0/13.4; 3,380 kWh (0% hydr)	Exports: $52,729	Imports: $60,787
1. Republic 2. Chinese Malay Tamil English 3. Dollar 4. $1 = 1.74 £1 = 2.88	1. Area 620 sq km 2. Population 2,723,000 3. Density 4,918 per sq km 4. BR 16; DR 6; AI 1.2% 5. Inf. Mort. 7 ; Med.30 6. Life Exp. M 72, F 77 7. Urban Population 100% 8. Singapore 9. Arable 3%; For .5%	2. Bananas 117 3. Pigs 320 4. Fish 15	2a. Meat 142 2b. Cement 2,000 Petrol 29.8 3. Telephones 44.2; Cars 236 Air: 24,947 pass-km; 1,257 ton-km Sea: goods loaded 81,564; unloaded 106,224	Office machines Petrol Telecomm. equip. Clothing Scientific instruments Exports to: USA Malaysia Japan Hong Kong	Oil Machinery Imports from: Japan USA Malaysia Saudi
				Services: Exports $16,464, Imports $11,328 Aid received: $95 (35)	

For detailed table headings and notes see the first page of this section

Country	Area and population	Production	Manufactures	Trade	
SOMALIA		1. GNP $1,035 mill (1.7%); $170 (-1.3%) Agric. 65%, Ind. 10% (Man. 5%), Serv. 26%	1. Energy 0/0.42; 60 kWh	Exports: $104	Imports: $132
1. Republic	1. Area 637,660 sq km	2. Bananas 117 Sorghum 250	2a. Butter 12 Meat168	Livestock Bananas	Petrol Foodstuffs
2. Somali Arabic	2. Population 7,497,000	3. Cattle 5,400 Goats 20,600 Sheep 13,900	3. Air: 298 pass-km; 4 ton-km	Myrrh	Construction mats
3. Shilling	3. Density 12 per sq km	4. Fish 18		Exports to: Saudi UAE	Imports from: Italy
4. $1 = 2606	4. BR 47; DR 18; AI 3.4%	5. Timber 6,758		Italy	UK Germany
£1 = 4,306.68	5. Inf. Mort. 122 ; Med.17				Kenya
	6. Life Exp. M 45, F 49				
	7. Urban Population 36%			Services: Exports $27, Imports $236	
	8. Mogadishu			Aid received: $440 (72)	
	9. Arable 2%; Past.46%; For.14%				
SOUTH AFRICA		1. GNP $86,029 mill (1.5%); $2,460 (-0.8%) Agric. 6%, Ind. 44% (Man. 24%), Serv. 50%	1. Energy 133.1/105.7; 25,870 kWh(2% hydr; 4% nucl)	Exports: $18,969	Imports: $17,075
1. Republic	1. Area 1,221,040 sq km	2. Cotton lint 56 Cotton seed 100 Grapes 1,463	2a. Beer 16.7 Butter 16 Cheese 40 Meat 1,325 Paper 1.6 Sugar 2,276 Wine 944	Gold Metals Minerals	Machinery and
2. English Afrikaans	2. Population 35,282,000	Maize 9,200 Oranges 520 Pineapples 265	2b. Aluminium 195 Cars 238 Cement 7,000	Precious stones	mechanical equip.
3. Rand	3. Density 29 per sq km	Potatoes 1,250 Sorghum 400	Commercial veh. 120 Copper 144	Precious stones	Transport equip.
4. $1 = 2.86	4. BR 31; DR 9; AI 2.2%	Sugar cane 19,000 Tobacco 34 Wheat 1,800	Fertilizers 478 Iron 6,500 Lead 369	Chemicals Textiles	Chemicals
£1 = 4.73	5. Inf. Mort. 62 ; Med.22	3. Cattle 11,850 Goats 5,862	Nickel 30 Petrol 13.8 Radios 776 Steel 9,500	Exports to: Japan USA	Imports from:
	6. Life Exp. M 60, F 66	Pigs 1,470 Sheep 32,605	TV 321 Tin 2.6 Zinc 86	UK Switzerland	Germany USA UK
	7. Urban Population 59%	4. Fish 1298	3. Telephones 14.3; Cars 3,107		Japan
	8. Pretoria	5. Timber 18,361	Rail: 21,408 pass-km; 92,184 ton-km		
	9. Arable 11%; Past.67%;For.4%	6. Antimony 5.2 Asbestos 145 Chromium 4,200	Air: 7,480 pass-km; 310 ton-km	Services: Exports $3,544, Imports $7,755	
		Coal 180 Cobalt 750 Copper 197	Sea: goods loaded 83,184; unloaded 10,440		
		Diamonds 8,500 Gold 610 Iron 19.9 Lead 78			
		Magnesite 110 Manganese 3,795 Nickel 34			
		Phosphates 3,086 Salt 700 Silver 178 Tin 1.3			
		Uranium 2,500 Vanadium 16,400 Zinc 77			
SPAIN		1. GNP $358,352 mill (2.9%); $9,150 (2.4%) Agric. 5%, Ind. 9%, Serv. 86%	1. Energy 29.13/97.11; 43,791 kWh (37% hydr; 17% nucl)	Exports: $55,640	Imports: $87,694
1. Kingdom	1. Area 504,780 sq km	2. Barley 9,325 Cotton lint 101 Cotton seed 160	2a. Beer 24.2 Butter 26 Cheese 162 Meat 3,297	Machinery and	Machinery and
2. Spanish	2. Population 39,187,000	Grapes 5,659 Lemons 575 Maize 3,170	Olive oil 592 Paper 3.4 Sugar 1,038	transport equip.	transport equip.
3. Peseta	3. Density 78 per sq km	Olives 3,012 Oranges 2,457 Peaches 751	Wine 3,030	Electrical goods	Petrol Chemicals
4. $1 = 112.1	4. BR 13; DR 9; AI 0.4%	Potatoes 5,444 Rye 274 Sugar beet 7,286	2b. Aluminium 431 Cars 1,696 Cement 28,092	Foods Fruit	Imports from:
£1 = 185.26	5. Inf. Mort. 9 ; Med.33	Tangerines 1,484 Tobacco 34	Commercial veh. 310 Copper 205 Fertilizers 2,144	Vegetables Iron and	Germany France
	6. Life Exp. M 74, F 80	Tomatoes 3,114 Wheat 4,759	Iron 6,000 Lead 114 Petrol 45.7 Ships 383	steel Chemicals	Italy USA UK
	7. Urban Population 78%	3. Cattle 5,300 Goats 3,200	Steel 12,000 TV 1,233 Tin 2.2 Zinc 258	Exports to: France	
	8. Madrid	Pigs 16,910 Sheep 27,400	3. Telephones 39.6; Cars 10,219	Germany UK Italy USA	
	9. Arable 41%; Past.20%; For.31%	4. Fish 1430	Rail: 15,480 pass-km; 11,256 ton-km		
		5. Timber 16,095	Air: 20,409 pass-km; 526 ton-km	Services: Exports $28,875, Imports $19,919	
		6. Coal 47 Copper 28 Iron 2.3 Lead 63	Sea: goods loaded 39,654; unloaded 119,904		
		Magnesite 450 Mercury 715 Oil 1.1			
		Potash 690 Salt 3,200 Silver 250 Tin 2			
		Tungsten 80 Uranium 150 Zinc 265			
SRI LANKA		1. GNP $7,268 mill (3.9%); $430 (2.4%) Agric. 26%, Ind. 27% (Man. 16%), Serv. 47%	1. Energy 0.33/1.87; 1,210 kWh (78% hydr)	Exports: $1,529	Imports: $2,088
1. Republic	1. Area 65,610 sq km	2. Cassava 490 Coconuts 1,698 Copra 170	2a. Meat 38	Tea Rubber Precious	Petrol Machinery
2. Sinhalese English Tamil	2. Population 17,217,000	Plantains 620 Rice 2,200 Rubber 120	3. Telephones 0.8; Cars 148	and semi-precious	and transport
3. Rupee	3. Density 263 per sq km	Tea 225 Tobacco 9	Rail: 1,680 pass-km; 168 ton-km	stones Coconuts	equip. Sugar
4. $1 = 39.94	4. BR 21; DR 6; AI 1.5%	3. Cattle 1,800 Goats 520	Air: 1,940 pass-km; 49 ton-km	Exports to: USA	Paper
£1 = 66.00	5. Inf. Mort. 24 ; Med.24	4. Fish 198	Sea: goods loaded 3,936; unloaded 6,525	Germany Japan UK	Imports from:
	6. Life Exp. M 70, F 74	5. Timber 8,882			Japan USA UK
	7. Urban Population 21%	6. Precious stones ... Graphite ...			Hong Kong
	8. Colombo			Services: Exports $401, Imports $774	
	9. Arable 29%; Past.7%; For.27%			Aid received: $558 (33)	
SUDAN		1. GNP $9,035 mill (1.1%); $380 (-1.8%) Agric. 54%, Ind. 9% (Man. 4%), Serv. 37%	1. Energy 0.06/1.52; 450 kWh (50% hydr)	Exports: $509	Imports: $1,060
1. Republic	1. Area 2,505,810 sq km	2. Cotton lint 125 Cotton seed 280 Dates 130	2a. Butter 13 Cheese 64 Meat 559 Sugar 376	Cotton Gum arabic	Machinery and
2. Arabic	2. Population 25,203,000	Groundnuts 400 Mangoes 134 Millet 112	3. Telephones 0.6; Cars 90	Sesame seeds	transport equip.
3. Pound	3. Density 11 per sq km	Seed cotton 480 Sorghum 1,503	Rail: 849 pass-km; 1860 ton-km	Exports to: Saudi	Manufactures
4. $1 = 11.39	4. BR 43; DR 14; AI 3.0%	3. Cattle 21,000 Goats 14,800 Pigs	Air: 471 pass-km; 21 ton-km	Thailand Egypt Italy	Petrol Food
£1 = 18.82	5. Inf. Mort. 99 ; Med.17	Sheep 20,300			Tobacco
	6. Life Exp. M 51, F 53	4. Fish 24			Chemicals Textiles
	7. Urban Population 22%	5. Timber 22,000			Imports from:
	8. Khartoum	6. Chromium 10			Saudi UK Germany
	9. Arable 5%; Past.24%; For.20%				USA
				Services: Exports $203, Imports $1,189	
				Aid received: $760 (31)	
SWEDEN		1. GNP $184,230 mill (2.2%); $21,710 (2.0%) Agric. 3%, Ind. 34% (Man. 23%), Serv. 63%	1. Energy 16.95/42.72; 32,783 kWh (48% hydr; 30% nucl)	Exports: $57,435	Imports: $54,568
1. Kingdom	1. Area 440,940 sq km	2. Apples 147 Barley 2,052 Oats 1,614	2a. Beer 4.1 Butter 69 Cheese 116 Meat 516	Machinery and	Machinery and
2. Swedish	2. Population 8,444,000	Potatoes 1,233 Rapeseed 401 Rye 340	Paper 8.2 Sugar 422	transport equip. Paper	transport equip
3. Krona	3. Density 21 per sq km	Sugar beet 2,500 Wheat 2,173	2b. Aluminium 130 Cars 400 Commercial veh. 70	and pulp Electrical	Chemicals
4. $1 = 6.48	4. BR 13; DR 12; AI 0.2%	3. Cattle 1,678 Pigs 2,175	Fertilizers 278 Iron 2,734 Lead 71.4	mach. Chemicals	Imports from:
£1 = 10.71	5. Inf. Mort. 6 ; Med.39	4. Fish 251	Petrol 15.5 Ships 27 Steel 4,452 TV 354	Exports to: Germany	Germany UK USA
	6. Life Exp. M 75, F 81	5. Timber 21,581	3. Telephones ...; Cars 3,367	UK USA Norway	Finland
	7. Urban Population 84%	6. Gold 3.9 Iron 12.9 Lead 82 Silver 200	Rail: 5,952 pass-km; 17,484 ton-km		
	8. Stockholm	Tungsten 250 Zinc 168	Air: 7,275 pass-km; 169 ton-km	Services: Exports $15,991, Imports $23,332	
	9. Arable 7%; Past.1%; For.70%		Sea: goods loaded 44,580; unloaded 54,976	Aid given: $1,799	
SWITZERLAND		1. GNP $197,984 mill (2.1%); $30,270 (1.8%) Agric. 4%, Ind. 38% (Man. 27%), Serv. 64%	1. Energy 6.47/24.22; 15,320 kWh (76% hydr; 19% nucl)	Exports: $63,884	Imports: $69,869
1. Republic	1. Area 41,290 sq km	2. Apples 293 Pears 120 Potatoes 800	2a. Beer 4.1 Butter 37 Cheese 130 Meat 486	Machinery	Machinery
2. German French Italian	2. Population 6,609,000	Wheat 565	Paper 1.2 Sugar 152 Wine 111	Pharmaceuticals	Chemicals Textiles
3. Franc	3. Density 169 per sq km	3. Cattle 1,848 Pigs 1,857	2b. Aluminium 103 Cement 5,500 Petrol 2.9	Jewellery Watches	Clothing
4. $1 = 1.55	4. BR 12; DR 10; AI 0.5%	4. Fish 5	Steel 1,000	Exports to: Germany	Imports from:
£1 = 2.56	5. Inf. Mort. 7 ; Med.38	5. Timber 4,521	3. Telephones 86; Cars 2,679	France USA Italy	Germany France
	6. Life Exp. M 75, F 81	6. Salt 260	Rail: 10,884 pass-km; 8,304 ton-km		Italy UK USA
	7. Urban Population 60%		Air: 13,834 pass-km; 734 ton-km	Services: Exports $35,324, Imports $20,697	
	8. Bern			Aid given: $558	
	9. Arable 10%; Past.40%; For.26%				

For detailed table headings and notes see the first page of this section

Country	Area and population	Production	Manufactures	Trade

SYRIA

1. Republic
2. Arabic
3. Pound
4. $1 = 20.89
   £1 = 34.52

1. Area 185,180 sq km
2. Population 12,530,000
3. Density 66 per sq km
4. BR 43; DR 6; AI 3.6%
5. Inf. Mort. 39 ; Med.16
6. Life Exp. M 65, F 69
7. Urban Population 52%
8. Damascus
9. Arable 30%; Past .45%; For. 3%

1. GNP $12,444 mill (5.4%); $1,020 (-2.1%)
   Agric. 22%, Ind. 23%, Serv. 55%
2. Barley 846 Cotton lint 180 Cotton seed 248
   Grapes 520 Olives 450 Oranges 150
   Seed cotton 431 Sugar beet 120 Tobacco 20
   Tomatoes 580 Wheat 2,069
3. Goats 1,078 Sheep 14,125
4. Fish 6
5. Timber 48
6. Oil 19.2 Phosphates 1,670

1. Energy 27.07/11.59; 3,555 kWh (25% hydr)
2a. Butter 17 Cheese 62 Meat 190
    Olive oil 98
2b. Cement 4,000 Fertilizers 126 Petrol 10.5
3. Telephones 5.9; Cars 125
   Rail: 1,029 pass-km; 1,508 ton-km
   Air: 792 pass-km; 14 ton-km
   Sea: goods loaded 12,084; unloaded 4,308

Exports: $3,006
Oil Natural gas
Chemicals Textiles
Clothing Foodstuffs
Exports to: USSR Italy
Romania France Iran

Imports: $2,097
Machinery
Foodstuffs
Chemicals Metals
Textiles
Imports from:
Japan France
USSR Germany

Services: Exports $570, Imports $857
Aid received: $139 (12)

TAIWAN

1. Republic
2. Chinese
3. Dollar
4. $1 = 27.07
   £1 = 44.74

1. Area 36,000 sq km
2. Population 20,100,000
3. Density 569 per sq km
4. BR 17; DR 5; AI 1.2%
5. Inf. Mort. ... ; Med...
6. Life Exp. M 71, F 76
7. Urban Population 70%
8. Taipei
9. Arable 13%; Past .11%; For. 52%

1. GNP $ 75,000mill (8.0%); $3,800 (7.0%)
   Agric. 6%, Ind. 46% (Man. 38%), Serv. 48%
2. Bananas 230 Pineapples 230 Rice 1,700
   Soybeans 15 Sugar cane 7,000 Tea 23
3. Cattle 176 Goats 196 Pigs 7,000
4. Fish 1,361
5. Timber 253
6. Coal 1 Gold 0.2 Salt 110

1. Energy ...; 71,643 kWh (10% hydr; 48% nucl.)
2a. Paper 0.9 Sugar 600
2b. Aluminium 67 Cars 275 Cement 17,000
    Copper 74.5 Lead 58.2 Nickel 10 Petrol 20
    Radios 1,199 Steel 8,000 TV 6,442
3. Telephones 3.6; Cars 1,578
   Rail: 8,233 pass-km; 2,278 ton-km

Exports: $60,316
Plastic goods
Calculators Clothing
Exports to: USA Japan
Hong Kong Germany
UK

Imports: $45,511
Petrol Cotton
Electrical goods
Imports from:
Japan USA
Germany Hong
Kong

TANZANIA

1. Republic
2. Swahili English
3. Shilling
4. $1 = 227.5
   £1 = 375.97

1. Area 945,090 sq km
2. Population 27,318,000
3. Density 29 per sq km
4. BR 50; DR 13; AI 3.8%
5. Inf. Mort. 97 ; Med.15
6. Life Exp. M 53, F57
7. Urban Population 70%
8. Dar es Salaam (Dodoma)
9. Arable 6%; Past .40%;For. 48%

1. GNP $3,079 mill (1.8%); $120 (-1.6%)
   Agric. 66%, Ind. 7% (Man. 4%), Serv. 27%
2. Bananas 1,380 Cassava 6,300 Coffee 50
   Cotton lint 59 Cotton seed 114 Maize 2,445
   Mangoes 186 Millet 200 Plantains 1,350
   Sisal 38 Sorghum 368 Tobacco 15
3. Cattle 14,500 Goats 6,650 Sheep 5,200
4. Fish 340
5. Timber 31,958
6. Diamonds 60 Gold 0.2

1. Energy 0.08/0.95;1,336 kWh (34% hydr)
2a. Meat 246 Sugar 110
3. Telephones 0.5; Cars 48
   Rail: 3,391 pass-km; 1,288 ton-km

Exports: $337
Coffee Cotton Sisal
Exports to: Germany
UK India Indonesia

Imports: $1,495
Machinery and
industrial goods
Imports from: UK
Japan Germany
Bahrain

Services: Exports $113, Imports $119
Aid received: $918 (38)

THAILAND

1. Kingdom
2. Thai
3. Baht
4. $1 = 24.81
   £1 = 41.00

1. Area 513,120 sq km
2. Population 57,196,000
3. Density 112 per sq km
4. BR 20; DR 7; AI 1.8%
5. Inf. Mort. 24 ; Med.23
6. Life Exp. M 65, F 69
7. Urban Population 23%
8. Bangkok
9. Arable 39%; Past .1%; For. 28%

1. GNP $64,437 mill (6.5%); $1,170 (4.5%)
   Agric. 15%, Ind. 38% (Man. 21%), Serv. 47%
2. Bananas 1,613 Cassava 23,460 Coconuts
   1,437 Cotton lint 33 Cotton seed 65
   Grapefruit 250 Jute 160 Maize 4,296
   Palm oil 180 Pineapples 1,745 Rice 18,500
   Rubber 930 Silk 1,250 Soybeans 686
   Sugar cane 33,561 Tobacco 74
3. Cattle 5,367 Pigs 4,650
4. Fish 2350
5. Timber 38,214
6. Antimony 0.7 Coal 10 Lead 24 Manganese 8
   Natural gas ... Salt 170 Tin 14.7
   Tungsten 650 Zinc 87

1. Energy 14.21/34.99; 7,872 kWh (29% hydr)
2a. Beer 1 Meat1,152 Paper 0.5 Sugar 4,052
2b. Cars 70 Cement 15,000 Commercial veh. 155
    Lead 18.7 Petrol 11.7 TV 544 Tin 14.6
    Tyres 4,296 Zinc 69
3. Telephones 2; Cars 770
   Rail: 11,172 pass-km; 3,264 ton-km
   Air: 13,448 pass-km; 492 ton-km
   Sea: goods loaded 22,176; unloaded 25,272

Exports: $20,059
Food and live animals
Machinery and
transport equip.
Manufactured goods
Exports to: USA Japan
Singapore Neths.

Imports: $32,746
Machinery and
transport equip.
Manufactures
Chemicals Fuels
Foodstuffs
Imports from:
Japan USA
Singapore
Germany

Services: Exports $7,109, Imports $6,859
Aid received: $697 (13)

TRINIDAD & TOBAGO

1. Commonwealth
2. English
3. Dollar
4. $1 = 4.23
   £1 = 6.99

1. Area 5,130 sq km
2. Population 1,281,000
3. Density 240 per sq km
4. BR 23; DR 6; AI 1.7%
5. Inf. Mort. 14 ; Med.24
6. Life Exp. M 70, F 75
7. Urban Population %
8. Port of Spain
9. Arable 23%; Past .2%; For. 43%

1. GNP $4,000 mill (-5.6%); $3,160 (-7.3%)
   Agric. 3%, Ind. 41% (Man. 8%), Serv. 56%
2. Cocoa beans 2 Copra 5 Grapefruit 3
   Sugar cane 1,250
3. Cattle 78 Goats 50 Pigs 84
4. Fish 3
5. Timber 60
6. Natural gas 140 Oil 7.8

1. Energy 16.06/7.13; 985 kWh (0% hydr)
2a. Meat 29 Sugar 97
2b. Fertilizers 248 Iron 550 Petrol 4.6
3. Telephones ...; Cars 244
   Air: 2,354 pass-km; 11 ton-km

Exports: $2,049
Mineral fuels
Chemicals Food and
live animals
Exports to: USA
Barbados UK Jamaica

Imports: $1,222
Machinery and
transport equip.
Foodstuffs
Chemicals
Imports from: USA
UK Canada Japan

Services: Exports $347, Imports $823
Aid received: $6 (5)

TUNISIA

1. Republic
2. Arabic French
3. Dinar
4. $1 = 0.98
   £1 = 1.62

1. Area 163,610 sq km
2. Population 8,180,000
3. Density 53 per sq km
4. BR 27; DR 6; AI 2.5%
5. Inf. Mort. 44 ; Med.21
6. Life Exp. M 67, F 69
7. Urban Population 54%
8. Tunis
9. Arable 31%; Past .20%; For. 4%

1. GNP $10,089 mill (3.1%); $1,260 (0.6%)
   Agric. 14%, Ind. 38% (Man. 16%), Serv. 53%
2. Barley 478 Dates 73 Grapes 130 Olives 330
   Oranges 150 Tomatoes 460 Wheat 1,122
3. Goats 1,200 Sheep 5,000
4. Fish 103
5. Timber 3,078
6. Iron 0.1 Lead 2 Natural gas 16 Oil 4.8
   Phosphates 6,566 Salt 450 Zinc 10

1. Energy 7.66/5.71; 1,414 kWh (5% hydr)
2a. Meat 134 Olive oil 165 Wine 23
2b. Cars 1 Cement 4,140 Commercial veh. 1
    Fertilizers 944 TV 58 Tyres 504
3. Telephones 4.1; Cars 166
   Rail: 1,032 pass-km; 2,052 ton-km
   Air: 1,389 pass-km; 19 ton-km
   Sea: goods loaded 6,552; unloaded 9,900

Exports: $2,933
Petroleum Clothing
Fertilizers
Exports to: France
Germany Italy Belgium

Imports: $4,378
Machinery Petrol
Cotton Iron and
steel
Imports from:
France Germany
Italy USA

Services: Exports $1,629, Imports $1,281
Aid received: $247 (31)

TURKEY

1. Republic
2. Turkish
3. Lira
4. $1 = 4286
   £1 = 7,083.04

1. Area 779,450 sq km
2. Population 58,687,000
3. Density 76 per sq km
4. BR 27; DR 8; AI 2.3%
5. Inf. Mort. 62 ; Med.22
6. Life Exp. M 65, F 68
7. Urban Population 48%
8. Ankara
9. Arable 36%; Past .11%; For. 26%

1. GNP $74,731 mill (5.3%); $1,360 (3.0%)
   Agric. 17%, Ind. 35% (Man. 23%), Serv. 48%
2. Barley 7,000 Cotton lint 628
   Cotton seed 1,005 Grapes 3,350 Maize 2,000
   Olives 1,000 Oats 274 Oranges 750
   Potatoes 4,100 Rye 285 Seed cotton 1,443
   Sugar beet 12,500 Tobacco 210
   Tomatoes 5,700 Wheat 20,000
3. Cattle 11,600 Goats 13,100 Sheep 31,500
4. Fish 628
5. Timber 16,809
6. Aluminium 562 Antimony 1 Asbestos 20
   Chromium 600 Coal 38 Copper 45 Iron 2.9
   Lead 14 Magnesite 1,200 Manganese 7
   Mercury 100 Molybdenum 100 Oil 3.8
   Phosphates 200 Salt 1,300 Tungsten 300
   Zinc 37

1. Energy 21.91/52.48; 15,806 kWh (42% hydr)
2a. Beer 2.4 Butter 119 Cheese 143 Meat 941
    Olive oil 90 Paper 0.4 Sugar 1,314
    Wine 23
2b. Aluminium 62 Cars 116 Cement 24,000
    Commercial veh. 31 Copper 75.6 Fertilizers 1,359
    Iron 3,800 Lead 9 Petrol 19.3 Radios 441
    Ships 8 Steel 7,500 TV 694 Tyres 6,636
    Zinc 24
3. Telephones 9.1; Cars 983
   Rail: 6,408 pass-km; 7,908 ton-km
   Air: 3,296 pass-km; 47 ton-km
   Sea: goods loaded 138,156; unloaded 43,476

Exports: $11,627
Textiles Agricultural
prods Metals
Foodstuffs
Exports to: Germany
Iraq Italy USA

Imports: $15,799
Fuels Machinery
Chemicals Iron
and steel
Pharmaceuticals
Imports from:
Germany USA Iraq
Italy

Services: Exports $7,083, Imports $5,474
Aid received: $122 (2)

For detailed table headings and notes see the first page of this section

Country	Area and population	Production	Manufactures	Trade	

UGANDA

1. Republic
2. English Swahili
3. Shilling
4. $1 = 698.67
   £1 = 1,154.62

1. Area 235,880 sq km
2. Population 18,794,000
3. Density 94 per sq km
4. BR 52; DR 14; AI 3.7%
5. Inf. Mort. 94 ; Med.15
6. Life Exp. M 51, F 55
7. Urban Population 10%
8. Kampala
9. Arable 34%; Past . 25%; For. 28%

1. GNP $4,254 mill (2.2%); $250 (-1.0%)
   Agric. 67%, Ind. 7% (Man. 5%), Serv. 26%
2. Cassava 2,500 Bananas 480 Coffee 168
   Cotton lint 6 Cotton seed 12 Groundnuts 120
   Maize 300 Millet 420 Plantains 6,700
   Sorghum 300
3. Cattle 3,950 Goats 3,000 Sheep 1,800
4. Fish 241
5. Timber 13,873
6. Tungsten 50

1. Energy 0.09/0.49; 163 kWh (96% hydr)
2a. Meat 149 Sugar 20
3. Telephones 0.4; Cars 10
   Rail: 231 pass-km; 61 ton-km
   Air: 98 pass-km; 15 ton-km

Exports: $250
  Coffee Cotton Tea
  Exports to: UK USA
  Spain

Imports: $544
  Sugar Vehicles
  Clothing
  Construction
  materials Food
  Imports from: UK
  Germany Kenya

Services: Exports $35, Imports $166
Aid received: $397 (24)

UNITED KINGDOM

1. Kingdom
2. English Welsh
3. Pound
4. $1 = 0.61
   £1 = 1.01

1. Area 244,880 sq km
2. Population 57,237,000
3. Density 237 per sq km
4. BR 14; DR 12; AI 0.2%
5. Inf. Mort. 8 ; Med.36
6. Life Exp. M 73, F 79
7. Urban Population 93%
8. London
9. Arable 29%; Past . 48%; For. 10%

1. GNP $834,166 mill (3.1%); $14,570 (2.9%)
   Agric. 2%, Ind. 37% (Man. 20%), Serv. 62%
2. Apples 249 Barley 7,985 Oats 535
   Potatoes 6,300 Rapeseed 1,231
   Sugar beet 8,000 Wheat 13,900 Wool 58
3. Cattle 11,933 Goats Pigs 7,383 Sheep 29,521
4. Fish 946
5. Timber 6,400
6. Coal 92.9 Lead 1 Natural gas 1,842
   Oil 87.9 Potash 455 Salt 5,900 Tin 4

1. Energy 280.18/288.16; 69,879 kWh (6% hydr;
   11% nucl)
2a. Beer 59.9 Butter 134 Cheese 307 Meat 3,371
   Paper 4.3 Sugar 1,304
2b. Aluminium 552 Cars 1,296 Cement 15,600
   Commercial veh. 274 Copper 248 Fertilizers 1,793
   Iron 12,492 Lead 350 Nickel 26.1 Petrol 80.9
   Radios 408 Ships 78 Steel 18,000 TV 3,022
   Tin 16.8 Tyres 29,376 Zinc 124
3. Telephones 60; Cars 18,859
   Rail: 33,528 pass-km; 15,096 ton-km
   Air: 77,161 pass-km; 2863 ton-km
   Sea: goods loaded 114,825; unloaded 168,336

Exports: $185,976
  Machinery and
  transport equip.
  Rubber Paper Textile
  manufactures
  Chemicals Fuels
  Beverages Food
  Tobacco
  Exports to: USA
  Germany Greece
  Neths. Belgium

Imports: $224,938
  Machinery and
  transport equip.
  Manufactures
  Chemicals Foods
  Mineral fuels
  Imports from:
  Germany USA
  France Neths.
  Japan

Services: Exports $172,010, Imports $157,790
Aid given: $2,587

UNITED STATES

1. Republic
2. English Spanish
3. Dollar
4. $1 = 1
   £1 = 1.65

1. Area 9,372,610 sq km
2. Population 249,224,000
3. Density 27 per sq km
4. BR 14; DR 9; AI 0.9%
5. Inf. Mort. 8 ; Med.33
6. Life Exp. M 73, F. 80
7. Urban Population 74%
8. Washington
9. Arable 21%; Past . 26%; For. 29%

1. GNP $5,237,707 mill (3.2%); $21,100 (2.2%)
   Agric. 2%, Ind. 29% (Man. 17%), Serv. 69%
2. Apples 4,400 Barley 9,121 Butter 572 Cheese
   3,132 Cotton lint 3,245 Cotton seed 5,140
   Grapefruit 1,772 Grapes 4,970
   Groundnuts 1,567 Maize 201,508 Oats 5,184
   Oranges 7,084 Peaches 1,205 Pineapples 545
   Potatoes 17,835 Rice 7,027 Seed cotton 6,986
   Sorghum 14,516 Soybeans 52,303
   Sugar beet 25,032 Sugar cane 22,407
   Tobacco 680 Wheat 74,534
3. Cattle 99,337 Goats 1,870 Pigs 53,852
   Sheep 11,368
4. Fish 5966
5. Timber 532,932
6. Aluminium 670 Antimony 2.5 Asbestos 20
   Coal 944 Copper 1,498 Gold 295 Iron 34.9
   Lead 419 Magnesite 100 Manganese 4
   Mercury 104 Molybdenum 43,000
   Natural gas 17,480 Oil 366 Phosphates 38,000
   Potash 1,635 Salt 35,700 Silver 2,007
   Tungsten 400 Uranium 3,500 Vanadium 1,500
   Zinc 288

1. Energy 2055.75/2504.66; 757,593 kWh(12%
   hydr; 14% nucl; 1% geo)
2a. Beer 230 Butter 572 Cheese 3,132 Meat 28,344
   Paper 69.5 Sugar 6,464 Wine 1,775
2b. Aluminium 6,084 Cars 6,052 Cement 70,944
   Commercial veh. 3,720 Copper 2,868
   Fertilizers 23,324 Iron 51,000 Lead 1,169
   Nickel 0.3 Petrol 688 Radios 4,315 Steel 87,000
   TV 12,871 Tin 16.2 Tyres 212,868 Zinc 606
3. Telephones 70; Cars 137,323
   Rail: 9,864 pass-km; 1,513776 ton-km
   Air: 642,121 pass-km; 11,938 ton-km
   Sea: goods loaded 381,096; unloaded 493,860

Exports: $393,893
  Machinery and
  transport equip.
  Chemicals Food
  Beverages Tobacco
  Crude materials
  Exports to: Canada
  Japan Mexico UK

Imports: $516,575
  Machinery and
  transport equip.
  Clothing
  Manufactures
  Metals Foodstuffs
  Imports from:
  Japan Canada
  Germany Mexico

Services: Exports $242,710, Imports $223,140
Aid given: $7,676

URUGUAY

1. Republic
2. Spanish
3. Peso
4. $1 = 2004
   £1 = 3,311.81

1. Area 177,410 sq km
2. Population 3,094,000
3. Density 18 per sq km
4. BR 17; DR 10; AI 0.6%
5. Inf. Mort. 20 ; Med.31
6. Life Exp. M 69, F 76
7. Urban Population 86%
8. Montevideo
9. Arable 8%; Past . 77%; For. 4%

1. GNP $8,069 mill (-0.2%); $2,620 (-0.8%)
   Agric. 11%, Ind. 28% (Man. 22%), Serv. 61%
2. Grapes 124 Maize 20 Oranges 95 Wheat 375
3. Cattle 8,723 Sheep 26,000
4. Fish 107
5. Timber 3,295

1. Energy 0.48/2.41; 1,681 kWh (71% hydr)
2a. Beer 0.7 Butter 12 Cheese 17 Meat 470
   Wine 74
3. Telephones 14.9; Cars 221
   Rail: 140 pass-km; 210 ton-km
   Air: 459 pass-km; 2 ton-km

Exports: $1,599
  Hides Leather Meat
  Wool
  Exports to: Brazil USA
  China Germany Arg.

Imports: $1,203
  Fuels Metals
  Machinery Vehicles
  Imports from: Brazil
  Argentina USA
  Germany

Services: Exports $593, Imports $912
Aid received: $38 (12)

USSR

1. Republic
2. Russian and others
3. Rouble
4. $1 = 1.79
   £1 = 2.96

1. Area 22,402,200 sq km
2. Population 288,595,000
3. Density 13 per sq km
4. BR 17; DR 10; AI 0.8%
5. Inf. Mort. 20 ; Med.31
6. Life Exp. M 67, F 75
7. Urban Population 68%
8. Moscow
9. Arable 10%; Past . 17%; For. 42%

1. GNP $1,064,000 mill (...); $3,800 (...)
2. Barley 59,000 Cotton lint 2,613
   Cotton seed 4,900 Flax 326 Grapes 4,700
   Hemp 24 Linseed 230 Maize 16,000
   Millet 5,000 Oats 18,800 Potatoes 62,000
   Rye 21,000 Seed cotton 8,600 Silk 4,400
   Soybeans 920 Sugar beet 90,000
   Tomatoes 7,300 Wheat 108,000
3. Cattle 118,400 Goats 6,480 Pigs 78,900
   Sheep 137,000
4. Fish 11332
5. Timber 391,800
6. Aluminium 5,750 Antimony 5.8 Asbestos 2,600
   Chromium 3,800 Coal 629 Cobalt 2,900
   Copper 950 Diamonds 15,000 Gold 280
   Iron 34.9 Lead 500 Magnesite 5,000
   Manganese 9,000 Mercury 2,500
   Molybdenum 11,500 Natural gas 27,580
   Nickel 210 Oil 570 Phosphates 35,000
   Potash 9,100 Salt 14,900 Silver 1,500 Tin 14
   Tungsten 7,000 Uranium 31,230
   Vanadium 9,600 Zinc 940

1. Energy 2356.77/1875.23; 333,100 kWh (19%
   hydr; 11% nucl)
2a. Beer 50.7 Butter 1,780 Cheese 2,112 Meat 19,970
   Paper 10.2 Sugar 9,565 Wine 1,900
2b. Aluminium 2,380 Cars 1,259 Cement 137,328
   Commercial veh. 860 Copper 1,355
   Fertilizers 36,059 Iron 51,000 Lead 750 Nickel 225
   Petrol 433 Radios 8,143 Steel 154,416 TV 9,081
   Tin 16.5 Tyres 51,396 Zinc 1,020
3. Telephones 13; Cars 13,000
   Rail: 417,444 pass-km; 3,717,012 ton-km
   Air: 200,123 pass-km; 28,272 ton-km
   Sea: goods loaded 90,308; unloaded 34,800

Exports: $104,640
  Machinery Iron and
  steel Oil Metals Petrol
  Timber Cotton
  Vehicles
  Exports to: Germany
  Czechoslovakia
  Poland Bulgaria

Imports: $120,867
  Machinery
  Clothing Iron and
  steel Ships
  Minerals
  Imports from:
  Germany
  Czechoslovakia
  Bulgaria Poland

For detailed table headings and notes see the first page of this section

Country	Area and population	Production	Manufactures	Trade

VENEZUELA

1. Republic
2. Spanish
3. Bolivar
4. $1 = 54.67
   £1 = 90.35

1. Area 912,050 sq km
2. Population 19,735,000
3. Density 22 per sq km
4. BR 28; DR 5; AI 2.8%
5. Inf. Mort. 33 ; Med.21
6. Life Exp. M 67, F 74
7. Urban Population 91%
8. Caracas
9. Arable 4%; Past . 20%; For. 35%

1. GNP $47,164 mill (0.5%); $2,450 (-2.3%)
   Agric. 6%, Ind. 46%(Man. 28%). Serv. 48%
2. Bananas 1,160 Cocoa beans 14 Coffee 70
   Cotton lint 31 Cotton seed 54 Maize 1,200
   Mangoes 127 Oranges 427 Plantains 475
   Sorghum 500 Tomatoes 195
3. Cattle 13,819 Goats 1,530 Pigs 2,326
4. Fish 294
5. Timber 1,464
6. Diamonds 160 Iron 11 Natural gas 919
   Oil 118 Phosphates 200 Salt 500

1. Energy 181.69/54.33; 17,733 kWh(39% hydr)
2a. Beer 5 Butter 2 Cheese 82 Meat 760
    Paper 0.7 Sugar 556
2b. Aluminium 576 Cement 6,072 Fertilizers 680
    Iron 3,580 Lead 14.1 Petrol 43.3 Steel 3,600
    Tyres 4,176
3. Telephones 9.2; Cars 1718
   Rail: 60 pass-km; 48 ton-km
   Air: 5,040 pass-km; 118 ton-km

Exports: $12,983 — Imports: $6,881
Oil Bauxite Iron ore — Food Chemicals
Exports to: USA Japan — Manufactures
Colombia — Imports from: USA
Germany Italy
Japan

Services: Exports $2,680, Imports $5,855
Aid received: $21 (1)

VIETNAM

1. Republic
2. Vietnamese
3. Dong
4. $1 = 8653
   £1 = 14,299.95

1. Area 331,690 sq km
2. Population 66,693,000
3. Density 202 per sq km
4. BR 30; DR 8; AI 2.2%
5. Inf. Mort. 54 ; Med.20
6. Life Exp. M 62, F 66
7. Urban Population 22%
8. Hanoi
9. Arable 20%; Past . 1%; For. 28%

1. GNP $30,000 mill (...); $450 (...)
2. Cassava 2,900 Copra 131 Groundnuts 210
   Maize 850 Pineapples 490 Rice 19,150
   Silk 450 Tea 34 Tobacco 30
3. Cattle 3,100 Pigs 11,700
4. Fish 874
5. Timber 26,620
6. Chromium 5 Coal 5.5 Phosphates 500
   Salt 300 Tin 0.8 Zinc 10

1. Energy 6.32/7.17; 1,270 kWh (26% hydr)
2a. Beer 0.9 Meat 886 Sugar 390
2b. Cement 2,000 Fertilizers 72 Tin 0.5 Zinc 10
3. Telephones 0.2; Cars ...

Exports:... — Imports:...
Raw materials — Fuels Raw mats
Handicrafts — Machinery
Agricultural products — Foodstuffs

Main trade with socialist countries Japan France
Singapore Hong Kong
Aid received: $138 (2)

YEMEN

1. Republic
2. Arabic
3. Rial/Dinar
4. $1 = 11.98/0.43
   £1 = 19.80/0.71

1. Area 527,970 sq km
2. Population 11,282,000
3. Density 21 per sq km
4. BR 51; DR 14; AI 1.8%
5. Inf. Mort. 107 ; Med.15
6. Life Exp. M 33, F 34
7. Urban Population 30 %
8. Sana
9. Arable 3%; Past . 30%; For. 6%

1. GNP $7,203 mill (...); $640 (...)
2. Coffee 5 Dates 28 Millet 52 Sorghum 500
   Wheat 150
3. Cattle 1,200 Goats 3,300 Sheep 3,700
4. Fish 73
5. Timber 312
6. Oil 9.9

1. Energy 12.05/3.84; 715 kWh (0% hydr)
2a. Cheese 13 Meat 139
2b. Petrol 3.5
3. Air: 525 pass-km; 7 ton-km

Exports: $1,283 — Imports: $2,899
Cotton Coffee Fish — Food and live
Exports to: Italy Saudi — animals
China — Imports from:
Japan Saudi UK
Italy

YUGOSLAVIA

1. Republic
2. Macedonian Serbo-Croat Slovene
3. Dinar
4. $1 = 23.13
   £1 = 38.22

1. Area 255,800 sq km
2. Population 23,807,000
3. Density 93 per sq km
4. BR 14; DR 9; AI 0.7%
5. Inf. Mort. 21 ; Med.33
6. Life Exp. M 70, F 76
7. Urban Population 50%
8. Belgrade
9. Arable 30%; Past . 25%; For. 37%

1. GNP $59,080 mill (0.0%); $2,490 (-0.7%)
   Agric. 10%, Ind. 42%, Serv. 48%
2. Grapes 1,022 Maize 6,600 Oats 260
   Potatoes 2,200 Sugar beet 5,000 Tobacco 54
   Wheat 6,348
3. Cattle 4,705 Goats 65 Pigs 7,231 Sheep 7,596
4. Fish 72
5. Timber 15,186
6. Antimony 0.8 Asbestos 9 Coal 64.8 Copper
   119 Gold 4.6 Iron 1.4 Lead 79 Magnesite 375
   Manganese 40 Mercury 69.9 Natural gas 89
   Nickel 6.3 Oil 3.1 Salt 370 Silver 133 Zinc 75

1. Energy 36.1/60.43; 16,470 kWh (43% hydr; 4% nucl; 4% geo)
2a. Beer 12.1 Butter 11 Cheese 137 Meat 1,503
    Olive oil 3 Paper 1.4 Sugar 930 Wine 486
2b. Aluminium 386 Cars 292 Cement 7,956
    Commercial veh. 49 Copper 203 Fertilizers 995
    Iron 2,628 Lead 117 Nickel 6.3 Petrol 13.7
    Radios 198 Ships 462 Steel 4,500 TV 591
    Tyres 12,636 Zinc 123
3. Telephones 16; Cars 3,024
   Rail: 11,748 pass-km; 24,576 ton-km
   Air: 5,229 pass-km; 99 ton-km
   Sea: goods loaded 8,100; unloaded 25,260

Exports: $12,612 — Imports: $13,171
Machinery and — Machinery and
transport equip. — transport equip.
Manufactures — Mineral fuels
Chemicals Foodstuffs — Chemicals
Exports to: USSR Italy — Manufactures Raw
Germany USA — materials Food
Imports from:
Germany USSR
Italy USA

Services: Exports $5,844, Imports $10,117
Aid received: $43 (2)

ZAIRE

1. Republic
2. French Swahili Lingala
3. Zaire
4. $1 = 4597
   £1 = 7,597.00

1. Area 2,345,410 sq km
2. Population 35,568,000
3. Density 16 per sq km
4. BR 45; DR 13; AI 3.1%
5. Inf. Mort. 75 ; Med.17
6. Life Exp. M 52, F 56
7. Urban Population 40%
8. Kinshasa
9. Arable 3%; Past . 7%; For. 77%

1. GNP $8,841 mill (1.5%); $260 (-1.6%)
   Agric. 30%, Ind. 32% (Man. 10%), Serv. 38%
2. Bananas 350 Coffee 96 Cassava 16,300
   Cotton lint 26 Cotton seed 50 Groundnuts 400
   Maize 756 Mangoes 160 Oranges 153
   Palm oil 180 Pineapples 190 Plantains 1,530
   Rubber 20
3. Cattle 1,500 Goats 3,050
4. Fish 166
5. Timber 34,239
6. Cobalt 25,400 Copper 441 Diamonds 24,000
   Gold 3.7 Silver 60 Tin 1.6 Tungsten 20
   Zinc 73

1. Energy 2.62/2.21; 2,827 kWh (98% hydr)
2a. Beer 3.7 Meat 200
2b. Copper 255 Tin 0.1 Zinc 54
3. Telephones 0.1; Cars 103
   Rail: 504 pass-km; 1,599 ton-km
   Sea: 487 pass-km; 53 ton-km

Exports: $1,249 — Imports: $849
Copper Coffee — Mining equip.
Diamonds Oil Cobalt — Foodstuffs
Zinc — Machinery and
Exports to: Belgium — transport equip.
France Switzerland — Imports from:
Belgium USA
France Germany

Services: Exports $193, Imports $1,324
Aid received: $637 (19)

ZAMBIA

1. Republic
2. English
3. Kwacha
4. $1 = 65.14
   £1 = 107.65

1. Area 752,610 sq km
2. Population 8,452,000
3. Density 11 per sq km
4. BR 50; DR 12; AI 4.0%
5. Inf. Mort. 72 ; Med.15
6. Life Exp. M 54, F 57
7. Urban Population 56%
8. Lusaka
9. Arable 7%; Past . 47%; For. 39%

1. GNP $3,060 mill (-0.2%); $390 (-3.8%)
   Agric. 13%, Ind. 47% (Man. 24%), Serv. 40%
2. Groundnuts 25 Maize 1,093 Millet 32
   Tobacco 5
3. Cattle 2,800 Goats 5304. Fish 68
5. Timber 12,149
6. Cobalt 6,700 Copper 500 Lead 12
   Silver 20 Zinc 24

1. Energy 1.16/1.61; 2,436 kWh (92% hydr)
2a. Beer 3 Meat 87 Sugar 132
2b. Copper 464 Lead 4.4 Zinc 13
3. Telephones 1.1; Cars 75
   Rail: 496 pass-km; 1,407 ton-km
   Air: 609 pass-km; 26 ton-km

Exports: $1,249 — Imports: $873
Copper Cobalt Zinc — Machinery Fuel
Tobacco — Chemicals
Exports to: Japan — Imports from:
Germany UK — S Africa UK USA
Japan

Services: Exports $108, Imports $687
Aid received: $388 (50)

ZIMBABWE

1. Republic
2. English
3. Dollar
4. $1 = 3.23
   £1 = 5.34

1. Area 390,580 sq km
2. Population 9,709,000
3. Density 24 per sq km
4. BR 40; DR 9; AI 3.1%
5. Inf. Mort. 55 ; Med.17
6. Life Exp. M 59, F 63
7. Urban Population 28%
8. Harare
9. Arable 7%; Past . 13%; For. 52%

1. GNP $6,076 mill (2.8%); $640 (-0.8%)
   Agric. 13%, Ind. 39% (Man. 25%), Serv. 49%
2. Groundnuts 119 Maize 1,993 Millet 143
   Soybeans 110 Tobacco 142
3. Cattle 6,550 Goats 2,500
4. Fish 18
5. Timber 7,832
6. Antimony 0.1 Asbestos 180 Chromium 600
   Coal 5.2 Copper 16 Gold 15 Iron 600
   Magnesite 35 Nickel 12.7 Phosphates 12.5
   Silver 22 Tin 0.8

1. Energy 5.44/6.55; 1,634 kWh (39% hydr)
2a. Beer 4 Meat 138 Sugar 460
2b. Copper 24 Fertilizers 764 Iron 800
    Nickel 18.6 Steel 1,000 Tin 0.8
3. Telephones 3.2; Cars 260
   Rail: ... pass-km; 5,592 ton-km
   Air: 626 pass-km; 67 ton-km

Exports: $1,420 — Imports: $1,043
Food Tobacco — Machinery and
Cotton Asbestos — transport equip.
Copper Nickel Iron — Petrol Chemicals
and steel — Manufactures
Exports to: S Africa UK — Imports from:
Germany USA — S Africa UK USA
Germany

Services: Exports $230, Imports $665
Aid received: $266 (28)

Population

thousands

	1950	1960	1970	1980	1990	2000
World	2,516,443	3,019,653	3,697,849	4,448,037	5,292,195	6,260,800
Africa	221,984	279,316	361,768	477,232	642,111	866,585
Asia	1,377,259	1,668,343	2,101,869	2,583,436	3,112,695	3,712,542
Europe	392,523	425,070	459,942	484,429	498,371	510,015
North America	220,361	269,565	321,036	373,767	427,226	479,393
Oceania	12,647	15,782	19,329	22,799	26,481	30,144
South America	111,594	147,242	191,138	240,829	296,716	354,759
Afghanistan	8,958	10,775	13,623	16,063	16,557	26,511
Albania	1,230	1,611	2,138	2,671	3,245	3,795
Algeria	8,753	10,800	13,746	18,740	24,960	32,904
American Samoa	19	21	27	32	38	39
Andorra	6	8	20	31	47	49
Angola	4,131	4,816	5,588	7,723	10,020	13,295
Anguilla	5	6	6	7	7	8
Antigua & Barbuda	46	55	66	75	76	79
Argentina	17,150	20,616	23,962	28,237	32,322	36,238
Aruba	57	59	61	64	60	58
Australia	8,219	10,315	12,552	14,695	16,873	18,855
Austria	6,935	7,048	7,467	7,549	7,583	7,613
Bahamas	79	113	171	210	253	295
Bahrain	116	156	220	347	516	683
Bangladesh	41,783	51,419	66,671	88,219	11,5593	150,589
Barbados	211	231	239	249	255	265
Belgium	8,639	9,153	9,656	9,852	9,845	9,832
Belize	67	91	120	145	187	230
Benin	2,046	2,237	2,693	3,459	4,630	6,369
Bermuda	39	45	55	56	58	62
Bhutan	734	868	1,045	1,245	1,516	1,906
BIOT	2	2	2	2	2	2
Bolivia	2,766	3,428	4,325	5,570	7,314	9,724
Botswana	389	481	623	902	1,304	1,822
Br. Virgin Is	6	7	10	12	13	15
Brazil	53,444	72,594	95,847	121,286	150,368	179,487
Brunei	46	90	133	228	266	333
Bulgaria	7,251	7,867	8,490	8,862	9,010	9,071
Burkina Faso	3,654	4,452	5,550	6,957	8,996	12,092
Burma	18,038	21,780	27,346	34,818	41,675	51,129
Burundi	2,456	2,948	3,522	4,132	5,472	7,358
Cambodia	4,346	5,433	6,938	6,400	8,246	10,046
Cameroon	4,467	5,297	6,610	8,653	11,833	16,701
Canada	13,737	17,909	21,324	24,043	26,521	28,488
Cape Verde Is	146	196	267	289	370	515
Cayman Is	6	9	10	17	25	35
Central African Rep.	1,314	1,534	1,849	2,320	3,039	4,074
Chad	2,658	3,064	3,652	4,477	5,678	7,337
Chile	6,082	7,614	9,504	11,145	13,173	15,272
China	554,760	657,492	830,675	996,134	1,139,060	1,299,180
Christmas Is	1	3	3	3	3	3
Cocos Is	1	1	1	1	1	1
Colombia	11,946	15,939	21,360	26,906	32,978	39,397
Comoros	173	215	274	392	550	789
Congo	808	988	1,263	1,669	2,271	3,167
Cook Is	15	18	21	19	18	17
Costa Rica	862	1,236	1,731	2,284	3,015	3,711
Cuba	5,850	6,985	8,520	9,679	10,608	11,504
Cyprus	494	573	615	629	701	762
Czechoslovakia	12,389	13,654	14,334	15,311	15,667	16,179
Denmark	4,271	4,581	4,929	5,123	5,143	5,153
Djibouti	60	80	168	304	409	552
Dominica	51	60	71	73	82	87
Dominican Rep.	2,353	3,231	4,423	5,697	7,170	8,621
Ecuador	3,310	4,413	6,051	8,123	10,587	13,319
Egypt	20,330	25,922	33,053	40,875	53,153	64,210
El Salvador	1,940	2,570	3,588	4,525	5,252	6,739
Equatorial Guinea	226	252	291	217	352	455
Ethiopia	19,573	24,191	30,623	38,750	49,240	66,364
Falkland Is	2	2	2	2	2	2
Faroe Is	31	35	39	41	47	49
Fiji	289	394	520	634	764	883
Finland	4,009	4,430	4,606	4,780	4,975	5,077
France	41,829	45,684	50,772	53,880	56,138	58,145
French Guiana	25	31	48	66	98	130
French Polynesia	62	82	109	147	206	268
Gabon	469	486	504	806	1,172	1,612
Gambia, The	294	352	464	641	861	1,119
Germany	68,376	72,673	77,709	78,303	79,479	76,962
Ghana	4,900	6,774	8,612	10,736	15,028	20,564
Gibraltar	23	24	26	29	30	32
Greece	7,566	8,327	8,793	9,643	10,047	10,193
Greenland	23	33	47	52	56	60
Grenada	76	90	94	107	85	83
Guadeloupe	210	275	320	327	343	365
Guam	59	67	86	108	118	128
Guatemala	2,969	3,964	5,246	6,917	9,197	12,222
Guinea	2,550	3,136	3,900	4,461	5,755	7,830
Guinea-Bissau	505	542	525	795	964	1,197
Guyana	423	569	709	759	796	891
Haiti	3,261	3,807	4,535	5,370	6,513	8,003
Honduras	1,401	1,935	2,627	3,662	5,138	6,846
Hong Kong	1,974	3,075	3,942	5,039	5,851	6,336
Hungary	9,338	9,984	10,338	10,711	10,552	10,531
Iceland	143	176	204	228	253	274
India	357,561	442,344	554,911	688,856	827,057	1,041,543
Indonesia	79,538	96,194	120,280	150,958	179,300	218,661
Iran	16,913	21,554	28,429	38,900	54,607	68,759
Iraq	5,158	6,847	9,356	13,291	18,920	26,339
Ireland	2,969	2,834	2,954	3,401	3,720	4,086
Israel	1,258	2,114	2,974	3,878	4,600	5,321
Italy	47,104	50,200	53,822	56,434	57,061	57,195
Ivory Coast	2,775	3,799	5,515	8,194	11,997	17,600
Jamaica	1,403	1,629	1,869	2,133	2,456	2,735
Japan	83625	94,096	104,331	116,807	123,460	128,470
Jordan	1,237	1,695	2,299	2,923	4,009	5,558
Kenya	6,265	8,332	11,498	16,632	24,031	35,060
Kiribati	32	41	49	58	66	72
Korea, North	9,726	10,789	14,619	18,260	21,773	26,117
Korea, South	20,357	25,003	31,923	38,124	42,793	46,403
Kuwait	152	278	744	1,375	2,039	2,639
Laos	1,755	2,177	2,713	3,205	4,139	5,463
Lebanon	1,443	1,857	2,469	2,669	2,701	3,327
Lesotho	734	870	1,064	1,339	1,774	2,370
Liberia	824	1,039	1,385	1,876	2,575	3,575
Libya	1,029	1,349	1,986	3,043	4,545	6,500
Liechtenstein	14	16	21	26	28	28
Luxembourg	296	314	339	364	373	377
Macau	188	169	245	287	479	656
Madagascar	4,230	5,309	6,742	8,785	12,004	16,627
Malawi	2,881	3,529	4,518	6,183	8,754	12,458
Malaysia	6,110	8,140	10,853	13,763	17,891	21,983
Maldives	82	92	114	154	215	283
Mali	3,520	4,375	5,484	6,863	8,156	12,685
Malta	312	329	326	364	353	366
Martinique	222	282	326	326	341	362
Mauritania	825	991	1,221	1,551	2,024	2,702
Mauritius	493	660	826	966	1,082	1,201
Mexico	28,012	38,020	52,771	70,416	86,154	107,233
Monaco	22	23	24	26	28	30
Mongolia	761	959	1,256	1,663	2,190	2,847
Montserrat	13	12	11	12	12	13
Morocco	8,953	11,626	15,310	19,382	25,061	31,559
Mozambique	6,198	7,461	9,395	12,095	15,656	20,493
Namibia	666	817	1,016	1,306	1,781	2,437
Nauru	4	5	7	7	9	10
Nepal	8,182	9,404	11,488	14,858	19,143	24,084
Neths. Antilles	162	192	222	247	188	203
Netherlands	10,114	11,480	13,032	14,144	14,951	15,829
New Caledonia	59	79	110	142	167	195
New Zealand	1,908	2,372	2,820	3,113	3,392	3,662
Nicaragua	1,098	1,493	2,053	2,771	3,871	5,261
Niger	2,400	3,028	4,165	5,586	7,731	10,752
Nigeria	32,935	42,305	56,581	78,430	108,542	149,621
Niue	4	4	5	4	3	2
Norfolk I.	1	1	1	2	2	2
Norway	3,265	3,581	3,877	4,086	4,212	4,331
Oman	413	505	654	984	1,502	2,176
Pacific Is	57	78	103	136	142	
Pakistan	39,513	49,955	65,706	85,299	112,050	162,409
Panama	893	1,148	1,531	1,956	2,418	2,893
Papua New Guinea	1,613	1,920	2,422	3,086	3,874	4,845
Paraguay	1,351	1,774	2,351	3,147	4,277	5,538
Peru	7632	9,931	13,193	17,295	22,330	26,276
Philippines	20,988	27,561	37,540	48,317	62,413	77,473
Pitcairn	0.06	0.06	0.06	0.06	0.06	0.06
Poland	24,824	29,561	32,526	35,574	38,423	40,366
Portugal	8,405	8,826	9,044	9,766	10,285	10,587
Puerto Rico	2,219	2,358	2,718	3,206	3,480	3,836
Qatar	25	45	111	229	368	499
Réunion	257	339	441	508	598	692
Romania	16,311	18,407	20,253	22,201	23,272	24,346
Rwanda	2,120	2,742	3,728	5,163	7,237	10,200
Samoa	82	111	143	157	168	171

Population of countries *continued*

	1950	1960	1970	1980	1990	2000
San Marino	13	15	18	21	23	25
São Tomé & Príncipe	60	64	74	85	121	151
Saudi Arabia	3,201	4,075	5,745	9,372	14,134	20,697
Senegal	2,500	3,187	4,158	5,538	7,327	8,716
Seychelles	34	42	52	65	69	75
Sierra Leone	1,944	2,241	2,656	3,263	4,151	5,437
Singapore	1,022	1,634	2,075	2,415	3,003	2,997
Solomon Is	104	123	163	231	320	429
Somalia	2,423	2,935	3,668	5,345	7,497	9,736
South Africa	13,683	17,396	22,458	28,270	35,282	43,666
Spain	28,009	30,455	33,779	37,542	39,187	40,667
Sri Lanka	7,678	9,889	1,2514	1,4819	17,217	19,416
St Christopher/N.	50	57	52	52	44	44
St Helena	5	5	5	5	7	10
St Lucia	79	88	101	120	150	177
St Pierre & M.	5	5	5	6	6	6
St Vincent/G.	67	80	88	99	116	128
Sudan	9,190	11,165	13,859	18,681	25,203	33,625
Surinam	215	290	372	352	422	497
Swaziland	264	326	419	563	788	1,121
Sweden	7,014	7,480	8,043	8,310	8,444	8,560
Switzerland	4,694	5,362	6,187	6,319	6,609	6,762
Syria	3,495	4,561	6,258	8,800	12,530	17,826
Taiwan	7,647	10,792	14,676	17,805	20,100	22,000
Tanzania	7,886	10,026	13,513	18,867	25,635	39,639
Thailand	20,010	26,392	35,745	46,718	57,196	63,670

	1950	1960	1970	1980	1990	2000
Togo	1,329	1,514	2,020	2,615	3,531	4,861
Tokelau	2	2	2	2	2	2
Tonga	50	65	85	96	95	92
Trinidad & Tobago	636	843	971	1,082	1,281	1,484
Tunisia	3,530	4,221	5,127	6,384	8,180	9,924
Turkey	20,809	27,509	35,321	44,438	58,687	66,789
Turks/Caicos	6	6	6	7	10	12
Tuvalu	5	5	6	7	9	11
USSR	180,075	214,335	242,766	265,546	288,595	308,363
US Virgin Is	27	33	64	98	116	135
Uganda	4,762	6,562	9,806	13,120	18,794	26,958
United Arab Em.	70	90	223	1,015	1,589	1,951
United Kingdom	50,616	52,372	55,632	56,330	57,237	58,393
United States	152,271	180,671	205,051	227,757	249,224	266,096
Uruguay	2,239	2,538	2,808	2,914	3,094	3,274
Vanuatu	52	65	86	118	158	206
Vatican City	1	1	1	1	1	1
Venezuela	5,009	7,502	10,604	15,024	19,735	24,715
Vietnam	29,954	34,743	42,729	53,700	66,693	82,427
Wallis & Futuna	7	8	9	10	17	26
Western Sahara	14	32	76	135	178	228
Yemen	4,316	5,247	6,332	7,675	11,282	13,219
Yugoslavia	16,346	18,402	20,371	22,304	23,807	24,900
Zaïre	12,184	15,310	19,769	26,225	35,568	49,190
Zambia	2,440	3,141	4,189	5,738	8,452	12,267
Zimbabwe	2,730	3,812	5,260	7,126	9,709	13,123

The estimates for 2000 are those of the UN. They use the 'medium variants' of the population controls.

Population change

percentage change in the decade

	1950-60	1960-70	1970-80	1980-90	1990-2000
World	20	22.5	20.3	19	18.3
Africa	25.8	29.5	31.9	34.5	35
Asia	21.1	26	22.9	20.5	19.3
Europe	8.3	8.2	5.3	2.9	2.3
North America	22.3	19.1	16.4	14.3	12.2
Oceania	24.8	22.5	18	16.1	13.8
South America	31.9	29.8	26	23.2	19.6
Afghanistan	20.3	26.4	17.9	3.1	60.1
Albania	31	32.7	24.9	21.5	16.9
Algeria	23.4	27.3	36.3	33.2	31.8
American Samoa	10.5	28.6	18.5	18.8	2.6
Andorra	33.3	150	55	51.6	4.3
Angola	16.6	16	38.2	29.7	32.7
Anguilla	20	0	16.7	0	14.3
Antigua & Barbuda	19.6	20	13.6	1.3	3.9
Argentina	20.2	16.2	17.8	14.5	12.1
Aruba	3.5	3.4	4.9	-6.3	-3.3
Australia	25.5	21.7	17.1	14.8	11.7
Austria	1.6	5.9	1.1	0.5	0.4
Bahamas	43	51.3	22.8	20.5	16.6
Bahrain	34.5	41	57.7	48.7	32.4
Bangladesh	23.1	29.7	32.3	31	30.3
Barbados	9.5	3.5	4.2	2.4	3.9
Belgium	5.9	5.5	2	-0.1	-0.1
Belize	35.8	31.9	20.8	29	23
Benin	9.3	20.4	28.4	33.9	37.6
Bermuda	15.4	22.2	1.8	3.6	6.9
Bhutan	18.3	20.4	19.1	21.8	25.7
BIOT	0	0	0	0	0
Bolivia	23.9	26.2	28.8	31.3	33
Botswana	23.7	29.5	44.8	44.6	39.7
Br. Virgin Is	16.7	42.9	20	8.3	15.4
Brazil	35.8	32	26.5	24	19.4
Brunei	95.7	47.8	71.4	16.7	25.2
Bulgaria	8.5	7.9	4.4	1.7	0.7
Burkina Faso	21.8	24.7	25.4	29.3	34.4
Burma	20.7	25.6	27.3	19.7	22.7
Burundi	20	19.5	17.3	32.4	34.5
Cambodia	25	27.7	-7.8	28.8	21.8
Cameroon	18.6	24.8	30.9	36.8	41.1
Canada	30.4	19.1	12.8	10.3	7.4
Cape Verde Is	34.2	36.2	8.2	28	39.2
Cayman Is	50	11.1	70	47.1	40
Central African Rep.	16.7	20.5	25.5	31	34.1
Chad	15.3	19.2	22.6	26.8	29.2
Chile	25.2	24.8	17.3	18.2	15.9
China	18.5	26.3	19.9	14.3	14.1
Christmas Is	200	0	0	0	0
Cocos Is	0	0	0	0	0

	1950-60	1960-70	1970-80	1980-90	1990-2000
Colombia	33.4	34	26	22.6	19.5
Comoros	24.3	27.4	43.1	40.3	43.5
Congo	22.3	27.8	32.1	36.1	39.5
Cook Is	20	16.7	-9.5	-5.3	-5.6
Costa Rica	43.4	40	31.9	32	23.1
Cuba	19.4	22	13.6	9.6	8.4
Cyprus	16	7.3	2.3	11.4	8.7
Czechoslovakia	10.2	5	6.8	2.3	3.3
Denmark	7.3	7.6	3.9	0.4	0.2
Djibouti	33.3	110	81	34.5	35
Dominica	17.6	18.3	2.8	12.3	6.1
Dominican Rep.	37.3	36.9	28.8	25.9	20.2
Ecuador	33.3	37.1	34.2	30.3	25.8
Egypt	27.5	27.5	23.7	30	20.8
El Salvador	32.5	39.6	26.1	16.1	28.3
Equatorial Guinea	11.5	15.5	-25.4	62.2	29.3
Ethiopia	23.6	26.6	26.5	27.1	34.8
Falkland Is	0	0	0	0	0
Faroe Is	12.9	11.4	5.1	14.6	4.3
Fiji	36.3	32	21.9	20.5	15.6
Finland	10.5	4	3.8	4.1	2.1
France	9.2	11.1	6.1	4.2	3.6
French Guiana	24	54.8	37.5	48.5	32.7
French Polynesia	32.3	32.9	34.9	40.1	30.1
Gabon	3.6	3.7	59.9	45.4	37.5
Gambia, The	19.7	31.8	38.1	34.3	30
Germany	6.3	6.9	0.8	1.5	-3.2
Ghana	38.2	27.1	24.7	40	36.8
Gibraltar	4.3	8.3	11.5	3.4	6.7
Greece	10.1	5.6	9.7	4.2	1.5
Greenland	43.5	42.4	10.6	7.7	7.1
Grenada	18.4	4.4	13.8	-20.6	-2.4
Guadeloupe	31	16.4	2.2	4.9	6.4
Guam	13.6	28.4	25.6	9.3	8.5
Guatemala	33.5	32.3	31.9	33	32.9
Guinea	23	24.4	14.4	29	36.1
Guinea-Bissau	7.3	-3.1	51.4	21.3	24.2
Guyana	34.5	24.6	7.1	4.9	11.9
Haiti	16.7	19.1	18.4	21.3	22.9
Honduras	38.1	35.8	39.4	40.3	33.2
Hong Kong	55.8	28.2	27.8	16.1	8.3
Hungary	6.9	3.5	3.6	-1.5	-0.2
Iceland	23.1	15.9	11.8	11	8.3
India	23.7	25.4	24.1	20.1	25.9
Indonesia	20.9	25	25.5	18.8	22
Iran	27.4	31.9	36.8	40.4	25.9
Iraq	32.7	36.6	42.1	42.4	39.2
Ireland	-4.5	4.2	15.1	9.4	9.8
Israel	68	40.7	30.4	18.6	15.7
Italy	6.6	7.2	4.9	1.1	0.2
Ivory Coast	36.9	45.2	48.6	46.4	46.7
Jamaica	16.1	14.7	14.1	15.1	11.4

	1950-60	1960-70	1970-80	1980-90	1990-2000
Japan	12.5	10.9	12	5.7	4.1
Jordan	37	35.6	27.1	37.2	38.6
Kenya	33	38	44.7	44.5	45.9
Kiribati	28.1	19.5	18.4	13.8	9.1
Korea, North	10.9	35.5	24.9	19.2	20
Korea, South	22.8	27.7	19.4	12.2	8.4
Kuwait	82.9	167.6	84.8	48.3	29.4
Laos	24	24.6	18.1	29.1	32
Lebanon	28.7	33	8.1	1.2	23.2
Lesotho	18.5	22.3	25.8	32.5	33.6
Liberia	26.1	33.3	35.5	37.3	38.8
Libya	31.1	47.2	53.2	49.4	43
Liechtenstein	14.3	31.3	23.8	7.7	0
Luxembourg	6.1	8	7.4	2.5	1.1
Macau	-10.1	45	17.1	66.9	37
Madagascar	25.5	27	30.3	36.6	38.5
Malawi	22.5	28	36.9	41.6	42.3
Malaysia	33.2	33.3	26.8	30	22.9
Maldives	12.2	23.9	35.1	39.6	31.6
Mali	24.3	25.3	25.1	18.8	55.5
Malta	5.4	-0.9	11.7	-3	3.7
Martinique	27	15.6	0	4.6	6.2
Mauritania	20.1	23.2	27	30.5	33.5
Mauritius	33.9	25.2	16.9	12	11
Mexico	35.7	38.8	33.4	22.4	24.5
Monaco	4.5	4.3	8.3	7.7	7.1
Mongolia	26	31	32.4	31.7	30
Montserrat	-7.7	-8.3	9.1	0	8.3
Morocco	29.9	31.7	26.6	29.3	25.9
Mozambique	20.4	25.9	28.7	29.4	30.9
Namibia	22.7	24.4	28.5	36.4	36.8
Nauru	25	40	0	28.6	11.1
Nepal	14.9	22.2	29.3	28.8	25.8
Neths. Antilles	18.5	15.6	11.3	-23.9	8
Netherlands	13.5	13.5	8.5	5.7	5.9
New Caledonia	33.9	39.2	29.1	17.6	16.8
New Zealand	24.3	18.9	10.4	9	8
Nicaragua	36	37.5	35	39.7	35.9
Niger	26.2	37.5	34.1	38.4	39.1
Nigeria	28.4	33.7	38.6	38.4	37.8
Niue	0	25	-20	-25	-33.3
Norfolk I.	0	0	100	0	0
Norway	9.7	8.3	5.4	3.1	2.8
Oman	22.3	29.5	50.5	52.6	44.9
Pacific Is	36.8	32.1	32	4.4	-100
Pakistan	26.4	31.5	29.8	31.4	44.9
Panama	28.6	33.4	27.8	23.6	19.6
Papua New Guinea	19	26.1	27.4	25.5	25.1
Paraguay	31.3	32.5	33.9	35.9	29.5
Peru	30.1	32.8	31.1	29.1	17.7
Philippines	31.3	36.2	28.7	29.2	24.1
Poland	19.1	10	9.4	8	5.1

Population change (continued)

	1950-60	1960-70	1970-80	1980-90	1990-2000
Portugal	5	2.5	8	5.3	2.9
Puerto Rico	6.3	15.3	18	8.5	10.2
Qatar	80	146.7	106.3	60.7	35.6
Réunion	31.9	30.1	15.2	17.7	15.7
Romania	12.9	10	9.6	4.8	4.6
Rwanda	29.3	36	38.5	40.2	40.9
Samoa	35.4	28.8	9.8	7	1.8
San Marino	15.4	20	16.7	9.5	8.7
São Tomé & Principe	6.7	15.6	14.9	42.4	24.8
Saudi Arabia	27.3	41	63.1	50.8	46.4
Senegal	27.5	30.5	33.2	32.3	19
Seychelles	23.5	23.8	25	6.2	8.7
Sierra Leone	15.3	18.5	22.9	27.2	31
Singapore	59.9	27	16.4	24.3	-0.2
Solomon Is	18.3	32.5	41.7	38.5	34.1
Somalia	21.1	25	45.7	40.3	29.9
South Africa	27.1	29.1	25.9	24.8	23.8
Spain	8.7	10.9	11.1	4.4	3.8
Sri Lanka	28.8	26.5	18.4	16.2	12.8
St Christopher/N.	14	-8.8	0	-15.4	0
St Helena	0	0	0	40	42.9
St Lucia	11.4	14.8	18.8	25	18
St Pierre & M.	0	0	20	0	0
St Vincent/G.	19.4	10	12.5	17.2	10.3
Sudan	21.5	24.1	34.8	34.9	33.4
Surinam	34.9	28.3	-5.4	19.9	17.8
Swaziland	23.5	28.5	34.4	40	42.3
Sweden	6.6	7.5	3.3	1.6	1.4
Switzerland	14.2	15.4	2.1	4.6	2.3
Syria	30.5	37.2	40.6	42.4	42.3
Taiwan	41.1	36	21.3	12.9	9.5
Tanzania	27.1	34.8	39.6	35.9	54.6
Thailand	31.9	35.4	30.7	22.4	11.3
Togo	13.9	33.4	29.5	35	37.7
Tokelau	0	0	0	0	0
Tonga	30	30.8	12.9	-1	-3.2
Trinidad & Tobago	32.5	15.2	11.4	18.4	15.8
Tunisia	19.6	21.5	24.5	28.1	21.3
Turkey	32.2	28.4	25.8	32.1	13.8
Turks/Caicos	0	0	16.7	42.9	20
Tuvalu	0	20	16.7	28.6	22.2
USSR	19	13.3	9.4	8.7	6.8
US Virgin Is	22.2	93.9	53.1	18.4	16.4
Uganda	37.8	49.4	33.8	43.2	43.4
United Arab Em.	28.6	147.8	355.2	56.6	22.8
United Kingdom	3.5	6.2	1.3	1.6	2
United States	18.7	13.5	11.1	9.4	6.8
Uruguay	13.4	10.6	3.8	6.2	5.8
Vanuatu	25	32.3	37.2	33.9	30.4
Vatican City	0	0	0	0	0
Venezuela	49.8	41.3	41.7	31.4	25.2
Vietnam	16	23	25.7	24.2	23.6
Wallis & Futuna	14.3	12.5	11.1	70	52.9
Western Sahara	128.6	137.5	77.6	31.9	28.1
Yemen	21.6	20.7	21.2	47	17.2
Yugoslavia	12.6	10.7	9.5	6.7	4.6
Zaire	25.7	29.1	32.7	35.6	38.3
Zambia	28.7	33.4	37	47.3	45.1
Zimbabwe	39.6	38	35.5	36.2	35.2

The estimates for 2000 are those of the UN. They use the 'medium variants' of the population controls.

Birth and death rates

births/deaths per 1,000 population

	Birth Rate					Death Rate				
	1950–55	1960–65	1970–75	1980–85	1990–95	1950–55	1960–65	1970–75	1980–85	1990–95
World	38	35	32	28	26	20	15	12	10	9
Africa	49	49	47	45	44	27	23	19	16	13
Asia	43	40	35	28	27	24	18	12	10	8
Europe	20	19	16	13	13	11	10	10	10	11
Latin America	43	41	35	31	27	15	12	10	8	7
North America	25	22	16	16	14	9	9	9	9	9
Oceania	28	27	24	20	19	12	11	10	8	8
Afghanistan	48	53	52	49	52	32	30	26	23	22
Albania	38	40	32	27	22	14	10	7	6	6
Algeria	51	50	48	41	35	24	19	15	10	7
Andorra	...	18	...	15	13	10	6	5	...	4
Angola	50	49	48	47	47	35	30	25	22	19
Anguilla	...	...	...	...	25	...	...	...	...	10
Antigua & Barbuda	35	32	20	15	15	12	9	7	6	5
Argentina	25	23	23	23	20	9	9	9	9	9
Australia	23	22	20	16	14	9	9	9	7	8
Austria	15	19	14	12	12	12	13	13	12	12
Bahamas	35	31	28	23	19	12	8	6	5	5
Bahrain	45	47	36	31	25	16	14	8	5	3
Bangladesh	47	47	49	45	41	24	22	21	18	14
Barbados	33	29	21	17	16	13	9	9	9	9
Belgium	17	17	14	12	12	12	12	12	11	12
Belize	41	46	40	39	37	12	8	6	5	4
Benin	43	48	49	49	49	37	31	26	21	18
Bermuda	28	26	20	18	16	9	7	7	7	7
Bhutan	44	42	41	39	38	27	24	21	18	16
Bolivia	47	46	45	44	41	24	22	19	16	12
Botswana	49	53	52	49	44	23	20	17	14	10
Br. Virgin Is	40	33	26	22	20	11	9	7	6	6
Brazil	45	42	34	31	26	15	12	10	8	8
Brunei	54	43	35	32	29	15	8	5	4	3
Bulgaria	21	17	16	14	12	10	8	9	11	12
Burkina Faso	51	51	50	47	47	32	27	24	20	17
Burma	42	41	38	34	30	24	20	14	11	9
Burundi	49	45	44	47	47	25	22	21	19	16
Cambodia	45	45	40	46	37	24	20	23	20	15
Cameroon	44	45	46	47	47	27	24	20	17	13
Canada	28	25	16	15	13	9	8	7	7	8
Cape Verde Is	51	49	39	39	41	19	15	12	10	7
Cayman Is	31	32	28	20	15	7	7	6	6	4
Central African Rep.	44	43	43	46	45	29	25	22	19	16
Chad	45	46	45	44	43	32	29	25	21	18
Chile	37	37	28	24	23	14	12	9	6	6
China	44	38	31	19	21	25	17	9	7	7
Christmas Is	49	26	...	...	16	6	5	5	...	...
Cocos Is	...	30	21	...	20	...	9	7	...	...
Colombia	47	44	35	29	26	17	12	9	6	6
Comoros	47	52	49	48	47	24	21	18	14	12
Congo	44	45	46	46	46	25	22	19	16	13
Cook Is	41	47	32	28	24	19	9	5	5	5
Costa Rica	47	45	32	30	26	13	9	6	4	4
Cuba	30	35	27	16	17	11	9	7	6	7
Cyprus	27	25	18	20	17	11	11	10	8	8
Czechoslovakia	22	16	18	15	14	11	10	11	12	11
Denmark	18	17	15	10	11	9	10	10	11	11
Djibouti	50	50	50	48	46	31	27	23	19	16
Dominica	39	43	...	21	21	17	12	...	6	5
Dominican Rep.	51	49	39	34	28	20	15	10	8	6
Ecuador	47	46	41	35	31	19	14	11	8	7
Egypt	49	45	38	39	31	24	20	16	13	9
El Salvador	48	48	43	38	36	20	15	11	11	7
Equatorial Guinea	42	41	42	43	44	32	28	24	21	18
Ethiopia	52	50	48	45	48	32	27	23	24	18
Falkland Is	23	22	19	15	9	12	11	12	8	6
Faroe Is	25	23	20	16	18	8	7	7	7	8
Fiji	44	39	32	31	24	14	11	9	8	6
Finland	23	18	13	14	12	10	9	10	9	10
France	20	18	16	15	13	13	11	11	11	10
French Guiana	28	30	29	28	28	16	13	8	7	6
French Polynesia	43	45	37	30	28	12	10	9	8	5
Gabon	30	31	31	34	43	27	23	20	18	16
Gambia, The	47	51	49	48	45	34	31	27	23	20
Germany	16	18	11	11	11	11	12	13	13	12
Ghana	48	48	46	45	44	22	19	16	14	12
Gibraltar	23	25	20	18	17	10	9	8	7	7
Greece	19	18	16	14	12	7	8	9	9	10
Greenland	44	47	43	28	22	19	9	6	7	8
Grenada	40	39	26	25	25	14	10	8	7	7
Guadeloupe	39	37	29	21	20	13	8	8	7	7
Guam	29	26	34	30	27	5	5	4	4	4
Guatemala	51	48	45	43	39	22	18	13	11	8
Guinea	55	52	52	51	51	34	30	27	24	20
Guinea-Bissau	41	40	41	43	43	30	28	27	25	21
Guyana	43	41	35	29	24	18	14	10	9	7
Haiti	44	42	39	37	35	28	22	18	15	12
Honduras	51	51	49	42	37	22	18	14	9	7
Hong Kong	38	33	20	17	12	9	6	5	5	6
Hungary	21	14	16	13	12	11	10	12	14	13
Iceland	28	26	21	18	15	8	7	7	7	7
India	44	42	38	35	31	25	19	16	13	10
Indonesia	43	43	38	32	27	26	22	17	11	9
Iran	48	47	44	42	33	25	20	15	10	7
Iraq	49	49	47	44	41	22	19	15	9	7
Ireland	21	22	22	20	18	13	12	11	9	9
Israel	33	26	27	24	21	7	6	7	7	7
Italy	18	19	16	11	11	10	10	10	10	11
Ivory Coast	53	53	51	50	50	28	24	19	16	13
Jamaica	35	40	33	27	22	12	9	8	6	6
Japan	24	17	19	13	12	9	7	7	6	7
Jordan	47	53	50	38	39	26	22	14	8	5
Kenya	53	53	53	51	47	25	21	17	13	10
Korea, North	37	41	36	22	5	32	13	8	6	5
Korea, South	32	13	9	6	15	32	13	9	6	6
Kuwait	45	45	44	35	26	11	9	5	3	2
Laos	46	45	44	45	44	25	23	23	19	15
Lebanon	41	43	32	29	30	19	13	9	9	8
Lesotho	42	43	42	42	40	27	23	19	14	11
Liberia	48	50	48	47	47	27	24	20	17	14
Libya	48	49	46	43		23	18	15	11	8
Liechtenstein	22	22	16	14	13	9	8	8	7	6
Luxembourg	15	16	12	12	12	12	12	12	11	11

Birth and death rates   *continued*

	Birth Rate					Death Rate				
	1950–55	1960–65	1970–75	1980–85	1990–95	1950–55	1960–65	1970–75	1980–85	1990–95
Macau	...	21	11	...	17	13	8	6	5	3
Madagascar	48	48	46	46	45	27	23	19	15	13
Malawi	52	55	57	57	55	31	28	24	22	19
Malaysia	45	43	35	32	28	20	13	9	6	5
Maldives	34	37	40	41	41	29	27	20	10	8
Mali	53	52	51	51	51	32	29	25	22	19
Malta	29	23	18	17	13	10	9	9	9	9
Martinique	40	36	26	18	18	13	9	7	7	7
Mauritania	48	48	47	47	46	31	28	24	21	18
Mauritius	47	43	26	22	17	16	9	7	6	6
Mexico	47	46	43	32	27	16	11	9	6	5
Monaco	14	19	8	...	20	14	16	12	15	17
Mongolia	44	43	42	38	34	22	17	13	10	8
Montserrat	32	27	24	20	17	14	10	10	10	10
Morocco	50	50	46	37	33	26	20	16	11	8
Mozambique	46	47	46	46	44	30	25	22	20	17
Namibia	45	46	45	44	42	25	23	18	14	11
Nauru	27	32	...	...	23	7	7	4	5	5
Nepal	46	46	47	43	36	27	25	21	17	13
Neths. Antilles	35	32	21	19	18	5	5	5	5	6
Netherlands	22	21	15	12	13	8	8	8	8	9
New Caledonia	27	33	33	29	25	10	9	9	7	6
New Zealand	26	26	21	16	16	9	9	8	8	8
Nicaragua	54	50	47	44	39	23	17	13	10	7
Niger	54	53	52	52	51	32	28	25	22	19
Nigeria	51	52	49	49	47	27	24	20	17	14
Niue	37	42	29	26	21	15	9	10	7	5
Norfolk I.	15	12	11	11	11	14	16	6	7	8
Norway	19	17	17	12	13	8	10	10	10	11
Oman	51	50	50	48	43	32	26	20	13	6
Pacific Is	29	34	35	33	31	6	7	5	5	4
Pakistan	50	48	48	50	42	29	22	18	14	11
Panama	40	41	36	28	25	13	10	7	5	5
Papua New Guinea	44	44	41	35	33	29	21	17	13	11
Paraguay	47	42	37	36	33	9	8	7	7	6
Peru	47	46	41	34	29	22	18	13	11	8
Philippines	49	44	37	36	30	20	13	11	9	7
Pitcairn	45	8	15	...	...	16	10	...	...	...
Poland	30	20	18	19	15	11	8	8	10	10
Portugal	24	24	20	15	13	12	11	11	10	10
Puerto Rico	37	31	24	21	18	9	7	7	7	8
Qatar	46	41	31	35	28	22	17	12	5	4
Reunion	39	40	31	24	21	13	10	7	6	6
Romania	25	17	19	16	15	12	9	9	10	11
Rwanda	47	51	53	52	50	23	21	21	19	16

	Birth Rate					Death Rate				
	1950–55	1960–65	1970–75	1980–85	1990–95	1950–55	1960–65	1970–75	1980–85	1990–95
Samoa	37	33	29	30	31	7	5	4	6	7
San Marino	18	18	16	14	10	8	8	8	8	8
São Tomé & Príncipe	41	50	45	40	35	26	19	11	11	10
Saudi Arabia	49	49	48	43	42	26	21	17	9	7
Senegal	49	50	49	47	44	28	27	24	19	16
Seychelles	31	41	32	28	24	12	12	9	9	8
Sierra Leone	48	48	49	48	48	34	32	29	25	22
Singapore	44	34	21	17	16	11	7	5	5	6
Solomon Is	...	...	...	...	42	15	15	14	12	10
Somalia	49	48	48	53	47	31	27	23	22	18
South Africa	43	42	36	33	31	20	17	13	11	9
Spain	20	22	20	13	13	10	9	8	8	9
Sri Lanka	39	35	29	27	21	12	9	8	6	6
St Christopher/N.	37	36	18	24	21	14	11	8	10	8
St Helena	28	28	25	...	9	9	10	10	8	7
St Lucia	36	42	39	29	21	15	12	8	6	6
St Vincent/G.	42	46	30	26	23	16	12	9	7	6
Sudan	47	47	47	46	43	27	25	21	17	14
Surinam	44	44	35	29	25	13	10	8	7	6
Swaziland	50	49	48	47	47	28	23	18	14	11
Sweden	16	15	14	11	13	10	10	10	11	12
Switzerland	17	19	14	12	12	10	10	9	9	10
Syria	47	47	47	46	43	21	17	12	9	6
Taiwan	45	36	24	21	17	10	6	5	5	5
Tanzania	49	52	51	51	50	27	23	19	15	13
Thailand	47	44	35	28	20	19	13	9	8	7
Togo	47	48	46	45	45	29	24	19	16	13
Tokelau	...	...	...	...	22	14	12	9	8	8
Tonga	37	33	32	31	29	9	4	3	3	4
Trinidad & Tobago	38	38	26	28	23	11	8	7	7	6
Tunisia	46	47	37	34	27	23	18	12	8	6
Turkey	48	43	35	31	27	24	16	12	9	8
Turks & Caicos	39	39	30	26	26	13	11	8	6	4
Tuvalu	34	28	23	...	29	19	9	7	...	4
USSR	26	22	18	19	17	9	7	9	11	10
Uganda	51	49	50	53	52	25	20	19	17	14
United Arab Em.	48	44	33	27	20	23	17	10	4	4
United Kingdom	16	18	15	13	14	12	12	12	12	12
United States	24	22	16	16	14	10	9	9	9	9
Uruguay	21	22	21	18	17	11	10	10	10	10
Venezuela	47	44	36	33	28	12	9	7	6	5
Vietnam	42	41	38	35	30	29	21	14	11	8
Yemen	53	54	55	54	51	32	27	22	18	14
Yugoslavia	29	22	18	17	14	12	9	9	9	9
Zaire	47	47	47	45	45	26	22	19	16	13
Zambia	50	49	49	51	50	26	21	18	15	12
Zimbabwe	52	52	49	43	40	23	19	15	12	9

The figures are yearly averages for the period shown. The figures for 1990-95 are estimates of the UN.

Infant mortality

deaths under one year per thousand live births

	1950-55	1970-75	1990-95
Afghanistan	227	194	162
Albania	145	58	32
Algeria	185	132	61
Angola	231	173	127
Argentina	64	49	29
Australia	24	17	7
Austria	53	24	9
Bahamas	82	...	21
Bahrain	175	55	12
Bangladesh	180	140	108
Barbados	132	33	10
Belgium	45	19	8
Benin	210	136	85
Bhutan	197	153	118
Bolivia	176	151	93
Botswana	130	95	58
Brazil	135	91	57
Brunei	112	20	7
Bulgaria	92	26	14
Burkina Faso	226	173	127
Burma	183	100	59
Burundi	166	143	110
Cambodia	165	181	116
Cameroon	190	119	86
Canada	36	16	7
Cape Verde Is	129	82	37
Central African Rep.	198	132	95
Chad	211	166	122
Chile	126	70	19
China	195	61	27
Colombia	123	73	37
Comoros	180	135	89
Congo	170	90	65
Costa Rica	94	51	17
Cuba	82	36	13
Cyprus	53	29	10
Czechoslovakia	54	21	13
Denmark	28	12	6
Djibouti	207	154	112
Dominica	130	20	17
Dominican Rep.	149	94	57
Ecuador	150	95	57
Egypt	200	150	57
El Salvador	175	110	53
Equatorial Guinea	204	157	117
Ethiopia	190	155	122
Fiji	88	45	24
Finland	34	12	5
France	45	16	7
Gabon	194	132	94
Gambia, The	231	179	132
Germany	53	20	6
Ghana	149	107	81
Greece	60	34	13
Greenland	120	...	27
Grenada	80	40	15
Guadeloupe	68	42	12
Guatemala	141	95	48
Guinea	222	177	134
Guinea-Bissau	211	183	140
Guyana	119	79	48
Haiti	220	135	86
Honduras	169	110	57
Hong Kong	79	17	6
Hungary	71	34	17
Iceland	21	12	5
India	190	135	88
Indonesia	160	114	65
Iran	190	122	40
Iraq	165	96	56
Ireland	41	18	8
Israel	41	23	10
Italy	60	26	9
Ivory Coast	186	129	87
Jamaica	85	42	14
Japan	51	12	5
Jordan	160	82	36
Kenya	150	98	64
Kiribati	121	89	59
Korea, North	115	47	24
Korea, South	115	47	21
Kuwait	125	43	15
Laos	180	145	97
Lebanon	87	48	40
Lesotho	160	130	89
Liberia	194	181	126
Libya	185	117	68
Luxembourg	43	16	9
Madagascar	245	172	110
Malawi	212	191	138
Malaysia	99	42	20
Maldives	...	...	73
Mali	213	203	159
Malta	75	22	9
Martinique	65	35	10
Mauritania	207	160	117
Mauritius	99	55	20
Mexico	114	71	36
Mongolia	148	98	60
Morocco	180	122	68
Mozambique	205	168	130
Namibia	168	134	97
Nepal	197	153	118
Netherlands	24	12	7
New Caledonia	40	38	34
New Zealand	26	16	9
Nicaragua	167	100	50
Niger	207	166	124
Nigeria	207	135	96
Norway	23	12	6
Oman	231	145	34
Pakistan	190	140	98
Panama	93	43	21
Papua New Guinea	190	105	53
Paraguay	106	53	39
Peru	159	110	76
Philippines	100	64	40
Poland	95	27	17
Portugal	91	45	13

Infant Mortality *continued*

	1950-55	1970-75	1990-95
Puerto Rico	63	25	13
Qatar	180	57	26
Reunion	141	41	12
Romania	101	40	19
Rwanda	160	140	112
São Tomé & Príncipe	...	60	45
Saudi Arabia	200	120	58
Senegal	184	122	80
Seychelles	...	...	18
Sierra Leone	231	193	143
Singapore	66	19	7
Solomon Is	...	...	50

	1950-55	1970-75	1990-95
Somalia	190	155	122
South Africa	152	110	62
Spain	62	21	9
Sri Lanka	91	56	24
St Christopher/N.	86	...	24
St Lucia	116	...	18
St Vincent/G.	115	...	22
Sudan	185	145	99
Surinam	89	49	28
Swaziland	160	144	107
Sweden	20	10	6
Switzerland	29	13	7
Syria	160	88	39
Taiwan	34	...	...

	1950-55	1970-75	1990-95
Tanzania	160	130	97
Thailand	132	65	24
Togo	204	129	85
Tonga	57	59	23
Trinidad & Tobago	79	30	14
Tunisia	175	120	44
Turkey	233	138	62
Turks/Caicos	125	...	...
Tuvalu	123	...	...
USSR	73	26	20
Uganda	160	116	94
United Arab Em.	180	57	22
United Kingdom	28	17	8
United States	28	18	8

	1950-55	1970-75	1990-95
Uruguay	57	46	20
Vanuatu	...	...	85
Venezuela	106	49	33
Vietnam	180	120	54
Western Samoa	41	...	47
Yemen	231	168	107
Yugoslavia	128	45	21
Zaire	182	117	75
Zambia	150	100	72
Zimbabwe	120	93	55

Infant mortality is the number of deaths per year of children under one year per 1,000 births. The figures are yearly averages for the period shown. The figures for 1990-95 are estimates of the UN.

Age

percentage

	1950 0-4	5-14	15-24	25-59	60+	1990 0-4	5-14	15-24	25-59	60+
World	14	21	18	39	8	12	20	19	40	9
Africa	17	25	19	34	5	18	27	19	31	5
Asia	15	22	19	37	7	12	21	21	38	8
Europe	9	16	16	46	13	6	13	15	47	19
North America	10	19	16	45	10	9	18	17	43	13
Oceania	12	18	16	43	11	9	17	18	43	13
South America	16	24	19	36	5	13	23	19	37	8
Afghanistan	20	27	19	31	3	18	24	20	34	5
Albania	14	25	19	32	10	11	22	19	40	8
Algeria	16	24	19	35	7	15	28	21	30	5
Angola	17	24	19	35	5	18	27	19	32	5
Argentina	11	19	18	44	7	10	20	16	41	13
Australia	11	16	15	46	13	7	15	16	46	15
Austria	7	15	14	48	15	6	12	15	48	20
Bahrain	18	25	19	34	5	13	20	15	49	3
Bangladesh	15	23	18	38	6	17	27	21	31	5
Barbados	12	21	18	41	9	8	17	19	41	15
Belgium	8	13	15	48	16	6	12	14	48	21
Benin	15	21	16	34	14	19	27	19	30	5
Bhutan	16	24	19	36	6	15	24	19	36	6
Bolivia	17	25	19	34	5	17	27	19	32	5
Botswana	18	28	20	28	6	20	29	18	27	5
Brazil	17	25	20	34	4	13	23	19	39	7
Bulgaria	10	17	19	44	10	6	14	14	47	19
Burkina Faso	18	26	19	33	5	18	26	19	32	5
Burma	16	22	18	39	6	13	24	21	36	6
Burundi	16	25	19	35	5	19	27	19	31	5
Cambodia	17	26	19	34	5	17	18	20	40	5
Cameroon	16	24	19	36	6	19	27	18	30	5
Canada	12	18	16	43	11	7	14	14	49	16
Cape Verde Is	11	24	24	36	7	18	27	22	27	6
Central African Rep.	14	22	18	38	8	18	27	18	32	6
Chad	15	23	18	37	7	17	25	19	33	6
Chile	14	23	18	38	7	11	19	19	42	9
China	14	20	18	41	8	10	17	22	42	9
Colombia	18	25	19	33	6	13	24	21	37	6
Comoros	18	26	19	32	5	19	28	20	29	4
Congo	16	24	19	36	6	19	27	19	30	5
Costa Rica	18	25	19	32	6	13	23	19	38	6
Cuba	13	22	17	40	7	8	14	21	44	12
Cyprus	14	21	18	39	9	9	17	15	45	14
Czechoslovakia	10	16	16	47	12	7	16	15	46	17
Denmark	10	17	14	47	13	5	12	15	48	20
Djibouti	17	24	18	36	5	18	27	18	32	4
Dominican Rep.	19	26	19	31	5	14	24	21	36	6
Ecuador	17	25	18	33	7	14	25	21	34	6
Egypt	16	24	19	36	5	15	25	19	36	6
El Salvador	17	26	19	33	5	16	29	21	29	6
Equatorial Guinea	13	21	18	39	9	17	25	18	33	6
Ethiopia	18	26	19	32	5	19	27	19	31	5
Fiji	19	28	19	30	5	12	25	19	39	5
Finland	13	18	16	44	10	6	13	13	49	18
France	10	13	15	46	16	7	13	15	46	19
Gabon	12	19	17	40	12	15	18	22	37	9
Gambia, The	16	23	19	37	5	18	26	18	33	5
Germany	7	14	17	51	11	6	10	14	65	4
Ghana	19	27	19	32	4	18	27	19	31	5
Greece	10	19	20	42	10	6	14	15	46	20
Guadeloupe	17	22	17	37	7	10	17	20	41	12
Guatemala	18	26	20	31	4	18	28	20	30	5
Guinea	18	25	19	34	4	19	27	19	30	4
Guinea-Bissau	15	22	19	38	6	17	24	18	35	7
Guyana	18	23	18	34	7	12	22	23	38	6

	1950 0-4	5-14	15-24	25-59	60+	1990 0-4	5-14	15-24	25-59	60+
Haiti	15	22	17	38	8	15	25	20	34	6
Honduras	18	27	20	32	3	17	28	21	30	5
Hong Kong	16	14	23	43	4	6	15	16	50	13
Hungary	9	16	17	47	11	6	14	14	47	19
Iceland	13	18	17	42	11	8	17	17	44	14
India	15	24	19	36	6	13	23	19	37	7
Indonesia	14	25	20	35	6	12	23	21	37	6
Iran	17	22	17	35	8	15	29	20	31	6
Iraq	19	27	19	31	4	18	28	19	30	4
Ireland	11	18	15	41	15	9	19	18	40	14
Israel	14	18	17	45	6	11	21	18	39	12
Italy	9	17	17	45	12	5	12	16	47	20
Ivory Coast	17	26	20	33	4	20	28	18	30	4
Jamaica	14	23	19	39	6	11	22	23	35	9
Japan	13	22	20	37	8	6	13	15	49	17
Jordan	19	27	17	30	7	17	27	22	29	4
Kenya	17	22	19	35	6	20	30	19	26	4
Korea, North	16	26	19	34	6	11	18	25	40	7
Korea, South	16	26	19	34	5	8	18	21	46	8
Kuwait	15	21	22	37	5	13	23	16	46	3
Laos	16	26	20	34	5	18	26	19	33	5
Lebanon	14	21	19	36	10	14	22	22	34	8
Lesotho	16	24	18	34	7	17	26	18	33	6
Liberia	17	25	19	35	5	18	27	19	31	5
Libya	17	25	19	32	7	18	28	19	31	4
Luxembourg	6	14	16	49	15	6	11	13	51	19
Madagascar	17	25	19	34	5	18	27	19	31	5
Malawi	19	26	19	31	5	21	28	18	29	4
Malaysia	17	24	18	34	7	15	24	20	37	6
Mali	18	26	19	33	4	20	27	19	30	4
Malta	15	20	17	38	9	7	16	14	48	14
Martinique	15	22	18	37	8	9	15	19	43	14
Mauritania	17	25	19	35	5	18	26	19	32	5
Mauritius	18	27	19	31	5	9	21	19	44	8
Mexico	17	27	19	32	5	13	24	22	35	6
Mongolia	17	25	19	34	6	16	26	20	34	5
Morocco	18	27	19	32	5	15	25	20	33	6
Mozambique	17	25	19	35	5	18	26	19	32	5
Namibia	17	24	19	34	6	18	27	18	31	5
Nepal	15	25	20	34	7	16	26	19	34	5
Netherlands	12	17	16	43	12	6	12	16	49	17
New Zealand	12	17	14	44	13	8	15	17	45	15
Nicaragua	18	26	20	32	4	18	28	20	30	4
Niger	18	26	19	33	4	20	27	19	30	4
Nigeria	19	27	19	31	4	20	28	19	30	4
Norway	10	15	13	48	14	6	13	16	44	21
Oman	17	25	19	34	5	19	27	17	33	4
Pakistan	15	23	19	35	8	19	27	19	31	4
Panama	17	25	18	35	6	12	23	21	37	7
Papua New Guinea	15	24	19	35	6	15	25	21	34	5
Paraguay	16	27	20	32	5	15	25	20	35	5
Peru	17	25	19	34	6	13	24	21	36	6
Philippines	18	26	18	33	6	15	25	20	35	5
Poland	12	18	19	43	8	8	17	14	46	15
Portugal	11	19	19	42	11	7	15	17	44	18
Puerto Rico	17	27	19	32	6	9	17	19	41	14
Qatar	17	25	19	34	6	14	21	16	46	4
Reunion	16	25	19	36	4	11	21	22	39	8
Romania	10	18	20	43	9	8	16	17	44	16
Rwanda	19	27	19	31	4	20	29	19	28	4
Saudi Arabia	17	25	18	34	6	18	27	18	33	4
Senegal	17	25	19	34	6	18	27	19	31	5
Sierra Leone	16	23	19	37	5	18	26	19	32	5
Singapore	16	24	18	38	4	9	15	17	51	9
Somalia	17	24	19	36	5	19	28	17	32	4

Age *continued*

	1950					1990				
	0-4	5-14	15-24	25-59	60+	0-4	5-14	15-24	25-59	60+
South Africa	15	23	18	37	6	14	23	19	37	6
Spain	10	17	18	44	11	6	14	17	45	18
Sri Lanka	16	25	20	34	6	11	22	19	41	8
Sudan	18	26	19	32	5	18	27	19	31	5
Surinam	17	23	18	34	8	13	21	22	37	7
Swaziland	18	25	19	33	5	19	28	19	29	5
Sweden	9	15	12	49	15	6	11	14	46	23
Switzerland	9	15	14	48	14	6	11	14	50	20
Syria	17	25	19	33	7	19	29	19	28	4
Taiwan	19	24	20	34	4	8	20	19	44	9
Tanzania	20	27	19	29	5	20	29	19	28	4
Thailand	17	26	20	32	5	10	23	22	39	6
Togo	17	24	18	34	7	18	27	19	31	5
Trinidad & Tobago	17	24	16	37	6	12	22	18	40	8
Tunisia	16	23	18	35	8	14	24	20	35	6

	1950					1990				
	0-4	5-14	15-24	25-59	60+	0-4	5-14	15-24	25-59	60+
Turkey	15	24	21	35	6	13	22	20	38	7
USSR	11	20	21	40	9	9	17	15	45	15
Uganda	19	25	19	32	5	21	29	19	28	4
United Arab Em.	17	25	19	34	6	11	20	13	53	3
United Kingdom	9	14	14	49	16	7	12	15	46	21
United States	11	16	15	46	12	7	14	15	47	17
Uruguay	10	18	18	43	12	8	17	16	42	17
Venezuela	19	25	19	34	3	14	24	20	36	6
Vietnam	13	22	19	40	7	14	25	21	33	7
Yemen	18	25	19	34	6	21	29	20	26	4
Yugoslavia	11	20	21	40	9	7	16	15	47	15
Zaire	18	26	19	32	6	19	28	19	31	4
Zambia	19	26	19	32	4	21	28	19	28	4
Zimbabwe	18	27	19	32	5	18	27	21	30	4

This table shows the percentage of the total population in each of the five age-groups for 1950 and 1990

Life expectancy

	1950-55		1970-75		1990-95	
	male	female	male	female	male	female
Afghanistan	31	32	38	38	43	44
Albania	54	56	66	70	70	75
Algeria	42	44	54	56	65	67
Angola	29	32	37	40	45	48
Argentina	60	65	64	71	68	75
Australia	67	72	68	75	74	80
Austria	63	68	67	74	72	79
Bahrain	50	53	62	65	70	74
Bangladesh	38	35	46	44	53	53
Barbados	55	60	67	72	73	78
Belgium	65	70	68	75	72	79
Benin	31	34	39	42	46	50
Bhutan	37	36	43	42	51	49
Bolivia	39	43	45	49	54	58
Botswana	41	44	49	53	58	64
Brazil	49	53	58	62	64	69
Bulgaria	62	66	69	74	70	76
Burkina Faso	32	35	40	43	48	51
Burma	39	41	51	54	61	64
Burundi	38	41	41	45	48	51
Cambodia	38	41	39	42	50	52
Cameroon	35	38	44	47	54	57
Canada	67	72	70	77	74	81
Cape Verde Is	47	50	56	59	67	69
Central African Rep.	33	38	40	46	48	53
Chad	31	34	38	41	46	49
Chile	52	56	61	67	69	76
China	39	42	63	64	69	73
Colombia	49	52	60	63	66	72
Comoros	40	41	47	48	56	57
Congo	36	41	44	49	52	57
Costa Rica	56	59	66	70	73	78
Cuba	58	61	69	73	74	78
Cyprus	65	69	70	73	74	79
Czechoslovakia	64	68	67	74	69	76
Denmark	70	72	71	76	73	79
Djibouti	32	35	39	43	47	51
Dominican Rep.	45	47	58	62	65	70
Ecuador	47	50	57	61	65	69
Egypt	41	44	51	53	60	63
El Salvador	44	47	57	61	64	69
Equatorial Guinea	33	35	39	42	46	50
Ethiopia	31	34	39	43	45	49
Fiji	49	52	56	60	64	68
Finland	63	70	67	75	72	80
France	64	70	69	76	73	81
Gabon	37	40	43	47	52	55
Gambia, The	29	32	36	39	43	47
Germany	65	70	68	74	72	78
Ghana	40	44	48	52	54	58
Greece	64	68	71	74	74	79
Guadeloupe	55	58	65	71	71	78
Guatemala	42	42	53	56	62	67
Guinea	31	32	37	38	44	45
Guinea-Bissau	31	34	35	38	42	45
Guyana	51	54	58	62	62	68
Haiti	36	39	47	50	55	58
Honduras	41	44	52	56	64	68
Hong Kong	57	65	69	76	75	80
Hungary	62	66	67	73	68	75
Iceland	70	74	71	77	75	81
India	39	38	51	49	60	61
Indonesia	37	38	48	51	61	65
Iran	46	46	56	56	67	68
Iraq	43	45	56	58	65	67
Ireland	66	68	69	74	73	78
Israel	64	66	70	73	74	78
Italy	64	68	69	75	73	80
Ivory Coast	35	38	44	47	53	56
Jamaica	56	59	67	71	71	76
Japan	62	66	71	76	76	82
Jordan	42	44	55	58	66	70
Kenya	39	43	49	53	59	63
Korea, North	46	49	59	64	68	74
Korea, South	46	49	59	64	68	74
Kuwait	54	58	65	69	72	77
Laos	37	39	39	42	50	53
Lebanon	54	58	63	67	65	69
Lesotho	34	41	44	53	54	63
Liberia	36	39	46	49	54	57
Libya	42	44	51	55	62	65
Luxembourg	63	69	67	74	72	79
Madagascar	36	39	45	48	54	57
Malawi	36	37	40	42	48	50
Malaysia	47	50	61	65	69	73
Mali	31	34	37	40	44	48
Malta	64	68	69	73	72	76
Martinique	55	58	66	72	73	79
Mauritania	32	35	39	42	46	50
Mauritius	50	52	61	65	68	73
Mexico	49	52	60	65	67	74
Mongolia	41	44	53	55	62	65
Morocco	42	44	51	55	62	65
Mozambique	32	35	41	44	47	50
Namibia	38	40	48	50	58	60
Nepal	37	36	44	43	54	53
Netherlands	71	73	71	77	74	81
New Zealand	68	72	69	75	73	79
Nicaragua	41	44	54	56	65	68
Niger	32	35	38	41	45	48
Nigeria	35	38	43	46	51	54
Norway	71	75	71	78	74	81
Oman	36	37	48	50	66	70
Pakistan	40	38	50	48	59	59
Panama	54	56	65	68	71	75
Papua New Guinea	36	35	48	48	55	57
Paraguay	61	65	64	68	65	70
Peru	43	45	54	57	63	67
Philippines	46	49	56	59	63	67
Pitcairn						
Poland	59	64	67	74	68	76
Portugal	57	62	65	71	71	78
Puerto Rico	63	67	69	76	73	79
Qatar	47	49	61	64	68	73
Reunion	50	56	60	68	68	76
Romania	59	63	67	71	69	74
Rwanda	39	42	43	46	49	52
Saudi Arabia	39	41	52	56	64	68
Senegal	36	38	39	41	48	50
Sierra Leone	29	32	34	37	41	45
Singapore	59	62	67	72	72	77
Somalia	32	35	39	43	45	49
South Africa	44	46	51	57	60	66
Spain	62	66	70	76	74	80
Sri Lanka	58	56	64	66	70	74
Sudan	36	38	41	44	51	53
Surinam	54	58	62	67	68	73
Swaziland	34	37	45	50	56	60
Sweden	70	73	72	78	75	81
Switzerland	67	72	71	77	75	81
Syria	45	47	55	59	65	69
Thailand	45	49	58	62	65	69
Togo	35	38	44	47	53	57
Trinidad & Tobago	56	59	64	69	70	75
Tunisia	44	45	55	56	67	69
Turkey	45	45	56	60	65	68
Uganda	39	42	45	49	51	55
United Arab Em.	47	49	61	64	70	74
United Kingdom	67	72	69	75	73	79
United States	66	72	68	75	73	80
Uruguay	63	69	66	72	69	76
USSR	60	69	64	74	67	75
Venezuela	54	57	64	69	67	74
Vietnam	39	42	48	53	62	66
Yemen	33	33	43	43	52	52
Yugoslavia	57	59	66	71	70	76
Zaire	38	41	44	48	52	56
Zambia	36	39	46	49	54	57
Zimbabwe	40	43	50	53	59	63

Life expectancy is the age to which a baby can expect to live with the rate of mortality pertaining at the time of birth.

Median age

years

	1950	1970	1990
Afghanistan	17	19	19
Albania	21	19	24
Algeria	20	16	18
Angola	19	19	18
Argentina	26	27	28
Australia	30	28	32
Austria	36	34	36
Bahrain	19	17	26
Bangladesh	22	18	18
Barbados	25	21	28
Belgium	36	35	37
Benin	24	18	17
Bhutan	20	20	20
Bolivia	19	18	18
Botswana	17	14	15
Brazil	19	19	23
Bulgaria	27	33	37
Burkina Faso	18	18	18
Burma	22	19	21
Burundi	20	17	17
Cambodia	19	18	23
Cameroon	20	19	17
Canada	28	26	33
Cape Verde Is	21	16	17
Central African Rep.	23	20	18
Chad	22	19	19
Chile	22	20	25
China	24	20	26
Colombia	19	17	22
Comoros	18	17	16
Congo	20	18	17
Costa Rica	18	17	22
Cuba	23	22	28
Cyprus	24	26	30
Czechoslovakia	30	32	34
Denmark	32	33	37
Djibouti	20	19	17
Dominican Rep.	18	16	21
Ecuador	19	17	20
Egypt	20	19	21
El Salvador	19	17	17
Equatorial Guinea	24	21	19
Ethiopia	18	18	17
Fiji	17	18	22
Finland	28	30	36
France	35	32	35
Gabon	26	25	23
Gambia, The	20	20	18
Germany	35	34	37
Ghana	17	17	17
Greece	26	33	36
Guadeloupe	21	18	26
Guatemala	18	17	17
Guinea	19	17	17
Guinea-Bissau	22	22	20
Guyana	20	16	22
Haiti	22	20	20
Honduras	18	16	17
Hong Kong	24	21	32
Hungary	30	34	37
Iceland	27	25	30
India	20	20	22
Indonesia	20	19	22
Iran	21	17	18
Iraq	17	17	17
Ireland	30	28	28
Israel	26	23	26
Italy	29	33	37
Ivory Coast	18	17	16
Jamaica	22	17	22
Japan	22	29	37
Jordan	17	17	17
Kenya	20	16	15
Korea, North	19	17	24
Korea, South	19	19	27
Kuwait	22	19	24
Laos	19	19	18
Lebanon	23	18	21
Lesotho	20	20	19
Liberia	19	19	17
Libya	19	18	17
Luxembourg	35	35	37
Madagascar	19	18	17
Malawi	17	17	16
Malaysia	20	18	21
Mali	18	17	17
Malta	24	26	33
Martinique	22	19	28
Mauritania	19	19	18
Mauritius	17	18	26
Mexico	18	17	20
Mongolia	19	18	19
Morocco	18	16	19
Mozambique	19	19	18
Namibia	20	18	17
Nepal	20	20	19
Netherlands	28	29	35
New Zealand	29	26	31
Nicaragua	18	16	17
Niger	18	17	16
Nigeria	17	17	16
Norway	33	33	36
Oman	19	18	17
Pakistan	21	17	17
Panama	20	18	22
Papua New Guinea	20	19	19
Paraguay	18	17	20
Peru	19	18	21
Philippines	18	17	20
Poland	26	28	32
Portugal	26	29	33
Puerto Rico	18	21	28
Qatar	19	23	25
Réunion	19	18	24
Romania	26	31	33
Rwanda	17	17	16
Saudi Arabia	19	18	18
Senegal	19	18	17
Sierra Leone	20	19	18
Singapore	20	20	30
Somalia	20	18	17
South Africa	21	19	22
Spain	28	30	33
Sri Lanka	20	19	24
Sudan	18	18	17
Surinam	20	16	22
Swaziland	18	17	16
Sweden	34	35	39
Switzerland	33	32	38
Syria	19	16	16
Tanzania	17	17	15
Thailand	18	17	23
Togo	19	18	17
Trinidad & Tobago	21	19	24
Tunisia	21	17	21
Turkey	20	19	22
Uganda	18	16	15
United Arab Em.	19	23	30
United Kingdom	35	34	36
United States	30	28	33
Uruguay	28	30	31
USSR	25	29	31
Venezuela	18	17	21
Vietnam	23	18	20
Yemen	19	18	16
Yugoslavia	24	29	33
Zaire	18	18	17
Zambia	18	17	15
Zimbabwe	18	15	17

The median age is the age which divides the population into two groups of equal size, one that is younger than the median and the other older.

Population density

inhabitants per square kilometre

	1950	1960	1970	1980	1990	2000
World	18	22	27	33	39	46
Africa	7	9	12	16	21	29
Asia	50	60	76	94	113	134
Europe	76	82	89	93	96	98
North America	8	9	11	12	13	14
Oceania	1	2	2	3	3	4
South America	6	8	11	14	17	20
Afghanistan	14	17	21	25	25	41
Albania	43	56	74	93	113	132
Algeria	4	5	6	8	10	14
American Samoa	95	106	136	161	191	196
Andorra	13	18	44	68	104	108
Angola	3	4	4	6	8	11
Anguilla	55	66	66	77	77	88
Antigua & Barbuda	115	138	165	188	190	198
Argentina	6	7	9	10	12	13
Aruba	295	306	316	332	311	301
Australia	1	1	2	2	2	2
Austria	83	84	89	90	90	91
Bahamas	6	8	12	15	18	21
Bahrain	171	230	324	512	761	1,008
Bangladesh	290	357	463	613	803	1,046
Barbados	491	536	555	579	593	616
Belgium	283	300	316	323	323	322
Belize	3	4	5	6	8	10
Benin	18	20	24	31	41	57
Bermuda	780	900	1,100	1,120	1,160	1,240
Bhutan	16	18	22	26	32	41
BIOT	33	33	33	33	33	33
Bolivia	3	3	4	5	7	9
Botswana	1	1	1	2	2	3
Br. Virgin Is	40	47	67	80	87	100
Brazil	6	9	11	14	18	21
Brunei	9	17	25	43	50	65
Bulgaria	65	71	77	80	81	82
Burkina Faso	13	16	20	25	33	44
Burma	26	32	40	50	62	76
Burundi	88	106	127	148	197	264
Cambodia	24	30	38	35	46	55
Cameroon	9	11	14	18	25	35
Canada	1	2	2	2	3	3
Cape Verde Is	36	49	66	72	92	128
Central African Rep.	2	2	3	4	5	7
Chad	2	2	3	3	4	6
Chile	8	10	13	15	17	20
China	58	69	87	104	119	135
Christmas Is	7	22	22	22	22	22
Cocos Is	71	71	71	71	71	71
Colombia	10	14	19	24	29	35
Comoros	77	96	123	175	246	353
Congo	2	3	4	5	7	9
Cook Is	75	90	105	95	90	85
Costa Rica	17	24	34	45	59	73
Cuba	53	63	77	87	96	104
Cyprus	53	62	66	68	76	82
Czechoslovakia	97	107	112	120	123	127
Denmark	99	106	114	119	119	120
Djibouti	3	3	7	13	18	24
Dominica	68	80	95	97	109	116
Dominican Rep.	48	66	91	117	147	177
Ecuador	12	16	21	29	37	47
Egypt	20	26	33	41	52	64
El Salvador	92	122	171	215	250	320
Equatorial Guinea	8	9	10	8	13	16
Ethiopia	16	20	25	32	40	54
Falkland Is	0.2	0.2	0.2	0.2	0.2	0.2
Faroe Is	22	25	28	29	34	35
Fiji	16	22	28	35	42	48
Finland	12	13	14	14	15	15
France	76	83	92	98	102	105
French Guiana	0.3	0.3	0.5	0.7	1.1	1.5
French Polynesia	17	22	29	40	56	72
Gabon	2	2	2	3	4	6
Gambia, The	26	31	41	57	78	99
Germany	192	204	218	220	218	216
Ghana	21	28	36	45	63	86
Gibraltar	3,538	3,692	4,000	4,462	4,615	4,923
Greece	57	63	67	73	76	77
Greenland	0.07	0.1	0.1	0.2	0.2	0.2
Grenada	221	262	273	311	247	241
Guadeloupe	123	161	188	192	201	214
Guam	109	124	159	199	218	236
Guatemala	27	36	48	64	84	112
Guinea	10	13	16	18	23	32
Guinea-Bissau	14	15	15	22	27	30
Guyana	2	3	3	4	4	4
Haiti	118	137	163	194	235	288
Honduras	13	17	23	33	46	61
Hong Kong	1,889	2,943	3,772	4,822	5,599	6,063
Hungary	100	107	111	115	113	113
Iceland	1	2	2	2	2	3
India	109	135	169	210	259	317
Indonesia	42	51	63	79	97	115
Iran	10	13	17	24	33	42
Iraq	12	16	21	30	43	60
Ireland	42	40	42	48	53	58
Israel	60	100	141	184	218	253
Italy	156	167	179	187	189	190
Ivory Coast	9	12	17	25	37	55
Jamaica	128	148	170	194	223	249
Japan	221	249	276	309	327	340
Jordan	13	17	24	30	41	57
Kenya	11	14	20	29	41	60
Kiribati	45	57	68	81	92	100
Korea, North	81	90	121	151	181	217
Korea, South	206	253	322	385	432	469
Kuwait	9	16	42	77	114	148
Laos	7	9	11	14	17	23
Lebanon	139	179	237	257	260	320
Lesotho	24	29	35	44	58	78
Liberia	7	9	12	17	23	32
Libya	1	1	1	2	3	4
Liechtenstein	88	100	131	163	175	175
Luxembourg	114	121	131	141	144	146
Macau	8,174	7,348	10,652	12,478	20,826	28,522
Madagascar	7	9	11	15	20	28
Malawi	24	30	38	52	74	105
Malaysia	19	25	33	42	54	67
Maldives	273	307	380	513	717	943
Mali	3	4	4	6	7	10

Population density

continued

	1950	1960	1970	1980	1990	2000
Malta	1	1	1	1	1	1
Martinique	201	256	295	296	309	329
Mauritania	1	1	1	2	2	3
Mauritius	242	324	405	474	530	589
Mexico	14	19	27	36	45	55
Monaco	11579	12105	12632	13684	14737	15789
Mongolia	1	1	1	1	1	2
Montserrat	130	120	110	120	120	130
Morocco	20	26	34	43	56	71
Mozambique	8	9	12	15	20	26
Namibia	1	1	1	2	2	3
Nauru	200	250	350	350	450	500
Nepal	58	67	82	106	136	171
Neths. Antilles	169	200	231	257	196	211
Netherlands	248	281	319	346	366	388
New Caledonia	3	4	6	7	9	10
New Zealand	7	9	10	11	13	14
Nicaragua	8	11	16	21	30	40
Niger	2	2	3	4	6	8
Nigeria	36	46	61	85	117	162
Niue	15	15	19	15	12	8
Norfolk I.	29	29	29	57	57	57
Norway	10	11	12	13	13	13
Oman	2	2	3	5	7	10
Pacific Is	81	111	147	194	202	210
Pakistan	50	63	83	107	154	204
Panama	12	15	20	25	31	38
Papua New Guinea	3	4	5	7	8	10
Paraguay	3	4	6	8	11	14
Peru	6	8	10	13	17	20
Philippines	70	92	125	161	208	258

	1950	1960	1970	1980	1990	2000
Pitcairn	12	12	12	12	12	12
Poland	79	95	104	114	123	129
Portugal	91	96	98	106	111	115
Puerto Rico	249	265	305	360	391	431
Qatar	2	4	10	21	33	45
Reunion	102	135	176	202	238	276
Romania	69	78	85	93	98	103
Rwanda	80	104	142	196	275	387
Samoa	32	43	56	61	65	66
San Marino	213	246	295	344	377	410
São Tomé & Príncipe	62	66	77	88	126	157
Saudi Arabia	1	2	3	4	7	10
Senegal	13	16	21	28	37	49
Seychelles	75	93	115	143	152	166
Sierra Leone	27	31	37	45	58	76
Singapore	1654	2644	3357	3907	4407	4850
Solomon Is	4	4	6	8	11	15
Somalia	4	5	6	8	12	15
South Africa	11	14	18	23	29	36
Spain	55	60	67	74	78	81
Sri Lanka	117	151	191	226	262	296
St. Christopher/N.	191	218	198	198	168	168
St. Helena	41	41	41	41	57	82
St. Lucia	128	143	164	195	244	287
St. Pierre & M.	21	21	21	25	25	25
St. Vincent/G.	173	206	227	255	299	330
Sudan	4	4	6	7	10	13
Surinam	1	2	2	2	3	3
Swaziland	15	19	24	32	45	65
Sweden	16	17	18	18	19	19
Switzerland	114	130	150	153	160	164
Syria	19	25	34	48	68	96

	1950	1960	1970	1980	1990	2000
Taiwan	211	298	406	492	556	608
Tanzania	8	11	14	20	29	42
Thailand	39	51	70	91	109	124
Togo	23	27	36	46	62	86
Tokelau	200	200	200	200	200	200
Tonga	67	87	114	128	127	123
Trinidad & Tobago	124	164	189	211	250	289
Tunisia	22	26	31	39	50	61
Turkey	27	35	45	57	72	86
Turks/Caicos	14	14	14	16	23	28
Tuvalu	21	21	25	29	38	46
Uganda	20	28	42	56	80	114
United Arab Em.	1	1	3	12	19	23
United Kingdom	207	215	228	231	234	239
United States	16	19	22	24	27	28
Uruguay	13	14	16	16	17	18
US Virgin Is	77	94	182	278	330	384
USSR	8	10	11	12	13	14
Vanuatu	4	4	6	8	11	14
Vatican City	2273	2273	2273	2273	2273	2273
Venezuela	5	8	12	16	22	27
Vietnam	90	105	129	162	201	249
Wallis & Futuna	26	29	33	36	62	78
Western Sahara	0.05	0.1	0.3	0.5	0.7	0.9
Yemen	17	21	25	33	47	68
Yugoslavia	64	72	80	87	93	97
Zaire	5	7	8	11	15	21
Zambia	3	4	6	8	11	16
Zimbabwe	7	10	13	18	25	34

The figures use the total area of the country for the calculations. This in some cases includes water areas. The densities given in the country-by-country section use land area only.

Urban population

percentage of total population

	1950	1970	1990	2000
Afghanistan	6	11	22	29
Albania	20	34	35	40
Algeria	22	40	45	51
Angola	8	15	28	36
Antigua & Barbuda	46	34	32	39
Argentina	65	78	86	89
Australia	75	85	86	86
Austria	49	52	58	61
Bahamas	62	58	59	65
Bahrain	64	78	83	85
Bangladesh	4	8	14	18
Barbados	34	37	45	51
Belgium	92	94	97	98
Belize	57	51	52	58
Benin	7	16	42	53
Bermuda	100	100	100	100
Bhutan	2	3	5	8
Bolivia	38	41	51	59
Botswana	0	8	24	33
Brazil	36	56	77	83
Brunei	27	62	58	59
Bulgaria	26	52	70	76
Burkina Faso	4	6	9	12
Burma	16	23	25	28
Burundi	2	2	7	12
Cambodia	10	12	12	15
Cameroon	10	20	49	60
Canada	61	76	76	79
Cape Verde Is	8	19	62	70
Cayman Is	100	100	100	100
Central African Rep.	16	30	47	55
Chad	4	11	33	44
Chile	58	75	86	89
China	11	20	21	25
Colombia	37	57	70	75
Comoros	3	11	28	34
Congo	31	35	42	50
Cook Is	32	30	35	42
Costa Rica	34	40	54	61
Cuba	49	60	75	80
Cyprus	30	41	53	60
Czechoslovakia	37	55	69	74
Denmark	68	80	86	89

	1950	1970	1990	2000
Djibouti	41	62	81	84
Dominican Rep.	24	40	60	68
Ecuador	28	40	57	65
Egypt	32	42	49	55
El Salvador	37	39	44	50
Equatorial Guinea	16	39	65	71
Ethiopia	5	9	13	17
Faroe Is	18	28	31	36
Fiji	24	35	44	51
Finland	32	50	68	74
France	56	71	74	76
French Guiana	54	67	75	79
French Polynesia	28	56	65	70
Gabon	11	26	46	54
Gambia, The	11	15	23	29
Germany	72	78	84	86
Ghana	15	29	33	38
Greece	37	53	63	68
Guadeloupe	42	41	49	55
Guatemala	31	36	42	48
Guinea	6	14	26	33
Guinea-Bissau	10	15	20	25
Guyana	28	29	34	42
Haiti	12	20	30	37
Honduras	18	29	44	52
Hong Kong	89	90	93	94
Hungary	37	46	60	67
Iceland	74	83	91	92
India	17	20	28	34
Indonesia	12	17	29	37
Iran	28	41	55	61
Iraq	35	56	74	79
Ireland	41	52	59	64
Israel	65	84	92	94
Italy	54	64	69	72
Ivory Coast	13	27	47	55
Jamaica	27	42	52	59
Japan	50	71	77	78
Jordan	35	51	68	74
Kenya	6	10	24	32
Kiribati	10	26	36	43
Korea, North	31	50	67	73
Korea, South	21	41	72	81
Kuwait	59	72	96	97
Laos	7	10	19	25
Lebanon	23	59	84	87
Lesotho	1	9	20	32

	1950	1970	1990	2000
Liberia	13	26	44	52
Libya	19	36	70	76
Liechtenstein	20	21	28	33
Luxembourg	59	62	84	87
Macau	97	97	99	99
Madagascar	8	14	25	32
Malawi	4	6	15	21
Malaysia	20	27	42	50
Maldives	11	14	21	24
Mali	9	14	19	23
Malta	61	78	87	90
Martinique	28	54	75	79
Mauritania	3	14	42	54
Mauritius	29	42	42	46
Mexico	43	59	73	77
Monaco	100	100	100	100
Mongolia	19	45	51	55
Montserrat	22	11	12	16
Morocco	26	35	49	56
Mozambique	2	6	27	41
Namibia	15	34	57	66
Nepal	2	4	10	14
Neths. Antilles	49	50	55	61
Netherlands	83	86	89	89
New Caledonia	40	49	81	87
New Zealand	73	81	84	85
Nicaragua	35	47	60	66
Niger	5	9	20	27
Nigeria	10	20	35	43
Niue	24	21	23	30
Norway	32	65	74	77
Oman	2	5	11	15
Pacific Is	22	41	60	66
Pakistan	18	25	32	38
Panama	36	48	55	61
Papua New Guinea	1	10	16	20
Paraguay	35	37	48	54
Peru	36	57	70	75
Philippines	27	33	42	49
Poland	39	52	63	67
Portugal	19	26	33	39
Puerto Rico	41	58	74	79
Qatar	63	80	90	91
Reunion	24	44	64	70
Romania	28	42	50	53
Rwanda	2	3	8	11
Samoa	13	12	23	27

	1950	1970	1990	2000
São Tomé & Príncipe	13	23	42	51
Saudi Arabia	16	49	77	82
Senegal	31	33	38	45
Seychelles	27	26	59	69
Sierra Leone	9	18	32	40
Singapore	100	100	100	100
Solomon Is	8	9	11	14
Somalia	13	23	36	44
South Africa	43	48	59	65
Spain	52	66	78	83
Sri Lanka	14	22	21	24
St. Christopher/N.	22	34	49	56
St. Lucia	38	40	46	53
St. Vincent/G.	13	15	21	27
Sudan	6	16	22	27
Surinam	47	46	48	54
Swaziland	1	10	33	45
Sweden	66	81	84	86
Switzerland	44	55	60	62
Syria	31	43	52	57
Tanzania	4	7	33	47
Thailand	11	13	23	29
Togo	7	13	26	33
Tonga	13	21	21	25
Trinidad & Tobago	23	39	69	75
Tunisia	31	44	54	59
Turkey	21	38	48	54
Turks/Caicos	40	41	51	58
USSR	39	57	68	71
Uganda	3	8	10	14
United Arab Em.	25	42	78	78
United Kingdom	84	89	93	94
United States	64	74	74	75
Uruguay	78	82	86	87
Vanuatu	5	14	22	85
Venezuela	53	72	91	94
Vietnam	12	18	22	27
Western Sahara	68	43	57	63
Yemen	5	10	32	38
Yugoslavia	22	35	50	58
Zaire	19	30	40	46
Zambia	9	30	56	65
Zimbabwe	11	17	28	35

Population of cities

The population of all towns with over 100,000 inhabitants is given. As far as possible figures for the urban agglomeration, as opposed to the city proper, have been given. In these cases the suburbs of a large city may be shown elsewhere in the list.

The capital city is shown by * and even if it does not have a population of over 100,000 its population is shown. Figures are based on census figures or more recent estimates where available, and are given in thousands.

The local spelling or an accepted transcription of the local form of the name is used in the list.

The estimates for the United Kingdom have been made using 1981 data for the geographic cities and updated in the light of the 1991 census. Populations for administrative districts can be found in the table 'Population of the United Kingdom'.

Afghanistan (1988)

Charikar	100
Herat	177
*Kabul	1,424
Mazar-e-Sharif	131
Qandahar	226
Qonduz	108

Albania (1987)

*Tiranë	233

Algeria (1989)

Al Asnam	119
*Algiers	1,722
Annaba	348
Batna	122
Bejaia	124
Blida	191
Constantine	449
Oran	664
Sétif	187
Sidi-Bel-Abbès	186
Skikda	141
Tizi-Ouzou	101
Tlemcen	146

American Samoa (1980)

*Pago Pago	3

Andorra (1986)

*Andorra La Vella	16

Angola (1988)

*Luanda	1,200

Anguilla (1985)

*The Valley	2

Antigua (1982)

*St. John City	30

Argentina (1988)

Bahia Blanca	257
*Buenos Aires	11,126
Catamarca	108
Comodoro Rivadavia	116
Córdoba	1,134
Corrientés	215
Formosa	119
La Plata	630
Mar del Plata	504
Mendoza	707
Nequén	127
Paranà	189
Posadas	181
Resistencia	281
Rio Cuarto	127
Rosario	1,071
Salta	328
San Juan	346
San Miguel de Tucumàn	603
San Nicolas	125
San Salvador de Jujuy	158
Santa Fé	330
Santiago del Estero	183

Australia (1987)

Adelaide	1,013
Brisbane	1,215
*Canberra	289
Geelong	149
Gold Coast	220
Hobart	180
Melbourne	2,965
Newcastle	419
Perth	1,083
Sydney	3,531
Townsville	109
Wollongong	234

Austria (1984)

Graz	395
Innsbruck	235
Klagenfurt	135
Linz	434
Salzburg	267
*Vienna	2,044

Bahamas (1980)

*Nassau	135

Bahrain (1988)

*Al Manamah	152

Bangladesh (1987)

Barisal	159
Chittagong	1,840
Comilla	126
*Dacca	4,770
Jessore	149
Khulna	860
Mymensingh	108
Narayanganj	298
Pabna	101
Rajshahi	430
Rangpur	156
Saidpur	128
Sylhet	169

Barbados (1987)

*Bridgetown	7

Belgium (1988)

Antwerp	500
Bruges	118
*Brussels	970
Charleroi	209
Ghent	233
Liège	200
Namur	103

Belize (1987)

*Belmopan	4

Benin (1982)

Cotonou	487
*Porto-Novo	208

Bermuda (1987)

*Hamilton	3

Bhutan (1987)

*Thimphu	60

Bolivia (1985)

Cochabamba	404
*La Paz	977
Oruro	178
Potosí	113
Santa Cruz	529
*Sucre	106

Botswana (1989)

*Gaborone	120
Mahalapye	104

Brazil (1989)

Alagoinhas	117
Alvoraba	106
Americana	157
Anàpolis	227
Aracaju	362
Aracatuba	142
Arapiraca	148
Araraguara	145
Arequemes	102
Baglé	106
Barra Mansa	187
Baurú	221
Belém	1,296
Belo Horizonte	3,446
Blumenau	193
Braganca Paulista	105
*Brasilia	1,577
Cabo	122
Cachoeiro de Itapemirim	138
Camacari	108
Camaragibe	113
Campina Grande	281
Campinas	845
Campo Grande	386
Campos	367
Canoas	262
Carapicuiba	268
Caratinga	110
Cariacica	244
Caruarú	191
Cascavel	201
Caucaia	109
Caxias	149
Caxias do Sul	268
Chapeco	101
Codó	119
Colatina	106
Contagem	386
Criciúma	129
Cuiabà	283
Curitiba	1,926
Diadema	322
Divinópolis	140
Dourados	124
Duque de Caxias	666
Embu	120
Feira de Santana	357
Florianópolis	218
Fortaleza	2,169
Foz de Iguaçú	183
Franca	184
Goiânia	928
Governador Valadares	217
Gravataí	142
Guarapuava	149
Guarujà	187
Guarulhos	718
Ilhéus	146
Imperatriz	237
Ipatinga	214
Irece	107
Itaborai	145
Itabuna	179
Itaguai	106
Itajai	104
Itapetininga	105
Itapipoca	109
Jaboatao	411
Jacarei	150
Jacobina	121
Jeguié	127
Joao Pessoa	398
Joinville	304
Juazeiro	154
Juazeiro do Norte	160
Juiz de Fora	351
Jundiaí	315
Lajes	144
Limeira	188
Linhares	123
Londrina	348
Luziania	101
Macapà	170
Maceió	484
Magé	200
Manaus	834
Marabú	134
Marilia	137
Maringà	198
Mauà	271
Mogi das Cruzes	235
Montes Claros	215
Mossoró	159
Natal	512
Nilópolis	166
Niterói	443
Nova Friburgo	144
Nova Hamburgo	168
Nova Iguaçu	1,325
Olinda	336
Osasco	594
Parnaíba	117
Passo Fundo	138
Paulista	161
Pelotas	278
Petrolina	131
Petrópolis	275
Piracicaba	253
Pitanga	100
Pocos de Caldas	100
Ponta Grossa	224
Pôrto Alegre	2,924
Pôrto Velho	202
Presidente Prudente	156
Recife	2,945
Resende	103
Ribeirao Preto	385
Rio Branco	146
Rio Claro	130
Rio de Janeiro	11,141
Rio Grande	165
Rondonopolis	102
Salvador	2,362
Santa Cruz do Sol	115
Santa Luzia	117
Santa Maria	197
Santarem	227
Santo André	637
Santo Angelo	108
Santos	461
Sao Bernardo do Campo	566
Sao Caetano do Sul	171
Sao Carlos	140
Sao Gonçalo	731
Sao Joao de Meriti	459
Sao José	106
Sao José do Rio Prêto	230
Sao José dos Campos	375
Sao Leopoldo	114
Sao Luis	564
Sao Paulo	16,832
Sao Vicente	241
Serra	102
Sete Lagoas	121
Sobral	128
Sorocaba	329
Sumaré	151
Susano	130
Taboao da Serra	123
Taubaté	206
Teofilo Otoni	126
Teresina	476
Teresopolis	116
Uberaba	246
Uberlândia	314
Uruquaiana	106
Varzea Grande	103
Viamao	149
Vila Velha	253
Vitoria	254
Vitoria da Conquista	199
Vitoria de Santo Antao	101
Volta Redonda	220

British Virgin Islands (1980)

*Road Town	4

Brunei (1982)

*Bandar Seri Begawan	58

Bulgaria (1987)

Burgas	198
Choumen	105
Dobrich (Tolbukhin)	111
Pleven	134
Plovdiv	357
Ruse	190
Sliven	107
*Sofia	1,129
Stara Zagora	156
Sumen	106
Varna	306

Burkina Faso (1985)

Bobo-Dioulasso	231
*Ouagadougou	442

Burma (Myanmar) (1983)

Bassein	144
Insein	144
Kanbe	254
Mandalay	533
Monywa	107
Moulmein	220
Pegu	150
*Rangoon	2,513
Sittwe	108
Taunggye	108
Thingangyun	141

Burundi (1986)

*Bujumbura	273

Cambodia (1983)

*Phnom Penh	500

Cameroon (1986)

Douala	1,030
Maroua	104
Nkongsamba	123
*Yaoundé	654

Canada (1986)

Calgary	671
Chicoutimi-Jonguière	158
Edmonton	785
Halifax	296
Hamilton	557
Kingston	122
Kitchener	311
London	342
Moncton	102
Montréal	2,921
Oshawa	204
*Ottawa-Hull	819
Québec	603
Regina	187
St. Catharines-Niagara	343
Saint John	121
St. John's	162
Saskatoon	201
Sherbrooke	130
Sudbury	149
Sydney	119
Thunder Bay	122
Toronto	3,427
Trois-Rivières	129
Vancouver	1,381
Victoria	256
Windsor	254
Winnipeg	623

Cape Verde Islands (1980)

*Praia	58

Cayman Islands (1988)

*Georgetown	12

Central African Republic (1988)

*Bangui	597

Chad (1986)

*Ndjamena	512
Sarh	124

Chile (1987)

Antofagasta	205
Arica	123
Chillàn	124
Concepción	394
Iquique	132
La Serena	107
Los Angeles	108
Osorno	105
Puente Alto	126
Puerto Mont	113
Punta Arenas	112
Rancagua	172
San Bernardo	136
Santiago	4,858
Talca	164
Talcahuano	221
Temuco	218
Valdivia	116
Valparaiso	279
Vina del Mar	316

China (1986)

Aksu	341
Altay	169
Anda	423
Anqing	441
Anshan	2,517
Anshun	216
Anyang	550
Baicheng	286
Baiyin	325
Bangiao	115
Baoding	548
Baoji	368
Baoshan	697
Baotou	1,119
Beihai	180
*Beijing	9,750
Beipiao	605
Bei'an	439
Bengbu	623
Benxi	839
Binzhou	186
Bole	141
Bose	275
Botou	460
Bozhou	1,112
Cangzhou	303
Changchun	1,908
Changde	230
Changji	237
Changsha	1,193
Changshu	1,004
Changzhi	474
Changzhou, Jiangsu	538
Chaohu	741
Chaoyang	328
Chao'an	1,227
Chengde	337
Chengdu	4,025
Chengzhou	201
Chifeng	896
Chongqing	6,511
Chuxiong	383
Chuzhou	370
Dali	399
Dalian	4,619
Dandong	2,574
Danjiangkou	431
Daqing	854
Datong	1,020
Daxian	218
Deyang	768
Dezhou	283
Dingzhou	938
Dongchuan	277
Dongguang	1,230
Dongsheng	130
Dongying	540
Dukhou	546
Dunhua	450
Duyun	392
Enshi	686

Ezhou	922
Foshan	323
Fuling	986
Fushun	2,045
Fuxin	1,693
Fuyang	208
Fuzhou, Fujian	1,652
Fuzhou, Jiangxi	174
Gangzhou	351
Gejiu	349
Gongzhuling	910
Guangyuan	819
Guangzhou	5,669
Guilin	686
Guiyang	1,403
Haicheng	992
Haikou	300
Hailar	185
Haining	600
Hami	275
Hancheng	307
Handan	1,014
Hangzhou	5,234
Hanzhong	420
Harbin	2,668
Hebi	326
Hechi	274
Hefei	1,541
Hegang	599
Heihe	138
Hengshui	292
Hengyang	616
Heshan	112
Heze	1,017
Hohhot	1,206
Hotan	121
Houma	160
Huaibei	1,308
Huaihua	436
Huainan	1,519
Huaiyin	391
Huangshan	151
Huangshi	1,069
Huizhou	188
Hunjiang	694
Huzhou	974
Hyaying	321
Jiamusi	571
Jiangmen	240
Jiaojiang	391
Jiaozou	519
Jiaxing	697
Jiayuguan	103
Jilin	3,974
Jinan	3,376
Jinchang	142
Jincheng	621
Jingdezhen	581
Jingmen	957
Jinhua	269
Jining, Inner Mongolia	220
Jining, Shandong	166
Jinshi	218
Jinxi	638
Jinzhou	4,448
Jishou	199
Jiujiang	390
Jiuquan	280
Jixi	816
Ji'an	186
Kaifeng	636
Kaili	349
Kaiyuan	219
Karamay	189
Kashi	202
Korla	230
Kunming	1,976
Kuytun	221
Laiwu	1,054

Langfang	533
Lanxi	612
Lanzhou	2,340
Laohekou	423
Lengshuijiang	287
Lengshuitan	371
Leshan	1,039
Lhasa	108
Lianyungang	472
Liaocheng	737
Liaoyang	1,612
Liaoyuan	378
Lichuan	718
Linfen	544
Linhai	1,012
Linhe	347
Linqing	611
Linxia	153
Linyi	1,385
Lishui	303
Liuzhou	655
Longkou	567
Longyan	386
Loudi	266
Luohe	164
Lupanshui	2,247
Luzhou	369
Lu'an	171
Lyoyang	1,063
Macheng	1,010
Manzhouli	119
Maoming	450
Ma'anshan	375
Meihekou	541
Meixian	749
Mianyang	859
Mudanjiang	635
Nanchang	2,471
Nanchong	246
Nangong	400
Nanjing	3,682
Nanning	963
Nanping	427
Nantong	420
Nanyang	304
Neijiang	310
Ningbo	1,033
Pingdingshan	843
Pingling	367
Pingxiang	1,305
Puqi	413
Putian	272
Puyang	1,125
Qingdao	4,205
Qingtongxia	198
Qingzhou	794
Qinhuangdao	448
Qinzhou	944
Qiqihar	1,301
Qitaihe	327
Quanzhou	444
Qufu	545
Qujing	775
Quzhou	211
Renqiu	591
Rizhao	988
Sanmenxia	157
Sanming	217
Sanya	324
Shanghai	12,320
Shangqui	205
Shangrao	145
Shantou	774
Shaoguan	696
Shaowu	269
Shaoxing	258
Shaoyang	475
Shashi	558
Shenyang	5,055

Shenzhen	335
Shihezi	546
Shijiazhuang	1,187
Shishou	558
Shiyan	338
Shizuishan	543
Shuangyashan	434
Siping	365
Suihua	737
Suining	1,195
Suizhou	140
Suzhou, Anhui	225
Suzhou, Jiangsu	722
Tacheng	121
Taiyuan	2,177
Taizhou	214
Tai'an	210
Tangshan	1,410
Tianjin	7,790
Tianshui	965
Tiefa	182
Tieling	289
Tongchuan	404
Tonghua	373
Tongliao	261
Tongling	501
Tsuen Wan	690
Tumen	102
Tunxi	107
Turpan	203
Ulanhot	198
Ürümqi	1,084
Ulanhot	198
Wafangdian	965
Wanxian	287
Weifang	310
Weihei	226
Weinan	709
Wenzhou	5,948
Wiamen	558
Wuhai	264
Wuhan	4,273
Wuhu	944
Wuwei	814
Wuxi	650
Wuzhong	231
Wuzhou	267
Xiamen	962
Xi'an	2,911
Xiangfan	442
Xiangtan	616
Xianning	406
Xianyang	662
Xiaogan	1,219
Xichang	436
Xilin Hot	102
Xiling	237
Xingtai	361
Xining	927
Xinji	532
Xintai	1,167
Xintao	1,272
Xinxiang	547
Xinyang	240
Xinyu	662
Xinzhou	400
Xuchang	254
Xuzhou	841
Yakeshi	393
Yancheng	1,265
Yangquan	516
Yangzhou	400
Yanji	228
Yantai	734
Yan'an	268
Ya'an	282
Yibin	664
Yichang	418
Yichun, Heilongjiang	1,167
Yichun, Jiangxi	785

Yinchuan	658
Yingcheng	546
Yingkou	2,789
Yingtan	120
Yining	237
Yiyang	374
Yongzhou	517
Yong'an	272
Yuci	425
Yueyang	422
Yulin	1,255
Yumen	184
Yuncheng	442
Yuxi	298
Yuyao	778
Zalantun	390
Zhangjiakou	629
Zhangye	400
Zhangzhou	318
Zhanjiang	947
Zhaodong	754
Zhaoging	194
Zhaotong	560
Zhaozhuang	1,612
Zhengzhou	1,943
Zhenjiang	422
Zhongshan	1,073
Zhoukou	227
Zhuhai	165
Zhumadian	216
Zhuozhou	478
Zhuzhou	513
Zibo	2,329
Zigong	1,673
Zixing	340
Zunyi	354

Colombia (1985)

Armenia	192
Barrancabermeja	137
Barranquilla	920
Bello	211
*Bogotá	4,185
Bucaramanga	364
Buenaventura	160
Cali	1,397
Cartagena	560
Cúcuta	407
Floridablanca	142
Ibagué	306
Itagui	136
Manizales	310
Medellin	1,506
Montería	238
Neiva	180
Palmira	175
Pasto	252
Pereira	302
Popayan	166
Santa Marta	226
Sincelejo	140
Soledad	164
Tulua	103
Valledupar	209
Villavicencio	182

Comoros (1980)

*Moroni	20

Congo (1985)

*Brazzaville	596
Pointe-Noire	298

Costa Rica (1984)

*San José	245

Cuba (1987)

Bayamo	119
Camagüey	275
Cienfuegos	116
Guantánamo	193
*Havana	2,059
Holguín	218
Las Tunas	111
Matanzas	110
Pinar del Rio	112
Santa Clara	188
Santiago de Cuba	390

Cyprus (1988)

Limassol	120
*Nicosia	167

Czechoslovakia (1989)

Bratislava	432
Brno	389
Hrádec Králové	100
Košice	230
Liberec	104
Olomouc	107
Ostrava	330
Plzeň	175
*Prague	1,209
Ústí nad Labem	106

Denmark (1989)

Ålborg	155
Århus	258
*Copenhagen	1,339
Odense	174

Djibouti (1988)

*Djibouti	290

Dominica (1981)

*Roseau	20

Dominican Rep. (1981)

Santiago	279
*Santo Domingo	1,313

Ecuador (1987)

Ambato	221
Cuenca	272
Esmeraldas	141
Guayaquil	1,301
Machala	111
Manta	136
Milagro	103
Portoviejo	167
*Quito	1,110
Riobamba	150
Santo Domingo	104

Egypt (1986)

Alexandria	2,893
Aswan	196
Asyût	292
Benha	120
Beni-Suef	163
*Cairo	6,325
Damanhûr	226
Damietta	121
El Faiyûm	227
El Giza	1,858
El Mahalla el Kubra	385
El Mansûra	358
El Minyâ	203
Helwân	352
Ismâ'ilîya	236

Kafr el Dauwâr	240
Kafr el Sheikh	104
Luxor	148
Port Said	382
Qena	142
Shibîn el Kom	136
Shubra el Kheima	711
Sohâg	142
Suez	265
Tanta	374
Zagazig	274

El Salvador (1983)

Mejicanos	107
San Miguel	176
*San Salvador	973
Santa Ana	224

Equatorial Guinea (1988)

*Malabo	37

Ethiopia

*Addis Ababa	1,739
Asmera	344
Diredawa	122

Falkland Islands (1986)

*Stanley	1

Fiji (1986)

*Suva	141

Finland (1989)

Espoo	170
*Helsinki	994
Oulu	100
Tampere	261
Turku	265
Vantaa	152

France (1982)

Aix-en-Provence	121
Amiens	153
Angers	196
Angoulême	102
Annecy	110
Avignon	173
Bayonne	128
Besançon	121
Bordeaux	628
Boulogne	103
Brest	187
Bruay-en-Artois	109
Caen	182
Calais	101
Clermont-Ferrand	256
Dijon	209
Douai	202
Dunkerque	196
Grasse-Cannes	296
Grenoble	392
Hagondange-Briey	120
La Rochelle	100
Le Havre	255
Le Mans	191
Lens	323
Lille	935
Limoges	172
Lorient	104
Lyons	1,170
Mantes-la-Jolie	168
Marseilles	1,080
Metz	185
Montbéliard	128
Montpellier	221

Mulhouse	220
Nancy	278
Nantes	465
Nice	449
Nîmes	130
Orléans	220
*Paris	8,510
Pau	131
Perpignan	130
Reims	199
Rennes	234
Rouen	380
St-Étienne	317
St-Nazaire	121
Strasbourg	373
Thionville	128
Toulon	410
Toulouse	523
Tours	255
Trappes	142
Troyes	125
Valence	104
Valenciennes	337
Villeurbanne	116

French Guiana (1982)

*Cayenne	38

French Polynesia (1987)

*Papeete	30

Germany (1988)

Aachen	239
Augsburg	243
Bergisch Gladbach	102
*Berlin	3,301
Bielefeld	306
Bochum	386
Bonn	277
Bottrop	112
Braunschweig	252
Bremen	533
Bremerhaven	132
Chemnitz (Karl-Marx-Stadt)	312
Cologne	928
Cottbus	129
Darmstadt	134
Dessau	104
Dortmund	584
Dresden	516
Düsseldorf	563
Duisburg	525
Erfurt	220
Erlangen	100
Essen	623
Frankfurt	619
Freiburg bei Breisgau	186
Gelsenkirchen	288
Gera	132
Göttingen	134
Hagen	206
Halle	235
Hamburg	1,594
Hamm	166
Hanover	500
Heidelberg	136
Heilbronn	112
Herne	171
Hildesheim	101
Jena	107
Karlsruhe	261
Kassel	185
Kiel	238
Koblenz	110
Krefeld	217
Leipzig	545
Leverkusen	155
Ludwigshafen am Rhein	152

Lübeck	209
Magdeburg	291
Mainz	189
Mannheim	295
Mönchengladbach	250
Mülheim a.d. Ruhr	170
Münster	246
Munich	1,189
Neuss	144
Nuremberg	472
Oberhausen	222
Offenbach am Main	107
Oldenburg	139
Osnabrück	154
Paderborn	110
Pforzheim	105
Potsdam	143
Recklinghausen	118
Regensburg	124
Remscheid	121
Rostock	254
Saarbrücken	184
Salzgitter	105
Schwerin	131
Siegen	107
Solingen	158
Stuttgart	552
Ulm	101
Wiesbaden	252
Witten	102
Wolfsburg	122
Würzburg	127
Wuppertal	366
Zwickau	121

Gabon (1983)

*Libreville	830

Gambia (1983)

*Banjul	146

Ghana (1984)

*Accra	965
Kumasi	489
Sekondi-Takoradi	116
Tamale	168
Tema	191

Gibraltar (1988)

Gibraltar	30

Greece (1981)

*Athens	3,027
Iràklion	102
Làrisa	102
Pàtrai	142
Thessaloniki	872

Greenland (1987)

*Godthåb	12

Grenada (1988)

*St. George's	29

Guadeloupe (1982)

*Basse Terre	14

Guam (1980)

*Agana	29

Guatemala (1989)

*Guatemala	2,000
Puerto Barrios	338
Quezaltenango	246

Guinea (1983)

*Conakry	705

Guinea Bissau (1988)

*Bissau	125

Guyana (1983)

*Georgetown	188

Haïti (1988)

Cap Haïtien	133
Gonaïves	144
Hinche	122
Jacmel	217
Jérémie	152
Les Cayes	215
*Port-au-Prince	1,144
Port-de-Paix	135

Honduras (1986) **133**

San Pedro Sula	400
*Tegucigalpa	605

Hong Kong (1986)

Kowloon	2,302
*Hong Kong	1,176
Tsuen Wan	690

Hungary (1989)

*Budapest	2,115
Debrecen	220
Györ	132
Kecskemét	106
Miskolc	208
Nyíreguháza	119
Pécs	183
Szeged	189
Székesfehérvár	114

Iceland (1988)

*Reykjavik	96

India (1981)

Adoni	109
Agartala	132
Agra	747
Ahmadabad	2,548
Ahmadnagar	181
Ajmer	376
Akola	225
Aligarh	321
Allahabad	650
Alleppey	170
Alwar	146
Amaravati	261
Ambala	121
Amritsar	595
Amroha	113
Anantapur	120
Arrah	125
Asansol	366
Aurangabad	316
Baharampur	102
Bahraich	103
Balurghat	113
Bangalore	2,922
Barddhaman	167
Bareilly	449
Batala	102
Belgaum	300
Bellary	202
Bermo	102
Bhadravati	131
Bhagalpur	225

Bharatpur	105
Bharuch	121
Bhatinda	127
Bhavnagar	307
Bhilwara	123
Bhimavaram	102
Bhiwandi	115
Bhiwani	101
Bhopal	671
Bhubaneshwar	219
Bhusawal	132
Bihar	151
Bijapur	147
Bikaner	288
Bilaspur	187
Bokaro Steel City	264
Bombay	8,243
Brahmapur	163
Bulandshahr	103
Burhanpur	141
Calcutta	9,194
Calicut	546
Cannanore	158
Chandigarh	423
Chandrapur	116
Chapra	112
Cochin	686
Coimbatore	920
Cuddalore	128
Cuddapah	103
Cuttack	327
Darbhanga	176
Davangere	197
Dehra Dun	293
*Delhi	5,729
Dhanbad	678
Dharwad	527
Dhule	211
Dindigul	170
Durgapur	312
Durg-Bhilai	490
Elluru	168
Erode	276
Etawah	212
Faizabad	143
Faridabad	331
Farrukhabad-cum-Fatehgarh	161
Firozabad	203
Gadag-Betgeri	117
Ganganagar	124
Gaya	247
Ghaziabad	287
Gondia	100
Gorakhpur	308
Gulbarga	219
Guntur	368
Gurgaon	101
Gwalior	556
Habra	130
Hapur	103
Haridwar	146
Hisar	137
Hospet	115
Hyderabad	2,546
Ichalkaranji	134
Imphal	157
Indore	829
Jabalpur	757
Jaipur	1,015
Jalgaon	145
Jalna	122
Jammu	223
Jamnagar	317
Jamshedpur	670
Jaunpur	105
Jhansi	284
Jodhpur	506
Jullundur	408

Junagadh	120
Kakinada	226
Kanchipuram	145
Kanpur	1,639
Karaikkudi	101
Karnal	132
Katihar	122
Khandwa	115
Kharagpur	233
Kolar Gold Fields	144
Kolhapur	351
Kota	358
Kumbakonam	142
Kurnool	206
Latur	112
Lucknow	1,008
Ludhiana	606
Machilipatnam	139
Madras	4,289
Madurai	908
Malegaon	246
Mandya	100
Mangalore	306
Mathura	159
Meerut	537
Mirzapur	128
Moradabad	345
Munger	129
Murwara	123
Muzaffarnagar	172
Muzaffarpur	190
Mysore	479
Nadiad	143
Nagercoil	172
Nagpur	1,302
Nanded	191
Nasik	429
Navadwip	130
Navsari	129
Nellore	237
Nizamabad	183
Ondal	109
Palghat	118
Panipat	138
Parbhani	109
Patan	105
Pathankot	110
Patiala	206
Patna	919
Pollachi	115
Pondicherry	251
Porbandar	133
Proddatur	107
Pune	1,686
Puri	101
Purnia	110
Quilon	168
Raichur	125
Raipur	338
Rajahmundry	268
Rajapalayam	102
Rajkot	445
Rampur	205
Ranchi	503
Raniganj	119
Ratlam	156
Raurkela	323
Rewa	101
Rohtak	166
Sagar	207
Saharanpur	295
Salem	519
Sambalpur	162
Sambhal	108
Sangli	269
Shahjahanpur	205
Shiliguri	154
Shillong	175
Shimoga	152

Sikar	103
Solapur	515
Sonipat	109
Srinagar	606
Surat	914
Tenali	119
Thane	390
Thanjavur	184
Tiruchchirappalli	610
Tirunelveli	323
Tirupati	115
Tiruppur	216
Trichur	170
Trivandrum	520
Tumkur	109
Tuticorin	251
Udaipur	233
Ujjain	282
Ulhasnagar	649
Vadodara	744
Valparai	115
Varanasi	797
Vazianagaram	115
Vellore	247
Vijayawada	543
Vishakhapatnam	604
Wadhawan	130
Warangal	335
Yamunanagar	160

Indonesia (1980)

Ambon	209
Balikpapan	281
Bandung	1,567
Banjarmasin	381
Bogor	247
Cirebon	224
*Jakarta	7,886
Jambi	230
Kediri	222
Madiun	151
Magelang	123
Malang	512
Manado	217
Medan	1,806
Padang	481
Pakanbaru	186
Palembang	787
Pekalongan	133
Pematangsiantar	150
Pontianak	305
Probolinggo	100
Samarinda	265
Semarang	1,027
Sukabumi	110
Surabaya	2,224
Surakarta	470
Tanjung Karang	284
Tegal	132
Ujung Pandang	709
Yogyakarta	399

Iran (1986)

Abadan	308
Ahvaz	580
Amol	118
Arak	265
Ardabil	282
Babol	115
Bakhtaran	561
Bandar-e-Abbas	202
Borujerd	184
Bushehr	120
Dezful	151
Esfahan	987
Gorgan	114
Hamadan	272
Islam Shahr/Qasemabad	215

Karaj	275
Kashan	139
Kerman	257
Khomeini Shahr	105
Khoramabad	209
Khorramshahr	151
Khoy	115
Malayer	104
Maraqeh	101
Mashhad	1,464
Masjed Soleyman	139
Najafabad	139
Neyshabur	109
Orumiyeh	301
Qaem Shahr	109
Qazvin	249
Qom	543
Rajai Shahr	118
Rasht	291
Rezaiyeh	164
Sabzewar	108
Sabzewar	129
Sanandaj	205
Sari	141
Shiraz	848
Tabriz	971
*Tehran	6,043
Yazd	230
Zahedan	282
Zahedan	165
Zanjan	215

Iraq (1985)

Al Amarah	132
Al Hillah	215
An Najaf	243
An Nasiriyah	139
Ar Ramadi	137
As Sulaymaniyah	279
*Baghdad	4,649
Basra	617
Ba'qubah	115
Bene Beraq	109
Irbil	334
Karbala	185
Kirkuk	208
Mosul	571
Netanya	118
Rishon Leziyyon	126

Irish Republic (1986)

Cork	174
*Dublin	921

Israel (1987)

Bat-Yam	133
Be'er Sheva	113
Haifa	223
Holon	146
*Jerusalem	495
Petah Tiqwa	134
Ramat Gan	116
Tel Aviv-Jaffa	318

Italy (1987)

Alessàndria	101
Ancona	104
Bari	357
Bérgamo	119
Bologna	422
Bolzano	101
Bréscia	199
Càgliari	220
Catània	370
Catanzaro	103
Consenza	106
Ferrara	143

Florence	417	Fukushima	278	Kyoto	1,461
Fóggia	159	Fukuyama	366	Machida	349
Forlí	110	Funabashi	533	Maebashi	286
Genoa	715	Gifu	410	Matsubara	136
La Spézia	107	Habikino	115	Matsudo	456
Lecce	102	Hachinohe	241	Matsue	143
Leghorn	173	Hachioji	466	Matsumoto	201
Messina	272	Hadano	156	Matsusaka	119
Milan	1,464	Hakodate	307	Matsuyama	443
Módena	177	Hamamatsu	535	Minoo	122
Monza	123	Handa	100	Misato	128
Naples	1,203	Higashikurume	114	Mishima	105
Novara	103	Higashimurayama	134	Mitaka	166
Padua	222	Higashiosaka	518	Mito	235
Palermo	731	Himeji	454	Miyakonojo	130
Parma	175	Hino	166	Miyazaki	287
Perugia	148	Hikone	100	Moriguchi	157
Pescara	131	Hirakata	391	Morioka	235
Piacenza	105	Hiratsuka	246	Muroran	118
Pisa	104	Hirosaki	175	Musashino	139
Prato	165	Hiroshima	1,086	Nagano	347
Ravenna	136	Hitachi	202	Nagaoka	186
Reggio di Calabria	179	Hofu	118	Nagareyama	140
Réggio nell'Emilia	130	Ibaraki	254	Nagasaki	445
Rimini	131	Ichihara	258	Nagoya	2,155
*Rome	2,817	Ichikawa	437	Naha	305
Salerno	154	Ichinomiya	262	Nara	349
Sassari	120	Ikeda	104	Narashino	151
Syracuse	124	Ikoma	100	Neyagawa	257
Tàranto	245	Imabari	123	Niigata	486
Terni	111	Iruma	138	Niihama	129
Torre del Greco	105	Ise	104	Niiza	139
Trento	101	Isesaki	116	Nishinomiya	427
Trieste	235	Ishinomaki	122	Nobeoka	131
Turin	1,012	Itami	186	Noda	114
Udine	102	Iwaki	356	Numazu	212
Venice	324	Iwakuni	110	Obihiro	167
Verona	259	Iwatsuki	106	Odawara	193
Vicenza	110	Izumi (Miyagi)	146	Ogaki	148

Ivory Coast (1982)

*Abidjan	1,850
Bouaké	640
Daloa [1974]	100
Korhogo	280
Man-Danane	450
Yamoussoukro	120

Jamaica (1982)

*Kingston	525

Japan (1990)

Abiko	121	Izumi (Osaka)	138
Ageo	195	Joetsu	130
Aizuwakamatsu	119	Kadoma	142
Akashi	271	Kagoshima	537
Akishima	105	Kakamigahara	130
Akita	302	Kakogawa	240
Amagasaki	499	Kamakura	174
Anjo	142	Kanazawa	443
Aomori	288	Kariya	120
Asahikawa	359	Kashihara	116
Asaka	104	Kashiwa	305
Ashikaga	168	Kasugai	267
Atsugi	197	Kasukabe	189
Beppu	130	Katsuta	110
Chiba	829	Kawachinagano	109
Chigasaki	202	Kawagoe	305
Chofu	198	Kawaguchi	439
Daito	126	Kawanishi	141
Ebina	106	Kawasaki	1,174
Fuchu	209	Kiryu	126
Fuji	223	Kisarazu	123
Fujieda	120	Kishiwada	189
Fujinomiya	117	Kitakyushu	1,026
Fujisawa	350	Kitami	107
Fukui	253	Kobe	1,477
Fukuoka	1,237	Kochi	317
		Kodaira	164
		Kofu	201
		Koganei	106
		Kokubunji	101
		Komaki	124
		Komatsu	106
		Koriyama	315
		Koshigaya	285
		Kumagaya	152
		Kumamoto	579
		Kurashiki	415
		Kure	217
		Kurume	228
		Kushiro	206

Oita	409	*Tokyo	11,936
Okayama	594	Tomokomai	158
Okazaki	307	Tondabayashi	110
Okinawa	106	Tottori	142
Ome	126	Toyama	321
Omiya	404	Toyohashi	338
Omuta	150	Toyokawa	112
Onomichi	101	Toyonaka	410
Osaka	2,624	Toyota	332
Ota	140	Tsu	157
Otaru	163	Tsuchiura	127
Otsu	260	Tsukuba	143
Oyama	148	Tsuruoka	100
Saga	170	Ube	175
Sagamihara	532	Ueda	119
Sakai	808	Uji	177
Sakata	101	Urawa	418
Sakura	145	Urayasu	166
Sapporo	1,672	Utsunomiya	427
Sasebo	245	Wakayama	397
Sayama	157	Yachiyo	149
Sendai	918	Yaizu	112
Seto	126	Yamagata	249
Shimizu	242	Yamaguchi	129
Shimonoseki	263	Yamato	195
Shizuoka	472	Yao	278
Soka	206	Yatsushiro	108
Suita	345	Yokkaichi	274
Suzuka	174	Yokohama	3,220
Tachikawa	153	Yokosuka	433
Takamatsu	330	Yonago	131
Takaoka	175	Zama	112
Takarazuka	202		
Takasaki	236		
Takatsuki	360		
Tama	144		
Tamakomai	160		
Tokorozawa	303		
Tokushima	263		
Tokuyama	111		

Jordan (1986)

*Amman	1,160
Az-Zarqa	405
Irbid	680

Kenya (1986)

Kisumu	198
Mombasa	457
*Nairobi	1,429
Nakuru	112

Kiribati (1985)

*Tarawa	24

Korea, North (1984)

Chinnamp'o	691
Chongjin	754
Haeju	131
Hamhung	775
Kaesong	346
Kimchaek	281
*Pyongyang	2,639
Sinuiju	500
Wonsan	350

Korea, South (1985)

Andong	114
Anyang	362
Chechon	102
Cheju	203
Cheonan	170
Chinhae	121
Chinju	227
Chongju	350
Chonju	426
Chunchon	163
Ch'angwon	174
Ch'ungju	113
Inchon	1,387
Iri	192
Kangnung	133
Kumi	142
Kunsan	186
Kwangju	906
Kwangmyong	220
Kyongju	127
Masan	449
Mokpo	236
Punch'on	456
Pusan	3,517
P'ohang	261
*Seoul	9,646
Songnam	448
Sunch'on	122
Suwon	431
Taegu	2,031
Taejon	866
Tonghae	114
Uijongbu	163
Ulsan	551
Wonju	151
Yosu	172

Kuwait (1985)

Al Jahra	111
*Kuwait	189
Salmiya	153

Laos (1985)

*Vientiane	377

Lebanon (1980)

*Beirut	702
Tripoli	175

Lesotho (1986)

*Maseru	109

Liberia (1984)

*Monrovia	425

Libya (1982)

Benghazi	650
Misratah	285
Sabhah	113
*Tripoli	980

Liechtenstein (1987)

*Vaduz	5

Luxembourg (1987)

*Luxembourg	77

Macau (1986)

*Macau	395

Madagascar (1986)

*Antananarivo	703
Fianarantsoa	111
Mahajanga	111
Toamasina	139

Malawi (1985)

Blantyre	355
*Lilongwe	234

Malaysia (1980)

Ipoh	301
Johor Baharu	250
Kelang	193
Kota Baharu	171
Kota Kinabalu	109
*Kuala Lumpur	938
Kuala Trengganu	187
Kuantan	137
Kuching	120

Petaling Jaya	208	Toluca	357	
Pinang	251	Torreón	730	
Sandakan	113	Tuxtla Gutiérrez	166	
Seremban	136	Uruapan	147	
Taiping	146	Veracruz	305	
Tawau	114	Villahermosa	251	

Maldives (1988)

*Malé	46

Monaco (1989)

*Monaco	28

Mali (1987)

*Bamako	646

Mongolia (1988)

*Ulan Bator	515

Malta (1987)

*Valletta	101

Montserrat (1980)

*Plymouth	3

Martinique (1982)

*Fort-de-France	100

Morocco (1982)

Agadir	246
Béni-Mellal	204
Casablanca	2,409
El Jadida	164
Fès	562
Kénitra	450
Khemisset	100
Khouribga	230
Marrakesh	549
Meknès	487
Nador	115
Oujda	471
*Rabat-Salé	893
Safi	256
Settat	167
Tangier	312
Taza	147
Tétouan	371

Mauritania (1985)

*Nouakchott	500

Mauritius (1988)

*Port Louis	139

Mayotte (1985)

*Dzaoudzi	6

Mexico (1980)

Acapulco	409
Aguascalientes	359
Atizapàn de Zaragoza	188
Campeche	152
Celaya	219
Chihuahua	407
Ciudad Juárez	596
Ciudad Madero	132
Ciudad Obregón	166
Ciudad Victoria	153
Coatzacoalcos	186
Córdoba	126
Cuernavaca	232
Culiacàn Rosales	560
Durango	321
Ensenada	175
Gómez Palacio	117
Guadalajara	2,587
Hermosillo	341
Irapuato	246
Jalapa	213
León	947
Los Mochis	123
Matamoros	239
Mazatlàn	250
Mérida	580
Mexicali	511
*Mexico	18,748
Minatitlàn	145
Monclova	120
Monterrey	2,335
Morelia	353
Nuevo Laredo	203
Oaxaca	157
Orizaba	115
Pachuca	135
Poza Rica de Hidalgo	167
Puebla	1,218
Querétaro	294
Reynosa	211
Salamanca	160
Saltillo	322
San Luis Potosi	602
Tampico	268
Tepic	177
Tijuana	461

Mozambique (1986)

Beira	270
*Maputo	1,070
Nampula	183

Namibia (1988)

*Windhoek	115

Nauru (1976)

*Domaneab	8

Nepal (1981)

*Katmandu	235

Netherlands (1989)

*Amsterdam	1,038
Arnhem	299
Breda	157
Dordrecht - Zwijndrecht	204
Eindhoven	381
Enschede-Hengelo	250
Geleen-Sittard	179
Groningen	206
Haarlem	214
Heerlen-Kerkrade	267
Hilversum	103
Leiden	185
Maastricht	161
Nijmegen	242
Rotterdam	1,040
s-Hertogenbosch	195
The Hague	684
Tilburg	227
Utrecht	526
Velsen-Beverwijk	126
Zaanstreek	142

Netherlands Antilles (1981)

*Willemstad	94

New Caledonia (1989)

*Nouméa	98

New Zealand (1991)

Auckland	885
Christchurch	307
Dunedin	109
Hamilton	148
Napier-Hastings	110
*Wellington	325

Nicaragua (1985)

León	101
*Managua	682

Niger (1983)

*Niamey	399

Nigeria (1983)

Aba	216
Abeokuta	309
Ado-Ekiti	266
Akure	117
Benin City	166
Calabar	126
Ede	222
Effon-Alaiye	111
Enugu	228
Gusau	114
Ibadan	1,060
Ife	215
Ijebu-Ode	113
Ikare	102
Ikerre-Ekiti	177
Ila	190
Ilesha	273
Ilobu	144
Ilorin	344
Iseyin	157
Iwo	262
Jos	149
Kaduna	247
Kano	487
Katsina	149
Kumo	107
*Lagos	1,097
Maiduguri	231
Mushin	241
Offa	142
Ogbomosho	527
Oka-Akoko	104
Ondo	123
Onitsha	269
Oshogbo	345
Owo	133
Oyo	185
Port Harcourt	296
Sapele	101
Shaki	126
Shomolu	107
Sokoto	148
Zaria	274

Northern Marianas (1974)

*Saipan	10

Norway (1988)

Bergen	211
*Oslo	643
Trondheim	136

Oman (1985)

*Muscat	250

Pakistan (1981)

Chiniot	106
Dera Ghazi Khan	103
Faisalabad	1,104
Gujranwala	659
Gujrat	154
Hyderabad	752
*Islamabad	204
Jhang	195
Karachi	5,181
Kasur	155
Lahore	2,953
Mardan	148
Multan	722
Okara	128
Peshawar	556
Quetta	286
Rahimyar-Khan	119
Rawalpindi	795
Sahiwal	106
Sargodha	291
Shekhupura	141
Sialkot	302
Sukkur	159
Wah-Cantt	108

Panama (1984)

*Panama	625

Papua New Guinea (1987)

*Port Moresby	152

Paraguay (1983)

*Asunción	729
Presidente Stroessner	110

Peru (1989)

Arequipa	612
Callao	515
Chiclayo	410
Chimbote	287
Cuzco	264
Huancayo	203
Ica	149
Iquitos	253
Juliaca	128
*Lima (Lima-Callao)	6,233
Piura	310
Pucallpa	147
Tacna	143
Trujillo	513

Philippines (1984)

Angeles	238
Bacolod	329
Bago	131
Baguio	154
Basilan	179
Batangas	176
Butuan	218
Cabanatuan	171
Cadiz	145
Cagayan de Oro	332
Calbayog	113
Caloocan	602
Cavite	104
Cotabato	102
Dagupan	114
Cebu	613
Davao	819
General Santos	195
Iligan	225
Iloilo	288
Las Pinas	190
Legaspi	122
Lipa	149
Lucena City	141
Makati	409
Malabon	213
Mandaluyong	227
Mandaue	169
*Manila	6,720
Marikina	248
Muntinlupa	172
Naga	113
Navotas	147
Olongapo	198
Pagadjan	105
Ormoc	123
Paranaque	253
Pasay	366
Pasig	319
Quezon City	1,546
San Carlos (Negros Occ.)	101
San Carlos (Pangasinan)	118
San Juan	139
San Pablo	163
Silay	136
Tacloban	120
Taguig	131
Valenzuela	276
Zamboanga	433

Pitcairn Island (1987)

*Adamstown	0

Poland (1988)

Białystok	260
Bielsko-Biala	178
Bydgoszcz	373
Bytom	240
Chorzów	138
Częstochowa	253
Dabrowa Górnicza	140
Elblag	122
Gdansk	469
Gdynia	250
Gliwice	211
Gorzów Wielkopolski	120
Jastrzębie Zdrój	102
Kalisz	105
Katowice	369
Kielce	208
Koszalin	105
Kraków	744
Legnica	101
Łodz	852
Lublin	333
Olsztyn	155
Opole	128
Płock	118
Poznań	586
Radom	222
Ruda Şlaska	168
Rybnik	141
Rzeszów	147
Sosnowiec	260
Szczecin	397
Tarnów	118
Torun	197
Tychy	188
Wałbrzych	141
*Warsaw	1,651
Włocławek	119
Wodzisław Şlaski	112
Wrocław	640
Zabrze	199
Zielona Góra	113

Portugal (1987)

*Lisbon	1,612
Oporto	1,315

Puerto Rico (1984)

Aguadilla	156
Arecibo	163
Bayamón	217
Caguas	275
Carolina	163
Mayagüez	210
Ponce	235
*San Juan	1,816

Qatar (1986)

*Doha	217

Réunion (1986)

*Saint-Denis	117

Romania (1985)

Arad	186
Bǎcau	175
Baia Mare	136
Botoşani	105
Brǎila	235
Braşov	351
*Bucharest	2,014
Buzau	132
Cluj-Napoca	310
Constanţa	328
Craiova	275
Galaţi	293
Iaşi	313
Oradea	209
Piatra Neamţ	108
Piteşti	154
Ploieşti	234
Reşiţa	104
Satu Mare	128
Sibiu	177
Timişoara	325
Tîrgu Mures	157

Rwanda (1981)

*Kigali	157

St. Christopher-Nevis (1980)

*Basseterre	15

St. Helena (1976)

*Jamestown	2

St. Lucia (1980)

*Castries	53

St. Pierre and Miquelon (1982)

*St. Pierre	5

St. Vincent and The Grenadines (1982)

*Kingstown	34

San Marino (1987)

*San Marino	4

São Tomé and Príncipe (1984)

*São Tomé	35

Saudi Arabia (1986)

Abha	155
Al Hufuf	101
Buraidah	184
Damman	128
Haradh	100
Jedda	1,400
Mecca	618
Medina	500
*Riyadh	2,000
Taif	205

Senegal (1985)

*Dakar	1,382
Kaolack	132
Thiès	156

Seychelles (1977)

*Victoria	23

Sierra Leone (1988)

*Freetown	470

Singapore (1987)

*Singapore	2,704

Solomon Islands (1986)

*Honiara	34

Somali Republic (1986)

Baidoa	300
Burao	300
Hargeisa	400
Kismayu	200
Merca	100
*Mogadishu	1,000

South Africa (1980)

Benoni	198
Bloemfontein	233
Boksburg	157
*Cape Town	1,912
Carletonville	123
Dlepmeadow	193
Durban	982
East London/ King William's Town	194
East Rand	1,038
Free State Goldfields	320
Germiston	166
Johannesburg	1,726
Kathlehong	138
Kayamnandi	221
Kimberley	150
Klerksdorp	134
Krugersdorp	122
Lekoa	218
Mamelodi	127
Nyanga	149
Pietermaritzburg	192
Port Elizabeth	652
Pretoria	823
Roodepoort	169
Sasolburg	540
Soweto	522
Springs	170
Tembisa	149
Uitenhage	125
Vanderbijlpark-Vereeniging	540
West Rand	647

Spain (1987)

Albacete	126
Alcalà de Henares	145
Alcorcón	138
Alicante	258
Almería	155
Badajoz	120
Badalona	224
Baracaldo	114
Barcelona	1,704
Bilbao	382
Burgos	159
Càdiz	156
Cartagena	169
Castellón	129
Córdoba	298
Elche	178
Getafe	133
Gijón	259
Granada	257
Hospitalet	278
Huelva	136
Jaén	104
Jérez de la Frontera	179
La Coruña	242
La Laguna	108
Las Palmas	358
Leganés	168
León	136
Lérida	109
Logroño	116
*Madrid	3,101
Màlaga	566
Móstoles	177
Murcia	305
Oviedo	186
Palma de Mallorca	307
Pamplona	179
Sabadell	188
Salamanca	156
San Sebastiàn	177
Santa Coloma de Gramanet	135
Santa Cruz de Tenerife	211
Santander	187
Seville	655
Tarragona	107
Tarrasa	160
Valencia	732
Valladolid	329
Vigo	263
Vitoria	201
Zaragoza	575

Sri Lanka (1987)

*Colombo	1,412
Dehiwala	191
Galle	109
Jaffna	138
Kandy	107
Kotte	102
Moratuwa	143

Sudan (1983)

El Obeid	140
Juba	116
Kassala	149
*Khartoum	561
Khartoum North	341
Omdurman	526
Port Sudan	207
Wâd Medanî	153
Wau	116

Surinam (1971)

*Paramaribo	192

Swaziland (1986)

*Mbabane	38

Sweden (1988)

Borås	101
Göteborg	720
Helsingborg	107
Jönköping	110
Linköping	120
Malmö	466
Norrköping	120
Örebro	120
*Stockholm	1,471
Uppsala	162
Västerås	118

Switzerland (1988)

Basle	359
*Bern	299
Geneva	389
Lausanne	263
Luzern	161
St Gallen	126
Winterthur	108
Zürich	839

Syria (1986)

Al Kamishli	125
Al Rakka	127
Aleppo	1,308
*Damascus	1,361
Hamah	214
Homs	464
Lattakia	258

Taiwan (1987)

Changhua	186
Changhwa	207
Chiai	255
Chilung	349
Chungho	343
Chunli	248
Fengshan	276
Fengyuan	144
Hainchung	259
Hsinchu	310
Hsintien	205
Hualien	107
Kaohsiung	1,343
Keelung	349
Panchiao	506
Pate	122
Pingchen	135
Pingtung	205
Sanchung	362
Taichung	715
Tainan	657
*Taipei	2,637
Taiping	105
Taitung	109
Tali	107
Taoyüan	220
Tucheng	111
Yuanlin	117
Yungho	242
Yungkang	115

Tanzania (1985)

Dar-es-Salaam	1,100
*Dodoma	85
Mbeya	194
Mwanza	252
Tabora	214
Tanga	172
Zanzibar	133

Thailand (1980-87)

*Bangkok	5,609
Chiang Mai	102
Chon Buri	115
Nakhon Si Thammarat	102
Songklha	173

Togo (1983)

*Lomé	366

Tonga (1989)

*Nuku'alofa	29

Trinidad and Tobago (1982)

*Port of Spain	60

Tunisia (1984)

Nabeul	335
Sfax	232
*Tunis	1,395

Turkey (1985)

Adana	776
Adapazari	155
*Ankara	2,252
Antakya	109
Antalya	258
Balikesir	152
Bursa	614
Denizli	171
Diyarbakir	305
Elazig	181
Erzurum	253
Eskisehir	367
Gaziantep	466
Isparta	102
Istanbul	5,495
Izmir	1,490
Izmit	236
Kayseri	378
Konya	439
Kütahya	120
Malatya	251
Manisa	126
Maras	212
Mersin	314
Samsun	280
Sivas	197
Trabzon	156
Urfa	206
Van	121
Zonguldak	119

Turks & Caicos (1980)

*Grand Turk	3

Tuvalu (1973)

*Fongafala	3

Uganda (1980)

*Kampala	459

United Arab Emirates (1984)

*Abu Dhabi	243
Al'Ayn	102
Dubai	266
Sharjah	125

United Kingdom (1991)

Aberdeen	187
Basildon	100
Belfast	281
Birkenhead	116
Birmingham	1,400

Blackpool	145	Baton Rouge	241	Macon	118
Bolton	142	Beaumont	120	Madison	176
Bournemouth	153	Berkeley	104	Memphis	653
Bradford	274	Birmingham	278	Mesa	251
Brighton	132	Boise City	108	Miami	3,001
Bristol	367	Boston	3,736	Milwaukee	1,572
Cambridge	100	Bridgeport	142	Minneapolis-St Paul	2,388
Cardiff	273	Brownsville	102	Mobile	203
Coventry	293	Buffalo	1,176	Modesto	133
Derby	214	Cedar Rapids	108	Montgomery	194
Dudley	186	Charlotte	1,112	Nashville-Davidson	474
Dundee	161	Chattanooga	162	New Haven	123
Edinburgh	404	Chesapeake	134	New York	18,120
Exeter	100	Chicago	8,181	Newark (N.J.)	316
Glasgow	730	Chula Vista	119	Newport Mews	162
Havant and Waterloo	116	Cincinnati	1,729	Norfolk	1,380
Huddersfield	122	Cleveland	2,769	Oakland	357
Hull	242	Colorado Springs	273	Odessa (Tex.)	101
Ipswich	116	Columbus (Ga.)	180	Oklahoma	446
Leeds	432	Columbus (Oh.)	1,344	Omaha	349
Leicester	280	Concord	106	Ontario	114
Liverpool	1,060	Corpus Christi	264	Orange	101
*London	6,378	Dallas	3,766	Orlando	146
Luton	165	Dayton	179	Oxnard	127
Manchester	1,669	Denver	1,858	Pasadena (Ca.)	130
Middlesbrough	140	Des Moines	192	Pasadena (Tx.)	118
Milton Keynes	148	Detroit	4,620	Paterson	139
Newcastle-upon-Tyne	617	Durham	114	Peoria	110
Newport	103	El Paso	492	Philadelphia	5,963
Northampton	166	Elizabeth	107	Phoenix	2,030
Norwich	119	Erie	115	Pittsburgh	2,284
Nottingham	262	Eugene	105	Plano	111
Oldham	100	Evansville	129	Pomona	116
Oxford	108	Flint	146	Portland (Or.)	1,414
Peterborough	129	Fort Lauderdale	149	Portsmouth	111
Plymouth	236	Fort Wayne	173	Providence	157
Poole	130	Fort Worth	430	Pueblo	101
Portsmouth	173	Fremont	154	Raleigh	180
Preston	140	Fresno	285	Reno	110
Reading	114	Fullerton	109	Richmond	218
Rochdale	100	Garden Grove	135	Riverside	197
Salford	100	Garland	177	Roanoke	102
Sheffield	445	Gary	137	Rochester	236
Solihull	109	Glendale (Az.)	126	Rockford	136
Southampton	192	Glendale (Ca.)	154	St Louis	2,467
Southend-on-Sea	154	Grand Rapids	187	St Paul	264
Stockport	130	Greensboro	177	St Petersburg	239
Stockton-on-Tees	153	Hampton	126	Sacramento	1,385
Stoke-on-Trent	245	Hartford	1,108	Salt Lake C.	1,065
Sunderland	192	Hartford	138	San Antonio	1,323
Swansea	162	Hayward	102	San Bernardino	139
Swindon	100	Hialeah	162	San Diego	2,370
Teeside	363	Hollywood	121	San Francisco	6,042
Telford	115	Honolulu	372	San Jose	712
Thurrock	124	Houston	3,642	Santa Ana	237
Torbay	121	Huntington Beach	184	Savannah	147
Walsall	171	Huntsville	163	Scottsdale	111
Warley	138	Independence	113	Seattle	2,421
Warrington	144	Indianapolis	1,237	Shreveport	220
West Bromwich	140	Inglewood	103	South Bend	107
Wolverhampton	237	Irving	129	Spokane	173
York	101	Jackson (Miss.)	208	Springfield (Ill.)	100
		Jacksonville (Fl.)	610	Springfield (Ma.)	149
		Jersey City	219	Springfield (Mo.)	139
		Kansas City (Ks.)	162	Stamford	101
		Kansas City (Mo.)	1,575	Sterling Heights	112
		Knoxville	173	Stockton	183
		Lakewood	122	Sunnyvale	112
		Lansing	129	Syracuse	161
		Laredo	117	Tacoma	159
		Las Vegas	192	Tallahassee	119
		Lexington-Fayette	213	Tampa	1,995
		Lincoln	183	Tempe	136
		Little Rock	181	Toledo	341
		Livonia	101	Topeka	119
		Lorg Beach	396	Torrance	136
		Los Angeles	13,770	Tucson	359
		Louisville	286	Tulsa	374
		Lubbock	186	Virginia Beach	333

United States (1986)

Abilene	112	Waco	105
Akron	222	Warren	150
Albuquerque	367	*Washington	3,734
Alexandria (Va.)	108	Waterbury	102
Allentown	104	Wichita	289
Amarillo	166	Winston-Salem	148
Ananeim	241	Worcester	158
Anchorage	235	Yonkers	186
Ann Arbor	108	Youngstown	105
Arlington	250		
Atlanta	2,737		
Aurora	218		
Austin	467		
Bakersfield	150		
Baltimore	2,343		

United States Virgin Is. (1980)

*Charlotte Amalie	53

Uruguay (1985)

*Montevideo	1,248

USSR (1987-89)

Abakan	151	Engels = Pokrovsk	
Achinsk	121	Fergana	203
Aktyubinsk	248	Frunze = Bishkek	
Almalyk	119	Gomel	500
Alma-Ata	1,108	Gorki = Nizhni Novgorod	
Almetyevsk	128	Gorlovka	345
Andizhan	288	Grodno	263
Andropov = Rybinsk		Grozny	404
Angarsk	262	Guryev	150
Angren	131	Gyandzha (Kirovabad)	270
Anzhero-Sudzhensk	112	Irkutsk	626
Arkhangelsk	416	Ivanovo	479
Armavir	172	Ivano-Frankovsk	225
Arzamas	108	Izhevsk (Ustinov)	635
Ashkhabad	382	Kadiyevka (Stakhanov)	112
Astrakhan	509	Kalinin = Tver	
Baku	1,757	Kaliningrad	394
Balakovo	188	Kaliningrad (Moscow)	146
Balashikha	132	Kaluga	307
Baranovichi	154	Kamenets-Podolskiy	101
Barnaul	602	Kamensk-Uralskiy	204
Batumi	135	Kamyshin	119
Belaya Tserkov	194	Kansk	108
Belgorod	293	Karaganda	614
Belovo	118	Karaklis (Kirovakan)	169
Beltsy	157	Karshi	126
Bendery	130	Kaunas	417
Berdyansk	133	Kazan	1,094
Berezniki	200	Kemerovo	520
Bishkek	616	Kerch	173
Biysk	231	Khabarovsk	601
Blagoveshchensk	202	Kharkov	1,611
Bobruysk	232	Kherson	358
Borisov	140	Khimki	128
Bratsk	249	Khodzhent (Leninabad)	157
Brest	238	Khmelnitskiy	230
Bryansk	445	Kiev	2,587
Bukhara	220	Kineshma	106
Chardzhou	166	Kirov = Vyatka	
Cheboksary	414	Kirovabad = Gyandzha	
Chelyabinsk	1,179	Kirovakan = Karaklis	
Cherepovets	315	Kirovograd = Yelizavetgrad	
Cherkassy	287	Kishinev	663
Cherkessk	107	Kisilevsk	128
Chernigov	291	Kislovodsk	110
Chernovtsy	254	Klaipeda	201
Chimkent	389	Kokand	173
Chirchik	160	Kokchetav	127
Chita	349	Kolomna	159
Daugavpils	128	Kolpino	134
Dimitrovgrad	121	Kommunarsk	126
Dneprodzerzhinsk	279	Komsomolsk-na-Amur	314
Dnepropetrovsk	1,179	Konstantinovka	115
Donetsk	1,110	Kopeisk	134
Dushanbe	595	Kostroma	276
Dzerzhinsk	281	Kovrov	158
Dzhambul	315	Kramatorsk	198
Dzhezkazgan	105	Krasnodar	620
Ekibastuz	141	Krasnoyarsk	912
Elektrostal	150	Krasnyy Luch	112
		Kremenchug	230
		Krivoy Rog	713
		Kurgan	354
		Kursk	434
		Kustanay	212
		Kutaisi	220
		Kuybyshev = Samara	
		Kzyl-Orda	189
		Leninabad = Khodzhent	
		Leninakan	228
		Leningrad = St Petersburg	
		Leninsk-Kuznetskiy	169
		Liepaja	114
		Lipetsk	465
		Lisichansk	124
		Lugansk (Voroshilovgrad)	509
		Lutsk	185
		Lvov	790
		Lyubertsy	162

Magadan	148
Magnitogorsk	430
Makeyevka	455
Makhachkala	320
Margilan	127
Mariupol (Zhdanov)	529
Maykop	145
Melitopol	174
Mezhdurechensk	104
Miass	163
Michurinsk	103
Minsk	1,589
Mogilev	359
*Moscow	8,967
Murmansk	432
Murom	124
Mytishchi	152
Naberezhniye-Chelni (Brezhnev)	501
Nakhodka	152
Nalchik	236
Namangan	291
Navoi	106
Nevinnomyssk	116
Nikolayev	503
Nikopol	157
Nizhnekamsk	183
Nizhnevartovskoye	212
Nizhni Novgorod (Gorki)	1,438
Nizhniy Tagil	427
Noginsk	122
Norilsk	181
Novgorod	228
Novocheboksarsk	109
Novocherkassk	188
Novokuybyshevsk	112
Novokuznetsk	600
Novomoskovsk	147
Novorossiysk	179
Novoshakhtinsk	105
Novosibirsk	1,436
Novotroitsk	105
Nukus	152
Odessa	1,115
Odintsovo	120
Oktyabrskiy	106
Omsk	1,148
Ordzhonikidze = Vladikavkaz	
Orekhovo-Zuyevo	147
Orel	335
Orenburg	547
Orsha	123
Orsk	271
Osh	209
Panevezhys	122
Pavlodar	331
Pavlograd	126
Penza	543
Perm	1,091
Pervouralsk	139
Petropavlovsk	233
Petropavlovsk-Kamchatskiy	252
Petrozavodsk	264
Pinsk	116
Podolsk	209
Pokrovsk (Engels)	182
Poltava	309
Prokopyevsk	270
Pskov	202
Pyatigorsk	121
Riga	915
Rostov	1,020
Rovno	233
Rubtsovsk	168
Rudniy	118
Rustavi	147
Ryazan	515
Rybinsk (Andropov)	254

St Petersburg (Leningrad)	5,020
Salavat	153
Samara (Kuybyshev)	1,257
Samarkand	388
Saransk	323
Sarapul	111
Saratov	905
Semipalatinsk	330
Sergiyev Posad (Zagorsk)	113
Serov	103
Serpukhov	142
Sevastopol	350
Severodonetsk	127
Severodvinsk	239
Shakhty	225
Shchelkovo	107
Shevchenko	161
Siauliai	140
Simbirsk (Ulyanovsk)	625
Simferopol	338
Slavyansk	142
Smolensk	338
Sochi	317
Solikamsk	108
Stakhanov = Kadiyevka	
Staryy Oskol	167
Stavropol	306
Sterlitamak	251
Sukhumi	130
Sumgait	234
Sumy	268
Surgut	227
Sverdlovsk = Yekaterinburg	
Syktyvkar	224
Syzran	174
Taganrog	295
Taldy-Kurgan	113
Tallinn	478
Tambov	305
Tartu	113
Tashkent	2,073
Tashsuz	110
Tbilisi	1,194
Temirtau	228
Ternopol	197
Tiraspol	173
Togliatti	627
Tomsk	502
Tselinograd	276
Tula	540
Tver (Kalinin)	447
Tyumen	456
Ufa	1,083
Ukhta	105
Ulan-Ude	351
Ulyanovsk = Simbirsk	
Uralsk	201
Urgench	123
Usolye-Sibirskoye	108
Ussuriysk	158
Ust Ilimsk	105
Ust-Kamenogorsk	321
Uzhgorod	111
Velikiye Luky	113
Vilnius	582
Vinnitsa	383
Vitebsk	347
Vladikavkaz (Ordzhonikidze)	313
Vladimir	343
Vladivostok	648
Volgodonsk	179
Volgograd	999
Vologda	278
Volzhskiy	257
Vorkuta	112
Voronezh	887
Voroshilovgrad = Lugansk	

Votkinsk	101
Vyatka (Kirov)	421
Yakutsk	188
Yaroslavl	634
Yekaterinburg (Sverdlovsk)	1,367
Yelets	119
Yelizavetgrad (Kirovograd)	269
Yenakievo	117
Yerevan	1,199
Yevpatoriya	106
Yoshkar-Ola	243
Yuzhno-Sakhalinsk	166
Zagorsk = Sergiyev Posad	
Zaporozhye	884
Zelenogradsk	148
Zhdanov = Mariupol	
Zhitomir	287
Zhukovskiy	100
Zlatoust	206

Vanuatu (1987)

*Port Vila	18

Vatican City State (1988)

*Vatican	1

Venezuela (1987)

Acarigua-Araure	122
Barcelona-Puerto La Cruz	157
Barinas	166
Barquisimento	718
Cabimas	214
*Caracas	3,247
Ciudad Bolivar	259
Ciudad Guayana	466
Cumanà	247
Guarenas	101
Lagunillas	125
Los Teques	159
Maracaibo	1,295
Maracay	857
Maturín	253
Mérida	250
Punto Fijo	120
San Cristóbal	338
Valencia	1,135
Valera	182

Vietnam (1979)

Ban Me Thuot	176
Cam-Ranh	118
Can-Tho	247
Dalat	105
Bien Hoa	246
Da-Nang	500
Haiphong	1,279
*Hanoi	2,571
Ho Chi Minh City	3,420
Hon Gai	120
Hué	211
Mytho	135
Quang Nghia	221
Long Xuyen	185
Nha-Trang	212
Qui-Nhon	165
Thanh-Hoa	115
Rach-gia	107
Vungtau	108
Vinh	207
Viettri	113
Minh Hai	103
Nam Dinh	193
Nguyen	138

Wallis and Futuna (1983)

*Mata Utu	1

Western Sahara (1982)

*Al Aaiún	97

Western Samoa (1988)

*Apia	33

Yemen (1986)

Aden	264
Hodeida	156
Mukalla	154
*Sana'	427
Ta'izz	178

Yugoslavia (1981)

Banja Luka	124
*Belgrade	1,470
Ljubljana	305
Maribor	106
Nis	161
Novi Sad	170
Osijek	105
Pristina	108
Rijeka	159
Sarajevo	449
Skopje	505
Split	169
Subotica	101
Zagreb	1,175

Zaïre (1984)

Bukavu	171
Ilebo	142
Kalémié	172
Kamina	160
Kananga	291
Kikwit	147
*Kinshasa	2,654
Kisangani	283
Kolwezi	201
Likasi	194
Lubumbashi	543
Luluabourg	506
Matadi	145
Mbandaka	125
Mbuji Mayi	423

Zambia (1987)

Chingola	187
Kabwe	191
Kitwe	449
Luanshya	161
*Lusaka	900
Mufulira	192
Ndola	418

Zimbabwe (1983)

Bulawayo	500
Chitungwiza	202
*Harare	681

Census details

France

Date of census: 5 March, 1990
Total population: 56,556,000

Within each département in alphabetical order are listed all communes with a population of over 20,000 inhabitants.

Ain	471,019
Bourg-en-Brasse	42,955
Oyonnax	23,992

Aisne	537,259
Laon	28,670
Saint-Quentin	62,085
Soissons	32,144

Alpes-de-Haute-Provence	130,883

Allier	357,710
Gap	35,647

Alpes-Maritimes	971,829
Antibes	70,688
Cagnes-sur-Mer	41,303
Cannes	69,363
Le Cannet	42,005
Grasse	42,077
Menton	29,474
Nice	345,674
Vallauris	24,406
Saint-Laurent-du-Var	24,475

Ardèche	277,581

Ardennes	296,357
Sedan	22,407

Ariège	136,455
Charleville-Mézières	59,439

Aube	289,207
Troyes	60,755

Aude	298,712
Carcassonne	44,991
Narbonne	47,086

Aveyron	270,141
Millau	22,458
Rodez	26,794

Bas-Rhin	953,053
Haguenau	30,38467
Illkirch-Graffenstaden	23,73867
Schiltigheim	29,330
Strasbourg	255,937

Bouches-du-Rhône	1,759,371
Aix-en-Provence	126,854
Arles	52,593
Aubagne	41,187
La Ciotat	30,748
Istres	36,516
Marignane	32,542
Marseille	807,726
Martigues	42,922
Miramas	21,882
Salon-de-Provence	35,041
Vitrolles	35,617

Calvados	618,478
Caen	115,624
Hérouville-Saint-Clair	25,061
Lisieux	24,506

Cantal	158,723
Aurillac	32,654

Charente	341,993
Angoulême	46,194

Charente-Maritime	527,146
Rochefort	26,949
La Rochelle	73,744
Saintes	27,546

Cher	321,559
Bourges	78,773
Vierzon	32,900

Corrèze	237,908
Brive-la-Gaillarde	52,677

Corse-du-Sud	118,174
Ajaccio	59,318

Côte-d'Or	493,866
Beaune	22,171
Dijon	151,636

Côtes-d'Armor	538,395
Saint-Brieuc	47,370

Creuse	131,349

Deux-Sèvres	345,965
Niort	58,660

Dordogne	386,365
Bergerac	27,886
Périgueux	32,848

Doubs	484,770
Besançon	119,194
Montbéliard	30,639

Drôme	414,072
Montélimar	31,386
Romans-sur-Isère	33,546
Valence	65,026

Essonne	1,084,824
Athis-Mons	29,695
Brétigny-sur-Orge	20,069
Brunoy	24,594
Corbeil-Essonnes	40,768
Draveil	28,034
Etampes	21,547
Evry	45,854
Grigny	24,969
Massy	38,972
Montgeron	21,818
Palaiseau	29,398
Ris-Orangis	24,788
Saint-Michel-sur-Orge	20,845
Sainte-Geneviève-des-Bois	31,372
Savigny-sur-Orge	33,651
Les Ulis	27,207

Vigneux-sur-Seine	25,265
Viry-Châtillon	30,738
Yerres	27,268

Eure	513,818
Evreux	51,452
Vernon	24,943

Eure-et-Loir	396,073
Chartres	41,850
Dreux	35,866

Finistère	838,687
Brest	153,099
Quimper	62,541

Gard	585,049
Alès	42,296
Nîmes	133,607

Gers	174,587
Auch	24,728

Gironde	1,213,499
Bègles	22,735
Bordeaux	213,274
Le Bouscat	21,574
Cenon	21,726
Gradignan	22,115
Liboume	21,931
Lormont	21,771
Mérignac	58,684
Pessac	51,424
Saint-Médard-en-Jalles	22,121
Talence	36,172
La Teste	21,244
Villenave-d'Ornon	25,957

Guadeloupe	386,987
Les Abymes	62,809
Le Gosier	20,708
Pointe-À-Pitre	26,083
Saint-Martin	28,524

Guyane	114,678
Cayenne	41,659

Haut-Rhin	671,319
Colmar	64,889
Mulhouse	109,905

Haute-Corse	131,563
Bastia	38,728

Haute-Garonne	925,962
Colomiers	27,253
Toulouse	365,933

Haute-Loire	206,568
Le-Puy-en-Velay	23,434

Haute-Marne	204,067
Chaumont	28,900
Saint-Dizier	35,558

Haute-Saône	229,650

Haute-Savoie	568,286
Annecy	51,143
Annemasse	27,927
Thonon-les-Bains	30,667

Haute-Vienne	353,593
Limoges	136,407

Hautes-Alpes	113,300
Montluçon	46,660
Moulins	23,353
Vichy	28,048

Hautes-Pyrénées	224,759
Tarbes	50,228

Hauts-de-Seine	1,391,658
Antony	57,916
Asnières-sur-Seine	72,250
Bagneux	36,453
Bois-Colombes	24,500
Boulogne-Billancourt	101,971
Châtenay-Malabry	29,359
Châtillon	26,508
Clamart	47,755
Clichy	48,204
Colombes	79,058
Courbevoie	65,649
Fontenay-aux-Roses	23,534
La Garenne-Colombes	21,831
Gennevilliers	45,052
Issy-les-Moulineaux	46,734
Levallois-Perret	47,788
Malakoff	31,135
Meudon	46,173
Montrouge	38,333
Nanterre	86,627
Neuilly-sur-Seine	62,033
Le Plessis-Robinson	21,349
Puteaux	42,917
Rueil-Malmaison	67,323
Saint-Cloud	28,673
Sèvres	22,057
Suresnes	36,950
Vanves	26,160
Villeneuve-la-Garenne	23,872

Hérault	794,603
Béziers	72,362
Montpellier	210,866
Sète	41,916

Ille-et-Vilaine	798,718
Fougères	23,138
Rennes	203,533
Saint-Malo	49,274

Indre	237,510
Châteauroux	52,949

Indre-et-Loire	529,345
Joué-lès-Tours	37,114
Tours	133,403

Isère	1,016,228
Bourgoin-Jallieu	22,749
Echirolles	34,646
Fontaine	23,089
Grenoble	153,973
Saint-Martin-d'Hères	34,501
Vienne	30,386

Jura	**248,759**
Dole	27,860
Lons-le-Saunier	20,140
Landes	**311,461**
Dax	20,119
Mont-de-Marsan	31,864
Loir-et-Cher	**305,937**
Blois	51,549
Loire	**746,288**
Firminy	23,367
Saint-Chamond	39,262
Saint-Étienne	201,569
Roanne	42,848
Loire-Atlantique	**1,052,183**
Nantes	252,029
Orvault	23,327
Rezé	33,703
Saint-Herblain	43,439
Saint-Nazaire	66,087
Saint-Sébastien-sur-Loire	22,763
Loiret	**580,612**
Fleury-les-Aubrais	20,730
Orléans	107,965
Lot	**155,816**
Cahors	20,787
Lot-et-Garonne	**305,989**
Agen	32,223
Villeneuve-sur-Lot	23,760
Lozère	**72,825**
Maine-et-Loire	**705,882**
Angers	146,163
Cholet	56,540
Saumur	31,894
Manche	**479,636**
Cherbourg	28,773
Saint-Lô	22,819
Marne	**558,217**
Châlons-sur-Marne	51,533
Epernay	27,738
Reims	185,164
Martinique	**359,572**
Fort-de-France	101,540
Le Lamentin	30,596
Mayenne	**278,037**
Laval	53,479
Meurthe-et-Moselle	**711,822**
Lunéville	22,393
Nancy	102,410
Vandoeuvre-lès-Nancy	34,420
Meuse	**196,344**
Verdun	23,427

Morbihan	**619,838**
Lanester	23,163
Lorient	61,630
Vannes	48,454
Moselle	**1,011,302**
Forbach	27,357
Metz	123,920
Montigny-lès-Metz	23,482
Sarreguemines	23,684
Thionville	40,835
Nièvre	**233,278**
Nevers	43,889
Nord	**2,531,855**
Armentières	26,240
Cambrai	34,210
Coudekerque-Branche	23,820
Croix	20,308
Douai	44,195
Dunkerque	71,071
Grande-Synthe	24,489
Hazebrouck	21,115
Hem	20,254
Lambersart	28,462
Lille	178,301
Lomme	26,807
Loos	21,358
La Madeleine	21,788
Marcq-en-Baroeul	36,898
Maubeuge	35,225
Mons-en-Baroeul	23,626
Roubaix	98,179
Saint-Pol-sur-Mer	24,013
Tourcoing	94,425
Valenciennes	39,276
Villeneuve-d'Ascq	69,695
Wattrelos	43,784
Oise	**725,603**
Beauvais	56,278
Compiègne	44,703
Creil	32,501
Nogent-sur-Oise	20,053
Orne	**293,204**
Alençon	31,139
Paris	**2,152,423**
Paris	2,175,200
Pas-de-Calais	**1,433,203**
Arras	42,715
Béthune	25,261
Boulogne-sur-Mer	44,244
Bruay-la-Buissière	25,451
Calais	75,836
Hénin-Beaumont	26,494
Lens	35,278
Liévin	34,012
Puy-de-Dôme	**598,213**
Clermont-Ferrand	140,167
Pyrénées-Atlantiques	**578,516**
Anglet	33,956
Bayonne	41,846
Biarritz	28,887
Pau	83,928

Pyrénées-Orientales	**363,796**
Perpignan	108,049
Réunion	**597,823**
Le Port	34,806
Saint-André	35,375
Saint-Benoît	26,457
Saint-Denis	122,875
Saint-Joseph	25,852
Saint-Leu	20,987
Saint-Louis	37,798
Saint-Paul	71,952
Saint-Pierre	59,645
Sainte-Marie	20,334
Le Tampon	48,436
Rhône	**1,508,966**
Bron	40,514
Caluire-et-Cuire	41,513
Décines-Charpieu	24,608
Lyon	422,444
Meyzieu	28,212
Oullins	26,400
Rillieux-la-Pape	31,149
Saint-Priest	42,131
Sainte-Foy-lès-Lyon	21,550
Vaulx-en-Velin	44,535
Vénissieux	60,744
Villefranche-sur-Saône	29,889
Villeurbanne	119,848
Saône-et-Loire	**559,413**
Chalon-sur-Saône	56,259
Le Creusot	29,230
Mâcon	38,508
Montceau-les-Mines	23,308
Sarthe	**513,654**
Le Mans	148,465
Savoie	**348,261**
Aix-les-Bains	24,826
Chambéry	55,603
Seine-Maritime	**1,223,429**
Dieppe	36,600
Fécamp	21,143
Le Grand-Quevilly	27,909
Le Havre	197,219
Mont-Saint-Aignan	20,329
Le Petit-Quevilly	22,718
Rouen	105,470
Saint-Etienne-du-Rouvray	31,012
Sotteville-lès-Rouen	29,957
Seine-et-Marne	**1,078,166**
Champs-sur-Marne	21,762
Chelles	45,495
Combs-la-Ville	20,001
Dammarie-les-Lys	21,228
Meaux	49,409
Le Mée-sur-Seine	20,971
Melun	36,489
Pontault-Combault	26,834
Seine-Saint-Denis	**1,381,197**
Aubervilliers	67,836
Aulnay-sous-Bois	82,537
Bagnolet	32,739
Le Blanc-Mesnil	47,093
Bobigny	44,881
Bondy	46,880
Clichy-sous-Bois	28,280
La Courneuve	34,351

Drancy	60,928
Epinal-sur-Seine	48,851
Gagny	36,151
Les Lilas	20,532
Livry-Gargan	35,471
Montfermeil	25,695
Montreuil	95,038
Neuilly-sur-Marne	31,603
Noisy-le-Grand	54,112
Noisy-le-Sec	36,402
Pantin	47,444
Pierrefitte-sur-Seine	23,882
Romainville	23,615
Rosny-sous-Bois	37,779
Saint-Denis	90,806
Saint-Ouen	42,611
Sevran	48,564
Stains	35,068
Tremblay-en-France	31,432
Villemomble	27,000
Villepointe	30,412
Somme	**547,825**
Abbeville	24,588
Amiens	136,234
Tarn	**342,723**
Albi	48,707
Castres	46,292
Tarn-et-Garonne	**200,220**
Montauban	53,278
Territoire de Belfort	**134,097**
Belfort	51,913
Val-d'Oise	**1,049,598**
Argenteuil	94,162
Bezons	25,792
Cergy	48,524
Eaubonne	22,208
Ermont	28,073
Franconville	33,874
Garges-lès-Gonesse	42,236
Gonesse	23,346
Goussainville	24,971
Herblay	22,435
Montmorency	21,003
Pontoise	28,463
Sannois	25,658
Sarcelles	57,121
Taverny	25,191
Villiers-le-Bel	26,223
Val-de-Marne	**1,215,538**
Alfortville	36,240
Arcueil	20,420
Cachan	25,370
Champigny-sur-Marne	79,778
Charenton-le-Pont	21,991
Choisy-le-Roi	34,230
Créteil	82,390
Fontenay-sous-Bois	52,105
Fresnes	27,032
L'Hay-les-Roses	29,841
Ivry-sur-Seine	54,106
Maisons-Alfort	54,065
Nogent-sur-Marne	25,386
Orly	21,824
Le Perreux-sur-Marne	28,540
Saint-Maur-des-Fossés	77,492
Sucy-en-Brie	25,924
Thiais	27,933
Villejuif	49,671
Villeneuve-le-Roi	20,376
Villeneuve-Saint-Georges	27,476

Villiers-sur-Marne	22,815	**Vaucluse**	**467,075**	**Vosges**	**386,258**	Elancourt	22,635
Vincennes	42,651	Avignon	89,440	Epinal	39,480	Houilles	30,027
Vitry-sur-Seine	82,820	Carpentras	25,477	Saint-Dié	23,670	Maisons-Laffitte	22,553
		Cavaillon	23,470			Mantes-la-Jolie	45,254
Var	**815,449**	Orange	28,136	**Yonne**	**323,096**	Montigny-le-Bretonneux	31,744
Draguignan	32,851			Auxerre	40,597	Les Mureaux	33,365
Fréjus	42,613	**Vendée**	**509,356**	Sens	27,755	Plaisir	25,949
La Garde	22,662	La Roche-sur-Yon	48,568			Poissy	36,864
Hyères	50,122			**Yvelines**	**1,307,150**	Rambouillet	25,293
Saint-Raphaël	26,799	**Vienne**	**379,977**	La Celle-Saint-Cloud	22,884	Saint-Germain-en-Laye	41,710
La Seeyne-sur-Mer	60,567	Châtellerault	35,691	Chatou	28,077	Sartrouville	50,440
Six-Fours-les-Plages	29,178	Poitiers	82,507	Le Chesnay	29,611	Trappes	30,938
Toulon	170,167			Conflans-Sainte-Honorine	31,857	Vélizy-Villacoublay	22,034
La Valette-du-Var	20,863					Versailles	91,029

Japan

Date of census: 1 October 1990
Total population: 123,612,000

The first list gives information about the prefectures. The second list gives the cities of each prefecture.

Prefecture	Population 1980	Population 1990	Density persons per sq km	Change 1980-90 %
Aichi	6,222	6,690	1,300	7.5
Akita	1,257	1,227	106	-2.4
Aomori	1,524	1,483	154	-2.7
Chiba	4,735	5,555	1,078	17.3
Ehime	1,507	1,515	267	0.5
Fukui	794	824	197	3.8
Fukuoka	4,553	4,811	969	5.7
Fukushima	2,035	2,104	153	3.4
Gifu	1,960	2,067	195	5.5
Gumma	1,849	1,966	309	6.3
Hiroshima	2,739	2,850	336	4.1
Hokkaido	5,576	5,644	72	1.2
Hyogo	5,145	5,405	645	5.1

Prefecture	Population 1980	Population 1990	Density persons per sq km	Change 1980-90 %
Ibaraki	2,558	2,845	467	11.2
Ishikawa	1,119	1,165	278	4.1
Iwate	1,422	1,417	93	-0.4
Kagawa	1,000	1,023	546	2.3
Kagoshima	1,785	1,798	196	0.7
Kanagawa	6,924	7,980	3,310	15.3
Kochi	831	825	116	-0.7
Kumamoto	1,790	1,840	249	2.8
Kyoto	2,527	2,603	564	3.0
Mie	1,687	1,793	310	6.3
Miyagi	2,082	2,249	309	8.0
Miyazaki	1,152	1,169	151	1.5
Nagano	2,084	2,157	159	3.5
Nagasaki	1,591	1,563	382	-1.8
Nara	1,209	1,375	373	13.7
Niigata	2,451	2,475	197	1.0
Oita	1,229	1,237	195	0.7
Okayama	1,871	1,926	271	2.9
Okinawa	1,107	1,222	540	10.4

Prefecture	Population 1980	Population 1990	Density persons per sq km	Change 1980-90 %
Osaka	8,473	8,735	4,640	3.1
Saga	866	878	360	1.4
Saitama	5,420	6,405	1,687	18.2
Shiga	1,080	1,222	304	13.1
Shimane	785	781	118	-0.5
Shizuoka	3,447	3,671	472	6.5
Tochigi	1,792	1,935	302	8.0
Tokushima	825	832	201	0.8
Tokyo	11,618	11,855	5,430	2.0
Tottori	604	616	176	2.0
Toyama	1,103	1,120	264	1.5
Wakayama	1,087	1,074	228	-1.2
Yamagata	1,252	1,258	135	0.5
Yamaguchi	1,587	1,573	257	-0.9
Yamanashi	804	853	191	6.1
Japan	**117,060**	**123,612**	**332**	**5.6**

Japan – census

Prefecture and City	Population 1990	Change 1985-90
Aichi		
Anjo	142,217	6.9
Bisai	55,881	-0.6
Chiryu	54,061	7
Chita	75,434	7.7
Gamagori	84,819	-0.9
Handa	99,550	7.2
Hekinan	65,901	3.3
Ichinomiya	262,434	2
Inazawa	96,277	1.9
Inuyama	69,803	1.6
Iwakura	43,807	3.1
Kariya	120,121	6.9
Kasugai	266,599	3.7
Komaki	124,441	9.8
Konan	93,836	1.9
Nagoya	2,154,664	1.8
Nishio	95,918	3.6
Obu	69,721	4.5
Okazaki	306,821	7.7
Owariasahi	65,676	14.4
Seto	126,343	1.4
Shinshiro	35,633	0.7
Takahama	33,477	7.1
Tokai	97,359	2.2
Tokoname	51,784	-2.4
Toyoake	62,156	7.2
Toyohashi	337,988	4.9
Toyokawa	111,731	4
Toyota	332,336	7.9
Tsushima	59,345	1

Prefecture and City	Population 1990	Change 1985-90
Akita		
Akita	302,359	2
Honjo	44,442	0.3
Kazuno	42,407	-4.7
Noshiro	55,915	-5.5
Odate	68,196	-5
Oga	34,291	-7.2
Omagari	40,430	-2.7
Yokote	42,294	-2.2
Yuzawa	36,539	-1.5
Aomori		
Aomori	287,813	-2.1
Goshogawara	47,973	-3.2
Hachinohe	241,065	-0.2
Hirosaki	174,710	-0.8
Kuroishi	39,214	-3.2
Misawa	41,344	-0.2
Mutsu	48,470	-1.7
Towada	60,916	-0.6
Chiba		
Abiko	120,629	8
Asahi	38,907	3.7
Chiba	829,467	5.1
Choshi	85,138	-3.1
Funabashi	533,273	5.2
Futtsu	54,877	-3.3
Ichihara	257,717	8.5
Ichikawa	436,597	9.7
Kamagawa	31,226	-1.8
Kamagaya	95,052	10.9

Prefecture and City	Population 1990	Change 1985-90
Kashiwa	305,060	11.7
Katsuura	25,334	0.7
Kimitsu	89,243	5.9
Kisarazu	123,434	2.7
Matsudo	456,211	6.7
Mobara	83,437	8.5
Nagareyama	140,059	12.3
Narashino	151,472	11.1
Narita	86,708	12.3
Noda	114,476	8.1
Sakura	144,688	19.4
Sawara	49,547	-0.5
Tateyama	54,574	-2.6
Togane	45,179	17.3
Urayasu	115,675	23.4
Yachiyo	148,615	4.5
Yatsukaido	72,157	7.7
Yokaichiba	32,305	0.3
Ehime		
Hojo	29,420	-3.6
Imabari	123,114	-1.6
Iyo	29,803	-0.1
Iyomishima	38,352	-0.7
Kawanoe	38,991	1.2
Matsuyama	443,317	3.9
Niihama	129,151	-2.3
Ozu	39,850	-0.2
Saijo	56,823	0.5
Toyo	33,752	-1.7
Uwajima	68,035	-4.7
Yawatahama	38,550	-7.3

Prefecture and City	Population 1990	Change 1985-90
Fukui		
Fukui	252,750	1
Katsuyama	29,805	-2
Obama	33,774	-0.7
Ono	40,990	-2.2
Sabae	62,284	1.4
Takefu	70,188	1.5
Tsuruga	68,039	3.6
Fukuoka		
Amagi	43,036	-1.2
Buzen	31,089	-2.8
Chikugo	43,836	1.1
Chikushino	70,303	11.2
Dazaifu	62,408	8.1
Fukuoka	1,237,107	6.6
Iizuka	83,133	1.5
Kasuga	88,703	17.4
Kitakyushu	1,026,467	-2.8
Kurume	228,350	2.5
Munakata	68,267	12
Nakama	49,216	-2.1
Nogata	62,532	-3
Ogari	47,116	7.5
Okawa	45,705	-4.5
Omuta	150,461	-5.6
Onojo	75,217	8.3
Tagawa	57,701	-3.4
Yamada	13,266	-6.7
Yame	39,817	-1.2
Yanagawa	43,791	-2.6
Yukuhashi	65,713	0.3

Japan census *continued*

Prefecture and City	Population 1990	Change 1985-90
Fukushima		
Aizuwakama	119,084	0.8
Fukushima	277,526	2.5
Haramachi	49,057	1.3
Iwaki	355,817	1.5
Kitakata	37,288	-0.7
Koriyama	314,651	4.3
Nihommatsu	34,927	1.7
Shirakawa	45,646	2.2
Soma	39,135	-0.5
Sukagawa	60,697	3.3
Gifu		
Ena	35,026	-0.9
Gifu	410,318	-0.3
Hashima	61,460	2.8
Kakamigahara	129,682	4.2
Kani	80,012	14.9
Mino	26,023	-3.4
Minokamo	43,013	3.1
Mizunami	41,006	2.3
Nakatsugawa	53,722	0.8
Ogaki	148,281	1.6
Seki	68,386	6.6
Takayama	65,245	0.3
Tojimi	94,036	10.9
Toki	64,946	-0.6
Gumma		
Annaka	45,526	2.1
Fujioka	60,983	6.8
Isesaki	115,939	3.1
Kiryu	126,443	-3.7
Maebashi	286,261	3.2
Numata	46,856	-0.7
Ota	139,801	4.6
Shibukawa	49,064	2.6
Takasaki	236,463	2
Tatebayashi	76,223	1.4
Tomioka	49,025	1
Hiroshima		
Fuchu	45,738	-4.3
Fukuyama	365,615	1.5
Hatsukaichi	63,441	22
Higashihiroshima	94,206	11.2
Hiroshima	1,085,677	4
Innoshima	32,640	-12.3
Kure	216,717	-4.3
Mihara	85,518	-0.5
Miyoshi	39,465	1.3
Onomichi	97,104	-3.5
Otake	33,236	-4.4
Shobara	22,677	-0.6
Takehara	34,771	-4.2
Hokkaido		
Abashiri	44,416	0.3
Akabiro	19,409	-14.3
Asahikawa	359,069	-1.3
Ashibetsu	25,079	-16.5
Bibai	35,176	-6
Chitose	78,947	7.3
Date	34,507	-0.9
Ebetsu	97,201	7.6
Eniwa	55,613	15.1
Fukagawa	30,674	-9.3
Furano	26,665	-4.3
Hakodate	307,251	-3.7
Iwamizawa	80,423	-1.5
Kitami	107,247	0

Prefecture and City	Population 1990	Change 1985-90
Kushiro	205,640	-4.1
Mikasa	17,049	-20.7
Mombetsu	31,077	-3.4
Muroran	117,852	-13.5
Nayoro	30,776	-9.7
Nemuro	36,914	-9.2
Noboribetsu	55,575	-4.8
Obihiro	167,389	2.7
Otaru	163,215	-5.4
Rumoi	32,428	-8.8
Sapporo	1,671,765	8.3
Shibetsu	25,754	-7.1
Sunagawa	23,152	-6.8
Takikawa	49,591	-4.6
Tomakomai	160,116	1.3
Utashinai	8,279	-13.9
Wakkanai	48,232	-7
Yubari	20,969	-33.8
Hyogo		
Ako	51,131	-2.4
Aioi	36,870	-7.5
Akashi	270,728	2.8
Amagasaki	498,998	-2
Ashiya	87,528	0.5
Himeji	454,360	0.3
Itami	186,132	1.9
Kakogawa	239,803	5.5
Kasai	51,789	-0.6
Kawanishi	141,254	3.6
Kobe	1,477,423	4.7
Miki	76,509	2.7
Nishinomiya	426,919	1.3
Nishiwaki	38,230	-1.4
Ono	46,007	0.7
Sanda	64,560	58.6
Sumoto	43,815	-1.7
Takarazuka	201,863	3.9
Takasago	93,267	2
Tatsuno	40,843	-0.8
Toyooka	47,247	-1
Ibaraki		
Hitachi	202,145	-1.9
Hitachiota	37,623	2.7
Ishioka	50,617	3.2
Iwai	43,103	2.2
Kasama	30,813	-2.3
Katsuta	109,826	6.9
Kitaibaraki	51,092	0.1
Koga	58,227	1.2
Mito	234,970	2.6
Mitsukaido	42,340	1.5
Nakaminato	32,577	-1.3
Ryugasaki	57,237	17.2
Shimodate	66,030	3.2
Shimotsuma	33,731	3.3
Takahagi	35,320	4
Toride	81,667	3.9
Tsuchiura	127,470	6.1
Tsukuba	143,408	12.5
Ushiku	60,698	16.9
Yuki	53,290	1.9
Ishikawa		
Hakui	27,515	-4.4
Kaga	69,199	0.8
Kanazawa	442,872	2.9
Komatsu	106,072	0
Matto	58,140	10.6
Nanao	50,101	-1
Suzu	23,471	-9.2
Wajima	30,166	-5.3

Prefecture and City	Population 1990	Change 1985-90
Iwate		
Esashi	34,434	-1.7
Hanamaki	70,514	0.9
Ichinoseki	61,971	1.7
Kamaishi	52,483	-12.5
Kitakami	58,782	3.6
Kuji	38,746	-1
Miyako	58,505	-5.1
Mizusawa	58,189	1.6
Morioka	235,440	0
Ninohe	28,858	-5.4
Ofunato	37,853	-3.7
Rikuzentakata	27,242	-4.1
Tono	28,954	-4.4
Kagawa		
Kanonji	45,500	-0.2
Marugame	75,607	1.8
Sakaide	63,878	-3.3
Takamatsu	329,695	0.8
Zensuji	38,425	-0.5
Kagoshima		
Akune	27,868	-4.5
Ibusuki	29,009	-3.5
Izumi	39,730	-0.9
Kagoshima	536,685	1.2
Kanoya	77,652	2.1
Kaseda	25,089	-2.6
Kokubu	46,557	13.7
Kushikino	29,386	-2.9
Makurozaki	28,795	-4.3
Naze	46,309	-6.9
Nishinoomote	20,951	-7.7
Okuchi	25,696	-3.6
Sendai	71,736	0.4
Tarumizu	22,264	-5.3
Kanagawa		
Atsugi	197,292	12.4
Ayase	77,926	9.5
Chigasaki	201,672	9
Ebina	105,816	13.6
Fujisawa	350,335	6.7
Hadano	155,619	9.7
Hiratsuka	245,944	6.9
Isehara	89,568	15.2
Kamakura	174,299	-0.7
Kawasaki	1,173,606	7.8
Minamiashigara	42,600	2.1
Miura	52,441	3.9
Odawara	193,415	4
Sagamihara	531,562	10.1
Yamato	194,870	9.7
Yokohama	3,220,350	7.6
Yokosuka	433,361	1.5
Zama	112,100	12.1
Zushi	56,705	-1.6
Kochi		
Aki	23,740	-5.1
Kochi	317,090	1.6
Muroto	23,310	-7.9
Nakamura	35,814	-0.8
Nankoku	46,827	-1.5
Sukumo	25,826	-1.6
Susaki	30,296	-3.4
Tosa	31,564	-1.8
Tosashimizu	21,182	-8

Prefecture and City	Population 1990	Change 1985-90
Kumamoto		
Arao	59,508	-4.9
Hitoyoshi	40,176	-5
Hondo	41,226	-3.3
Kikuchi	28,167	-1.2
Kumamoto	579,305	4.2
Minamata	34,595	-5.3
Tamana	45,285	-1.8
Ushibuka	21,442	-7
Uto	33,388	-0.6
Yamaga	33,439	-0.6
Yatsushiro	108,135	-0.6
Kyoto		
Ayabe	40,594	-3.1
Fukuchiyama	66,506	0.8
Joyo	84,770	3.6
Kameoka	85,283	11.9
Kyoto	1,461,140	-1.2
Maizuru	96,329	-2.5
Miyazu	26,450	-5.2
Muko	52,932	1.4
Nagaokakyo	77,193	2.6
Uji	177,018	7
Yawata	75,761	4.7
Mie		
Hisai	39,680	1.4
Ise	104,162	-1.2
Kameyama	37,631	6
Kumano	23,719	-5.6
Kuwana	97,911	3.4
Matsusaka	118,727	1.6
Nabari	68,933	22.1
Owase	27,114	-8.8
Suzuka	174,103	5.6
Toba	27,317	-3.7
Tsu	157,178	4.3
Ueno	60,239	-0.9
Yokkaichi	274,184	4.3
Miyagi		
Furukawa	64,227	5.8
Ishinomaki	121,980	-0.6
Iwanuma	38,093	4.3
Kakuda	35,432	0.9
Kesennuma	65,578	-3.8
Natori	53,735	5.6
Sendai	918,378	7.1
Shiogama	62,025	0.3
Shiroishi	42,028	-0.6
Tagajo	58,456	7.4
Miyazaki		
Ebino	26,825	-4.3
Hyugo	58,448	-1.2
Kobayashi	41,048	0.2
Kushima	26,735	-5.6
Miyakonojo	130,155	-1.5
Miyazaki	287,367	3
Nichinan	49,178	-5.4
Nobeoka	130,615	-4.2
Saito	37,216	-3
Nagano		
Chino	50,064	5.9
Iida	91,859	-0.6
Iiyama	28,115	-3.2
Ina	60,063	1.8
Komagane	32,771	1.2
Komoro	44,885	2.7
Koshoku	36,921	0.2
Matsumoto	200,723	1.7

Japan census   *continued*

Prefecture and City	Population 1990	Change 1985-90
Nagano	347,036	3
Nakano	40,996	1.1
Okaya	59,854	-3.1
Omachi	31,597	-2.6
Saku	62,005	3.4
Shiojiri	57,331	2.4
Suwa	52,465	0.3
Suzaka	53,662	0.1
Ueda	119,435	2.8

Nagasaki

Fukue	29,709	-4
Hirado	26,864	-5.5
Isahaya	90,678	2.6
Matsuura	24,183	-2.3
Nagasaki	444,616	-1.1
Omura	73,437	5.7
Sasebo	244,693	-2.4
Shimabara	44,828	-2.7

Nara

Gojo	34,546	1.9
Gose	36,643	-0.1
Ikoma	99,598	15.4
Kashihara	115,556	2.4
Nara	349,356	6.6
Sakurai	60,261	2.3
Tenri	68,818	-0.4
Yamatokoriyama	92,948	3.7
Yamatotakada	68,236	4.6

Niigata

Arai	28,325	-0.6
Gosen	39,378	-2.2
Itoigawa	34,047	-4.9
Joetsu	130,114	-0.4
Kamo	34,865	-3
Kashiwazaki	88,309	2.4
Mitsuke	43,117	1.3
Murakami	32,173	-3.5
Nagaoka	185,938	1.2
Niigata	486,087	2.2
Niitsu	64,005	0.2
Ojiya	43,438	-1.7
Ryotsu	19,432	-4.8
Sanjo	85,824	-0.6
Shibata	78,168	1.2
Shirone	35,800	4.6
Tochio	27,809	-6.3
Tokamachi	46,279	-3.6
Toysaka	45,962	3.2
Tsubame	43,891	-1.7

Oita

Beppu	130,323	-3.3
Bungotakada	20,084	-2.1
Hita	64,694	-1.6
Kitsuki	21,936	-0.9
Nakatsu	66,383	0.2
Oita	408,502	4.7
Saiki	52,325	-4.4
Taketa	20,164	-8.2
Tsukumi	26,796	-7.1
Usa	50,830	-2.7
Usuki	37,870	-4.7

Okayama

Bizen	31,145	-3.4
Ibara	36,076	-3.1
Kasaoka	59,618	-1.6
Kurashiki	414,692	0.3

Prefecture and City	Population 1990	Change 1985-90
Niimi	27,291	-3.7
Okayama	593,742	3.7
Soja	52,724	2.9
Takahashi	26,006	-2.1
Tamano	73,240	-4.8
Tsuyama	89,405	3

Okinawa

Ginowan	75,899	9.7
Gushikawa	54,026	5.2
Hirara	32,605	-2.4
Ishigaki	41,241	0.2
Ishikawa	20,729	3
Itoman	49,638	8.1
Nago	51,149	4.3
Naha	304,896	0.4
Okinawa	105,852	4.6
Urasoe	89,993	10.3

Osaka

Daito	126,460	3.3
Fujiidera	65,924	1
Habikino	115,035	3.3
Higashiosaka	518,251	-0.9
Hirakata	390,790	2.2
Ibaraki	254,080	1.4
Ikeda	104,219	2.5
Izumi	146,105	6.2
Izumiotsu	67,037	-1.1
Izumisano	88,862	-2.9
Kadoma	142,288	1.2
Kaizuka	79,236	-0.4
Kashiwara	76,819	4.9
Katano	65,311	1.7
Kawachinagano	108,770	19.1
Kishiwada	188,553	1.5
Matsubara	135,921	-0.4
Minoo	122,133	6.4
Moriguchi	157,365	-1.3
Neyagawa	256,521	-0.7
Osaka	2,623,831	-0.5
Osakasayama	54,323	8.9
Sakai	807,859	-1.3
Sennan	60,054	0
Settsu	87,465	1.3
Shijonawate	50,036	-0.6
Suita	345,187	-1.1
Takaishi	65,084	-2.8
Takatsuki	359,867	3.2
Tondabayashi	110,444	7.6
Toyonaka	409,843	-0.8
Yao	277,724	0.5

Saga

Imari	60,887	-1.9
Karatsu	79,206	0.6
Kashima	34,337	-1.1
Saga	169,964	1
Takeo	34,488	-0.9
Taku	25,162	-2.6
Tasu	55,878	0.2

Saitama

Ageo	194,952	9.2
Asaka	103,621	9.7
Chichibu	60,916	-0.2
Fujimi	94,858	10.7
Fukaya	94,023	5.5
Gyoda	83,181	4.8
Hanno	73,216	10
Hanyu	53,766	4.4
Hasuda	59,703	10.6
Hatogaya	56,441	1.8
Higashimatsuyama	84,395	19.8

Prefecture and City	Population 1990	Change 1985-90
Honjo	59,094	4.6
Iruma	137,585	16
Iwatsuki	106,462	5.5
Kamifukuoka	58,753	1.9
Kasukabe	188,809	9.8
Kawagoe	304,860	6.8
Kawaguchi	438,667	8.8
Kazo	56,400	11.6
Kitamoto	63,933	10
Konosu	72,436	19.6
Koshigaya	285,280	12.5
Kuki	66,852	14
Kumagaya	152,122	6
Misato	128,377	18.9
Niiza	138,919	7.5
Okegawa	69,030	12.2
Omiya	403,779	8.2
Sakado	95,736	9.3
Satte	54,339	5.6
Sayama	157,307	9
Shiki	63,492	7.7
Soka	206,129	6.1
Toda	87,600	13.8
Tokorozawa	303,047	10.1
Urawa	418,267	10.9
Wako	56,891	3
Warabi	73,620	4.6
Yashio	72,474	7.2
Yono	79,058	10.4

Shiga

Hikone	99,518	5.6
Kusatsu	94,766	8.3
Moriyama	58,561	10.4
Nagahama	55,482	-0.1
Omihachiman	66,068	3.6
Otsu	260,004	10.9
Yokaichi	40,814	2.7

Shimane

Gotsu	27,748	-3
Hamada	49,139	-3.8
Hirata	30,632	-2.2
Izumo	82,680	2.4
Masuda	52,408	-3
Matsue	142,931	2.1
Oda	36,922	-3.5
Yasugi	32,440	-1.9

Shizuoka

Atami	47,290	-4.2
Fuji	222,500	3.8
Fujieda	119,815	7
Fujinomiya	117,093	4
Fukuroi	53,180	7.5
Gotemba	79,560	6.2
Hamakita	81,159	5.1
Hamamatsu	534,624	4
Ito	71,223	1.5
Iwata	83,521	3.4
Kakegawa	72,795	5.9
Kosai	43,055	4.1
Mishima	105,419	5.8
Numazu	211,731	0.6
Shimada	73,809	2
Shimizu	241,524	-0.3
Shimoda	30,081	-0.4
Shizuoka	472,199	0.8
Susono	49,039	8.6
Tenryu	24,519	-2
Yaizu	112,188	3.3

Prefecture and City	Population 1990	Change 1985-90
Tochigi		
Ashikaga	167,687	0
Imaichi	56,009	5.5
Kanuma	90,044	2.2
Kuroiso	52,346	5.2
Mooka	61,747	7.8
Nikko	20,128	-7.3
Otawara	52,547	6.1
Oyama	142,263	6
Sano	83,484	3.4
Tochigi	86,216	-0.1
Utsunomiya	426,809	5.3
Yaita	35,603	3

Tokushima

Anan	59,045	-2.8
Komatsushima	43,179	-1.9
Naruto	64,577	0.4
Tokushima	263,336	2.1

Tokyo

Akigawa	50,388	10.4
Akishima	105,375	8
Chofu	197,680	3.5
Fuchu	209,419	3.7
Fussa	58,053	12.6
Hachioji	466,373	9.3
Higashikurume	113,800	3.4
Higashimurayama	134,002	8.2
Higashiyamato	75,124	7.5
Hino	165,935	6.3
Hoya	95,148	3.9
Inagi	58,593	15.4
Kiyose	67,540	3.8
Kodaira	164,021	3.4
Koganei	105,888	1.2
Kokubunji	100,958	5.8
Komae	74,197	0.6
Kunitachi	65,830	1.5
Machida	349,030	8.7
Mitaka	165,555	-0.4
Musashimurayama	65,555	7.6
Musashino	139,069	0.2
Ome	125,945	13.6
Tachikawa	152,817	4.3
Tama	144,490	18.3
Tanashi	75,141	5.3
Tokyo	8,163,127	-2.3

Tottori

Kurayoshi	51,835	-1
Sakaiminato	37,291	-0.2
Tottori	142,477	4
Yonago	131,453	-0.3

Toyama

Himi	60,768	-2.2
Kurobe	36,493	1
Namerikawa	30,923	0.1
Oyabe	36,377	-0.9
Shimminato	39,435	-5.4
Takaoka	175,469	-0.2
Tonami	37,070	1.5
Toyama	321,259	2.3
Uozu	49,516	-0.6

Japan census — continued

Prefecture and City	Population 1990	Change 1985-90
Wakayama		
Arida	34,808	-1.7
Gobo	29,133	-4.3
Hashimoto	46,590	15.1
Kainan	48,598	-4.3
Shingu	35,927	-6
Tanabe	69,861	-1.4
Wakayama	396,554	-1.2

Prefecture and City	Population 1990	Change 1985-90
Yamagata		
Higashine	42,750	2.1
Kaminoyama	38,238	-1.5
Murayama	31,589	-1.9
Nagai	33,261	-0.7
Nanyo	36,977	-0.5
Obanazawa	23,910	-3.6
Sagae	42,076	0.6
Sakata	100,808	-0.6
Shinjo	43,124	0.2
Tendo	57,339	4
Tsuruoka	99,891	-0.3
Yamagata	249,493	1.8
Yonezawa	94,763	1.1

Prefecture and City	Population 1990	Change 1985-90
Yamaguchi		
Hagi	50,619	-4
Hikari	47,613	-3.3
Hofu	117,639	-0.4
Iwakuni	109,534	-2.1
Kudamatsu	53,029	-2.6
Mine	19,642	-6.6
Nagato	26,110	-5.2
Onada	46,491	0
Shimonoseki	262,643	-2.4
Shinnanyo	32,988	-2.7
Tokuyama	110,900	-1.5
Ube	175,052	0.2
Yamaguchi	129,467	4.2
Yanai	36,356	-2.8

Prefecture and City	Population 1990	Change 1985-90
Yamanashi		
Enzan	26,550	-0.6
Fujiyoshida	54,802	0
Kofu	200,630	-0.9
Nirasaki	29,761	5.6
Otsuki	34,941	0.1
Tsuru	33,903	2.2
Yamanashi	31,102	0.1

New Zealand

Date of census: 5 March 1991

Total population: 3,429,364

D. - District   C. - City

Area	Population 1986	Population 1991	Change %
Ashburton D.	24,856	24,334	-2.1
Auckland C.	301,428	315,925	4.8
Banks Peninsula D.	7,232	7,636	5.6
Buller D.	11,151	10,926	-2
Carterton D.	6,336	6,910	9.1
Central Hawke's Bay D.	13,054	12,551	-3.9
Central Otago D.	16,805	15,669	-6.8
Chatham Island County	775	754	-2.7
Christchurch C.	286,601	292,537	2.1
Clutha D.	19,201	18,300	-4.7
Dunedin C.	114,349	116,524	1.9
Far North D.	47,912	51,427	7.3
Franklin D.	37,328	42,177	13
Gisborne D.	45,758	44,281	-3.2
Gore D.	13,877	13,587	-2.1
Grey D.	14,300	13,976	-2.3
Hamilton C.	95,388	101,276	6.2
Hastings D.	64,371	64,558	0.3
Hauraki D.	15,904	16,893	6.2
Horowhenua D.	28,858	29,526	2.3
Hurunui D.	9,280	9,561	3
Invercargill C.	57,042	56,059	-1.7
Kaikoura D.	3,529	3,709	5.1
Kaipara D.	17,200	17,470	1.6
Kapiti Coast D.	29,754	35,136	18.1
Kawerau D.	8,311	8,143	-2
Lower Hutt C.	95,342	94,355	-1
Mackenzie D.	4,866	5,019	3.1
Manawatu D.	25,826	27,149	5.1
Manukau C.	206,741	226,496	9.6
Marlborough D.	34,854	36,386	4.4
Masterton D.	22,508	22,927	1.9
Matamata-Piako D.	29,409	29,369	-0.1
Napier C.	52,512	51,442	-2
Nelson C.	35,919	37,995	5.8

Area	Population 1986	Population 1991	Change %
New Plymouth D.	66,878	67,844	1.4
North Shore C.	144,149	151,330	5
Opotiki D.	8,134	8,780	7.9
Otorohanga D.	9,282	9,225	-0.6
Palmerston North C.	66,821	70,206	5.1
Papakura D.	32,765	36,579	11.6
Porirua C.	45,663	46,685	2.2
Queenstown-Lakes D.	12,024	14,349	19.3
Rangitikei D.	17,699	16,615	-6.1
Rodney D.	45,883	55,808	21.6
Rotorua D.	62,912	65,013	3.3
Ruapehu D.	19,461	18,062	-7.2
Selwyn D.	20,520	21,343	4
South Taranaki D.	30,770	29,511	-4.1
South Waikato D.	28,266	26,155	-7.5
South Wairarapa D.	8,747	8,998	2.9
Southland D.	34,570	33,615	-2.8
Stratford D.	10,086	9,852	-2.3
Tararua D.	19,884	19,510	-1.9
Tasman D.	33,729	36,300	7.6
Taupo D.	29,027	30,635	5.5
Tauranga D.	60,194	67,245	11.7
Thames-Coromandel D.	21,715	24,805	14.2
Timaru D.	43,394	42,505	-2
Upper Hutt C.	37,290	36,993	-0.8
Waikato D.	36,475	37,625	3.2
Waimakariri D.	25,400	27,805	9.5
Waimate D.	8,234	7,759	-5.8
Waipa D.	35,553	37,016	4.1
Wairoa D.	10,680	10,374	-2.9
Waitakere C.	122,581	136,600	11.4
Waitaki D.	23,268	22,988	-1.2
Waitomo D.	10,522	10,065	-4.3
Wanganui D.	44,019	44,933	2.1
Wellington C.	149,868	149,598	-0.2
Western Bay of Plenty D.	26,912	30,121	11.9
Westland D.	9,519	9,251	-2.8
Whakatane D.	31,185	32,044	2.8
Whangarei D.	62,542	62,574	0.1
New Zealand	3,307,084	3,429,364	3.7

Urban areas

Area	Population 1986	Population 1991	Change %
Ashburton	15,229	15,085	-0.9
Auckland	821,647	885,377	7.8
Blenheim	22,681	23,562	3.9
Christchurch	300,052	306,856	2.3
Dunedin	107,639	109,423	1.7
Feilding	12,802	13,355	4.3
Gisborne	32,238	31,413	-2.6
Gore	11,249	10,988	-2.3
Greymouth	11,261	10,863	-3.5
Hamilton	140,106	148,468	6
Hawera	11,375	11,137	-2.1
Invercargill	52,558	51,905	-1.2
Kapiti	23,203	27,332	17.8
Levin	18,962	18,967	0
Masterton	19,930	19,994	0.3
Napier-Hastings	110,785	109,875	-0.8
Nelson	44,593	47,429	6.4
New Plymouth	47,384	48,429	2.2
Oamaru	14,247	13,811	-3.1
Palmerston North	67,405	70,838	5.1
Pukekohe	13,931	14,924	7.1
Rotorua	51,991	53,611	3.1
Taupo	17,602	18,306	4
Tauranga	63,254	70,701	11.8
Timaru	28,676	27,653	-3.6
Tokoroa	18,193	16,610	-8.7
Wanganui	40,758	41,168	1
Wellington	325,711	324,792	-0.3
Whakatane	15,959	16,740	4.9
Whangarei	44,318	44,168	-0.3
Total urban areas	2,505,739	2,603,780	3.9

Norway

Date of Census: 3 November 1990

Total population: 4,247,553

The population totals are given for the Fylke and any Kommune with over 10,000 inhabitants. The population at 1 January 1985 is also shown with the percentage change 1985-90 indicated.
The 1985 and 1990 figures are not always comparable because of boundary changes/amalgamations. The Fylke are shown, as normally in Norwegian listings, in geographic order south to north.

	1985	1990	Change 1985-90 %
NORWAY	4,159,335	4,247,553	2.1
ØSTFOLD	234,941	238,296	1.4
Halden	25,876	25,873	0.0
Sarpsborg	12,069	11,790	-2.3
Fredrikstad	27,125	26,546	-2.1
Moss	24,830	24,683	-0.6
Borge	11,291	11,966	6.0
Skjeberg	13,495	14,279	5.8
Askim	12,500	12,864	2.9
Tune	18,511	18,324	-1.0
Onsoy	12,488	12,914	3.4
Rygge	11,540	12,037	4.3
AKERSHUS	393,217	417,653	6.2
Vestby	10,657	11,266	5.7
Ski	20,849	22,337	7.1
Ås	11,410	11,946	4.7
Frogn	9,223	10,298	11.7
Nesodden	11,243	13,189	17.3
Oppegård	18,603	20,669	11.1
Bærum	82,918	90,333	8.9
Asker	37,767	41,848	10.8
Aurskog-Holand	12,438	12,582	1.2
Sorum	9,905	11,265	13.7
Rælingen	13,387	13,781	2.9
Lorenskog	23,779	26,454	11.2
Skedsmo	32,339	34,110	5.5
Nittedal	14,945	16,177	8.2
Ullensaker	17,506	18,125	3.5
Nes	14,729	15,703	6.6
Eidsvoll	15,643	16,689	6.7
OSLO	447,351	459,292	2.7
HEDMARK	186,355	187,275	0.5
Hamar	15,693	16,315	4.0
Kongsvinger	17,467	17,469	0.0
Ringsaker	30,419	31,377	3.1
Stange	17,790	17,616	-1.0
Elverum	16,903	17,428	3.1

	1985	1990	Change 1985-90 %
OPPLAND	181,791	182,578	0.4
Lillehammer	22,012	22,850	3.8
Gjøvik	25,957	26,207	1.0
Østre Toten	14,131	14,336	1.5
Vestre Toten	13,626	13,366	-1.9
Gran	12,514	12,641	1.0
BUSKERUD	219,967	225,172	2.4
Drammen	50,749	51,880	2.2
Kongsberg	20,913	21,185	1.3
Ringerike	26,870	27,384	1.9
Modum	12,160	12,243	0.7
Øvre Eiker	14,113	14,801	4.9
Nedre Eiker	17,906	18,901	5.6
Lier	18,113	18,961	4.7
Røyken	13,687	14,393	5.2
VESTFOLD	191,600	198,399	3.5
Borre	8,940	22,568	152.4
Tønsberg	8,891	31,551	254.9
Sandefjord	35,011	36,095	3.1
Larvik	8,070	38,223	373.6
Nøtterøy	17,183	18,031	4.9
TELEMARK	162,547	162,907	0.2
Porsgrunn	31,402	31,268	-0.4
Skien	46,656	47,870	2.6
Notodden	12,622	12,426	-1.6
Bamble	13,162	13,784	4.7
Kragerø	10,904	10,817	-0.8
AUST-AGDER	94,688	97,333	2.8
Arendal	12,051	12,439	3.2
Grimstad	14,631	15,656	7.0
VEST-AGDER	140,232	144,917	3.3
Kristiansand	62,197	65,543	5.4
Mandal	12,218	12,496	2.3
Vennesla	11,092	11,544	4.1
ROGALAND	323,365	337,504	4.4
Eigersund	12,126	12,409	2.3
Sandnes	39,678	44,798	12.9
Stavanger	94,193	98,109	4.2
Haugesund	27,014	27,736	2.7
Hå	12,660	13,022	2.9
Klepp	11,289	11,854	5.0
Time	11,013	12,051	9.4
Sola	14,420	15,944	10.6
Karmøy	33,888	35,087	3.5

	1985	1990	Change 1985-90 %
HORDALAND	399,702	410,568	2.7
Bergen	207,416	212,944	2.7
Stord	13,591	14,632	7.7
Kvinnherad	13,159	13,093	-0.5
Voss	14,059	14,082	0.2
Os	11,499	12,768	11.0
Fjell	12,200	14,912	22.2
Askoy	18,036	18,598	3.1
Lindås	11,079	11,863	7.1
SOGN OG FJORDANE	106,116	106,659	0.5
Flora	9,344	10,049	7.5
MØRE OG ROMSDAL	237,290	238,408	0.5
Molde	21,310	22,251	4.4
Kristiansund	17,818	17,190	-3.5
Ålesund	35,008	35,862	2.4
Orsta	10,289	10,248	-0.4
SØR-TRØNDELAG	246,824	250,978	1.7
Trondheim	134,075	137,846	2.8
Orkdal	9,960	10,138	1.8
Melhus	11,834	12,505	5.7
NORD-TRØNDELAG	126,692	127,157	0.4
Steinkjer	20,590	20,665	0.4
Namsos	11,847	11,909	0.5
Stjordal	16,717	17,321	3.6
Levanger	16,278	16,829	3.4
Verdal	13,131	13,503	2.8
NORDLAND	242,268	239,311	-1.2
Bodø	34,013	36,890	8.5
Narvik	18,865	18,609	-1.4
Vefsn	13,282	13,410	1.0
Rana	25,251	24,650	-2.4
Fauske	10,093	10,001	-0.9
Vestvågoy	10,846	10,566	-2.6
TROMS	146,736	146,716	0.0
Harstad	21,760	22,375	2.8
Tromso	47,753	51,218	7.3
Lenvik	11,204	10,843	-3.2
FINNMARK	75,667	74,524	-1.5
Alta	13,928	15,170	8.9
Sor-Varanger	10,073	9,671	-4.0

United States – census

Date of census: 1 April 1990
Total population: 249,632,692

The first list gives information concerning the states and the second about cities with a population exceeding 100,000. All population totals are for resident population and exclude people normally resident in the state or city who were abroad at the time of the census.

State	Resident Population 1980	Resident Population 1990	Change 1980-90 %	Rank 1990	% of US total population 1990	Density 1990 per sq mile
Alabama	3,890,061	4,040,587	3.9	22	1.6	80
Alaska	400,481	550,043	37.3	50	0.2	1
Arizona	2,717,866	3,665,228	34.9	24	1.5	32
Arkansas	2,285,513	2,350,725	2.9	33	0.9	45
California	23,668,562	29,760,021	25.7	1	12	190
Colorado	2,888,834	3,294,394	14	26	1.3	32
Connecticut	3,107,576	3,287,116	5.8	27	1.3	675
D. of Columbia	637,651	606,900	-4.8	48	0.2	9633
Delaware	595,225	666,168	11.9	46	0.3	345
Florida	9,739,992	12,937,926	32.8	4	5.2	239
Georgia	5,464,265	6,478,216	18.6	11	2.6	112
Hawaii	,965,000	1,108,229	14.8	41	0.4	172
Idaho	,943,935	1,006,749	6.7	42	0.4	12
Illinois	11,418,461	11,430,602	0.1	6	4.6	205
Indiana	5,490,179	5,544,159	1	14	2.2	154
Iowa	2,913,387	2,776,755	-4.7	30	1.1	50
Kansas	2,363,208	2,477,574	4.8	32	1	30
Kentucky	3,661,433	3,685,296	0.7	23	1.5	93
Louisiana	4,203,972	4,219,973	0.4	21	1.7	95
Maine	1,124,660	1,227,928	9.2	38	0.5	40
Maryland	4,216,446	4,781,468	13.4	19	1.9	486
Massachusetts	5,737,037	6,016,425	4.9	13	2.4	769
Michigan	9,258,344	9,295,297	0.4	8	3.7	163
Minnesota	4,077,148	4,375,099	7.3	20	1.8	55
Mississippi	2,520,638	2,573,216	2.1	31	1	54
Missouri	4,917,444	5,117,073	4.1	15	2.1	74
Montana	786,690	799,065	1.6	44	0.3	5
Nebraska	1,570,006	1,578,385	0.5	36	0.6	21
Nevada	799,184	1,201,833	50.4	39	0.5	11
New Hampshire	920,610	1,109,252	20.5	40	0.4	123
New Jersey	7,364,158	7,730,188	5	9	3.1	1035
New Mexico	1,299,968	1,515,069	16.5	37	0.6	12

State	Resident Population 1980	Resident Population 1990	Change 1980-90 %	Rank 1990	% of US total population 1990	Density 1990 per sq mile
New York	17,557,288	17,990,455	2.5	2	7.2	380
North Carolina	5,874,429	6,628,637	12.8	10	2.7	136
North Dakota	652,695	638,800	-2.1	47	0.3	9
Ohio	10,797,419	10,847,115	0.5	7	4.4	265
Oklahoma	3,025,266	3,145,585	4	28	1.3	46
Oregon	2,632,663	2,842,321	8	29	1.1	30
Pennsylvania	11,866,728	11,881,643	0.1	5	4.8	265
Rhode Island	947,154	1,003,464	5.9	43	0.4	951
South Carolina	3,119,208	3,486,703	11.8	25	1.4	115
South Dakota	690,178	696,004	0.8	45	0.3	9
Tennessee	4,590,750	4,877,185	6.2	17	2	119
Texas	14,228,383	16,986,510	19.4	3	6.8	65
Utah	1,461,037	1,722,850	17.9	35	0.7	21
Vermont	511,456	562,758	10	49	0.2	61
Virginia	5,346,279	6,187,358	15.7	12	2.5	156
Washington	4,130,163	4,866,692	17.8	18	2	73
West Virginia	1,949,644	1,793,477	-8	34	0.7	74
Wisconsin	4,705,335	4,891,769	4	16	2	90
Wyoming	470,816	453,588	-3.7	51	0.2	5
USA	226,504,825	248,709,873	9.8		100	70

USA – cities

City	Population 1990	Change 1980-90 %
Abilene	106,654	8.5
Akron	223,019	-6
Albany	101,082	-0.6
Albuquerque	384,736	15.6
Alexandria	111,183	7.7
Allentown	105,090	1.3
Amarillo	157,615	5.6
Anaheim	266,406	21.4
Anchorage	226,338	29.8
Ann Arbor	109,592	1.5
Arlington	261,721	63.5
Atlanta	394,017	-7.3
Aurora	222,103	40.1
Austin	465,622	34.6
Bakersfield	174,820	65.5
Baltimore	736,014	-6.4
Baton Rouge	219,531	-0.4
Beaumont	114,323	-3.2
Berkeley	102,724	-0.6
Birmingham	265,968	-6.5
Boise	125,738	23
Boston	574,283	2
Bridgeport	141,686	-0.6
Buffalo	328,123	-8.3
Cedar Rapids	108,751	-1.4
Charlotte	395,934	25.5
Chattanooga	152,466	-10.1
Chesapeake	151,976	32.7
Chicago	2,783,726	-7.4
Chula Vista	135,163	61
Cincinnati	364,040	-5.5
Cleveland	505,616	-11.9
Colorado Springs	281,140	30.7
Columbus (Ga.)	179,278	5.4
Columbus (Oh.)	632,910	12
Concord	111,348	7.3
Corpus Christi	257,453	10.9
Dallas	1,006,877	11.3
Dayton	182,044	-5.9
Denver	467,610	-5.1
Des Moines	193,187	1.1
Detroit	1,027,974	-14.6
Durham	136,611	35.1
El Monte	106,209	
El Paso	515,342	21.2
Elizabeth	110,002	3.6
Erie	108,718	-8.7
Escondido	108,635	68.8
Eugene	112,669	6.6
Evansville	126,272	-3.2
Flint	140,761	-11.8
Fort Lauderdale	149,377	-2.5
Fort Wayne	173,072	0.4
Fort Worth	447,619	16.2
Fremont	173,339	31.4
Fresno	354,202	62.9
Fullerton	114,144	11.6
Garden Grove	143,050	16
Garland	180,650	30.1
Gary	116,646	-23.3
Glendale (Ar.)	148,134	52.4
Glendale (Ca.)	180,038	29.5
Grand Rapids	189,126	4
Greensboro	183,521	17.9
Hampton	133,793	9.1
Hartford	139,739	2.5
Hayward	111,498	19.1
Hialeah	188,004	29.4
Hollywood	121,697	0.3
Honolulu	365,272	0.1
Houston	1,630,553	2.2
Huntington Beach	181,519	6.5
Huntsville	159,789	12.1
Independence	112,301	0.5
Indianapolis	741,952	4.3
Inglewood	109,602	16.4
Irvine	110,330	77.6
Irving	155,037	41
Jackson	196,637	-3.1
Jacksonville	672,971	17.9
Jersey City	228,537	2.2
Kansas City (Ks.)	149,767	-7.1
Kansas City (Mo.)	435,146	-2.9
Knoxville	165,121	-5.7
Lakewood	126,481	11.1
Lansing	127,321	-2.4
Laredo	122,899	34.4
Las Vegas	258,295	56.9
Lexington-Fayette	225,366	10.4
Lincoln	191,972	11.7
Little Rock	175,795	10.5
Livonia	100,850	-3.8
Long Beach	429,433	18.8
Los Angeles	3,485,398	17.4
Louisville	269,063	-9.9
Lowell	103,439	11.9
Lubbock	186,206	6.8
Macon	106,612	-8.8
Madison	191,262	12.1
Memphis	610,337	-5.5
Mesa	288,091	89
Mesquite	101,484	51.3
Miami	358,548	3.4
Milwaukee	628,088	-1.3
Minneapolis	368,383	-0.7
Mobile	196,278	-2.1
Modesto	164,730	54
Montgomery	187,106	5.2
Moreno Valley	118,779	...
Nashville-Dav.	510,784	6.9
New Haven	130,474	3.5
New Orleans	496,938	-10.9
New York	7,322,564	3.5
Newark	275,221	-16.4
Newport News	170,045	17.4
Norfolk	261,229	-2.2
Oakland	372,242	9.7
Oceanside	128,398	67.4
Oklahoma	444,719	10.1
Omaha	335,795	7
Ontario	133,179	49.9
Orange	110,658	21
Orlando	164,693	28.4
Overland Park	111,790	36.7
Oxnard	142,216	31.4
Pasadena (Ca.)	131,591	11.4
Pasadena (Tx.)	119,363	6
Paterson	140,891	2.1
Peoria	113,504	-8.6
Philadelphia	1,585,577	-6.1
Phoenix	983,403	24.5
Pittsburgh	369,879	-12.8
Plano	128,713	77.9
Pomona	131,723	42
Portland	437,319	18.8
Portsmouth	103,907	-0.6
Providence	160,728	2.5
Raleigh	207,951	38.4
Rancho Cucamonga	101,409	83.5
Reno	133,850	32.8
Richmond	203,056	-7.4
Riverside	226,505	32.8
Rochester	231,636	-4.2
Rockford	139,426	-0.2
St. Louis	396,685	-12.4
St. Paul	272,235	0.7
St. Petersburg	238,629	0
Sacramento	369,365	34
Salem	107,796	21
Salinas	108,777	35.2
Salt Lake City	159,936	-1.9
San Antonio	935,933	19.1
San Bernardino	164,164	38.2
San Diego	1,110,549	26.8
San Francisco	723,959	6.6
San Jose	782,248	24.3
Santa Ana	293,742	44
Santa Clarita	110,642	...
Santa Rosa	113,313	37.1
Savannah	137,560	-2.9
Scottsdale	130,069	46.8
Seattle	516,259	4.5
Shreveport	198,525	-4.1
Simi Valley	100,217	29.3
Sioux Falls	100,814	23.9
South Bend	105,511	-3.8
Spokane	177,196	3.4
Springfield (Il.)	105,227	5.2
Springfield (Ma.)	156,983	3.1
Springfield (Mo.)	140,494	5.5
Stamford	108,056	5.5
Sterling Heights	117,810	8.1
Stockton	210,943	42.3
Sunnyvale	117,229	10
Syracuse	163,860	-3.7
Tacoma	176,664	11.5
Tallahassee	124,773	53
Tampa	280,015	3.1
Tempe	141,865	32.7
Thousand Oaks	104,352	35.4
Toledo	332,943	-6.1
Topeka	119,883	1
Torrance	133,107	2.5
Tucson	405,390	22.6
Tulsa	367,302	1.8
Vallejo	109,199	36
Virginia Beach	393,069	49.9
Waco	103,590	2.3
Warren	144,864	-10.1
Washington	606,900	-4.9
Waterbury	108,961	5.5
Wichita	304,011	8.6
Winston-Salem	143,485	8.8
Worcester	169,759	4.9
Yonkers	188,082	-3.7

Agriculture

Land use

	Land area th km²	Arable land %	Permanent crops %	Permanent grassland %	Forest %	Other land %	Irrigated land 100 km²
World	130,693	10.5	0.8	24.6	31	33.2	22,867.2
Africa	29,639	5.7	0.6	26.8	23.1	43.9	111.5
Asia	26,787	15.7	1.2	25.3	19.6	38.2	1,427.6
Europe	4,721	26.7	3	17.7	33.4	19.3	173
North America	21,318	12.5	0.3	17.3	32.2	37.7	258.1
Oceania	8,427	5.7	0.1	52.1	18.5	23.5	21.3
South America	17,529	6.6	1.5	27.3	51.1	13.5	87.6
Afghanistan	652	12.1	0.2	46	2.9	38.7	26.6
Albania	27.4	21.5	4.5	14.7	38.2	21	4.2
Algeria	2,382	2.9	0.2	12.8	2	82.1	3.7
Americ. Samoa	0.2	10	10	0	70	10	0
Andorra	0.5	2.2	0	55.6	22.2	20	0
Angola	1,247	2.4	0.4	23.3	42.5	31.3	0
Antigua & Barbuda	0.4	18.2	0	9.1	11.4	61.4	0
Argentina	2,737	9.5	3.6	52	21.7	13.2	17.4
Aruba	0.2	10.5	0	0	0	89.5	0
Australia	7,618	6.1	0	55.8	13.9	24.1	18.5
Austria	82.7	17.3	0.9	24	38.7	19.1	0
Bahamas	10	0.8	0.2	0.2	32.4	66.4	0
Bahrain	0.7	1.5	1.5	5.9	0	91.2	0
Bangladesh	130	69.1	2.1	4.6	15.1	9.1	22.2
Barbados	0.4	76.7	0	9.3	0	14	0
Belgium-Lux.	32.8	24.5	0.5	21	21.3	32.8	0
Belize	22.8	1.9	0.5	2.1	44.4	51.1	0
Benin	111	12.7	4.1	4	32.3	46.9	0.1
Bermuda	0.1	0	0	0	20	80	0
Bhutan	47	2.4	0.4	5.7	55.3	36.2	0.3
BIOT	0.1	0	0	0	0	100	0
Bolivia	1,084	3	0.2	24.6	51.4	20.8	1.7
Botswana	567	2.4	0	77.6	1.7	18.2	0
Br. Virgin Is	0.2	20	6.7	33.3	6.7	33.3	0
Brazil	8,457	7.9	1.4	20	65.7	5	26
Brunei	5.3	0.6	0.8	1.1	46.5	51	0
Bulgaria	111	34.7	2.7	18.3	35	9.3	12.4
Burkina Faso	274	13	0	36.5	24.5	25.9	0.2
Burma	658	14.5	0.7	0.6	49.3	34.9	0
Burundi	25.7	43.7	8.3	35.6	2.5	9.9	0.7
Cambodia	177	16.5	0.8	3.3	75.8	3.6	0.9
Cameroon	465	12.8	2.3	17.8	53.2	13.9	0.3
Canada	9,161	5	0	3.5	38.9	52.6	8.2
Cape Verde Is	4	9.4	0.5	6.2	0.2	83.6	0
Cayman Is	0.3	0	0	7.7	23.1	69.2	0
Central Afr. Rep.	623	3.1	0.1	4.8	57.5	34.5	0
Chad	1,259	2.5	0	35.7	10.2	51.5	0.1
Chile	749	5.5	0.3	17.9	11.8	64.5	12.6
China	9,326	10	0.4	34.2	12.6	42.9	449.4
Christmas Is	0.1	0	0	0	0	100	0
Cocos Is	0	0	0	0	0	100	0
Colombia	1,039	3.7	1.4	38.7	49	7.1	5.1
Comoros	2.2	34.5	9.9	6.7	15.7	33.2	0
Congo	342	0.4	0.1	29.3	62.1	8.1	0
Cook Is	0.2	4.3	21.7	0	0	73.9	0
Costa Rica	51.1	5.6	4.8	45.2	32.1	12.3	1.2
Cuba	111	23.6	6.5	25.3	24.8	19.9	8.7
Cyprus	9.2	11.3	5.7	0.5	13.3	69.2	0.3
Czechoslovakia	125	39.8	1.1	13.1	36.7	9.3	3.1
Denmark	42.4	60.6	0.1	5.1	11.6	22.6	4.3
Djibouti	23.2	0	0	8.6	0.3	91.1	0
Dominica	0.8	9.3	13.3	2.7	41.3	33.3	0
Dominican Rep.	48.4	23.2	7.3	43.2	12.8	13.5	2.3
Ecuador	277	6.2	3.4	18.2	41.5	30.6	5.5
Egypt	995	2.4	0.2	0	0	97.4	25.8
El Salvador	20.7	27.3	8.1	29.4	5	30.2	1.2
Equatorial Guinea	28.1	4.6	3.6	3.7	46.2	41.9	0
Ethiopia	1101	12	0.7	40.9	24.8	21.7	1.6
Falkland Is	12.2	0	0	98.6	0	1.4	0
Faroe Is	1.4	2.1	0	0	0	97.9	0
Fiji	18.3	8.3	4.8	3.3	64.9	18.7	0
Finland	305	8	0	0.4	76.2	15.3	0.6
France	550	33.2	2.3	21.3	26.7	16.4	13.7
French Guiana	88.2	0.1	0	0.1	82.8	17	0
French Polynesia	3.7	1.4	19.1	5.5	31.4	42.6	0
Gabon	258	1.1	0.6	18.2	77.6	2.4	0
Gambia, The	10	17.5	0	9	16.8	56.7	0.1
Germany	349	34.2	1.3	16.3	29.6	18.6	0
Ghana	230	5	7.5	14.7	35.7	37.1	0.1
Gibraltar	0	0	0	0	0	100	0
Greece	131	22	8	40.2	20	9.8	11.8
Greenland	342	0	0	0.7	0	99.3	0
Grenada	0.3	14.7	23.5	2.9	8.8	50	0
Guadeloupe	1.7	13.6	5.3	16.6	42	22.5	0
Guam	0.6	10.9	10.9	14.5	18.2	45.5	0
Guatemala	108	12.7	4.5	12.7	36.1	34	0.8
Guinea	246	2.5	0.5	12.2	40.1	44.7	0.2
Guinea-Bissau	28.1	10.7	1.2	38.4	38.1	11.6	0
Guyana	197	2.4	0.1	6.2	83.2	8.1	1.3
Haiti	27.6	20.1	12.7	17.9	1.8	47.5	0.7
Honduras	112	14.1	1.9	22.7	30.6	30.8	0.9
Hong Kong	1	6.1	1	1	12.1	79.8	0
Hungary	92.3	54.7	2.6	13.1	18.2	11.5	1.8
Iceland	100	0.1	0	22.7	1.2	76	0
India	2,973	55.8	1.2	4	22.4	16.6	417.9
Indonesia	1,812	8.7	3	6.5	62.6	19.2	75
Iran	1,636	8.6	0.4	26.9	11	53	57.5
Iraq	437	12	0.5	9.1	4.3	74.1	25.4
Ireland	68.9	13.9	0	68.1	4.9	13	0
Israel	20.6	16.6	4.4	7.2	5.3	66.5	2.2
Italy	294	31	10.4	16.7	22.9	19.1	30.8
Ivory Coast	318	7.6	3.9	9.4	18.5	60.6	0.6
Jamaica	10.8	19.1	5.7	18	17.3	39.9	0.3
Japan	377	11.1	1.4	1.7	66.7	19.2	28.9
Jordan	88.9	3.5	0.7	8.9	0.8	86.1	0.6
Kenya	570	3.4	0.9	6.5	6.4	82.8	0.5
Kiribati	0.7	0	52.1	0	2.8	45.1	0
Korea, North	120	19.2	0.8	0.4	74.5	5.1	11.9
Korea, South	98.7	20.2	1.4	0.9	65.7	11.7	13.6
Kuwait	17.8	0.2	0	7.5	0.1	92.1	0
Laos	231	3.8	0.1	3.5	55.9	36.7	1.2
Lebanon	10.2	20.3	9.1	1	7.8	61.8	0.9
Lesotho	30.4	10.5	0	65.9	0	23.6	0
Liberia	96.3	1.3	2.5	2.5	21.8	71.8	0
Libya	1,760	1	0.2	7.6	0.4	90.8	2.4
Liechtenstein	0.2	25	0	37.5	18.8	18.8	0
Macau	0	0	0	0	0	100	0
Madagascar	582	4.4	0.9	58.5	25.1	11.2	8.9
Malawi	94.1	25.1	0.3	19.6	44.5	10.5	0.2
Malaysia	329	3.2	11.7	0.1	58.9	26.2	3.4
Maldives	0.3	10	0	3.3	3.3	83.3	0
Mali	1,220	1.7	0	24.6	6.9	66.7	2.1
Malta	0.3	37.5	3.1	0	0	59.4	0
Martinique	1.1	10.4	8.5	18.9	35.8	26.4	0.1
Mauritania	1,025	0.2	0	38.3	14.6	46.9	0.1
Mauritius	1.9	54.1	3.2	3.8	30.8	8.1	0.2
Mexico	1909	12.1	0.8	39	22.8	25.2	51
Mongolia	1,567	0.9	0	79.1	8.9	11.2	0.5
Montserrat	0.1	20	0	10	40	30	0
Morocco	446	18.5	1.3	46.8	11.7	21.8	12.6
Mozambique	782	3.7	0.3	56.3	18.8	20.9	1.1
Namibia	823	0.8	0	64.3	22.4	12.6	0
Nauru	0	0	0	0	0	100	0
Nepal	137	17	0.2	14.6	16.9	51.3	6.6
Netherlands	33.9	26.6	0.9	31.9	8.8	31.8	5.5
Neths. Antilles	1	8.1	0	0	0	91.9	0
New Caledonia	18.3	0.5	0.5	15.3	38.7	44.9	0
New Zealand	268	1.8	0.1	51.4	27.2	19.5	2.8
Nicaragua	119	9.2	1.5	44.6	30.3	14.4	0.9
Niger	1,267	2.8	0	7.3	1.9	87.9	0.3
Nigeria	911	31.6	2.8	23	15.4	27.2	8.6
Niue	0.3	61.5	3.8	3.8	19.2	11.5	0
Norfolk I.	0	0	0	25	0	75	0
Norway	307	2.8	0	0.3	27.1	69.7	0.9
Oman	212	0.1	0.2	4.7	0	95.1	0.4

Land use
continued

	Land area (th km²)	Arable land %	Permanent crops %	Permanent grassland %	Forest %	Other land %	Irrigated land (100 km²)
Pacific Is	1.8	14	19.1	13.5	22.5	30.9	0
Pakistan	771	26.6	0.6	6.5	4.3	62.1	156.8
Panama	76	5.8	1.8	17.5	51.7	23.2	0.3
Papua New Guinea	453	0.1	0.8	0.2	84.4	14.5	0
Paraguay	397	5.3	0.3	51.6	37.6	5.2	0.7
Peru	1,280	2.7	0.3	21.2	53.8	22.1	12.4
Philippines	298	15.3	11.5	4.1	36.1	33.1	15
Poland	304	47.5	1	13.3	28.7	9.5	1
Portugal	92	22.2	7.7	5.8	39.6	24.7	6.3
Puerto Rico	8.9	7.7	6.8	37.7	20	27.9	0.4
Qatar	11	0.5	0	4.5	0	95	0
Réunion	2.5	20	1.6	4	35.2	39.2	0.1
Romania	230	43.8	2.6	19.1	27.5	6.9	34
Rwanda	25	33.3	11.6	15.8	19.9	19.4	0
Samoa	2.8	19.4	23.7	0.4	47.3	9.2	0
San Marino	0.1	16.7	0	0	0	83.3	0
Sao Tome & Principe	1	2.1	36.5	1	0	60.4	0
Saudi Arabia	2,150	0.5	0	39.5	0.6	59.3	4.3
Senegal	193	27.1	0	29.6	30.8	12.5	1.8
Seychelles	0.3	3.7	18.5	0	18.5	59.3	0
Sierra Leone	71.6	23.1	2	30.8	28.9	15.2	0.3
Singapore	0.6	3.3	0	0	4.9	91.8	0
Solomon Is	28	1.4	0.6	1.4	91.5	5.1	0
Somalia	627	1.6	0	46	13.9	38.4	1.1
South Africa	1,221	10.1	0.7	66.6	3.7	18.9	11.3
Spain	499	31.2	9.6	20.4	31.4	7.4	33.2
Sri Lanka	64.6	14.3	15.1	6.8	27	36.8	5.5
St Pierre M.	0.2	13	0	0	4.3	82.6	0
St Christopher/N.	0.4	22.2	16.7	2.8	16.7	41.7	0
St Helena	0.3	0	0	0	0	0	0
St Lucia	0.6	8.2	21.3	4.9	13.1	52.5	0
St Vincent/G.	0.4	33.3	12.8	5.1	35.9	12.8	0
Sudan	2,376	5.2	0	23.6	19.6	51.6	18.8
Surinam	156	0.4	0.1	0.1	95.2	4.2	0.6
Swaziland	17.2	9.3	0.2	68.3	6.3	15.9	0.6
Sweden	403	7.3	0	1.4	69.6	21.7	1.1
Switzerland	39.8	9.8	0.5	40.5	26.5	22.7	0.3
Syria	184	26.9	3.3	44.7	3	22.1	6.5
Taiwan	36	11.7	1.7	11.4	51.8	23.4	0
Tanzania	886	4.7	1.2	39.5	47.7	6.8	1.5
Thailand	511	35	4.4	1.5	27.7	31.3	40.5
Togo	54.4	25.2	1.3	3.7	23.9	46	0.1
Tokelau	0	0	0	0	0	100	0
Tonga	0.7	23.6	43.1	5.6	11.1	16.7	0
Trinidad & Tobago	5.1	14.4	9	2.1	43.3	31.2	0.2
Tunisia	155	21.2	9.8	19.6	3.6	45.7	2.8
Turkey	770	32.2	3.9	11.2	26.2	26.6	22
Turks/Caicos	0.4	2.3	0	0	0	97.7	0
Tuvalu	0	0	0	0	0	100	0
Uganda	200	25.1	8.5	25.1	28.4	13	0.1
United Arab Em.	83.6	0.3	0.1	2.4	0	97.1	0.1
United Kingdom	242	28.7	0.2	47.8	9.8	13.4	1.6
United States	9,167	20.5	0.2	26.3	28.9	24	181
Uruguay	175	7.3	0.3	77.3	3.8	11.3	1.1
US Virgin Is	0.3	14.7	5.9	26.5	5.9	47.1	0
USSR	22,272	10.2	0.2	16.7	42.4	30.4	207.8
Vanuatu	12.2	1.6	10.3	2.1	1.3	84.7	0
Venezuela	882	3.6	0.8	20	34.9	40.8	3.3
Vietnam	327	17.4	2.6	1	28.4	50.5	18.2
Wallis & Futuna	0.2	5	20	0	0	75	0
Western Sahara	267	-	-	19	-	81	0.7
Yemen	528	2.6	0.2	30.4	5.9	60.9	0
Yugoslavia	255	27.6	2.8	24.9	36.6	8.1	1.7
Zaïre	2,267	3.2	0.3	6.6	77.2	12.8	0.1
Zambia	743	7	0	47.1	39.1	6.7	0.3
Zimbabwe	387	7	0.2	12.6	51.5	28.6	2.2

The statistics are for 1989. The land area is the total area of the country less that covered by lakes and major rivers. Arable land is land under temporary crops including temporary grassland for mowing or pasture. It includes market gardens and fallow land. Land under permanent crops is land that is under the same crop for a long period. Rubber or coffee is an example. Permanent pasture includes both cultivated and natural grassland that has been so for five or more years. Forest and woodland includes both planted and natural forests. Other land includes built-up areas, wasteland and desert. Irrigated land also includes land which is occasionally flooded by rivers for crop production or pasture improvement. Note that the unit in this column is 100 sq km.

Agricultural population

percentage of the economically active population employed in agriculture

	1930-44	1945-64	1970	1980	1989
World	...	...	55	51	47
Africa	...	...	74	69	64
Asia	...	...	70	66	61
Europe	...	...	20	14	10
North America	...	...	14	12	11
Oceania	...	...	23	20	17
South America	...	...	38	29	24
Afghanistan	...	...	66	61	55
Albania	...	...	66	60	49
Algeria	...	75	47	31	25
Angola	...	...	78	74	70
Argentina	...	19	16	13	11
Australia	19	11	8	7	5
Austria	36	23	15	9	6
Bahamas	...	16	13	9	7
Bangladesh	...	...	81	75	69
Barbados	...	24	18	10	7
Belgium	17	6.0	5	3	2
Belize	...	35	...	...	...
Benin	...	...	81	70	62
Bermuda	...	6	...	...	...
Bhutan	...	...	94	92	91
Bolivia	...	63	52	46	42
Botswana	...	...	86	70	64
Brazil	67	54	45	31	25
Brunei	...	34	...	...	...
Bulgaria	80	64	35	18	13
Burkina Faso	...	...	88	87	85
Burma	...	...	59	53	47
Burundi	...	...	94	93	91
Cambodia	...	81	78	74	70
Cameroon	...	...	83	70	62
Canada	26	11	8	5	4
Cape Verde	...	40	64	52	44
Central Afr. Rep.	...	...	83	72	64
Chad	...	...	90	83	76
Chile	35	28	23	17	13
China	...	...	78	74	68
Colombia	72	54	39	34	28
Comoros	...	...	87	83	80
Congo	...	...	65	62	60
Costa Rica	...	49	43	31	25
Cuba	...	42	30	24	20
Cyprus	51	40	39	26	21
Czechoslovakia	37	38	17	13	10
Denmark	29	17	11	7	5
Dominica	...	50	...	...	...
Dominican Rep.	77	56	55	46	37
Ecuador	...	56	51	39	31
Egypt	71	57	52	46	41
El Salvador	75	60	56	43	37
Equat. Guinea	...	...	75	66	57
Ethiopia	...	...	85	80	75
Fiji	...	57	52	46	40
Finland	57	35	20	12	8
France	36	20	14	9	6
Fr. Polynesia	...	41	...	...	...
Gabon	...	84	80	75	69
Gambia	...	...	87	84	81
Germany, E.	22	18	13	11	8
Germany, W.	27	11	8	6	3
Ghana	...	58	58	56	51
Greece	...	54	42	31	25
Greenland	...	34	...	...	...
Grenada	...	40	...	...	...
Guadeloupe	...	29	15	10	...
Guam	...	2	...	...	...
Guatemala	71	68	61	57	52
Guinea	...	...	85	81	75
Guinea-Bissau	...	...	84	82	79
Guyana	51	34	32	27	23
Haiti	...	83	74	70	64
Honduras	...	67	65	61	56
Hong Kong	...	7	4	2	1
Hungary	53	38	25	18	12
Iceland	57	38	17	10	7
India	66	73	72	70	67
Indonesia	66	68	66	57	49
Iran	...	55	44	36	28
Iraq	...	48	47	30	21
Ireland	49	35	26	19	14
Israel	...	12	10	6	4
Italy	48	25	19	12	8
Ivory Coast	...	86	77	65	57
Jamaica	...	39	33	31	27
Japan	48	27	20	11	7
Jordan	...	35	28	10	6
Kenya	...	...	85	81	77
Korea, N.	...	57	53	43	34
Korea, S.	...	...	49	36	26
Kuwait	...	...	2	2	...
Laos	...	...	79	76	72
Lebanon	...	...	20	14	9
Lesotho	...	...	90	86	80
Liberia	...	81	78	74	70
Libya	...	...	29	18	14
Macau	...	5	...	...	...
Madagascar	...	...	84	81	77
Malawi	...	...	91	83	76
Malaysia	61	58	54	42	33
Mali	...	...	89	86	81
Malta	...	10	7	5	4
Martinique	...	39	24	13	8
Mauritania	...	...	85	69	65
Mauritius	...	40	34	28	23
Mexico	65	54	44	37	31
Mongolia	...	...	48	40	31
Morocco	74	56	58	46	38
Mozambique	...	75	86	85	82
Namibia	...	59	51	43	36

Agricultural population — *continued*

	1930-44	1945-64	1970	1980	1989
Nepal	...	93	94	93	92
Netherlands	21	11	7	6	4
Neths. Antilles	...	2	...	...	...
New Caledonia	...	55	...	...	...
New Zealand	23	14	12	11	9
Nicaragua	73	60	52	47	39
Niger	...	97	94	91	88
Nigeria	...	...	71	68	65
Norway	35	19	12	8	6
Oman	...	...	...	50	41
Pakistan	...	75	59	55	50
Panama	52	46	42	32	26
Papua New G.	...	...	84	76	68
Paraguay	...	52	53	49	47
Peru	62	50	47	40	35
Philippines	73	57	55	52	47
Poland	64	48	39	29	22
Portugal	49	42	32	26	17
Puerto Rico	52	24	14	4	3
Réunion	...	45	38	18	12
Romania	...	70	49	31	21
Rwanda	...	...	94	93	91
Saudi Arabia	...	...	64	48	40
Senegal	...	...	83	81	79
Sierra Leone	...	...	76	70	63
Singapore	...	8	3	2	1
Somalia	...	...	79	76	72
South Africa	64	30	33	17	14
Spain	52	35	26	17	11
Sri Lanka	...	52	55	53	52
St. Vincent	...	40	...	...	...
Sudan	...	86	77	71	61
Surinam	...	25	25	20	17
Swaziland	...	...	81	74	67
Sweden	33	14	8	6	4
Switzerland	21	11	8	6	4
Syria	...	50	50	32	25
Taiwan	...	50	48	28	24
Tanzania	...	...	90	86	81
Thailand	89	82	80	71	65
Togo	...	...	77	73	70
Trinidad & T.	...	20	19	10	8
Tunisia	...	58	42	35	25
Turkey	82	75	71	58	49
Uganda	...	...	89	86	81
UK	6	5	3	3	2
Uruguay	...	18	19	16	14
USA	...	7	4	4	2
USSR	50	39	26	20	14
Venezuela	50	32	26	16	11
Vietnam	...	...	77	68	61
Yemen, N.	...	...	76	69	33
Yemen, S.	...	...	51	41	63
Yugoslavia	78	57	50	32	23
Zaïre	...	86	79	72	66
Zambia	...	...	77	73	69
Zimbabwe	...	...	77	73	69

The figures also include people engaged in forestry, hunting and fishing. The economically active population includes people in work and seeking employment. The people working in agriculture does not include dependents unless the latter are working in the enterprises.

Food supply

calories per person per day

	1950	1960	1970	1980	1986-1988 average
World	...	...	2,488	2,600	2,671
Africa	...	...	2,239	2,327	2,261
Asia	...	...	2,133	2,320	2,450
Europe	...	...	3,302	3,388	3,459
North America	...	...	3,211	3,273	3,371
Oceania	...	...	3,096	3,099	3,149
South America	...	...	2,523	2,623	2,662
Afghanistan	...	2,040	...	...	...
Albania	...	...	2,533	...	...
Algeria	...	2,180	1,826	2,607	2,699
Angola	...	2,038	...	2,040	...
Antigua & Barb.	...	...	2,291	2,106	2,178
Argentina	2,970	2,810	3,351	3,239	3,168
Australia	3,240	3,140	3,226	3,287	3,347
Austria	2,670	2,970	3,477	3,370	3,474
Bahamas	...	...	2,623	2,512	2,680
Bangladesh	...	...	2,033	1,848	1,925
Barbados	...	...	2,894	3,122	3,188
Belgium	2,880	3,060	3,482	3,628	3,901
Belize	...	...	2,644	2,707	2,627
Benin	...	2,120	2,056	2,099	2,115
Bermuda	...	...	2,933	2,525	3,004
Bolivia	...	1,990	1,972	2,097	2,096
Botswana	...	...	2,176	2,148	2,251
Brazil	2,240	2,720	2,485	2,621	2,703
Brunei	...	...	2,305	2,810	2,839
Bulgaria	...	...	3,494	3,639	3,650
Burkina Faso	...	2,020	1,987	2,032	2,002
Burma	...	...	2,031	2,333	2,545
Burundi	...	...	2,363	2,344	2,320
Cameroon	...	2,130	2,151	2,145	2,142
Canada	3,110	3,020	3,369	3,310	3,451
Cape Verde	...	...	1,992	2,553	2,500
Central Afr. Rep.	...	2,120	2,158	2,119	1,965
Chad	...	2,180	2,135	...	...
Chile	2,450	2,480	2,697	2,637	2,581
China	...	1,870	2,092	2,328	2,637
Colombia	...	2,370	2,133	2,499	2,544
Comoros	...	...	2,223	2,071	2,059
Congo	...	2,120	2,181	2,441	2,519
Costa Rica	...	2,420	2,406	2,615	2,781
Cuba	...	...	2,576	2,839	3,103
Czechoslovakia	...	...	3,429	3,414	3,540
Denmark	3,160	3,260	3,385	3,524	3,605
Dominica	...	...	2,237	2,385	2,884
Dominican Rep.	...	1,930	1,971	2,323	2,359
Ecuador	...	1,990	1,986	2,059	2,302
Egypt	2,360	2,690	2,510	3,031	3,196
El Salvador	...	1,890	1,840	...	...
Ethiopia	...	2,110	...	...	...
Fiji	...	...	2,439	2,764	2,785
Finland	2,980	3,110	3,128	3,083	3,120
Fr. Guiana	...	...	2,509	2,553	2,778
Fr. Polynesia	...	...	2,841	2,845	2,856
France	2,800	3,090	3,391	3,232	3,312
Gabon	...	1,910	2,224	...	...
Gambia	...	2,300	2,348	2,176	2,339
Germany	2,730	2,990	3,300	3,450	3,650
Ghana	...	2,160	2,230	1,753	2,167
Greece	2,500	2,940	3,190	3,543	3,702
Grenada	...	...	2,342	2,294	2,959
Guadeloupe	...	...	2,329	2,502	2,713
Guatemala	...	2,050	2,062	2,194	2,327
Guinea	...	...	2,041	1,833	2,007
Guinea-Bissau	...	...	2,107	2,230	...
Guyana	...	...	2,293	2,409	2,423
Haiti	...	...	1,918	1,887	1,992
Honduras	...	2,160	2,151	2,176	2,138
Hong Kong	...	...	2,690	2,727	2,883
Hungary	...	3,030	3,336	3,493	3,635
Iceland	...	...	2,963	3,005	3,361
India	1,700	2,020	1,992	2,117	2,104
Indonesia	...	...	1,872	2,441	2,645
Iran	...	2,050	2,199	...	...
Iraq	...	1,920	...	...	...
Ireland	3,430	3,480	3,591	3,707	3,688
Israel	2,680	2,810	3,019	2,976	3,133
Italy	2,350	2,690	3,459	3,606	3,571
Ivory Coast	...	2,290	2,484	2,541	2,405
Jamaica	...	2,230	2,473	2,570	2,579
Japan	...	2,330	2,758	2,833	2,822
Jordan	...	2,220	2,374	...	...
Kenya	...	2,120	2,259	2,206	2,016
Kiribati	...	...	2,207	2,912	2,952
Korea, N.	...	...	2,501	3,059	3,172
Korea, S.	...	2,090	2,456	2,828	2,867
Kuwait	...	...	2,780	3,094	3,127
Laos	...	...	2,061	...	...
Lebanon	...	2,160	2,503	...	...
Lesotho	...	...	2,032	2,346	2,275
Liberia	...	...	2,167	2,380	2,344
Libya	...	1,770	2,366	3,664	3,393
Macau	...	...	2,181	2,185	2,233
Madagascar	...	2,300	2,469	2,485	2,174
Malawi	...	...	2,281	2,421	2,057
Malaysia	...	...	2,417	2,595	2,665
Mali	...	2,120	2,067	1,702	2,114
Malta	...	...	3,085	2,961	3,258
Martinique	...	...	2,363	2,646	2,844
Mauritania	...	...	1,975	2,052	2,465
Mauritius	...	2,330	2,343	2,723	2,690
Mexico	...	2,500	2,641	3,051	3,123
Mongolia	...	...	2,434	2,716	2,481
Morocco	...	2,080	2,453	2,763	2,808
Mozambique	...	...	2,089	1,801	1,604
Namibia	...	2,289	...	...	...
Nepal	...	...	2,020	2,000	2,034
Netherlands	2,950	3,160	3,431	3,306	3,303
Neths Antilles	...	...	2,448	2,765	2,794
New Caledonia	...	...	3,309	2,936	2,919
New Zealand	3,360	3,490	3,475	3,352	3,476
Nicaragua	2,300	...	...	...	...
Niger	...	...	2,074	2,370	2,321
Nigeria	...	2,180	2,232	2,246	2,083
Norway	3,110	2,930	3,101	3,327	3,266
Pakistan	...	2,090	2,018	2,244	2,167
Panama	...	2,350	2,443	2,338	2,484
Paraguay	...	2,520	2,753	2,770	2,784
Peru	2,270	2,260	2,251	2,166	2,277
Philippines	1,700	1,880	2,206	2,343	2,238
Poland	...	...	3,331	3,431	3,434
Portugal	2,270	2,530	3,069	3,034	3,284
Réunion	...	...	2,497	2,847	2,661
Romania	...	3,040	3,065	3,339	3,327
Rwanda	...	...	1,967	2,055	1,817
Samoa	...	...	2,088	2,403	2,474
Sao Tome	...	...	2,167	2,262	2,529
Saudi Arabia	...	...	1,906	2,819	2,805
Senegal	...	2,280	2,300	2,397	2,162
Seychelles	...	...	2,408	2,302	2,117
Sierra Leone	...	...	2,071	2,052	1,813
Singapore	...	1,700	2,682	2,691	2,882
Solomon Is.	...	...	2,114	2,146	2,140
Somalia	...	...	2,163	2,080	1,781
South Africa	2,640	2,820	2,767	2,932	2,963
Spain	2,400	2,820	2,867	3,332	3,494
Sri Lanka	1,900	2,080	2,308	2,225	2,297
St. Christ.-N.	...	...	2,145	2,308	2,822
St. Lucia	...	...	2,132	2,319	2,760
St. Vincent	...	...	2,247	2,491	2,764
Sudan	...	...	2,100	2,379	1,981
Surinam	...	1,920	2,337	2,588	2,775
Swaziland	...	...	2,203	2,504	2,554
Sweden	3,110	2,990	3,038	3,013	3,031
Switzerland	3,170	3,210	3,479	3,501	3,623
Syria	...	2,350	2,363	2,945	3,142
Taiwan	1,980	2,350	2,660	2,800	...
Tanzania	...	2,080	1,948	2,279	2,186
Thailand	...	...	2,160	2,312	2,288
Togo	...	...	2,191	2,173	2,110
Tonga	...	...	2,589	2,922	2,964
Trinidad & T.	...	...	2,385	2,882	2,983
Tunisia	...	1,730	2,264	2,759	2,911
Turkey	2,510	3,110	2,829	3,113	3,084
UAE	2,360	2,690	3,130	3,595	3,489
Uganda	...	2,090	2,260	2,135	2,034
UK	3,130	3,270	3,356	3,257	3,259
Uruguay	2,900	3,200	2,982	2,793	2,746
USA	3,200	3,120	3,497	3,529	3,644
USSR	...	...	3,348	3,362	3,382
Vanuatu	...	...	2,408	2,417	2,533
Venezuela	2,030	2,300	2,335	2,662	2,534
Vietnam	2,121	2,172	2,000	...	...
Yemen	...	...	1,900	2,210	2,300
Yugoslavia	2,600	2,970	3,340	3,598	3,570
Zaïre	...	1,920	2,218	2,126	2,079
Zambia	...	...	2,194	2,204	2,028
Zimbabwe	2,450	...	2,055	2,152	2,193

Tractors

thousands

	1950	1970	1980	1987	1988
World	6,046	15,483	21,742	25,535	25,865
Africa	95	330	455	547	557
Asia	38	805	3,550	4,937	5,101
Europe	969	6,104	8,465	10,118	10,2840
North America	4,041	5,382	5,621	5,675	5,689
Oceania	162	429	429	419	418
South America	86	455	660	1,102	1,124
Afghanistan	...	0.6	0.8	0.8	0.8
Albania	0.3	6.2	11	11	12
Algeria	12	47	44	86	90
Amer. Samoa	...	0.02	0.01	0.01	0.01
Angola	...	7.2	10	10	10
Antigua & Barb.	0.05	0.1	0.2	0.2	0.2
Argentina	50	172	167	208	210
Australia	121	329	332	332	332
Austria	18	249	320	326	326
Bahamas	...	0.03	0.07	0.08	0.08
Bangladesh	...	1.9	4.2	5	5.1
Barbados	0.2	0.4	0.6	0.6	0.6
Belgium	14	98	116	120	117
Belize	...	0.2	1.3	1	1.1
Benin	...	0.07	0.1	0.1	0.1
Bermuda	0.01	0.05	0.04	0.05	0.05
Bolivia	0.1	0.4	4.5	4.9	4.7
Botswana	0.07	1.6	2.1	3.1	3.3
Brazil	21	168	330	694	715
Brunei	0.1	0.01	0.03	0.07	0.07
Bulgaria	8.9	54	62	54	54
Burkina Faso	...	0.04	0.07	0.1	0.1
Burma	0.05	5.1	9.3	11	12
Burundi	...	0.004	0.05	0.05	0.05
Cambodia	0.06	1.3	1.4	1.4	1.4
Cameroon	...	0.2	0.5	0.9	1
Canada	400	596	657	742	756
Cape Verde	...	0.02	0.03	0.02	0.02
Central Afr. Rep.	...	0.07	0.2	0.2	0.2
Chad	...	0.09	0.2	0.2	0.2
Chile	7.2	26	35	41	37
China	1.3	135	745	892	876
Colombia	14	23	28	34	35
Congo	...	0.6	0.7	0.7	0.7
Cook Is.	...	0.09	0.1	0.1	0.1
Costa Rica	0.5	5.1	6	6.3	6.4
Cuba	8.9	48	68	59	56
Cyprus	0.5	6.8	11	14	14
Czechoslovakia	27	136	137	140	141
Denmark	22	174	189	165	168
Dominica	...	0.07	0.09	0.09	0.09
Dominican Rep.	0.5	2.5	3.2	2.3	2.3
Ecuador	1.2	3.1	6.2	8.2	8.4
Egypt	...	17	36	45	46
El Salvador	...	2.5	3.3	3.4	3.4
Equat. Guinea	...	0.07	0.1	0.1	0.1
Ethiopia	...	2.9	3.9	3.9	3.9
Falkland Is.	0.01	0.09	0.1	0.1	0.1
Fiji	0.06	1.2	1.6	4.3	4.3
Finland	15	156	212	240	240
Fr. Guiana	0.01	0.04	0.1	0.2	0.2
Fr. Polynesia	0.02	0.1	0.1	0.2	0.2
France	140	1,239	1,504	1,520	1,518

	1950	1970	1980	1987	1988
Gabon	...	0.8	1.3	1.4	1.4
Gambia	...	0.06	0.04	0.04	0.04
Germany	180	1,516	1,598	1,635	1,628
Ghana	0.02	2.7	3.5	3.9	3.9
Greece	7.1	62	140	186	187
Greenland	...	0.2	0.08	0.09	0.09
Grenada	0.02	0.02	0.03	0.03	0.03
Guadeloupe	0.1	0.5	0.9	1.6	1.6
Guam	0.06	0.07	0.08	0.08	0.08
Guatemala	...	3.2	4	4.1	4.2
Guinea	...	0.05	0.1	0.2	0.3
Guinea-Bissau	...	0.02	0.03	0.05	0.05
Guyana	0.3	3.3	3.5	3.6	3.6
Haiti	...	0.4	0.5	0.6	0.6
Honduras	0.3	0.6	3.3	3.4	3.4
Hungary	13	67	55	54	53
Iceland	1.9	9.9	13	13	13
India	8.1	111	418	698	751
Indonesia	0.1	8.5	13	13	14
Iran	...	20	75	111	113
Iraq	0.4	14	23	42	43
Ireland	16	84	140	162	164
Israel	3.1	16	27	25	25
Italy	61	619	1,072	1,315	1,363
Ivory Coast	...	1.4	3.1	3.4	3.5
Jamaica	...	1.7	2.8	3	3
Japan	0.7	278	1,471	1,904	1,985
Jordan	0.1	2.8	4.5	5.7	5.7
Kenya	2.9	6.4	6.5	9.4	9.5
Kiribati	...	0.01	0.02	0.02	0.02
Korea, N.	0.2	20	30	75	77
Korea, S.	...	0.1	2.6	20	25
Kuwait	...	...	0.02	0.1	0.1
Laos	...	0.3	0.5	0.8	0.8
Lebanon	0.1	2.5	3	3	3
Lesotho	...	0.4	1.4	1.7	1.8
Liberia	...	0.2	0.3	0.3	0.3
Libya	...	3.9	14	31	32
Liechtenstein	0.2	0.5	0.4	0.4	0.4
Madagascar	0.7	2.4	2.7	2.8	3
Malawi	0.1	0.8	1.2	1.4	1.4
Malaysia	0.1	4.6	8.1	12	12
Mali	...	0.4	0.8	0.8	0.8
Malta	0.04	0.1	0.4	0.4	0.4
Martinique	0.1	0.4	0.9	0.8	0.8
Mauritania	...	0.3	0.3	0.3	0.3
Mauritius	0.3	0.3	0.3	0.3	0.4
Mexico	23	91	115	163	165
Mongolia	0.1	5.4	9.7	12	12
Montserrat	...	0.02	0.01	0.01	0.01
Morocco	6.2	12	25	33	34
Mozambique	0.5	4.2	5.8	5.8	5.8
Namibia	0.3	2.1	2.6	2.9	3
Nepal	...	0.4	0.5	2.9	2.9
Netherlands	23	134	178	192	194
Neths Antilles	0.03	0.1	0.1	0.2	0.2
New Caledonia	0.06	0.5	1	1.3	1.3
New Zealand	40	96	92	78	78
Nicaragua	0.5	0.5	2.2	2.5	2.5
Niger	...	0.04	0.2	0.2	0.2
Nigeria	...	2.9	8.6	10.1	11
Niue	...	0.007	0.01	0.01	0.01
Norfolk I.	...	0.01	0.01	0.01	0.01

	1950	1970	1980	1987	1988
Norway	12	91	131	151	150
Oman	...	...	0.1	0.1	0.1
Pacific Is.	...	0.04	0.05	0.05	0.05
Pakistan	1.8	21	71	175	180
Panama	0.4	2.4	4	6.2	6.2
Papua New G.	...	1.4	1.4	1.2	1.2
Paraguay	0.1	2.2	3.2	10	11
Peru	2.9	11	14	16	16
Philippines	1.4	7.8	17	20	21
Poland	30	222	620	1,044	1,101
Portugal	2.9	29	72	79	77
Puerto Rico	2.1	5.4	3.7	2	2
Qatar	...	...	...	0.09	0.09
Réunion	0.06	0.3	1.2	2.1	2.2
Romania	12	108	147	184	183
Rwanda	...	0.06	0.08	0.1	0.1
Samoa	...	0.02	0.03	0.04	0.04
Sao Tome	...	0.1	0.1	0.1	0.1
Saudi Arabia	...	0.6	1.2	1.8	1.9
Senegal	...	0.3	0.5	0.5	0.5
Seychelles	...	0.02	0.03	0.04	0.04
Sierra Leone	...	0.1	0.3	0.5	0.5
Singapore	...	0.02	0.04	0.06	0.06
Somalia	0.3	0.9	1.7	2.1	2.1
South Africa	48	155	180	183	183
Spain	15	261	524	702	720
Sri Lanka	0.3	13	24	29	29
St. Christ.-N.	0.09	0.2	0.2	0.2	0.2
St. Lucia	0.02	0.04	0.04	0.09	0.09
St. Vincent	0.01	0.06	0.08	0.08	0.08
Sudan	...	5	11	20	21
Surinam	0.2	0.9	1.4	1.2	1.3
Swaziland	0.1	1.2	2.9	3.3	3.3
Sweden	63	179	181	183	182
Switzerland	18	73	95	108	109
Syria	0.9	9.1	28	52	55
Taiwan	0.3	...	...	...	...
Tanzania	0.9	5	18.6	19	19
Thailand	0.6	8	73	136	142
Togo	...	0.06	0.2	0.4	0.4
Tonga	...	0.06	0.06	0.1	0.1
Trinidad & T.	0.3	1.9	2.4	2.6	2.6
Tunisia	5.2	21	34	26	26
Turkey	20	106	436	636	655
Uganda	0.1	1.3	2.2	4	4.2
UK	302	456	512	519	518
Uruguay	16	27	33	35	35
US Virgin Is.	...	0.4	0.3	0.08	0.08
USA	3,640	4,617	4,740	4,670	4,670
USSR	595	1,977	2,562	2,735	2,692
Vanuatu	...	...	0.06	0.06	0.06
Venezuela	3.9	19	38	46	47
Vietnam	...	3.6	37	36	36
W. Sahara	...	0.01	0.01	0.01	0.01
Yemen	1.5	1.6	3.3	2.5	5.4
Yugoslavia	6.3	82	426	1,017	1,066
Zaïre	0.2	1	1.9	2.3	2.3
Zambia	1	3.1	4.6	4.4	4.4
Zimbabwe	4.2	17	20	20	20

The figures refer to all tractors, except garden tractors, used in agriculture.

Fertilizer consumption

kg per hectare of agricultural land

	1970	1980	1987	1988
World	15.1	25.2	30	31.1
Africa	1.6	3.4	3.6	3.8
Asia	11.5	27.8	41.5	45.9
Europe	106	137	142	144
North America	28.5	40.8	35.9	35.7
Oceania	2.8	3.4	3.5	3.8
South America	3	9	9.9	9.4
Afghanistan	1.4	1.3	2.1	1.4
Albania	37.8	69	85.5	86.4
Algeria	2.5	5.4	7.5	4.5
Angola	0.4	0.5	0.3	0.3
Argentina	0.5	0.6	0.9	0.9
Australia	2	2.5	2.8	3.1
Austria	105	111	85.5	91.5
Bahamas	...	109	41.7	41.7

	1970	1980	1987	1988
Bahrain	...	...	250	66.7
Bangladesh	14.8	42.8	72.3	81
Barbados	154	157	83.8	83.8
Belgium	298	292	276	276
Belize	50.8	19.7	43.7	43.1
Benin	1.7	0.4	4	2.8
Bhutan	...	0.3	0.3	0.3
Bolivia	0.1	0.1	0.4	0.2
Brazil	5.1	18.1	16.8	15.1
Brunei	...	...	108	30.8
Bulgaria	106	134	121	147
Burkina Faso	0	0.3	1.3	1.1
Burma	2.1	9.6	14.9	10.3
Burundi	0.3	0.5	1.2	1.2
Cambodia	1	2.2	0.2	...
Cameroon	1.3	2.3	3.3	2.3
Canada	11.9	27.6	28.4	27.5
Central Afr. Rep.	0.4	0.3	0.2	0.1

	1970	1980	1987	1988
Chad	...	...	0.1	0.1
Chile	9.1	7.6	16.7	18.3
China	12.6	39.7	55.1	60.9
Colombia	6.5	8.8	11.1	11.5
Congo	0.5	...	0.2	0.1
Costa Rica	26.6	28.2	33.6	35.6
Cuba	78.9	92.5	107	94.6
Cyprus	52.3	28.3	123	133
Czechoslovakia	181	253	230	236
Denmark	201	216	216	220
Djibouti	...	6	0.5	...
Dominica	...	158	158	158
Dominican Rep.	16.2	14.7	23	21.3
Ecuador	5.7	11.7	11.1	9.3
Egypt	131	271	351	400
El Salvador	52.1	45.2	68.9	60.8
Equat. Guinea	5.8	0.3	0.3	1.1
Ethiopia	0.1	0.7	0.9	1.1

Fertilizer consumption *continued*

	1970	1980	1987	1988
Fiji	20.5	55.4	71.7	84
Finland	172	191	206	184
Fr. Guiana	...	14.3	76.9	70.3
Fr. Polynesia	...	7.5	9.5	9.5
France	143	178	186	194
Gabon	...	...	0.3	0.2
Gambia	0.5	8.2	10	13.2
Germany	250	280	265	265
Ghana	0.2	1.9	1.7	2
Greece	36.8	57.4	73.1	70.6
Guadeloupe	124	125	83.3	190
Guatemala	18.7	27.7	39.1	39.4
Guinea	0.3	0.1	0.2	0.2
Guinea-Bissau	...	0.2	0.2	...
Guyana	7.3	3.5	7.7	8.3
Haiti	0.2	0.3	1.6	1.6
Honduras	8.5	5.5	9.5	8.7
Hungary	122	211	211	225
Iceland	10.2	12.8	10.2	9.6
India	12.2	28.9	50	60.9
Indonesia	8.4	37.2	68.6	72.5
Iran	3.5	9.9	16.6	17.5
Iraq	1.9	9.8	22.9	26
Ireland	87.3	104	118	122
Israel	46.7	64.4	78	174
Italy	66.3	120	135	123
Ivory Coast	1.3	7.7	3.3	6.2
Jamaica	45	37.2	53	63.1
Japan	355	333	382	365
Jordan	1.9	28.9	12	23.3
Kenya	8.9	10.2	16	20.2
Korea, N.	151	318	305	331
Korea, S.	240	358	406	384
Kuwait	...	3.3	2.4	5.6
Laos	1	2.4	0.3	0.2

	1970	1980	1987	1988
Lebanon	131	74.3	65	72.7
Lesotho	0.1	2	1.5	1.9
Liberia	3.8	5.1	4.4	5.5
Libya	1.3	5.2	4	5.7
Madagascar	0.4	0.2	0.5	0.3
Malawi	2.8	8.0	11.5	12.1
Malaysia	53.4	105	159	150
Mali	0.1	0.4	0.9	0.4
Malta	45.6	115	45.8	41.8
Martinique	271	247	426	509
Mauritius	196	233	288	249
Mexico	6.1	12.7	18.7	17.7
Mongolia	...	0.1	0.2	0.2
Morocco	4.4	12.7	10.3	11
Mozambique	0.1	0.6	0.1	...
Nepal	1.3	5.2	12.4	13
Netherlands	296	336	316	301
New Caledonia	...	5.2	4	4
New Zealand	33.2	31.8	25.8	26.1
Nicaragua	9.7	8.8	8.4	11
Niger	...	0.2	0.2	0.1
Nigeria	0.2	3.4	5.9	6
Norway	209	277	242	226
Oman	...	1	4.2	3.9
Pakistan	11.6	42.8	66.8	67.2
Panama	12.5	17.6	19.9	20
Papua New G.	4.6	11.4	31.1	29.7
Paraguay	0.6	0.4	0.4	0.4
Peru	2.8	3.9	7.5	7
Philippines	25.7	26.3	56.4	54.9
Poland	132	185	174	193
Portugal	30	63.5	86	65
Qatar	...	15.4	12.8	10.9
Réunion	180	70.7	200	205
Romania	39.8	81.7	92.1	89.7
Rwanda	0.1	0.1	1.5	0.3
Saudi Arabia	0.1	0.5	5	6.4

	1970	1980	1987	1988
Senegal	1.0	1.8	1.9	2.4
Sierra Leone	0.4	0.5	0.1	0.1
Singapore	250	550	1,833	2,800
Somalia	0.1	...	0.1	0.1
South Africa	5.6	11.2	7.5	8.7
Spain	37.9	53.2	65.7	65.2
Sri Lanka	38.8	63.9	91.8	90.4
St. Christ.-N.	140	140	194	207
St. Lucia	142	43	81	81
St. Vincent	137	205	200	190
Sudan	1.1	1.2	0.5	1.2
Surinam	45.5	22.3	125	73.9
Swaziland	4.2	11.5	4.9	5.6
Sweden	146	131	114	107
Switzerland	67.7	89.7	87.8	87.9
Syria	2.9	9	16.4	21.1
Taiwan	0.7	1.5	...	...
Tanzania	0.3	0.9	1.2	1
Thailand	5.7	15.9	28.2	37.2
Togo	0.2	1.6	6.6	6.1
Trinidad & T.	60.7	47.2	48.1	16.4
Tunisia	4.5	8.2	13.4	12.9
Turkey	7.8	38.1	47.1	44.4
UAE	...	14.3	14.2	41.4
Uganda	0.7	0.1	0.1	...
UK	99.5	111	134	130
Uruguay	3.3	5.4	4	4.5
US Virgin Is.	60	68.8	81.3	81.3
USA	35.9	50.2	41.1	41.2
USSR	15.4	31	45.4	45
Venezuela	2.8	11.5	28.5	32.2
Vietnam	31.4	20.4	60	76.8
Yemen	...	0.8	0.9	2.1
Yugoslavia	43.2	57.7	73	72
Zaïre	0.1	0.5	0.2	0.1
Zambia	1	2	2.4	2.1
Zimbabwe	14.8	23.5	18.3	19.9

Total consumption of fertilizers

thousand tonnes

	1970	1980	1987	1988	Rank	%
World	69,245	116,473	139,518	145,642		
Africa	1,616	3,316	3,476	3,752		
Asia	11,740	30,665	46,227	51,851		
Europe	24,886	31,191	31,811	32,272		
North America	17,614	25,604	23,106	22,909		
Oceania	1,421	1,651	1,757	1,869		
South America	1,656	5,290	5,737	5,829		
Algeria	112	236	241	170		
Argentina	87	116	161	167		
Australia	964	1,162	1,349	1,456		
Austria	408	407	334	319		
Bangladesh	143	417	706	799		
Belgium	517	447	417	416		
Brazil	1,002	4,201	3,759	3,729	7	2.6
Bulgaria	639	830	745	903		
Burma	22	100	125	107		
Canada	802	1,906	2,200	2,159		
Chile	153	132	303	324		
China	4,220	15,335	22,688	25,322	2	17.4
Colombia	144	312	502	525		
Cuba	396	530	654	600		

	1970	1980	1987	1988	Rank	%
Czechoslovakia	1,282	1,730	1,556	1,593		
Denmark	598	627	606	614		
Egypt	373	664	897	1,034		
Finland	486	489	522	473		
France	4,651	5,609	5,818	5,998	5	4.1
Germany	4,763	5,169	4,810	4,876	6	3.3
Greece	337	527	607	648		
Hungary	837	1,399	1,373	1,463		
India	2,177	5,231	8,734	11,052	4	7.6
Indonesia	240	1,173	2,266	2,393	10	1.6
Iran	95	572	976	1,031		
Iraq		92	216	246		
Ireland	423	601	670	691		
Italy	1,338	2,111	2,303	2,093		
Japan	2,139	1,816	2,037	1,943		
Korea, N.	309	729	746	812		
Korea, S.	563	803	840	855		
Malaysia	193	453	699	735		
Mexico	538	1,238	1,859	1,758		
Morocco	88	259	318	326		
Netherlands	650	679	636	605		
New Zealand	448	464	370	372		
Nigeria	7	174	293	312		
Norway	119	259	232	218		

	1970	1980	1987	1988	Rank	%
Pakistan	283	1,080	1,720	1,740		
Peru	84	118	232	215		
Philippines	201	334	486	505		
Poland	2,572	3,499	3,277	3,625	8	2.5
Portugal	129	259	283	295		
Romania	594	1,223	1,390	1,355		
Saudi Arabia	5	41	434	548		
South Africa	558	1,064	713	827		
Spain	1,216	1,662	2,021	1,994		
Sri Lanka	105	165	207	211		
Sweden	503	484	401	366		
Switzerland	148	181	177	178		
Syria	40	126	227	291		
Taiwan	679	1,360	1,286	1,338		
Thailand	81	296	587	778		
Turkey	431	1,456	1,778	1,614		
UK	1,894	2,054	2,484	2,416	9	1.7
Uruguay	69	81	61	67		
USA	15,535	21,480	17,792	17,772	3	12.2
USSR	10,312	18,756	27,403	27,187	1	18.7
Venezuela	100	241	611	692		
Vietnam	311	196	421	530		
Yugoslavia	631	824	1,031	1,015		
Zimbabwe	106	173	140	152		

Energy

million tonnes of coal-equivalent
tonnes of c-e per person

	Production		Imports		Exports		Consumption per capita		Consumption total
	1970	1989	1970	1989	1970	1989	1970	1989	1989
World		10,611.08		3,404.61		3,339.79		1.96	10,176.64
Africa		645.76		73.99		426.05		0.41	254.88
Asia		2,908.61		970.18		1,256.78		0.79	2,414.4
Europe		1,367.37		1,479.34		568.34		4.43	2,200.46
North America		2,695.21		738.81		413.72		7.09	2,989.93
Oceania		207.12		28.68		92.47		5.41	141.33
South America		430.26		83.48		175.69		1.03	300.45
Afghanistan	3.44	4.18	0.33	0.89	3.08	1.37	0.05	0.23	3.64
Albania	2.65	6.31	0.03	0.25	0.12	0.08	0.6	1.28	4.07
Algeria	72.45	132.01	0.96	1.22	70.16	101.93	0.37	0.94	22.7
Amer. Samoa	0	0	0.06	0.26	0	0	2.44	3.21	0.12
Angola	7.49	32.73	0.39	0.02	6.38	30.35	0.16	0.09	0.87
Antigua & Barb.	0	0	0.88	0.2	0.28	0.01	2	1.75	0.14
Argentina	37.29	64.99	4.53	6.13	0.15	2.91	1.58	1.94	61.86
Australia	59.65	192.99	26.8	19.32	17.2	91.12	4.83	7.24	120.6
Austria	11.15	8.52	15.48	22.9	0.84	1.2	3.27	4.02	30.45
Bahamas	0	0	4.58	4.31	2.95	2.98	4.07	2.55	0.64
Bahrain	6.03	9.97	12.64	14.61	14.31	13.01	2.64	15.6	7.79
Bangladesh	...	5.41	...	3.13	0	0	...	0.07	7.74
Barbados	0.01	0.11	0.47	0.49	0	0.004	0.82	1.61	0.41
Belgium	11.46	8.45	64.98	83.76	12.87	27.37	5.61	5.81	57.22
Belize	0	0	0.06	0.1	0	0	0.49	0.46	0.09
Benin	...	0.42	...	0.26	0	0.43	0.05	0.05	0.22
Bermuda	0	0	0.31	0.43	0	0	1.96	6.57	0.38
Bolivia	1.76	5.61	0.02	0.001	0.86	2.94	0.23	0.38	2.7
Br. Virgin Is.	0	0	0.01	0.02	0	0	0.9	1.77	0.02
Brazil	18.69	79.1	28.05	59.78	1.5	8.02	0.44	0.8	117.61
Brunei	10.07	22.64	0.02	0.05	9.79	19.23	2.18	13.78	3.56
Bulgaria	16.15	18.02	16.58	33.82	0.4	3.78	3.77	4.89	44.03
Burkina Faso	0	0	...	0.26	0	0	0	0.03	0.25
Burma	1.3	2.91	0.49	0.05	0	0	0.06	0.06	2.53
Burundi	0	0.02	0	0.08	0	0	0.01	0.02	0.1
Cameroon	0.14	11.77	0.37	0.02	0	9.32	0.07	0.25	2.9
Canada	204.98	369.51	74.8	61.24	84.45	150.47	8.81	10.91	287.06
Cape Verde	0	0	0	0.04	0	0	0.05	0.1	0.04
Cayman Is.	0	0	0.02	0.13	0	0	2.1	4.32	0.11
Central Afr. Rep.	0.005	0.009	0.11	0.11	0	0	0.05	0.05	0.14
Chad	0	0	0.09	0.13	0	0	0.02	0.02	0.1
Chile	6.11	7.35	4.48	8.79	0.02	0.26	1.15	1.21	15.63
China	304	971.91	0.62	7.66	2.15	69.99	0.35	0.81	892.48
Colombia	21.79	56.54	0	1.88	7.99	29.32	0.61	0.82	26.45
Comoros	...	0	0.01	0.03	0	0	0.04	0.05	0.03
Congo	0.05	10.58	0.17	0.03	0.03	9.8	0.15	0.36	0.78
Costa Rica	0.12	0.41	0.64	1.23	0.07	0.12	0.37	0.5	1.46
Cuba	0.24	1.08	8.74	19.91	0	3.4	0.99	1.53	16.09
Cyprus	0	0	0.81	2.29	0	0	1.21	2.52	1.75
Czechoslovakia	63.75	63.84	22.21	44.96	6.69	8.68	5.39	5.96	93.13
Denmark	0.05	11.54	33.61	23.45	3.22	9.29	5.44	4.41	22.68
Djibouti	0	0	1.07	0.74	0	0	0.3	0.38	0.1
Dominica	0	0.002	0.01	0.03	0	0	0.18	0.37	0.03
Dominican Rep.	0.1	0.12	1.35	2.78	0	0	0.32	0.4	2.77
Ecuador	0.38	21.6	1.45	0.38	0.06	14.66	0.28	0.66	6.84
Egypt	24.55	73.09	5.89	2.6	21.36	32.96	0.26	0.75	38.27
El Salvador	0.06	0.23	0.64	1.04	0	0.04	0.19	0.23	1.16
Equat. Guinea	0	0	0.02	0.05	0	0	0.06	0.15	0.05
Ethiopia	0.03	0.08	0.02	1.59	0	0.3	0.06	0.02	1.15
Faeroe Is.	0	0.009	0.12	0.28	0	0	3.28	6.06	0.29
Fiji	0	0.04	0.45	0.6	0.02	0.12	0.45	0.48	0.36
Finland	1.18	5.85	22.46	26.41	0.3	1.57	4.07	5.78	28.68
Fr. Guiana	0	0	0.05	0.24	0	0	0.96	2.34	0.22
Fr. Polynesia	0	0.008	0.21	0.39	0	0	0.82	1.44	0.29
France	60.8	66.58	175	184.15	14.42	22.72	3.84	3.95	220.9
Gabon	7.91	15.77	0.01	0.05	7.33	12.74	0.46	1.25	1.42
Gambia	0	0	0.02	0.09	0	0.001	0.05	0.11	0.09
Germany	254.96	243.07	228.64	257.11	40.2	34.74	5.4	5.86	453.95
Ghana	0.35	0.59	1.32	1.78	0.13	0.13	0.17	0.11	1.58
Gibraltar	0	0	0.24	0.69	0	0	0.77	0.6	0.02

Production, trade and consumption

continued

	Production		Imports		Exports		Consumption per capita		Consumption total
	1970	1989	1970	1989	1970	1989	1970	1989	1989
Greece	2.03	11.41	10.03	29.12	0.23	8.13	1.11	3.12	31.31
Greenland	0.02	0	0.16	0.27	0	0.01	3.68	4.63	0.26
Grenada	0	0	0.02	0.04	0	0	0.22	0.43	0.04
Guadeloupe	0	0	0.15	0.62	0	0	0.45	1.8	0.62
Guam	0	0	1.09	0.84	0.13	0	6.49	5.8	0.68
Guatemala	0.04	0.51	1.14	1.63	0	0.14	0.19	0.2	1.76
Guinea	0	0.02	0.36	0.49	0	0	0.09	0.09	0.49
Guinea-Bissau	0	0	0.03	0.08	0	0	0.05	0.07	0.07
Guyana	0	0.001	0.74	0.32	0	0	1.04	0.39	0.31
Haiti	...	0.04	0.17	0.32	0	0	0.04	0.05	0.33
Honduras	0.02	0.11	1.11	0.81	0.5	0	0.23	0.2	0.89
Hong Kong	0	0	5.49	19.09	0.07	3.5	0.96	2	11.54
Hungary	20.66	20.76	11.72	22.04	1.53	6.11	2.81	3.67	38.76
Iceland	0.18	0.56	0.75	1.02	0	0	4.11	5.66	1.43
India	67.76	236.1	17.95	40.37	0.42	0.34	0.14	0.31	256.88
Indonesia	63.91	138.7	0.5	12.01	44.78	91.69	0.12	0.27	49.58
Iran	295	235.27	0	6.52	257	146.7	0.96	1.53	81.84
Iraq	112	204.48	0.05	0.01	107	174.16	0.61	1.06	19.36
Ireland	1.76	4.74	7.86	9.19	1.02	0.66	2.58	3.63	13.39
Israel	7.46	0.08	1.55	16.58	0.29	1.47	2.14	3.03	13.72
Italy	25.8	32.01	182	201.82	34.89	19.89	2.65	3.81	217.76
Ivory Coast	0.03	0.43	1.23	3.63	0.12	0.36	0.2	0.18	2.04
Jamaica	0.02	0.01	3.04	2.41	0.22	0.1	1.21	0.85	2.07
Japan	...	47.12	...	488.48	...	8.48	...	4.03	495.9
Jordan	...	0.03	0.77	4.5	0	0	0.27	1.01	3.91
Kenya	0.04	0.34	3.59	3.26	1.38	0.8	0.1	0.11	2.45
Kiribati	0	0	...	0.01	0	0	...	0.15	0.01
Korea, N.	26.63	52.2	1.92	7.24	0.09	0.05	2.05	2.8	59.93
Korea, S.	9.53	19.74	13.05	92.52	0.63	5.43	0.67	2.2	93.11
Kuwait	224	118.38	0	4.79	208	98.41	4.77	8.28	16.36
Laos	0	0.14	0.27	0.11	0	0.09	0.09	0.04	0.16
Lebanon	0.11	0.06	3.18	4.04	0.58	0	0.68	1.45	3.88
Liberia	0.03	0.04	0.76	0.46	0	0.03	0.48	0.15	0.38
Libya	232.37	88.62	0.7	0.05	232.03	67.59	0.67	4.01	17.6
Luxembourg	0.11	0.1	5.84	4.81	0	0.12	17.05	12.48	4.64
Macau	0	0	0.09	0.49	0	0	0.36	1.06	0.49
Madagascar	0.01	0.04	0.87	0.48	0.34	0.09	0.07	0.04	0.45
Malawi	0.02	0.07	0.2	0.28	0	0	0.04	0.04	0.34
Malaysia	1.5	60.35	15.09	9.82	8.53	45.99	0.56	1.38	24.07
Mali	0	0.02	0.11	0.22	0	0	0.02	0.03	0.22
Malta	0	0	0.46	0.84	0	0	0.92	2.05	0.72
Martinique	0	0	0.18	0.9	...	0.27	0.53	1.74	0.59
Mauritania	0	0.003	0.2	1.56	0	0	0.16	0.72	1.41
Mauritius	0.01	0.01	0.3	0.82	0	0	0.33	0.44	0.47
Mexico	53.81	250.89	3.65	7.83	5.27	101.22	1.04	1.69	146.49
Mongolia	0.72	3.12	0.37	1.13	0	0.26	0.87	1.87	3.99
Morocco	0.73	0.79	2.21	9.43	0.08	0.002	0.19	0.37	8.99
Mozambique	0.03	0.05	1.66	0.55	0.53	0	0.14	0.03	0.5
Nauru	0	0	0.03	0.07	0	0	4.43	5.9	0.06
Nepal	0	0.07	0.15	0.38	0	0.04	0.01	0.02	0.44
Netherlands	45.22	83.24	104	130.73	58.32	105.03	4.59	6.64	98.62
Neths Antilles	0	0	71.55	16.26	57.51	11.16	33.81	9.79	1.83
New Caledonia	0.03	0.06	1.05	0.69	0	0.007	9.44	4.28	0.71
New Zealand	3.7	13.96	5.04	4.43	0.08	1.21	2.79	5.06	17.02
Nicaragua	0.04	0.07	0.66	1.03	0	0.001	0.32	0.28	1.05
Niger	...	0.16	0.09	0.32	0	0	0.02	0.06	0.45
Nigeria	79.19	127.83	1	5.31	74.5	109.07	0.05	0.19	20.19
Norway	7.48	166.29	16.01	5.95	3.39	140.1	4.55	7.18	30.18
Oman	24.11	48.96	0.63	0.11	24.11	43.81	0.17	3.43	4.97
Pacific Is.	0	0.004	0.08	0.13	0	0	0.62	0.64	0.11
Pakistan	...	20.47	...	12.49	...	0.38	...	0.27	31.74
Panama	...	0.27	...	4.9	...	0.46	...	0.59	1.4
Papua New Guinea	0.02	0.06	0.35	1.12	0	0.01	0.14	0.29	1.12
Paraguay	0	0.34	0	0.77	0	0.15	0	0.22	0.91
Peru	6.43	12.11	2	2.25	0.34	3.24	0.61	0.5	10.64
Philippines	0.29	2.76	13.3	17.69	1.12	0.38	0.26	0.3	17.98
Poland	131	167.63	15.42	36.56	28.03	28.46	3.56	4.53	173.11
Portugal	0.99	0.87	7.85	22.46	0.69	2.65	0.74	1.81	18.59
Puerto Rico	0.03	0.03	14.35	10.1	5.05	0.61	3.46	2.81	9.66
Qatar	26.6	36.71	0.08	0	25.29	28.07	13.26	24.1	8.53
Reunion	0.01	0.07	0.13	0.45	0	0.001	0.31	0.84	0.49
Romania	64.85	78.74	6.18	47.89	7.65	19.66	2.99	4.49	103.9

Production, trade and consumption continued

	Production		Imports		Exports		Consumption per capita		Consumption total
	1970	1989	1970	1989	1970	1989	1970	1989	1989
Rwanda	0.01	0.02	0.03	0.2	0	0	0.01	0.03	0.21
Samoa	0	0.002	0.02	0.06	0	0	0.11	0.36	0.06
Saudi Arabia	277	417.1	0.73	0.57	251	295.47	0.47	6.36	86.49
Senegal	0	0	2.36	1.85	0.05	0.01	0.12	0.2	1.39
Seychelles	0	0	0.03	0.22	0	0	0.29	0.93	0.06
Sierra Leone	0	0	0.61	0.48	0	0.003	0.12	0.08	0.31
Singapore	0	0	29.42	76.84	17.38	43.24	2.01	4.98	13.4
Solomon Is.	0	0	0.02	0.08	0	0	0.11	0.24	0.08
Somalia	0	0	0.11	0.58	0	0.06	0.04	0.06	0.42
South Africa	48.52	133.1	16.61	23.33	1.84	42.87	2.19	2.64	105.7
Spain	14.12	29.13	49.75	93.56	17.94	13.89	1.47	2.49	97.11
Sri Lanka	0.09	0.33	2.97	2.77	0.14	0.2	0.14	0.11	1.87
St Christopher-Nevis	...	0	...	0.03	0	...	...	0.71	0.03
St. Lucia	...	0	...	0.08	0	0	0.32	0.53	0.08
St. Pierre & M.	...	0	...	0.13	0	0	3.2	7	0.05
St. Vincent	...	0.004	...	0.04	...	0	0.17	0.36	0.04
Sudan	0.01	0.06	2.35	1.81	...	0.05	0.15	0.06	1.52
Surinam	0.12	0.43	0.74	0.42	0	0.07	2.32	1.26	0.52
Sweden	5.14	16.95	49.06	37.83	2.78	12.97	5.77	5.07	42.72
Switzerland	3.91	6.47	20.04	22.02	1.57	2.56	3.21	3.66	24.22
Syria	6.18	27.07	1.86	1.42	5.15	13.96	0.38	0.96	11.59
Tanzania	0.04	0.08	1.68	1.01	0.83	0.06	0.06	0.04	0.95
Thailand	0.37	14.21	7.34	22.6	0.05	0.96	0.18	0.64	34.99
Togo	0	0.001	0.13	0.25	0	0.02	0.06	0.07	0.24
Tonga	0	0	0.01	0.04	0	0	0.14	0.37	0.04
Trinidad & Tobago	13.01	16.06	23.14	1.26	27.99	9.52	4.4	5.65	7.13
Tunisia	6.05	7.66	0.86	3.68	4.7	5.15	0.29	0.71	5.71
Turkey	11.96	21.91	6.01	37.75	0.3	2.98	0.48	0.96	52.48
UAE	55.12	160.85	0.28	0.003	54.8	128.66	2.52	19.79	30.61
Uganda	0.09	0.09	0.64	0.42	0.04	0.02	0.07	0.03	0.49
UK	145	280.18	173	107.55	28.67	97.34	4.87	5.03	288.16
Uruguay	0.15	0.48	2.85	2.34	0	0.15	0.89	0.78	2.41
US Virgin Is	0	0	21.28	20.98	13.06	15.56	42.42	17.15	1.97
USA	2,102	2055.75	276	593.01	74	128.81	10.82	10.13	2,504.66
USSR	1,216	2356.77	17.04	30.03	163	406.74	4.13	6.55	1,875.23
Vanuatu	0	0	0.04	0.06	0	0	0.22	0.2	0.03
Venezuela	298	181.69	0.31	0.16	258	113.99	2.24	2.82	54.33
Vietnam	3.07	6.32	9.76	1.99	0.24	1.08	0.3	0.11	7.17
Wake I.	0	0	0.52	0.59	0	0	38	19	0.04
Yemen	0	12.05	10.05	7.03	8.19	14.17	0.18	0.91	3.84
Yugoslavia	21.44	36.1	9.68	28.19	1.04	1.33	1.41	2.55	60.43
Zaïre	0.49	2.62	1.48	1.41	0.19	1.56	0.07	0.06	2.21
Zambia	0.61	1.16	1.01	0.86	0	0.25	0.39	0.2	1.61
Zimbabwe	3.61	5.44	0.7	1.36	0.54	0.13	0.71	0.7	6.55

South Africa includes Botswana, Lesotho, Namibia and Swaziland.

World trade

The total value of the imports and exports for the year stated is shown in the left-hand column of figures. These figures are in million $US. This total figure is divided into percentages for the ten basic groups of the Standard International Trade Classification

(SITC). These are the latest figures divided into the SITC groupings that are available. In the next column the latest total figure for imports and exports is given. This also is in million $US. This figure is usually for 1990, but if earlier than this, the numeral indicates the year in the

eighties, 7 is 1987, 8 is 1988 etc. In the extreme right-hand column the latest trade figure quoted in the column to the left has been divided by the population showing the value of imports and exports per capita. This figure is in $US.

	Year		Total	Percentages Food and live animals	Beverages and tobacco	Crude materials inedible excl. fuels	Mineral fuels, lubricants & related materials	Animal and vegetable oils and fats	Chemicals	Manufactured goods	Machinery and transport equipment	Misc. manufactured goods	Misc. transactions and goods	1990 Total or latest available figure million $US	Latest year	Trade per capita
			million $US													$US
Afghanistan	1981	Imports	622	12.5	0.7	1.3	17.9	4.2	7.2	25.3	23.9	3.5	3.5	798	9	50
		Exports	694	33.9	–	38.9	12.6	–	0.6	13.5	–	0.5	–	238	9	15
Algeria	1987	Imports	7,028	24.7	0.3	4.7	2.3	2.4	13.7	19.5	28.6	2.9	0	7,396	8	313
		Exports	8,185	–	0.2	0.5	97.5	–	0.6	0.3	–	–	–	8,164	8	345
Angola	1981	Imports	1,678	24.5	1.4	1.6	0.8	4.3	12.1	20.2	20.9	9.7	–	451	7	49
		Exports	1,874	5.2	–	0.4	82.1	–	–	11.9	–	–	0.4	2,190	7	237
Argentina	1987	Imports	5,817	4.3	–	7.7	11.5	–	20	13.7	37.8	4.6	–	4,204	9	132
		Exports	6,360	46.6	0.8	9.1	1.5	8.6	6.4	18.3	6.4	2.4	–	9,579	9	300
Australia	1987	Imports	26,914	4.3	0.8	3.1	4.9	0.3	9.9	17.2	40.2	13.8	5.3	38,843		2,273
		Exports	25,361	22.1	0.5	29.4	20	0.4	1.9	11.8	4.9	2.1	7	39,539		2,314
Austria	1988	Imports	36,636	5	0.4	5.3	5.7	0.2	10.2	19.5	36.7	16.8	–	50,017		6,487
		Exports	28,120	3.3	0.3	5.1	1.4	–	6.4	40	33.9	14.4	–	41,881		5,432
Bahamas	1986	Imports	329	5	0.9	0.5	73.8	–	4	4.9	6.1	4.8	–	3,001	9	12,004
		Exports	2,702	0.8	0.4	0.6	87.5	–	9.8	0.3	0.5	–	–	2,786	9	11,144
Bahrain	1985	Imports	3,106	6.8	1.3	1.2	46.6	–	5.2	11.2	20.6	6.6	–	2,866	9	5,849
		Exports	2,901	–	–	–	88	–	0.3	8.7	2.2	0.4	–	2,689	9	5,488
Bangladesh	1987	Imports	2,572	22	–	5.6	13.8	6.3	8.3	23.2	17.2	2.9	0.4	3,524	9	33
		Exports	1,194	17	0.1	8.1	1.3	–	–	36.1	1.4	35.5	0.2	1,305	9	12
Barbados	1987	Imports	515	16.1	2.3	3.1	10.8	0.9	10.4	17.9	24.2	11.8	2.4	705		2,716
		Exports	155	23.4	3.4	1.1	18.6	–	10.7	9.6	21.5	10.5	1.2	209		804
Belgium/Lux.	1988	Imports	91,942	8.7	1.1	6.7	7.3	0.4	11.4	24	23.9	10.4	6.2	120,067		12,202
		Exports	89,426	8.7	0.8	2.7	3.6	0.4	14.3	33.6	26.7	8.1	1.1	118,295		12,022
Belize	1985	Imports	128	23.8	2.7	0.5	17.1	0.3	7.8	12.5	17.6	17.4	0.4	216	9	1,200
		Exports	90	60.9	1.3	3	–	1.5	1.4	7.4	22.8	0.7	–	124	9	689
Benin	1982	Imports	475	11.5	13.3	1.1	4.7	–	5.3	31	22.2	9.9	0.9	288	4	73
		Exports	426	23.9	0.4	13.1	3.9	11	0.4	3.6	3.7	39	0.9	167	4	42
Bermuda	1984	Imports	413	16.4	3.5	0.9	12.2	–	8.8	12.8	23.2	19.7	2.9	488	8	8,133
		Exports	40	0.1	0.9	0.4	–	–	57	0.8	20.4	4.1	16.5	31	8	517
Bolivia	1987	Imports	766	15.5	0.7	2	0.4	1.8	9.9	19.5	41.2	8.4	0.6	715	9	97
		Exports	569	7.6	–	35.5	44.9	–	0.3	10.3	–	–	0.2	900	9	122
Brazil	1987	Imports	16,577	7.1	–	5.8	32.6	0.3	15.6	8.5	26.1	3.9	–	18,281		209
		Exports	26,228	26.7	1.7	12.7	3.6	1.7	6.1	19.7	20.3	6.7	0.8	34,392		
Brunei	1985	Imports	610	14.6	5.3	1.3	1.8	0.6	7.2	21.7	34.1	10.8	2.8	656	6	2,852
		Exports	2,972	–	–	–	98.5	–	–	–	0.8	0.2	–	1,797	6	7,813
Burkina Faso	1984	Imports	252	29.8	2.6	3.3	14.8	2.9	10.8	18.6	18.3	1.2	–	489	8	57
		Exports	78	15.1	–	73.2	–	2.8	–	6.2	2.5	–	–	142	8	17
Burma	1984	Imports	299	3	–	1.3	2.5	6	10.6	31.6	40	4.6	0.4	261	6	193
		Exports	193	64.2	–	31.2	1	–	0.5	3.1	–	–		325		8
Burundi	1985	Imports	193	13.6	0.6	1.2	17.6	1.7	8.4	19.2	28.2	6.1	3.4	236		52
		Exports	109	89.5	0.1	1.2	–	–	–	4.8	–	–	4.4	75		16
Cameroon	1987	Imports	1,749	11.6	2.7	2.4	1.4	0.4	14.8	21.7	35.8	8.5	0.7	1,271	8	115
		Exports	829	37.9	0.9	15.8	17.5	1.1	2.4	13.4	9.9	0.9		924	8	83
Canada	1988	Imports	106,777	5.1	0.4	3.5	3.9	–	6	13.3	55.1	10.5	2.2	116,461		4,391
		Exports	113,145	8	0.5	15	9	–	5.5	16.6	37.9	3.6	3.6	126,995		4,789
Cape Verde Is.	1980	Imports	67	35.4	2.4	2.1	9.1	1.9	6.6	17.2	13.9	7	4.4	112	9	311
		Exports	49	4.4	–	0.5	86.1	–	–	1.3	7.3	–	–	7	9	19
Central Afr. Republic	1980	Imports	80	14	6.4	2.6	1.8	0.5	11.8	19.1	33.9	10	–	150	9	52
		Exports	115	27.5	2	41.3	–	0.3	–	25.1	–	–	3.6	134	9	47
Chad	1975	Imports	110	11.6	2	2.1	14.2	0.8	16.4	18.4	28.9	5.4	0.2	419	8	78
		Exports	40	14.9	0.3	68.6	7.9	–	0.5	1.6	5.4	–	0.8	141	8	26
Chile	1986	Imports	2,964	3.8	0.4	4.8	14.9	0.8	16.4	15.2	35.7	6.5	1.6	7,272		522
		Exports	4,157	27.5	0.5	29.7	0.1	0.5	2.5	36	1.6	0.7	1	8,580		652

57

	Year		Total	**Percentages** Food and live animals	Beverages and tobacco	Crude materials inedible excl. fuels	Mineral fuels, lubricants & related materials	Animal and vegetable oils and fats	Chemicals	Manufac- tured goods	Machinery and transport equipment	Misc. manufac- tured goods	Misc. trans- actions and goods	1990 Total or latest available figure	Latest year	Trade per capita
			million $US											million $US		$US
China	1987	Imports	43,393	5.6	0.6	7.7	1.2	0.8	11.6	22.5	33.9	4.3	11.7	466,040		409
		Exports	39,541	12.1	0.4	9.3	11.5	0.2	5.7	21.7	4.4	15.9	18.8	450,560		396
Colombia	1988	Imports	5,005	6	0.3	6.2	3.6	1.2	24.2	15.7	36.8	4	2.1	5,010	9	155
		Exports	5,026	43	0.4	5.4	25.8	–	4.2	10.3	1.2	8.6	1.2	5,739	9	177
Comoros	1980	Imports	30	33.2	–	–	21.2	–	–	–	–	–	45.6	67	7	134
		Exports	11	60.9	–	3.4	–	–	17.2	–	–	–	18.5	15	7	30
Congo	1985	Imports	580	16.4	2	1	3.1	0.7	8.4	26	35.3	7	–	544	8	255
		Exports	1,087	1.3	–	1.8	93.3	–	–	3.3	0.2	–	–	751	8	353
Costa Rica	1985	Imports	1,098	6.8	0.5	2.1	16.6	0.8	21.4	22	22.6	6.3	0.9	1,743	9	589
		Exports	941	71.4	–	2.7	1.8	–	6.5	9.3	2.7	4.3	1.2	1,362	9	460
Cuba	1987	Imports	7,611	9.7	–	3.9	34.7	0.9	5.8	10.8	30.7	3.4	–	7,579	8	728
		Exports	5,401	82	1.9	6.9	5.8	–	0.5	1	0.6	0.5	–	5,518	8	530
Cyprus	1987	Imports	1,483	10.4	2.1	2	12.3	1.3	9.6	27.9	24.5	8.6	1.5	2,281	9	3,306
		Exports	621	23.7	6.5	3	5.2	1.9	4.7	8.3	9.8	36.1	0.7	793	9	1,410
Czechoslovakia	1987	Imports	23,283	5.1	0.8	7.3	27.9	0.2	6.7	8.4	35.7	4.9	2.9	13,106		837
		Exports	23,016	2.2	0.4	3	3.6	–	6.4	16.4	55.2	11.6	1.2	11,882		759
Denmark	1988	Imports	26,468	10.7	1.2	4.5	6.4	0.5	11.2	20.3	28.7	13.1	3.2	31,743		6,176
		Exports	27,821	25.6	1.1	5.9	2.2	0.4	8.7	11.3	24.8	14.8	5	35,087		6,826
Djibouti	1984	Imports	227	15.9	5.6	12.7	8.4	1.3	5.3	14.2	24.7	9	3	188	6	409
		Exports	11	5.2	0.8	–	–	–	0.1	0.5	–	–	93.4	20	6	43
Dominica	1985	Imports	55	19.3	3.4	1.9	10.9	3.6	11.4	20.5	22.5	6.5	–	107	9	1,338
		Exports	28	54.3	0.8	0.3		4.5	28.6	3.4	5.4	2.7	–	45	9	563
Dominican Rep.	1985	Imports	1,247	6.4	0.2	3.5	35.2	3.9	11.7	13.5	23.2	2.3	–	1,964	9	1,338
	1983	Exports	648	71.4	3.8	0.6	–	–	4	14.1	4.5	1.5	–	924	9	563
Ecuador	1987	Imports	1,890	5.3	0.4	3.1	1.7	0.7	20.3	21	42.2	5.2	–	1,862		690
	1986	Exports	2,184	52.2	–	4.2	41.8	–	0.5	0.4	–	0.3	–	2,722		238
Egypt	1988	Imports	23,297	23.4	1.1	7.7	2.6	1.8	13.1	20.6	26.8	3.1	–	7,434	9	144
		Exports	5,706	8.9	0.1	9.8	33.2	–	3.1	39.8	0.3	4.8	–	2,565	9	50
El Salvador	1985	Imports	1,104	12.2	0.3	2.4	24.3	2.9	20.1	18.1	13.7	5.9	–	902		172
		Exports	610	66.5	0.3	0.8	2.5	–	6.1	14.8	1.6	4.9	2.4	412		78
Ethiopia	1985	Imports	988	24.5	0.4	3.3	14.8	4.8	7.4	13.5	28.7	2.6	–	1,075	8	22
		Exports	337	69.6	–	19.1	9.8	0.2	0.6	–	–	0.2	–	446	8	9
Fiji	1986	Imports	435	15.7	0.7	0.6	16.6	1.2	8.4	20.9	23.5	9	3.3	633	9	844
		Exports	273	55.8	–	2	15.1	1.4	0.8	4.7	3.1	3.2	13.7	386	9	515
Finland	1988	Imports	20,918	4.9	0.4	6.3	9.6	–	10.5	16.1	39.2	12.1	0.8	27,108		5,432
		Exports	21,658	1.6	–	12.3	1.7	–	5.5	43	27.4	7.9	–	26,743		5,359
France	1988	Imports	177,186	9.3	1	4.6	8.3	0.3	10.9	18.3	33.2	13.9	0.3	233,140		4,131
		Exports	162,089	11.9	3.4	4.1	2.1	0.2	14.4	18.2	35.1	10.3	0.4	209,958		3,720
French Guiana	1987	Imports	394	16.7	5.1	0.3	10.1	0.4	5.9	14.6	31.9	13.9	1	697		6,970
		Exports	53	69.6	–	7.4	0.1	–	0.9	4.7	5.6	4	7.6	85		850
Gabon	1983	Imports	685	13.9	3.4	1.8	1.8	0.7	7.5	22.1	38.5	10	0.5	732	7	691
		Exports	1,475	0.4	–	14.2	79.5	–	1.2	3.8	0.6	0.3	–	1,288	7	1,215
Germany, West	1988	Imports	250,293	9.4	1	6.4	7.7	0.3	9.5	18.6	29.2	14.9	3	342,586		5,418
		Exports	323,196	4.2	0.6	1.8	1.3	0.3	13.2	18.2	48	11.1	1.2	398,446		6,302
Ghana	1983	Imports	542	11.1	2.5	2.5	10.1	1.1	10.3	16.2	32.5	6.3	7.5	907	8	64
	1982	Exports	717	60.5	–	4.2	9.7	–	–	24	0.1	–	1.3	1,014	8	72
Greece	1988	Imports	12,002	15.3	1.3	6.1	5.1	0.5	11.6	22	30.2	7.5	0.5	16,126	9	1,608
		Exports	5,155	18.2	5.6	6.3	5.3	1.4	3.6	28.9	3.3	26	1.4	7,543	9	751
Grenada	1987	Imports	507	24.2	2.6	4.5	5.8	0.3	9.5	19.3	23.2	10.4	–	88	7	880
	1986	Exports	390	89.8	1	–	–	–	0.1	0.5	5.6	3.1	–	31	7	310
Guadeloupe	1987	Imports	1,040	17.7	4.2	1.9	4.7	0.6	9.2	16	29.6	15.5	0.5	1,650		4,853
		Exports	92	67.8	9.9	0.7	0.2	–	3	4.9	9.8	3.5	–	118		347
Guatemala	1985	Imports	1,296	5.5	0.2	2.2	35.8	2.6	20	14.8	14.9	4.1	–	1,654	9	185
		Exports	991	60	1.5	16.5	1.7	–	9.2	7.1	0.9	3.1	–	1,108	9	124
Guinea–Bissau	1980	Imports	55	17	2.2	2.3	6.2	0.8	5.6	23.9	36.4	3.7	1.9	52	0	65
		Exports	11	39.9	1.6	44.7	–	1.2	0.3	6.9	–	–	5.4	16	0	20
Guyana	1979	Imports	290	14.5	0.6	1	21.8	2.7	11.1	22.8	19.7	5.5	0.5	248	5	314
		Exports	289	45.3	2.1	46.5	–	–	1.6	0.9	1.7	1.2	0.5	227	9	287
Haiti	1984	Imports	472	18.4	1.9	2.8	12.9	7.1	9.1	19.6	19.5	8	0.6	344	8	55
		Exports	178	35.8	–	1.4	–	–	3.2	15.5	–	37.6	6.5	165	9	26
Honduras	1985	Imports	873	8.2	0.3	0.8	25.9	0.7	17.6	19.3	21.4	5.3	0.5	933	8	194
		Exports	699	80.9	1.9	11	0.9	1.4	1.1	1.7	–	0.8	–	869	8	181

	Year		Total	Percentages										1990	Latest	Trade
				Food and live animals	Beverages and tobacco	Crude materials inedible excl. fuels	Mineral fuels, lubricants & related materials	Animal and vegetable oils and fats	Chemicals	Manufac-tured goods	Machinery and transport equipment	Misc. manufac-tured goods	Misc. trans-actions and goods	Total or latest available figure million $US	year	per capita
			million $US													$US
Hong Kong	1988	Imports	63,937	6.3	1.5	3.4	1.9	0.2	8.9	26.4	28.8	22	0.5	82,496		14,223
		Exports	63,175	2.7	1.2	2.9	0.4	–	5.8	17.8	26.9	41.4	0.9	58,333		10,057
Hungary	1988	Imports	9,345	6.6	0.8	7.1	13.8	–	16.1	17.1	30.8	6.2	1.4	8,764		831
		Exports	9,930	17.5	1.3	4.4	2.9	1	12.2	15.3	32.5	11	1.8	9,707		920
Iceland	1988	Imports	1,587	6.7	1.5	4.5	6.4	0.3	7.2	17.9	36.7	18.5	–	1,395	9	5,580
		Exports	1,431	72.4	–	1.8	–	2.2	–	16.6	4.8	1.6	0.5	1,401	9	5,604
India	1985	Imports	16,223	4.5	–	7.6	26.5	3.9	14.6	19.2	20.8	2.5	–	23,382		28
		Exports	8,988	23	1.6	10.3	6	0.4	3.6	34.1	6.5	14.4	–	17,786		22
Indonesia	1987	Imports	12,370	5	–	7.7	9.2	0.8	18.9	14.6	38.9	3.8	0.7	21,837		122
		Exports	17,135	9.8	0.4	11.2	50	1.7	1.5	19.1	0.3	4.3	1.6	25,675		143
Iran	1977	Imports	14,448	10.5	0.9	3.1	–	1.3	7.2	29.1	44.3	3.4	0.2	9,738	9	180
		Exports	25,943	0.7	–	0.9	97.6	–	–	0.4	–	–	0.4	12,378	5	259
Iraq	1978	Imports	4,213	10.8	0.4	2	–	1.1	4.7	22.9	53.7	4.2	0.2	3,854	7	237
		Exports	10,091	0.7	–	–	98.5	–	0.4	–	–	–	0.4	21,431	9	1,172
Ireland	1988	Imports	15,568	10.4	1.2	2.8	5.6	0.4	12.3	16.2	34.4	13.5	3.2	20,716		5,919
		Exports	18,738	23.5	2	4.3	0.5	–	12.8	8.5	31.2	13.3	3.8	23,788		6,797
Israel	1988	Imports	12,959	6.8	0.5	3.9	7.4	0.3	8.9	36.1	27.3	6.8	1.9	15,104		3,241
		Exports	9,739	8.3	–	4.2	0.6	–	13.4	36.6	25.5	11.2	–	11,576		2,484
Italy	1988	Imports	138,639	11.6	1.1	8.6	8.5	0.6	11.4	16.9	28.9	7.8	4.8	180,105		3,124
		Exports	128,561	5.1	1.1	1.7	1.7	0.3	8	22.7	35.6	23.8	0.1	168,680		2,925
Ivory Coast	1985	Imports	1,734	15	1.9	1.8	22	–	12.8	17.6	22.2	5.8	0.7	2,100	8	189
		Exports	2,669	64.7	0.2	12.7	9.7	2.4	2.5	4.3	1.8	0.9	0.9	2,792	8	251
Jamaica	1988	Imports	1,434	15.6	1	3.3	13.6	1.1	11.4	21.4	21.3	9.8	1.5	1,809	9	757
		Exports	833	20.6	4.4	50.8	2.3	–	2.5	1.9	2.4	15.2	–	1,029	9	431
Japan	1988	Imports	187,353	14.3	1.2	15.2	20.8	–	7.5	14.3	12.4	10.8	3.3	234,806		1,901
		Exports	264,916	0.6	–	0.7	0.3	–	5.1	13.4	70	8.5	1.4	286,948		2,323
Jordan	1988	Imports	2,786	16.8	0.7	3.7	15.5	1.2	9.9	17.3	22.6	7.2	5.1	2,119	9	546
		Exports	884	9.3	0.5	45.1	–	0.2	28.3	11.2	1.1	4.2	–	926	9	239
Kenya	1986	Imports	1,649	5.6	–	2.4	18.2	3.5	16.5	12.6	37.6	3.5	–	2,147	9	89
		Exports	1,216	67.9	1.1	7.5	11.1	–	3.2	4.8	1.5	1.9	1.1	969	9	40
Korea, South	1987	Imports	41,017	4	–	14.4	14.7	0.3	11.2	15.3	34.4	5.3	0.4	68,771		1,607
		Exports	47,207	4.4	0.2	1	1.6	–	2.8	21.6	35.8	32.5	–	64,933		1,532
Kuwait	1984	Imports	6,896	15.7	1.2	1.4	0.6	0.3	4.8	21.7	38.7	15.2	0.3	6,303	9	3,075
		Exports	12,274	1	–	0.4	82.8	–	8.8	2.4	3.5	1.2	–	11,476	9	5,598
Lebanon	1977	Imports	1,692	16.9	1.1	9.4	6.6	0.6	4.6	19.7	21.1	5.8	14.2	3,567	2	1,364
		Exports	394	13.8	4.2	2.7	–	–	9.6	24.7	11.4	33.1	0.5	923	2	334
Liberia	1984	Imports	363	21.6	2.1	1.4	19.7	1.3	6.7	13.8	26.8	6.2	0.4	308	7	131
		Exports	449	6.9	–	90.4	0.1	1.5	0.1	0.2	0.2	1.2	0.6	382	7	163
Libya	1987	Imports	4,722	15	0.5	1.2	0.4	2.3	9	19.4	34.9	16.1	–	5,879	8	1,560
		Exports	8,766	–	–	–	96.9	–	3.2	–	–	–	–		8	
Macau	1985	Imports	776	8.9	2.8	6.7	6.4	0.3	4.4	49.7	10.3	8.5	2	1,534		3,196
		Exports	902	12.6	–	1	–	–	0.6	14.6	5.4	75.6	0.2	1,694		3,529
Madagascar	1986	Imports	373	13.6	0.5	2.2	23.1	1.2	11.8	14.4	28.8	4.3	0.2	340	9	29
		Exports	316	80.8	–	8.9	2.2	–	1.3	4	1.9	0.08	–	312	9	26
Malawi	1985	Imports	295	4.7	0.7	2.7	13.3	2.2	20.2	21.6	29.4	4.9	0.4	573		69
		Exports	242	44.3	45.2	5.5	–	–	0.5	4.2	0	0.2	0.1	418		50
Malaysia	1987	Imports	12,679	9.3	0.6	4	7.5	0.6	10.3	15.5	45.1	6.1	1	22,541	9	1,299
		Exports	17,920	5.4	–	23.5	19.9	9.2	1.6	8.2	25.9	5.8	–	25,113	9	1,456
Mali	1982	Imports	401	18.6	1.8	1.3	27.5	0.4	9.2	18.8	18.4	3.6	0.4	500	9	63
		Exports	233	36.2	–	58.1	–	2.9	–	1.9	–	0.8	–	271	9	34
Malta	1987	Imports	1,116	10.8	2.3	2	6.5	0.4	8	26.2	33.3	9.9	0.7	1,505	9	4,300
		Exports	551	2.2	2.4	0.8	–	–	1.5	12.1	28.8	52.1	–	858	9	2,451
Martinique	1987	Imports	1,120	16.9	3.1	1.6	8.2	0.5	8.7	14.2	31.1	14.9	1	1,708		5,024
		Exports	194	54.4	14.1	0.9	14.5	–	3.4	5.1	3.9	3.7	–	272		800
Mauritania	1975	Imports	165	28.7	0.6	2.3	8	1	3.5	15	35	5	0.9	222	9	113
		Exports	180	9.3	–	90.6	–	–	–	–	–	–	0.1	437	9	222
Mauritius	1987	Imports	1,012	11.3	0.4	3.1	7.5	1.1	6.4	40	22.1	8	–	1,326	9	1,287
		Exports	900	40.4	–	0.4	–	–	0.6	3.8	0.6	54.2	–	987	9	958
Mexico	1987	Imports	9,427	8.2	–	13.2	5.1	1.2	16.4	9.4	40	6	0.8	23,633	9	280
		Exports	19,353	12.3	1.5	4.3	44.3	–	5	11.6	18.6	2.5	2.5	22,818	9	270
Morocco	1988	Imports	4,772	9.7	1	14.4	13.2	2	11.9	19.9	24.1	4	–	5,492	9	225
		Exports	3,625	25.4	–	20.8	2.1	–	25.6	8.3	1.6	16	–	3,308	9	135

	Year		Total	Percentages Food and live animals	Beverages and tobacco	Crude materials inedible excl. fuels	Mineral fuels, lubricants & related materials	Animal and vegetable oils and fats	Chemicals	Manufactured goods	Machinery and transport equipment	Misc. manufactured goods	Misc. transactions and goods	1990 Total or latest available figure	Latest year	Trade per capita
			million $US											million $US		$US
Mozambique	1984	Imports	487	25	–	–	18.7	–	4.6	3.5	17.3	–	30.9	715	8	48
		Exports	86	66.8	–	13.9	6.3	0.2	–	1.1	–	–	11.9	103	8	7
Nepal	1986	Imports	494	11.6	1	2.4	12.1	1.1	13.3	31.6	20.6	6.3	0.1	580	9	31
		Exports	153	25.7	0.1	7.4	–	2	3.2	30.6	–	31.1	0	156	9	8
Netherlands	1988	Imports	99,801	12.3	1.3	5.7	9.4	0.6	10.5	17.2	29	13.2	0.8	126,195		8,447
		Exports	103,559	17.9	1.9	6.1	8.6	0.7	16.7	14.2	22	8.3	3.7	131,839		8,825
Netherlands Antilles	1984	Imports	4,024	3.7	0.4	–	85.2	–	2.3	2.4	3.2	2.6	0.3	1,404	8	7,389
		Exports	3,727	0.1	–	0.4	98.4	–	0.8	–	–	–	0.4	1,133	8	5,963
New Caledonia	1986	Imports	458	15.2	3.7	0.5	13.8	0.5	–	14	31.8	13.1	7	883		5,194
		Exports	189	1.1	–	22.8	–	–	–	56.4	–	–	19.6	449		2,641
New Zealand	1988	Imports	7,305	6	1.2	4.2	5.5	0.4	12.7	18.9	38	12.7	0.3	9,489		2,833
		Exports	8,806	42.1	0.4	24.8	1.4	0.8	4.9	14.6	5.8	3.2	2.1	9,435		2,816
Nicaragua	1985	Imports	1,066	7.2	–	1.1	31.8	1.9	17.8	13.7	22.9	3.5	–	923	7	264
		Exports	274	52.8	1.7	35.9	0.1	–	3.2	2.2	–	1.5	2.6	300	7	86
Niger	1982	Imports	465	25.2	1.8	4.2	11.6	1.1	9	23	23.4	–	0.7	345	5	52
		Exports	331	4.9	0.1	84.4	1.5	–	–	8.1	0.8	–	–	209	5	32
Nigeria	1986	Imports	4,028	13.4	–	3.2	0.5	2.1	17.4	20.7	38.1	4.1	–	3,419	9	33
		Exports	5,899	4.8	–	0.6	93.1	–	–	–	–	–	1.3	8,138	9	77
Norway	1988	Imports	23,221	4.9	0.6	7.4	3.6	–	8.2	18.6	39.6	16.7	0.5	26,905		6,346
		Exports	22,511	8	–	4.4	36.5	0.3	8.2	23.9	14.9	3.6	–	34,072		8,036
Oman	1987	Imports	1,820	18.6	2.1	1.3	3	0.4	6.8	17.4	37.1	9.9	2.9	2,255	9	1,555
		Exports	322	20.2	1	2.6	2.2	–	2.2	12.4	49	3.2	7.1	3,933	9	2,712
Pakistan	1987	Imports	5,829	7.9	–	7.1	17.7	5.7	17.2	12.4	28.5	3.3	–	7,356		66
		Exports	4,177	13.1	0.3	13.3	0.7	–	0.4	52.1	1.3	18.5	0.2	5,522		49
Panama	1985	Imports	1,383	10	0.7	0.7	21.2	1.3	12.3	17.9	23.5	10.6	1.9	1,489		615
		Exports	301	74.3	1.7	1.3	7.2	2	3.3	5	–	5	–	321		133
Papua New Guinea	1986	Imports	931	18.1	1.1	0.8	10.4	0.4	9	16.6	34.1	8.4	1.2	1,335	9	368
		Exports	1,048	27.8	–	43.9	–	4	–	–	–	–	24.3	1,281	9	353
Paraguay	1987	Imports	595	1.5	7.6	0.9	24	–	8.2	13.9	36.7	7	–	695	9	167
		Exports	353	13.7	2.8	70.6	–	2.7	3.2	5.5	–	1.6	–	1,163	9	280
Peru	1987	Imports	3,002	18	0.2	4.1	4.3	1.1	20.7	12.6	33.8	5.3	0	1,839	9	84
		Exports	2,203	22.4	–	30.2	12.8	0	2.2	24.2	1.2	4.6	2.3	3,562	9	163
Philippines	1988	Imports	8,731	9	1.1	5.2	13.3	–	12.9	15.2	19.8	2.3	20.9	10,732	9	179
		Exports	7,074	14.9	0.5	10	2.2	6	3.6	9.8	9.8	13.5	29.7	7,747	9	129
Poland	1987	Imports	10,843	9.3	1.1	8.7	17.3	0.6	11.4	13.3	32.2	5.7	0.3	8,160		214
		Exports	12,204	9.9	0.5	6.1	11.3	–	7.2	18.3	33.6	7.9	5.1	13,627		357
Portugal	1987	Imports	13,441	10	0.5	8.2	11.6	0.2	10.6	19.1	33.4	6	0.4	25,072		2,381
		Exports	9,166	3.8	3.6	9.2	1.9	0.7	5.5	25.3	16.5	32.9	0.7	16,348		1,553
Qatar	1982	Imports	1,945	8.7	1.6	2.7	6.6	0.4	4.7	20.3	49	11.9	–	1,267	8	3,726
	1981	Exports	5,389	–	–	–	93.9	–	3.9	2.2	–	–	–	3,541	5	11,803
Réunion	1987	Imports	1,464	17.5	2.4	2	4.5	0.6	9.3	16.8	30.3	15.3	1.3	2,049		3,415
		Exports	148	84.3	4.1	0.4	0.2	–	2.9	1.7	4.7	1.3	0.3	184		307
Romania	1973	Imports	352	7.6	0.8	9.8	6.9	–	3.8	22.4	42.3	3.1	3.3	9,156		395
		Exports	117	16.2	2.3	8	8.9	1.8	6.7	13.1	24.4	16.1	2.5	5,962		257
Rwanda	1986	Imports	352	9.3	1.5	4.9	13.7	3.4	5.3	25.1	29.6	6.2	0	369	8	55
		Exports	117	87.2	–	10.3	–	–	–	–	–	–	2.2	101	8	15
St. Lucia	1986	Imports	154	19.9	3.2	2.5	7.7	0.3	12.2	21.5	19.8	12.9	–	155	6	1,107
		Exports	82	69.3	3	1	–	2.3	0.3	8.3	4.3	11	–	83	6	593
St. Vincent/ Grenadines	1980	Imports	57	30	5.1	3.2	8.9	–	9.6	19.1	15.7	8.3	–	61	2	555
		Exports	15	79.1	2.2	0.4	1.6	2.7	–	2.6	4.4	6.8	–	32	2	291
Saudi Arabia	1987	Imports	20,110	13.9	1.3	1.4	0.3	0.4	8	19.9	32.5	14.6	7.7	21,153	9	1,466
	1985	Exports	27,487	0.4	–	–	94.4	–	2.8	0.6	1.4	–	–	28,369	9	1,966
Senegal	1987	Imports	1,023	18.2	2.1	3.7	17.4	1.1	10.7	16.2	25.8	4.9	–	1,023	7	148
		Exports	606	34.9	–	10.7	18.8	10	12.5	6.3	4.6	1.7	0.3	606	7	88
Seychelles	1986	Imports	104	15.2	3.4	1.2	17	1	5.7	18.8	27.5	9.9	0.3	164	9	2,343
		Exports	18	8.6	1.9	3.7	71.9	–	0.4	1.5	7.6	4.5	–	31	9	443
Sierra Leone	1986	Imports	151	26.2	2.8	1.5	11.1	2.5	8.8	15.8	27.1	3.7	0.6	189	9	47
	1984	Exports	147	5.4	0.7	5	20.7	1.8	5.8	7.4	38.6	8.3	6.3	138	9	34
Singapore	1988	Imports	43,861	5.3	0.8	3.4	14.1	1.1	6.6	14.7	43.4	9.1	1.6	49,676	9	18,536
		Exports	39,304	4.2	0.7	5.1	12.8	1.1	6.6	8.3	48	9	4.3	44,678	9	16,671
Solomon Islands	1987	Imports	67	15	3.1	0.6	14.7	0.6	6.6	20.5	29.1	9.2	0.4	114	9	368
	1984	Exports	92	27.6	–	55.1	–	14.5	–	–	–	–	2.7	75	9	242

	Year		Total million $US	Percentages Food and live animals	Beverages and tobacco	Crude materials inedible excl. fuels	Mineral fuels, lubricants & related materials	Animal and vegetable oils and fats	Chemicals	Manufactured goods	Machinery and transport equipment	Misc. manufactured goods	Misc. transactions and goods	1990 Total or latest available figure million $US	Latest year	Trade per capita $US
Somalia	1981	Imports	512	18	0.7	6.4	2.2	1.5	2.1	16.1	49.9	2.4	0.7	132	7	19
	1981	Exports	152	97.5	–	1.9	0.2	–	–	–	–	–	0.4	104	7	15
South Africa	1985	Imports	10,311	3.9	0.9	3.9	0.6	1.5	12.2	12.2	39.5	7.4	17.9	17,075		484
		Exports	16,419	4.7	–	8.9	8.9	–	2.8	19.6	2.5	0.9	51.3	18,969		496
Spain	1988	Imports	60,556	8.8	1.2	8.1	11.4	0.3	10.1	12.7	39.1	8	0.2	87,694		2,250
		Exports	40,466	13.8	1.6	4	4.7	1.7	8.8	21.7	33.5	9.8	0.4	55,640		1,428
Sri Lanka	1986	Imports	1,831	16.8	0.4	2.1	12.6	0.5	11.1	29	22.7	4.8	0.1	2,088	9	124
		Exports	1,194	35.7	–	12.1	6.9	2.1	0.9	9.4	3.3	29.2	–	1,529	9	91
Sudan	1982	Imports	1,284	16.3	1.4	0.1	27.1	0.9	8.2	0.3	28.3	1.7	15.7	1,060	8	45
	1981	Exports	509	26.2	–	64.5	5	2.5	–	0.8	0.9	–	–	509	8	21
Surinam	1974	Imports	230	9.1	1.3	2.1	19.9	0.5	12.9	21.4	18.7	7.9	6.2	294	7	735
		Exports	240	12.8	–	30.1	–	–	39.3	11.6	–	–	6.2	301	7	753
Sweden	1988	Imports	45,819	5.5	0.8	4.6	6.9	0.2	9.7	17	39.5	15.1	0.9	54,568		6,374
		Exports	49,979	1.6	–	9.2	2.2	–	7	26.9	42.8	7.8	2.2	57,435		6,710
Switzerland	1988	Imports	56,385	5.4	1.3	3	3.7	–	11.6	21.7	32.3	20.6	0.3	69,869		10,413
		Exports	50,733	2.3	0.5	1.3	0.1	–	21.6	19.9	32.4	21.2	0.7	63,884		9,521
Syria	1987	Imports	2,481	14.8	0.1	2.1	19.8	1.5	12.5	20.9	25.1	2.9	0.3	2,097	9	179
		Exports	1,350	3.8	–	13	51.8	–	10.7	13.4	0.5	6.6	–	3,006	9	256
Tanzania	1981	Imports	867	5.6	–	1.5	30.8	1	10	12.3	35	3.8	–	1,495	8	62
		Exports	564	57.1	3.7	25.9	0.2	0.4	0.7	8.5	2.5	0.9	–	337	8	14
Thailand	1987	Imports	12,972	4.7	0.5	7.1	13.4	0.1	14.4	19.4	32.1	4	4.5	25,768	9	465
		Exports	11,659	36.3	0.5	8.9	0.8	1.7	0.3	16.5	11.9	22.5	0.9	20,059	9	363
Togo	1985	Imports	288	18.1	–	–	–	–	11.4	–	–	–	70.5	288	5	97
	1983	Exports	162	23	–	59.3	1.3	–	–	13.1	1.4	0.7	1.2	190	5	64
Tonga	1986	Imports	39	22.8	5.8	3.5	12.3	0.2	8.7	23.3	15.3	7.9	0.2	57	9	633
		Exports	6	54	–	2.6	–	21.2	–	3.8	3.6	11.4	3.3	9	9	100
Trinidad & Tobago	1988	Imports	1,126	16.7	0.5	5.8	11.9	0.6	12	19.5	25.1	7.6	0.3	1,222		993
		Exports	1,411	4.5	1.5	0.6	60.5	–	21	9.1	1.3	1.4	–	2,049		1,666
Tunisia	1987	Imports	3,023	9.6	0.7	9.9	10.8	1.6	11.2	26.9	22	6.9	0.5	4,378	9	553
		Exports	2,152	8.5	0.5	2.9	23.6	3.7	18.1	8.9	5.8	27.6	0.5	2,933	9	371
Turkey	1988	Imports	14,335	1.3	1.3	9.4	21.3	2	15.6	16.9	29.4	3.2	–	15,799		269
		Exports	11,662	21.3	2.4	6.7	2.9	0.7	8.2	29.8	6.4	21	–	11,627		198
Uganda	1977	Imports	182	7.1	0.4	1.5	29.7	0.4	11.1	17.9	26.8	5.1	–	544	8	31
		Exports	373	89.2	0.5	7	0.8	–	–	2.4	–	–	0.1	274	8	16
United Arab Emirates	1982	Imports	8,600	8.3	1	1.2	6	0.2	5.4	22.2	40.6	14	1.1	10,003	9	6,454
		Exports	9,078	0.9	0.3		94.9	–	–	1	1.4	1.1	0.1	15,837	6	11,312
United Kingdom	1988	Imports	189,465	8.5	1.4	5.3	4.8	0.4	8.8	18.4	37.6	13.8	1.1	224,938		3,918
		Exports	145,076	4.3	2.6	2.5	7.1	0.1	14.2	15.6	39.2	12.9	1.9	185,976		3,239
United States	1984	Imports	460,259	4.7	10	3.1	9.6		4.4	14.2	44	16	2.8	516,575		2,067
		Exports	321,971	8.3	1.4	8	2.6	0.5	9.9	7.6	44.3	8.3	9.2	393,893		1,576
Uruguay	1988	Imports	1,176	4.6	1.4	6.2	14.2	0.5	21.1	14.4	32.7	5	–	1,203	9	391
		Exports	1,443	31.7	0.2	29.3	0	0.5	7.1	16.1	2.8	12.1	0.3	1,599	9	519
USSR	1983	Imports	80,410	20.4	2.6	3	–	0.6	7.1	11.3	44.4	9.6	1	120,867		419
		Exports	91,340	–	–	7.6	66.7	–	3.2	5.9	15.2	0.9	0.5	104,640		363
Venezuela	1988	Imports	14,690	5.4	0.7	6	1.2	0.7	15	13	51.8	5.8	0.6	6,881	9	357
		Exports	1,892	3.8	0.8	4	1.3	–	7.7	64.5	4.7	1.7	11.6	12,983	9	674
Yemen, North	1985	Imports	1,290	28.5	2.4	0.8	6.7	2	9.2	21.9	23.2	5	0.4	1,378	7	131
		Exports	13	20.8	0.5	2.3	–	–	1.6	6	64.4	2	2.4	101	7	10
Yugoslavia	1987	Imports	12,549	5.6	–	9.2	17.4	0.2	16.3	16.6	30.5	4	–	18,890		793
		Exports	11,396	7.7	1	4.9	1.9	–	11.3	26.4	30.4	16	–	14,312		601
Zaïre	1978	Imports	796	19.5	1.1	2.9	7.6	–	10.3	21.1	31.7	5.1	0.7	849	9	25
		Exports	899	28	–	11.2	1.4	1	0.2	55.7	0.6	–	1.8	1,249	9	36
Zambia	1982	Imports	1,000	5.3	–	1.2	20.8	1.3	16	17.8	34.5	2.9	–	873	9	112
		Exports	1,021		0.2		–	–	–	95.9	–	–	–	1,347	9	172
Zimbabwe	1986	Imports	985	1.6	0.5	5.9	15.1	1.1	15.9	14.5	38	4.6	2.8	1,043	7	121
		Exports	1,018	19.2	25	16.1	1.1	–	1.4	31.6	1.6	2.9	1	1,420	7	164

Gross national product

	GNP 1989	GNP per capita	GDP share agriculture	GDP share industry	GDP share manufacture	Services	Average annual rate of change 1980-89	Consumer price index Jan 1991
	million $US	$US	%	%	%	%	%	1980=100
Afghanistan	...	450	...	...	...	...	...	...
Albania	...	800	...	...	...	...	...	...
Algeria	53,116	2,170	16	44	14	40	3.1	202 90
Americ Samoa	...	6,000	...	...	...	...	...	...
Angola	6,010	620	...	...	...	...	...	...
Antigua & Barbuda	...	3,000	6	...	...	...	6.8	...
Argentina	68,780	2,160	14	33	35	53	-0.3	142,000
Aruba	...	6,000	...	...	...	...	...	...
Australia	242,131	14,440	4	32	15	64	3.2	224
Austria	131,899	17,360	3	37	27	60	1.9	143
Bahamas	2,820	11,370	...	...	...	...	4.2	180
Bahrain	3,500	7,000	1	25	19	74	-0.4	...
Bangladesh	19,913	180	44	14	7	42	3.5	263
Barbados	1,622	6,370	7	...	...	...	1.8	179
Belgium	162,026	16,390	2	31	22	67	1.7	159
Belize	294	1,600	19	...	...	...	3.2	144 89
Benin	1,753	380	46	12	5	42	1.3	...
Bermuda	...	25,000	...	...	...	...	0.6	151 87
Bhutan	...	190	45	25	6	30	10	...
Bolivia	4,301	600	24	30	13	46	-0.8	1,294,159
Botswana	1,947	1,600	3	57	4	40	10.2	250
Brazil	375,146	2,550	9	43	31	48	3.1	93,752
Brunei	...	6,000	2	45	9	53	...	...
Bulgaria	20,860	2,320	11	59	...	30	...	...
Burkina Faso	2,716	310	32	26	15	42	5	113
Burma	...	500	51	10	9	39	...	363
Burundi	1,149	220	56	15	10	29	4.5	217
Cambodia	...	450	...	...	...	...	...	...
Cameroon	11,661	1,010	27	27	15	46	3.9	...
Canada	500,337	19,020	3	21	17	76	3.6	186
Cape Verde Is.	281	760	14	...	...	...	5.8	141 88
Central African Republic	1,144	390	42	15	8	43	1.2	135
Chad	1,038	190	36	20	16	44	6.3	...
Chile	22,910	1,770	8	30	21	62	2.7	714
China	393,006	360	32	48	34	20	9.6	...
Colombia	38,607	1,190	17	36	21	47	3	173
Comoros	209	460	41	10	4	49	3.1	...
Congo	2,045	930	14	35	9	51	3.6	185 88
Costa Rica	4,898	1,790	18	27	...	56	2.7	1,122
Cuba	16,200	1,500	12	36	...	52	...	...
Cyprus	4,892	7,050	7	27	15	66	5.8	167
Czechoslovakia	...	4,000	6	57	...	37	...	170
Denmark	105,263	20,510	5	29	20	66	2.2	179
Djibouti	...	1,000	5	...	...	...	...	...
Dominica	147	1,800	27	16	7	57	4.6	144 87
Dominican Rep.	5,513	790	15	26	11	59	2.2	804
Ecuador	10,774	1,040	15	39	21	46	2.2	2,348
Egypt	32,501	630	19	30	14	51	...	502
El Salvador	5,356	1,040	21	21	16	58	0.3	611
Equatorial Guinea	149	430	54	7	...	39	...	...
Ethiopia	5,953	120	42	16	11	42	1.8	177
Fiji	1,218	1,640	24	21	10	55	0.2	145
Finland	109,705	22,060	7	36	22	57	3.4	197
France	1,000,866	17,830	4	29	21	67	2	187
French Guiana	...	2,000	35	...	...	...	...	202
French Polynesia	...	6,000	...	...	...	...	...	200
Gabon	3,060	2,770	11	47	10	42	1	168 87
Gambia, The	196	230	34	12	7	54	2.2	495
Germany	1,272,959	16,500	3	37	32	60	2	132
Ghana	5,503	380	50	17	10	33	2.6	3,800
Gibraltar	...	4,000	...	...	...	...	...	169
Greece	53,626	5,340	16	29	18	55	1	627
Greenland	...	6,000	30	...	...	...	...	172 88
Grenada	179	1,900	21	19	5	60	5.9	116
Guadeloupe	...	6,000	...	...	...	...	...	188
Guatemala	8,205	920	18	26	...	56	0.2	416
Guinea	2,372	430	30	33	3	37	...	...
Guinea-Bissau	173	180	47	16	11	37	3.4	...
Guyana	248	310	25	31	15	44	-0.6	...
Haiti	2,556	400	31	38	15	31	1.1	211
Honduras	4,495	900	21	25	14	54	2.3	360
Hong Kong	59,202	10,320	0	28	21	72	7.2	230
Hungary	27,078	2,560	14	36	...	50	1.3	330
Iceland	5,351	21,240	10	15	...	75	2.8	1,846
India	287,383	350	32	29	18	39	5.4	255
Indonesia	87,936	490	24	37	17	39	5.7	118
Iran	160,000	3,190	25	15	7	60	...	...
Iraq	...	2,000	25	20	12	55	...	...
Ireland	30,054	8,500	10	10	3	80	1.2	214
Israel	44,131	9,750	5	...	...	...	3.2	71,328
Italy	871,955	15,150	4	34	22	62	2.4	258
Ivory Coast	9,305	790	46	24	17	30	0.9	167 89
Jamaica	3,011	1,260	6	45	18	49	-0.4	161
Japan	2,920,310	23,730	3	41	30	56	4.1	125
Jordan	5,291	1,730	6	29	16	65	0.6	213
Kenya	8,785	380	31	20	12	49	4.2	309 90
Kiribati	48	700	30	...	...	...	3.7	...
Korea, South	186,467	4,400	10	44	26	46	10.1	193
Kuwait	3,3082	16,380	1	56	9	43	2.2	133 89
Laos	693	170	...	...	...	...	3	...
Lebanon	...	2,000	8	...	...	...	...	...
Lesotho	816	470	24	30	14	46	2.2	128
Liberia	...	500	37	28	5	35	...	142 88
Libya	...	6,000	5	50	7	45	-0.6	...
Liechtenstein	...	33,000	...	...	...	...	...	...
Luxembourg	9,408	24,860	3	41	30	56	4	102
Macau	...	2,000	1	...	...	...	...	...
Madagascar	2,543	230	31	14	12	55	0.1	529
Malawi	1,475	180	35	19	11	46	3.3	460
Malaysia	37,005	2,130	23	42	22	35	4.6	102
Maldives	87	420	14	53	...	33	9.5	...
Mali	2,109	260	50	12	6	38	3.5	...
Malta	2,041	5,820	4	41	27	55	2.5	110
Martinique	...	4,000	...	...	...	...	...	200
Mauritania	953	490	37	24	...	39	0.4	...
Mauritius	2,068	1,950	13	32	24	55	6.4	152
Mexico	170,053	1,990	9	32	23	59	0.6	17,249
Mongolia	...	400	...	...	...	...	...	...
Montserrat	...	4,000	4	6	5	90	...	...
Morocco	22,069	900	16	34	17	50	4.1	208
Mozambique	1,193	80	64	22	...	14	-3.5	...
Namibia	...	1,300	11	38	5	51	...	...
Nepal	3,206	170	58	14	6	28	4.7	270
Neth. Antilles	...	6,000	1	10	9	89	...	151
Netherlands	237,248	16,010	5	31	20	64	1.8	129
New Caledonia	...	4,000	25	7	5	68	...	193
New Zealand	39,437	11,800	10	28	17	62	1.7	280
Nicaragua	2,803	1,000	29	23	19	48	-1.4	...
Niger	2,195	290	35	13	8	52	-1.7	128 88
Nigeria	28,314	250	31	44	10	25	-0.3	575 90
Norway	92,097	21,850	4	34	15	62	3.9	212
Oman	7,756	5,220	4	80	4	16	10.3	...
Pakistan	40,134	370	27	24	16	49	6.3	177
Panama	4,211	1,780	10	15	7	75	...	122
Papua New Guinea	3,444	900	28	30	10	42	1.8	181
Paraguay	4,299	1,030	29	22	16	49	1.6	411 88
Peru	23,099	1,090	8	30	21	62	0.6	12,528 90
Philippines	42,754	700	23	33	22	44	0.6	401
Poland	66,974	1,760	14	50	...	36	2.6	30,000
Portugal	44,058	4,260	9	37	...	54	2.7	515

	GNP 1989	GNP per capita	GDP share agriculture	GDP share industry	GDP share manufacture	Services	Average annual rate of change 1980-89	Consumer price index Jan 1991
	million $US	$US	%	%	%	%	%	1980=100
Puerto Rico	20,118	6,010	2	40	40	58	3.4	139
Qatar	...	10,000	2	39	12	59	-5.8	...
Reunion	...	5,000	6	9	9	85	...	187
Romania	...	2,000	14	53	53	33	...	...
Rwanda	2,157	310	37	23	15	40	1.3	...
Samoa	114	720	51	12	6	37	...	298 90
Sao Tome & Principe	43	360	31	11	...	58	-2.8	...
Saudi Arabia	89,986	6,230	8	45	8	47	-1.1	96 89
Senegal	4,716	650	22	31	20	47	3	174
Seychelles	285	4,170	7	21	11	72	2.5	139
Sierra Leone	813	200	46	11	6	43	-0.8	3,952 87
Singapore	28,058	10,450	0	37	26	63	6.9	127
Solomon Is	181	570	35	...	...	...	7.3	337
Somalia	1,035	170	65	10	5	25	1.7	1,112 87
South Africa	86,029	2,460	6	44	24	50	1.5	422
Spain	358,323	9,150	6	9	18	85	2.9	253
Sri Lanka	7,268	430	26	27	16	47	3.9	340
St Christopher/N.	120	3,000	10	25	15	65	5.6	...
St Lucia	267	1,810	16	22	7	62	6.6	110 88
St Vincent/G.	152	1,500	17	25	9	58	6	...
Sudan	9,035	380	36	15	8	49	1.1	...
Surinam	1,314	3,020	11	26	15	63	-3.3	276 89
Swaziland	683	900	23	40	26	37	4.1	283 88
Sweden	184,230	21,710	4	34	23	62	2.2	219
Switzerland	197,984	30,270	4	42	27	54	2.1	144
Syria	12,444	1,020	38	23	...	39	5.4	740
Taiwan	...	6,600	6	46	38	48	6.4	...
Tanzania	3,079	120	66	7	4	27	1.8	1,061 89
Thailand	64,437	1,170	17	38	21	45	6.5	162
Togo	1,364	390	34	23	8	43	1	142
Tonga	89	910	41	19	9	40	2.1	201 89
Trinidad & Tob.	4,000	3,160	3	41	8	56	-5.6	298
Tunisia	10,089	1,260	13	33	16	54	3.1	229
Turkey	74,731	1,360	17	35	23	48	5.3	3,338
Uganda	8,069	250	17	7	5	76	2.2	...
United Arab Em.	28,449	18,430	2	55	8	43	-4	...
United Kingdom	834,166	14,570	2	37	20	61	3.1	195
United States	5,237,707	21,100	2	29	17	69	3.2	163
Uruguay	8,069	2,620	11	28	22	61	-0.2	17,385
USSR	1,064,000	3,800	23	42	...	35	...	108 88
Vanuatu	131	860	34	11	5	55	0.7	174 90
Venezuela	47,164	2,450	6	46	28	48	-0.5	590
Vietnam	...	450	...	...	...	...	...	...
Yemen	7,203	640	24	...	...	...	...	...
Yugoslavia	59,080	2,490	14	42	...	44	0	1,026,255
Zaire	8,841	260	30	32	10	38	1.5	...
Zambia	3,060	390	14	47	24	39	-0.2	2874 90
Zimbabwe	6,076	640	13	39	25	47	2.8	...

The Gross National Product (GNP) is a measure of a country's total production of goods and services including the net income from overseas. Owing to difficulties in the use of exchange rates and individual nation's methods of calculating the GNP, the figures must be used cautiously. For socialist countries an equivalent, but not directly comparable measure, the Net Material Product, has been used. Services includes items which cannot be included in the other groups. Gross Domestic Product (GDP) is the value of domestic production and excludes net overseas income.

The consumer price index is for early 1991 unless stated otherwise in the last column.

Money exchange rates

national units per US dollar

	unit	1970	1980	Summer 1991
Afghanistan	afghani	45	45.85	60.06
Albania	lek	5	7	6.22
Algeria	dinar	4.94	3.97	18.05
Andorra	Spain-France			
Angola	kwanza	...	...	59.09
Antigua & Barb.	E. Carib. $	2	2.7	2.69
Argentina	austral		1.2	9906
Australia	dollar	0.9	0.85	1.29
Austria	schilling	25.88	13.81	12.59
Bahamas	dollar	1	1	1
Bahrain	dinar	0.48	0.38	0.37
Bangladesh	taka	0	16.25	34.49
Barbados	dollar	2	2.01	2
Belgium	franc	49.68	31.52	36.88
Belize	dollar	1.67	2	1.99
Benin	CFA franc	276	225.8	303.4
Bermuda	dollar	1	1	1
Bhutan	ngultrum	...	...	25.42
Bolivia	boliviano	11.88	24.51	3.57
Botswana	pula	0.72	0.74	2.08
Br. Virgin Is	US dollar	1	1	1
Brazil	cruzeiro	4.95	65.5	323.66
Brunei	dollar			1.74
Bulgaria	lev	1.17	0.85	18.77
Burkina Faso	CFA franc	276	225.8	303.4
Burma	kyat	4.8	6.76	6.53
Burundi	franc	87.5	90	172.77
Cambodia	riel	55.5		795.64
Cameroon	CFA franc	276	225.8	303.4
Canada	dollar	1.01	1.2	1.14
Cape Verde	escudo	28.75	42.49	77.7
Cayman Is	dollar	...	...	0.83
Central Afr. Rep.	CFA franc	276	225.8	303.4
Chad	CFA franc	276	225	303.4
Chile	peso	0.01	39	350.27
China	renminbi yuan	2.46	1.53	5.33
Colombia	peso	19.09	50.92	621.24
Comoros	CFA franc	276	225.8	303.4
Congo	CFA franc	276	225.8	303.4
Costa Rica	colon	6.64	8.57	128.29
Cuba	peso	1	0.71	0.79
Cyprus	pound	0.42	0.37	0.48
Czechoslovakia	koruna	16.2	10.94	29.8
Denmark	krone	7.49	6.02	6.93
Djibouti	franc	214.39	177.72	175.49
Dominica	E. Carib. $	2	2.7	2.69
Dominican Rep.	peso	1	1	12.73
Ecuador	sucre	25	25	1022
Egypt	pound	0.44	0.7	3.28
El Salvador	colon	2.5	2.5	7.99
Equat. Guinea	CFA franc	276	225.8	303.4
Ethiopia	birr	2.5	2.07	2.04
Falkland Is	pound	0.42	0.42	0.61
Fiji	dollar	0.87	0.79	1.5
Finland	markka	4.18	3.84	4.29
Fr. Guiana	franc	5.52	4.52	6.07
France	franc	5.52	4.52	6.07
Gabon	CFA franc	276	225	303.4
Gambia	dalasi	2.09	1.68	8.71
Germany	mark	3.65	1.96	1.79
Ghana	cedi	1.02	2.75	366.44
Gibraltar	pound	0.42	0.42	0.61

	unit	1970	1980	Summer 1991
Greece	drachma	30	46.54	194.73
Greenland	krone	7.49	6.02	6.93
Grenada	E. Carib. $	2.01	2.7	2.69
Guadeloupe	franc	5.52	4.52	6.07
Guatemala	quetzal	1	1	4.87
Guinea	franc	24.69	19.36	616.61
Guinea-Bissau	peso	28.75	34.49	646.45
Guyana	dollar	2.01	2.55	126.29
Haiti	gourde	5	5	5
Honduras	lempira	2	2	5.59
Hong Kong	dollar	6.05	5.14	7.73
Hungary	forint	60	32.21	76.24
Iceland	krona	0.88	6.24	62.62
India	rupee	7.58	7.93	25.42
Indonesia	rupiah	378	626.75	1,950
Iran	rial	76.38	72.32	67.78
Iraq	dinar	0.36	0.3	0.36
Ireland	punt	0.42	0.53	0.67
Israel	shekel	0.35	7.55	2.39
Italy	lira	623	930.5	1,331
Ivory Coast	CFA franc	276	225.8	303.4
Jamaica	dollar	0.83	1.78	9.96
Japan	yen	357.65	203	136.61
Jordan	dinar	0.36	0.31	0.68
Kenya	shilling	7.14	7.57	28.71
Kiribati	Australian $	0.9	0.85	1.29
Korea, N.	won	...	...	0.96
Korea, S.	won	316.65	659.9	729.73
Kuwait	dinar	0.36	0.27	0.29
Laos	kip	0.12	10	696.19
Lebanon	pound	3.25	3.65	892.59
Lesotho	maluti	0.72	0.75	2.87
Liberia	dollar	1	1	1
Libya	dinar	0.36	0.3	0.3
Liechtenstein	Swiss franc	4.32	1.76	1.55
Luxembourg	franc	49.68	31.53	36.88
Macau	pataca	...	...	7.96
Madagascar	franc	276	225.8	1,763
Malawi	kwacha	0.83	0.83	2.91
Malaysia	ringgit	3.01	2.22	2.77
Maldives	rufiya	4.75	7.55	9.92
Mali	CFA franc	276	225.8	303.4
Malta	pound	0.42	0.35	0.34
Martinique	franc	5.52	4.52	6.07
Mauritania	ouguiya	55.2	46.03	82.6
Mauritius	rupee	5.57	7.84	16.67
Mexico	peso	12.5	23.26	3,008
Monaco	French franc	5.52	4.52	6.07
Mongolia	tugrik	4	2.85	3.34
Montserrat	Fr.C.d	2.01	2.7	2.69
Morocco	dirham	5.03	4.33	9.08
Mozambique	metical	...	...	1519
Namibia	SA rand	0.72	0.75	2.87
Nepal	rupee	10.13	12	42.47
Netherlands	guilder	3.5	2.13	2.01
Neths Antilles	guilder	1.87	1.8	1.78
New Zealand	dollar	0.9	1.04	1.77
Nicaragua	cordoba	7.03	10.05	4.97
Niger	CFA franc	276	225.8	303.4
Nigeria	naira	0.36	0.54	10.69
Norway	krone	7.14	5.18	6.98
Oman	rial	0.42	0.35	0.38

	unit	1970	1980	Summer 1991
Pakistan	rupee	4.8	9.9	23.6
Panama	balboa	1	1	1
Papua New G.	kina	0.9	0.64	0.95
Paraguay	guarani	126	126	1317
Peru	sol	38.7	341.17	0.8
Philippines	peso	6.44	7.6	25.72
Poland	zloty	24	33.2	11,317
Portugal	escudo	28.75	53.04	153.4
Puerto Rico	US dollar	1	1	1
Qatar	riyal	4.76	3.64	3.62
Reunion	Fr.franc	5.52	4.52	6.07
Romania	leu	6	18	61.29
Rwanda	franc	100	92.84	127.11
Samoa	tala	0.72	0.93	2.35
San Marino	Ital.lira	623	930.5	1,331
Sao Tome	dobra	28.75	35.48	185.99
Saudi Arabia	riyal	4.5	3.33	3.75
Senegal	CFA franc	276	225.8	303.4
Seychelles	rupee	5.57	6.52	5.33
Sierra Leone	leone	0.84	1.06	272.86
Singapore	dollar	3.08	2.09	1.74
Solomon Is	dollar	0.9	0.8	2.71
Somalia	shilling	7.14	6.3	2,606
South Africa	rand	0.72	0.75	2.86
Spain	peseta	69.72	79.25	112.1
Sri Lanka	rupee	5.96	18	39.94
St. Christopher-N.	E.Carib.d	2	2.7	2.69
St. Helena	pound	0.42	0.42	0.61
St. Lucia	E.Carib.d	2.01	2.7	2.69
St. Pierre & M.	French franc	5.52	4.52	6.07
St. Vincent	E.Carib.d	2	2.7	2.69
Sudan	pound	0.35	0.5	11.39
Surinam	guilder	1.89	1.79	1.78
Swaziland	lilangeni	0.72	0.75	2.87
Sweden	krona	5.17	4.37	6.48
Switzerland	franc	4.32	1.76	1.55
Syria	pound	3.82	3.93	20.89
Taiwan	dollar	...	...	27.07
Tanzania	shilling	7.14	8.18	227.5
Thailand	baht	20.93	20.63	24.81
Togo	CFA franc	276	225.8	303.4
Trinidad & T.	dollar	2	2.4	4.23
Tunisia	dinar	0.52	0.42	0.98
Turkey	lira	14.93	90.15	4,286
Turks & Caicos	US dollar	1	1	1
UAE	dirham	4.76	3.67	3.65
UK	pound	0.42	0.42	0.61
US Virgin Is	US dollar	1	1	1
USA	dollar	1	1	1
USSR	rouble	0.9	0.66	1.79
Uganda	shilling	7.14	7.57	698.67
Uruguay	peso	0.25	10.03	2004
Vanuatu	vatu	91.93	75.21	110.44
Venezuela	bolivar	4.45	4.3	54.67
Vietnam	dong	0.24	2.1	8,653
Yemen (former N.)	rial	5.5	4.56	11.98
Yemen (former S.)	dinar	0.42	0.35	0.43
Yugoslavia	dinar	12.5	29.3	23.13
Zaïre	zaire	0.5	2.99	4,597
Zambia	kwacha	0.71	0.8	65.14
Zimbabwe	dollar	0.72	0.63	3.23

Production statistics

■ Agricultural, forest and fishing products

Apples
thousand tonnes

	1950	1970	1980	1989	1990	Rank	%
World	13,512	26,702	34,503	40,748	40,486		
Africa	61	260	460	951	907		
Asia	676	3,155	7,963	11,470	11,671		
Europe	9,508	13,966	13,221	13,874	13,257		
North America	2,758	3,508	4,474	5,514	5,325		
Oceania	247	561	533	714	695		
South America	273	718	1,406	2,225	2,330		
Algeria	8	8	20	50	54		
Argentina	195	435	945	964	1,050	9	2.6
Australia	199	430	329	344	315		
Austria	339	293	300	321	325		
Belgium-Lux.	286	279	257	274	223		
Brazil	6	15	59	478	598		
Bulgaria	176	389	357	398	305		
Canada	300	413	466	468	500		
Chile	45	142	252	660	665		
China	118	396	2,993	4,514	4,567	2	11.3
Czechoslovakia	206	205	216	55	50		
Denmark	212	121	107	40	30		
France	3,751	3,895	2,412	2,339	2,400	5	5.9
Germany	1,383	2,399	2,134	2,484	2,658	4	6.6
Greece	31	230	287	260	296		
Hungary	158	700	1,027	959	900		
India	...	277	734	950	980		
Iran	...	89	460	1,246	1,250	10	3.1
Iraq	...	36	104	75	80		
Ireland	38	18	10	10	10		
Israel	2	71	111	118	118		
Italy	741	1,923	1,892	1,925	1,950	6	4.8
Japan	382	1,037	887	1,045	1,069	8	2.6
Korea, N.	...	135	460	638	660		
Korea, S.	48	217	459	640	653		
Lebanon	14	98	122	200	205		
Mexico	49	128	280	499	416		
Morocco	2	168	70	300	300		
Netherlands	285	482	407	393	333		
New Zealand	48	131	204	370	380		
Norway	40	51	41	45	45		
Pakistan	3	34	100	220	236		
Peru	...	73	78	75	78		
Poland	138	636	898	1,312	740		
Portugal	36	89	111	80	145		
Romania	67	265	468	780	800		
South Africa	39	214	375	503	450		
Spain	193	535	1,052	815	621		
Sweden	162	137	125	165	147		
Switzerland	472	400	342	280	293		
Syria	6	25	86	175	200		
Tunisia	6	8	17	40	42		
Turkey	102	716	1,420	1,850	1,850	7	4.6
UK	597	524	322	395	249		
Uruguay	24	30	29	35	35		
USA	2,409	2,962	3,722	4,539	4,400	3	10.9
USSR	...	4,533	6,445	6,000	6,300	1	15.6
Yugoslavia	170	362	430	546	500		

Asses
thousand head

	1950	1970	1980	1988	1989	Rank	%
World	36,424	42,285	37,920	42,391	43,201		
Africa	9,130	10,893	11,534	12,500	12,763		
Asia	17,000	21,463	17,140	20,928	21,521		
Europe	2,877	1,775	1,263	1,057	1,048		
North America	3,097	3,779	3,648	3,656	3,658		
South America	3,302	3,743	3,940	3,941	3,901		
Afghanistan	1,000	1,283	1,303	1,300	1,300	9	3
Albania	51	58	52	52	52		
Algeria	302	335	511	327	300		
Argentina	106	94	90	90	90		
Bolivia	411	653	680	620	630		
Botswana	21	51	124	145	146		
Brazil	1,541	1,400	1,334	1,310	1,320	8	3.1
Bulgaria	177	301	338	333	329		
Burkina Faso	112	200	200	330	403		
Cameroon	69	84	33	39	39		
Chad	250	343	262	239	245		
China	10,650	11,623	7,567	10,846	11,296	1	26.1
Colombia	329	327	630	650	650		
Dominican Rep.	80	129	119	142	142		
Ecuador	46	179	207	279	220		
Egypt	816	1,361	1,706	1,950	1,960	5	4.5
Ethiopia	3,183	3,837	3,925	4,800	4,900	2	11.3
Greece	409	375	241	173	170		
Haiti	163	173	206	216	217		
India	1,249	1,000	1,000	1,328	1,400	7	3.2
Iran	1,230	2,021	1,917	1,760	1,740	6	4
Iraq	460	549	439	410	415		
Italy	762	287	135	86	86		
Lesotho	54	74	97	126	127		
Libya	36	95	60	61	62		
Mali	165	470	488	510	530		
Mauritania	80	215	144	149	150		
Mexico	2,683	3,380	3,212	3,185	3,186	3	7.4
Morocco	729	990	1,032	860	870		
Namibia	78	56	67	68	68		
Niger	225	328	467	512	512		
Nigeria	1,260	840	700	700	700		
Pakistan	922	1,900	2,423	3,022	3,108	4	7.2
Peru	407	507	488	490	490		
Portugal	267	168	182	170	170		
Saudi Arabia	22	95	106	110	110		
Senegal	45	178	235	210	210		
South Africa	701	218	210	210	210		
Spain	830	394	193	131	130		
Sudan	500	627	682	650	670		
Syria	256	240	240	177	177		
Tanzania	117	160	163	172	173		
Tunisia	137	180	204	220	224		
Turkey	1,696	1,910	1,349	1,200	1,200	10	2.8
USSR	900	626	384	300	300		
Venezuela	387	493	452	440	440		
Yemen	600	673	695	690	690		
Zimbabwe	91	85	95	101	102		

Bananas
thousand tonnes

	1950	1970	1980	1989	1990	Rank	%
World	12,121	30,837	39,201	44,005	45,036		
Africa	710	3,783	4,357	6,054	6,121		
Asia	3,905	9,283	14,695	17,615	18,770		
Europe	242	462	471	453	453		
North America	3,297	6,197	7,119	7,074	7,065		
Oceania	112	939	1,080	1,229	1,240		
South America	3,855	10,173	11,480	11,449	11,977		
Angola	...	283	303	280	280		
Argentina	10	196	157	240	260		
Australia	75	130	118	207	210		
Bangladesh	...	645	619	600	600		
Bolivia	50	214	213	360	391		
Brazil	2,084	4,809	6,522	5,502	5,552	2	12.3
Burundi	...	810	972	1,608	1,600	8	3.6
Cambodia	113	118	62	112	115		
Cameroon	119	94	90	68	70		
Central Afr. Rep.	142	56	80	87	88		
China	217	694	291	1,602	1,900	6	4.2
Colombia	386	788	1,075	1,350	1,340		
Comoros	...	87	31	47	48		
Costa Rica	434	1,119	1,105	1,400	1,530	9	3.4
Cuba	39	44	149	183	183		
Dominica	9	48	29	40	40		
Dominican Rep.	278	242	299	384	396		
Ecuador	360	2,895	2,192	2,376	2,817	4	6.3
Egypt	39	89	119	300	320		
El Salvador	...	46	52	36	36		
Ethiopia	23	53	73	77	78		
Guadeloupe	84	140	106	93	50		
Guatemala	185	487	619	420	478		
Guinea	57	82	99	110	111		
Haiti	247	47	207	225	230		
Honduras	802	1,426	1,320	1,040	1,050		
India	2,014	3,148	4,425	6,000	6,200	1	13.8
Indonesia	...	1,574	1,622	2,350	2,360	5	5.2
Israel	6	57	67	68	68		
Ivory Coast	19	185	164	130	130		
Jamaica	131	193	110	130	130		
Kenya	...	111	134	149	150		
Lebanon	16	35	19	23	24		
Liberia	...	60	74	80	80		
Madagascar	157	258	264	217	220		
Malawi	4	8	72	82	83		
Malaysia	171	375	455	500	505		
Martinique	75	187	118	200	200		
Mexico	412	934	1,543	1,185	1,065		
Mozambique	26	46	65	84	84		
Nicaragua	13	29	170	109	108		
Pakistan	569	76	130	205	217		
Panama	349	1,007	1,044	1,250	1,250		
Papua New G.	...	758	916	972	979		
Paraguay	140	250	304	425	440		
Philippines	462	893	4,046	3,190	3,250	3	7.2
Portugal	20	35	25	51	50		
Puerto Rico	209	114	108	86	72		
Samoa	39	26	21	24	24		
Somalia	26	136	66	116	117		
South Africa	22	71	103	180	182		
Spain	222	426	443	390	400		
St. Lucia	6	66	65	90	90		
St. Vincent	...	36	26	65	65		
Sudan	10	58	90	63	65		
Surinam	5	44	38	55	55		
Taiwan	107	462	214	228	230		
Tanzania	13	558	773	1,350	1,380	10	3.1
Thailand	250	1,200	2,014	1,610	1,613	7	3.6
Uganda	...	314	362	480	480		
Venezuela	756	969	975	1,134	1,160		
Vietnam	62	399	856	1,186	1,200		
Zaire	16	282	312	350	350		
Zimbabwe	...	37	54	72	72		

Barley
thousand tonnes

	1950	1970	1980	1989	1990	Rank	%
World	59,267	138,485	158,536	167,633	181,248		
Africa	3,486	3,724	3,810	5,633	5,208		
Asia	22,456	27,765	16,345	15,311	18,456		
Europe	15,014	48,479	48,799	71,650	70,630		
North America	10,247	19,740	20,416	20,900	22,822		
Oceania	580	2,594	3,514	4,422	4,012		
South America	1,130	1,052	850	1,208	1,121		
Afghanistan	269	363	330	250	250		
Algeria	808	470	667	700	700		
Argentina	656	497	257	358	350		
Australia	531	2,372	3,245	4,095	3,550		
Austria	210	954	1,288	1,422	1,353		
Belgium-Lux.	254	608	843	693	625		
Bulgaria	332	1,108	1,437	1,568	1,400		
Canada	4,245	10,024	11,034	11,666	13,232	3	7.3
China	12,360	15,934	3,433	3,200	3,200		
Czechoslovakia	1,046	2,543	3,560	3,550	3,650		
Denmark	1,708	5,175	6,239	4,959	4,923	9	2.7
Egypt	123	93	110	126	150		
Ethiopia	720	727	751	1,142	1,188		
Finland	201	943	1,421	1,630	1,641		
France	1,534	8,865	11,031	9,810	10,077	4	5.6
Germany	1,995	7,312	13,385	12,266	13,309	2	7.3
Greece	211	655	848	496	400		
Hungary	654	749	863	1,340	1,300		
India	2,384	2,642	2,003	1,721	1,600		
Iran	767	1,042	1,133	2,750	2,700		
Iraq	722	692	598	663	1,000		
Ireland	163	854	1,435	1,474	1,337		
Italy	258	326	918	1,691	1,700		
Japan	2,020	629	394	371	365		
Korea, N.	155	353	390	630	630		
Korea, S.	846	1,589	1,030	516	416		
Libya	84	70	83	120	130		
Mexico	160	240	512	433	468		
Morocco	1,483	2,190	1,712	2,999	2,224		

Barley
continued

	1950	1970	1980	1989	1990	Rank	%
Netherlands	201	366	265	251	213		
New Zealand	49	222	269	327	462		
Norway	109	545	652	585	600		
Pakistan	150	97	126	123	116		
Peru	208	164	147	125	100		
Poland	1,061	2,182	3,575	3,909	4,000	10	2.2
Portugal	96	64	44	84	35		
Romania	412	615	2,336	3,436	3,100		
Saudi Arabia	...	...	8	350	350		
South Africa	41	27	99	286	250		
Spain	1,909	3,922	6,432	9,308	9,325	5	5.1
Sweden	231	1,836	2,343	1,870	2,052		
Switzerland	55	147	218	360	320		
Syria	321	328	279	271	846		
Tunisia	218	126	279	200	478		
Turkey	2,270	3,720	5,480	4,500	7,000	8	3.9
UK	2,061	8,257	10,032	8,070	7,985	7	4.4
USA	5,843	9,476	8,869	8,800	9,121	6	5
USSR	6,354	35,132	31,901	48,509	59,000	1	32.6
Yugoslavia	323	442	726	702	650		

Beef and veal
thousand tonnes

	1950	1970	1980	1988	1989	Rank	%
World	20,621	38,871	45,507	49,140	49,436		
Africa	1,502	2,367	2,876	3,321	3,455		
Asia	2,393	3,090	3,887	3,509	3,532		
Europe	4,178	8,816	10,556	10,705	10,500		
North America	5,955	11,913	12,264	13,667	14,028		
Oceania	818	1,397	2,192	2,202	2,021		
South America	4,049	5,779	6,934	7,119	7,100		
Afghanistan	25	66	75	65	65		
Algeria	13	24	32	81	80		
Angola	4	40	50	56	56		
Argentina	1,972	2,503	2,989	2,650	2,600	3	5.3
Australia	628	997	1,688	1,588	1,491	8	3
Austria	85	155	202	220	213		
Bangladesh	...	141	183	137	140		
Belgium-Lux.	129	264	307	324	323		
Bolivia	31	51	85	131	135		
Brazil	1,092	1,822	2,149	2,581	2,478	4	5
Bulgaria	38	78	116	120	121		
Burma	1	67	76	84	86		
Cameroon	21	35	50	74	76		
Canada	503	864	979	973	985	10	2
Chile	132	165	167	197	200		
China	1,330	1,433	1,680	642	662		
Colombia	281	422	572	594	602		
Costa Rica	15	44	80	86	86		
Cuba	179	199	146	141	141		
Czechoslovakia	181	336	382	405	409		
Denmark	149	193	246	217	205		
Dominican Rep.	15	32	47	69	71		
Ecuador	23	45	92	95	116		
Egypt	130	119	125	185	188		
Ethiopia	166	243	214	206	237		
Finland	45	109	115	112	106		
France	964	1,577	1,832	1,826	1,716	7	3.5
Germany	662	1,633	1,901	2,047	2,005	5	4.1
Greece	9	86	98	82	85		
Guatemala	35	57	83	57	53		
Honduras	16	32	61	46	46		
Hungary	96	122	145	110	120		
India	173	66	74	232	234		
Indonesia	206	133	131	232	230		
Iran	50	75	171	169	180		
Iraq	21	42	52	47	48		
Ireland	193	217	383	452	441		
Italy	268	1,062	1,126	1,163	1,166	9	2.4
Ivory Coast	2	39	45	37	45		
Japan	61	270	430	569	550		
Kenya	10	131	198	238	228		
Korea, S.	7	37	88	140	120		
Libya	2	4	42	50	53		
Madagascar	122	114	127	137	137		
Mali	27	38	38	65	69		
Mexico	145	450	743	1,224	1,796	6	3.6
Mongolia	74	49	71	71	70		
Morocco	50	91	113	137	139		
Namibia	40	27	33	52	55		
Netherlands	135	337	419	506	478		

	1950	1970	1980	1988	1989	Rank	%
New Zealand	187	387	491	600	515		
Nicaragua	20	59	56	33	29		
Niger	33	29	37	35	35		
Nigeria	113	205	251	256	256		
Norway	41	57	74	77	78		
Pakistan	205	155	167	260	272		
Panama	13	35	41	51	57		
Paraguay	86	124	113	131	135		
Peru	65	96	84	117	112		
Philippines	42	73	77	68	67		
Poland	178	513	646	689	660		
Portugal	31	87	95	110	112		
Romania	101	214	310	230	235		
Somalia	13	45	45	50	50		
South Africa	245	398	579	650	650		
Spain	101	296	409	450	446		
Sudan	51	126	205	238	300		
Sweden	118	159	155	126	138		
Switzerland	75	133	165	157	163		
Tanzania	95	123	129	162	168		
Thailand	16	89	143	158	160		
Turkey	37	187	228	245	250		
Uganda	40	68	82	59	59		
UK	588	921	1,064	949	958		
Uruguay	292	338	336	313	362		
USA	4,785	10,062	10,085	10,879	10,655	1	21.6
USSR	1,726	5,508	6,797	8,616	8,800	2	17.8
Venezuela	67	208	340	307	357		
Vietnam	11	33	33	114	120		
Yugoslavia	97	244	339	301	290		
Zimbabwe	39	127	113	86	89		

Beer
million hectolitres

	1948	1970	1980	1986	1987	Rank	%
World	...	644	900	726	907		
Angola	...	0.7	1.3	...	...		
Argentina	3.4	3.6	2.3	5.5	5.9		
Australia	5.7	15.5	20.2	18.6	18.6		
Austria	1.7	7.4	7.7	9	8.6		
Belgium	11.3	13.0	14	13.7	14		
Bolivia	0.3	0.4	1.2	0.8	0.8		
Brazil	3.2	9.1	29.8	34	33.9	7	3.7
Bulgaria	0.3	3.0	5.3	6	6.2		
Cameroon	0.1	0.7	1.9	5.3	5.9		
Canada	8.2	16.9	22.7	23.5	23.5	10	2.6
Chile	0.9	2.1	1.8	2.1	2.5		
China	...	1.2	6.9	41.3	54	5	6
Colombia	2.3	7.1	12.9	16.9	15.4		
Cuba	1	1	2.4	2.9	3.3		
Czechoslovakia	8.2	21.2	23.4	22.8	22.2		
Denmark	2.5	7.1	9.2	9.1	8.8		
Dominican Rep.	0.1	0.4	1.2	1.1	1.3		
Ecuador	0.1	0.7	2	0.6	0.6		
Ethiopia	0	0.4	1	0.8	0.8		
Finland	0.7	2.4	2.9	3.2	3.5		
France	8.3	20.9	21.3	19	19		
Gabon	1	1	1	0.9	0.9		
Germany	15	98	113	113	111	2	12.2
Greece	0.1	0.8	2.6	3.3	3.3		
Hungary	0.5	5	7.8	9	9		
India	...	0.3	0.9	2	1.5		
Indonesia	1	1	1	0.7	0.8		
Ireland	2.7	3.9	5	4.5	4.4		
Italy	0.9	5.9	8.7	11.4	11.5		
Ivory Coast	0	0.2	1.6	1.4	1.4		
Japan	0.9	30.5	45.6	50.8	54.9	4	6.1
Kenya	0.1	0.8	2.3	2.9	3.1		
Korea, S.	0.1	0.9	5.8	8	8.8		
Mexico	3.4	14.3	26.9	27.4	31.5	8	3.5
Netherlands	1.5	8.8	15.6	18	17.6		
New Zealand	1.6	3.4	3.8	4.1	4.1		
Nigeria	0.1	1.3	4	10.2	6.7		
Norway	0.5	1.5	2	2	2.1		
Panama	1	1	1	0.9	1		
Paraguay	0	0	1	0.8	0.9		
Peru	0.3	2.3	5.3	7.5	8.6		
Philippines	0.3	3.5	5.2	8.8	11.1		
Poland	1.6	10.4	11.2	11.3	11.9		
Portugal	0.2	1.3	3.8	4.1	5.1		
Puerto Rico	0.1	0.9	0.9	0.2	0.3		
Romania	0.4	4.4	9.9	10.6	10.6		

	1948	1970	1980	1986	1987	Rank	%
South Africa	0.1	2.9	7.5	13.8	16.7		
Spain	0.5	12.3	19	23.5	24.2	9	2.7
Sweden	1.7	4.4	4	4	4.1		
Switzerland	1.7	4.7	4.1	4.1	4.1		
Thailand	0	0.4	1.2	0.9	1		
Turkey	0.2	0.5	2.5	1.9	2.4		
UK	46.1	55.2	64.8	59.4	59.9	3	6.6
Uruguay	0	0.8	0.7	0.7	0.7		
USA	107	158	215	230	230	1	25.4
USSR	7.1	41.9	61.3	48.9	50.7	6	5.6
Venezuela	0.6	4.9	5	5	5.0		
Vietnam	0.5	1.5	1	0.9	0.9		
Yugoslavia	1.2	6.7	11.7	11.6	12.1		
Zaire	0.2	3.4	5	3.7	3.7		
Zambia	...	2.6	2.8	...	...		
Zimbabwe	...	0.4	1	1.1	1.2		

Buffaloes
thousand head

	1950	1970	1980	1989	1990	Rank	%
World	88,611	123,487	120,091	138,092	140,863		
Africa	1,212	2,014	2,338	2,500	2,550		
Asia	90,600	120,499	116,438	133,665	136,274		
Europe	539	385	448	368	376		
South America	50	118	500	1,150	1,235		
Bangladesh	...	778	1,560	2,000	2,050	10	1.5
Brazil	50	118	500	1,150	1,235		
Bulgaria	287	80	52	24	24		
Burma	741	1,593	1,906	2,020	2,020		
Cambodia	317	903	386	730	750		
China	21,837	29,509	18,362	21,102	21,430	2	15.2
Egypt	1,212	2,014	2,338	2,650	2,650	9	1.9
India	40,831	56,119	61150	73,700	75,000	1	53.2
Indonesia	2,770	2,956	2,481	3,300	3,463	5	2.5
Iran	111	210	220	230	230		
Iraq	244	280	226	145	145		
Laos	152	932	831	1,050	1,085		
Malaysia	280	303	287	220	220		
Nepal	2,982	3,552	4,206	3,003	2,950	6	2.1
Pakistan	6,420	8,828	11,549	14,349	15,000	3	10.6
Philippines	2,108	4,452	2,841	2,842	2,765	8	2
Romania	84	182	228	210	212		
Sri Lanka	600	744	843	1,000	1,000		
Thailand	5,398	5,642	6,250	4,612	4,719	4	3.4
Turkey	939	1,184	1,030	540	540		
USSR	335	465	350	400	420		
Vietnam	1,049	2,333	2,300	2,907	2,900	7	2.1

Buffalo meat
thousand tonnes

	1970	1980	1988	1989	Rank	%
World	1,142	1,267	1,387	1,487		
Africa	96	119	165	170		
Asia	1,031	1,132	1,209	1,306		
Europe	15	16	13	11		
Bangladesh	5	6	3	3		
Bulgaria	5	4	2	2		
Burma	14	18	22	22		
Cambodia	5	6	10	10		
China	504	496	203	223	3	15
Egypt	96	119	165	170	4	11.4
India	112	122	290	355	1	23.9
Indonesia	32	35	47	40	9	2.7
Iran	9	9	10	10		
Iraq	4	4	3	3		
Laos	10	19	30	31	10	2.1
Macau	0	1	1	2		
Malaysia	8	7	6	6		
Nepal	15	65	91	91	5	6.1
Pakistan	139	177	289	301	2	20.2
Philippines	30	47	44	44	8	3
Romania	9	12	10	8		
Sri Lanka	6	6	...	...		
Thailand	58	72	68	68	7	4.6
Turkey	19	25	11	11		
Vietnam	56	62	82	84	6	5.6

Butter and ghee
thousand tonnes

	1950	1970	1980	1988	1989	Rank	%
World	...	5,957	6,950	7,495	7,611		
Africa	...	168	149	169	176		
Asia	...	1,029	1,234	1,561	1,635		
Europe	...	2,397	2,993	2,704	2,763		
North America	...	703	669	719	746		
Oceania	...	456	362	392	355		
South America	...	119	170	156	158		
Afghanistan	—	5	12	10	10		
Argentina	43	33	32	34	37		
Australia	159	208	99	110	116		
Austria	25	42	42	35	35		
Bangladesh	...	12	15	13	13		
Belgium-Lux.	72	98	101	80	90		
Brazil	23	46	94	79	80		
Bulgaria	4	14	21	24	21		
Burma	...	4	6	13	13		
Canada	140	150	108	108	110		
Chile	7	7	4	5	5		
China	...	71	32	60	63		
Colombia	...	11	12	14	15		
Cuba	1	0	10	9	10		
Czechoslovakia	33	91	125	148	150	10	2
Denmark	156	133	117	94	93		
Egypt	8	56	67	81	82		
Ethiopia	33	9	9	9	9		
Finland	50	91	73	61	60		
France	217	489	584	516	539	5	7.1
Germany	341	718	838	696	695	3	9.1
Greece	5	6	5	5	4		
Hungary	9	20	31	35	38		
India	447	492	640	800	840	2	11
Iran	...	43	69	73	71		
Iraq	...	6	8	8	8		
Ireland	52	75	128	125	132		
Italy	56	68	76	82	80		
Japan	3	44	66	69	85		
Korea, S.	...	...	9	39	33		
Mexico	...	18	22	32	33		
Mongolia	...	5	7	9	10		
Morocco	...	7	12	14	14		
Nepal	...	7	10	11	12		
Netherlands	81	119	189	170	177	9	2.3
New Zealand	177	246	262	280	238	8	3.1
Nigeria	...	7	8	8	8		
Norway	18	20	22	26	25		
Pakistan	...	209	217	312	332	6	4.3
Poland	90	198	296	267	290	7	3.8
Portugal	2	6	5	10	11		
Romania	4	37	37	44	41		
Somalia	...	1	6	11	12		
South Africa	30	50	18	12	16		
Spain	6	7	19	24	26		
Sudan	...	9	12	13	13		
Sweden	102	53	65	68	69		
Switzerland	19	30	36	36	37		
Syria	6	10	14	17	17		
Turkey	59	113	120	118	119		
UK	17	63	168	140	134		
Uruguay	3	7	7	13	12		
USA	697	517	508	548	572	4	7.5
USSR	452	1,085	1,375	1,794	1,780	1	23.4
Venezuela	2	5	9	4	2		
Yugoslavia	9	15	11	11	11		

Cabbages
thousand tonnes

	1950	1970	1980	1988	1989	Rank	%
World	...	27,751	34,507	36,864	36,640		
Africa	...	438	677	824	834		
Asia	...	10,535	14,685	15,897	15,910		
Europe	6,957	7,892	7,718	8,044	8,176		
North America	...	1,409	1,770	1,819	1,827		
Oceania	...	102	132	104	114		
South America	...	409	578	475	480		
Australia	84	71	97	80	88		
Austria	162	101	179	100	97		
Bangladesh	...	51	50	64	65		
Belgium	28	52	55	97	108		
Bulgaria	137	168	155	126	135		

	1950	1970	1980	1988	1989	Rank	%
Canada	57	70	141	126	135		
China	4,000	4,110	5,848	7,750	7,960	2	21.7
Colombia	...	241	455	335	335		
Cuba	...	15	32	40	43		
Czechoslovakia	299	268	261	281	306		
Denmark	63	55	49	50	50		
Ecuador	30	84	24	10	10		
Egypt	...	271	362	452	450		
Ethiopia	...	31	42	47	48		
France	611	406	285	274	233		
Germany	1,002	851	857	986	942	8	2.6
Greece	54	127	159	174	175		
Hong Kong	13	56	81	28	17		
Hungary	186	220	165	155	139		
India	...	380	463	500	550		
Indonesia	...	150	334	500	500		
Ireland	138	125	31	59	62		
Israel	10	15	29	38	38		
Italy	611	801	517	493	547		
Japan	1,040	3,753	3,126	2,874	2,900	3	7.9
Korea, North	...	208	330	405	410		
Korea, South	372	909	3,387	2,508	2,300	4	6.3
Mexico	...	29	63	145	145		
Netherlands	160	229	267	275	270		
New Zealand	30	30	33	23	25		
Norway	62	49	46	50	50		
Peru	30	42	37	48	48		
Philippines	25	44	66	70	76		
Poland	1,068	1,459	1,493	1,574	1,617	5	4.4
Portugal	...	120	143	170	170		
Romania	234	541	855	1,300	1,200	7	3.3
South Africa	...	109	217	250	260		
Spain	762	718	560	426	472		
Sri Lanka	3	14	42	46	40		
Sweden	30	30	36	35	35		
Switzerland	51	26	27	6	6		
Syria	...	19	80	91	43		
Thailand	169	180	200	210	210		
Turkey	300	543	573	655	641		
UK	878	926	796	880	850	9	2.3
USA	1,348	1,229	1,431	1,400	1,400	6	3.8
USSR	...	6,967	8,947	9,700	9,300	1	25.4
Vietnam	...	48	65	90	93		
Yugoslavia	335	605	753	496	678	10	1.9

Camels
thousand head

	1950	1970	1980	1988	1989	Rank	%
World	10,819	15,965	16,691	18,918	19,072		
Africa	7,079	11,348	12,399	14,142	14,267		
Asia	3,419	4,368	4,058	4,511	4,535		
Afghanistan	350	300	267	265	265		
Algeria	144	173	150	114	135		
Chad	266	397	432	518	516	9	2.7
China	...	693	597	475	475	10	2.5
Djibouti	15	22	51	58	58		
Egypt	165	120	84	75	77		
Ethiopia	855	981	980	1,060	1,070	4	5.6
India	638	1,110	1,050	1,390	1,400	3	7.3
Iran	450	135	29	27	27		
Iraq	181	262	72	55	58		
Kenya	152	507	608	790	800	7	4.2
Libya	324	163	134	185	190		
Mali	79	217	230	231	235		
Mauritania	160	707	734	810	810	6	4.2
Mongolia	741	636	601	547	542	8	2.8
Morocco	195	205	157	42	43		
Niger	224	340	404	417	420		
Oman	...	11	41	82	83		
Pakistan	466	700	851	957	972	5	5.1
Saudi Arabia	265	107	277	405	405		
Somalia	2,730	4,700	5,533	6,680	6,700	1	35.1
Sudan	1,550	2,495	2,611	2,850	2,900	2	15.2
Tunisia	186	233	173	184	185		
United Arab Em.	...	100	57	99	100		
USSR	306	248	235	265	270		
Yemen	123	224	158	144	144		

Carrots
thousand tonnes

	1970	1980	1988	1989	Rank	%
World	7,980	10,499	13,704	13,684		
Africa	214	382	505	524		
Asia	1,777	2,677	3,742	3,858		
Europe	3,049	3,591	4,141	4,267		
North America	1,099	1,338	1,921	1,807		
Oceania	117	141	178	186		
South America	281	456	667	642		
Algeria	27	42	150	160		
Argentina	94	106	180	180		
Australia	85	109	144	149		
Austria	20	22	21	19		
Belgium-Lux.	67	84	134	93		
Bolivia	15	19	33	35		
Bulgaria	32	28	24	26		
Canada	180	247	271	276		
Chile	34	84	153	130		
China	1,135	1,791	2,599	2,697	1	19.7
Colombia	88	153	160	160		
Czechoslovakia	136	126	130	160		
Denmark	50	54	72	68		
Ecuador	3	10	30	30		
Egypt	51	133	120	125		
Finland	15	26	39	35		
France	552	503	540	485	8	3.5
Germany	335	421	404	475	9	3.5
Greece	22	30	31	30		
Hungary	106	107	131	112		
Indonesia	...	45	60	80		
Iraq	9	13	22	20		
Ireland	...	38	42	40		
Israel	52	49	73	76		
Italy	246	275	300	508	7	3.7
Japan	498	597	679	690	5	5
Korea, South	4	78	95	90		
Libya	2	9	27	28		
Martinique	4	12	12	14		
Mexico	66	85	150	155		
Morocco	38	52	61	62		
Netherlands	137	163	291	300	10	2.2
New Zealand	31	32	34	37		
Norway	52	47	65	62		
Peru	24	38	55	55		
Poland	438	595	743	757	4	5.5
Portugal	75	82	80	82		
Saudi Arabia	1	5	25	25		
South Africa	65	106	89	90		
Spain	38	140	180	180		
Sweden	41	55	83	57		
Switzerland	22	49	55	57		
Tunisia	25	35	53	53		
Turkey	41	72	157	146		
UK	611	649	663	580	6	4.2
Uruguay	11	12	12	14		
USA	838	980	1,474	1,347	3	9.8
USSR	1,371	1,914	2,550	2,400	2	17.5
Venezuela	11	28	31	31		
Yugoslavia	55	96	105	118		

Cassava
thousand tonnes

	1950	1970	1980	1988	1989	Rank	%
World	55,200	96,695	121,472	141,110	147,500		
Africa	29,700	38,339	46,445	59,757	62,098		
Asia	9,720	22,943	43,970	52,069	54,378		
North America	570	783	900	894	941		
Oceania	580	2,594	3,514	174	180		
South America	15,100	34,444	30,000	28,215	29,902		
Angola	...	1,597	1,850	1,980	1,920		
Argentina	344	296	212	150	150		
Benin	700	533	894	780	1,002		
Bolivia	77	223	217	430	312		
Brazil	12,466	29,922	24,315	21,612	23,247	2	15.8
Burundi	700	843	412	567	600		
Cambodia	...	...	146	115	110		
Cameroon	1,000	637	1,273	1,500	1,530		
Central Afr. Rep.	...	767	950	533	540		
Chad	...	140	185	330	330		
China	...	1,938	3,390	3,277	3,185		
Colombia	825	1,380	2,070	1,282	1,396		

Cassava *continued*

	1950	1970	1980	1988	1989	Rank	%
Congo	...	461	631	761	760		
Cuba	179	217	280	305	305		
Dominican Rep.	148	173	98	126	156		
Ecuador	18	382	216	123	135		
Gabon	100	155	242	260	260		
Ghana	512	1,533	1,803	2,788	3,327	10	2.3
Guinea	218	482	480	400	358		
Haiti	104	205	250	260	280		
India	1,255	4,993	5,904	5,213	5,250	7	3.6
Indonesia	6,817	10,695	13,670	15,471	16,581	3	11.2
Ivory Coast	623	546	1,067	1,333	1,300		
Kenya	...	510	588	600	620		
Liberia	397	264	300	447	400		
Madagascar	866	1,227	1,641	2,200	2,250		
Malawi	545	150	292	135	155		
Malaysia	350	271	347	390	400		
Mozambique	...	2,549	3,100	3,400	3,400	9	2.3
Niger	21	182	191	212	212		
Nigeria	10,422	9,473	11,000	15,000	16,500	4	11.1
Papua New G.	...	...	99	110	111		
Paraguay	903	1,442	1,973	3,891	4,000	8	2.7
Peru	288	477	481	392	325		
Philippines	290	436	2,275	1,846	1,847		
Rwanda	300	333	578	390	360		
Senegal	61	159	26	15	12		
Sierra Leone	36	75	92	116	116		
Sri Lanka	220	376	524	492	490		
Sudan	...	133	125	65	15		
Tanzania	900	3,373	5,547	6,200	6,300	6	4.3
Thailand	269	3,208	15,128	22,307	23,460	1	15.9
Togo	233	562	404	413	403		
Uganda	2,024	1,058	2,122	2,502	2,500		
Venezuela	152	317	322	328	328		
Vietnam	...	950	3,378	2,810	2,900		
Zaïre	6,000	10,232	12,500	16,254	16,300	5	11.1
Zambia	60	159	183	240	260		
Zimbabwe	...	46	55	87	88		

Cattle

thousand head

	1950	1970	1980	1989	1990	Rank	%
World	799,216	1,097,120	1,201,961	1,273,735	1,282,862		
Africa	97,473	153,683	169,296	185,728	189,903		
Asia	276,312	346,311	359,123	388,430	393,289		
Europe	99,987	123,593	133,740	125,967	125,398		
North America	114,131	167,069	177,209	163,245	160,767		
Oceania	19,725	31,608	35,216	30,835	31,266		
South America	135,820	178,149	212,600	259,935	263,840		
Afghanistan	2,500	3,502	3,800	1,600	1,600		
Algeria	784	1,047	1,351	1,410	1,410		
Angola	1,263	2,514	3,150	3,100	3,100		
Argentina	42,320	48,841	55,620	50,782	50,581	6	3.9
Australia	14,534	22,382	26,165	22,434	22,602		
Austria	2,207	2,440	2,560	2,541	2,562		
Bangladesh	...	25,395	33,247	23,000	23,113	10	1.8
Belgium	2,017	2,882	3,104	3,174	3,069		
Bolivia	2,227	2,298	4,030	7,828	8,073		
Botswana	1,019	1,660	2,954	2,350	2,350		
Brazil	51,305	75,658	91,333	132,934	140,000	2	10.5
Bulgaria	1,688	1,277	1,782	1,613	1,577		
Burkina Faso	1,120	2,556	2,755	2,850	2,850		
Burma	4,494	6,949	8,479	9,111	9,150		
Cambodia	842	2,233	800	2,000	2,100		
Cameroon	1,540	2,308	3,200	4,582	4,600		
Canada	7,945	11,678	12,399	12,199	12,287		
Central Afr. Rep.	306	503	1,240	2,495	2,500		
Chad	3,260	4,500	3,900	4,115	4,200		
Chile	2,293	2,923	3,661	3,257	3,700		
China	46,404	63,184	52,839	74,800	76,980	5	6
Colombia	14,910	20,233	24,109	24,598	25,000	9	1.9
Costa Rica	601	1,498	2,183	1,735	1,735		
Cuba	4,333	6,146	5,900	4,927	4,920		
Czechoslovakia	3,966	4,253	4,935	5,075	5,129		
Denmark	2,998	2,855	2,976	2,232	2,190		
Dominican Rep.	711	1,176	2,153	2,245	2,200		
Ecuador	1,467	2,434	2,931	4,024	4,100		
Egypt	1,356	2,108	1,930	1,950	2,000		
Ethiopia	18,901	26,310	26,000	28,900	30,000	7	2.3
Finland	1,700	1,907	1,752	1,379	1,400		
France	15,606	21,674	23,793	21,780	21,200		
Germany	14,151	19,281	20,672	20,369	20,287		
Guatemala	977	1,474	1,653	2,023	2,023		
Guinea	1,060	1,300	1,753	1,800	1,800		
Haiti	582	800	1,100	1,550	1,500		
Honduras	884	1,578	2,277	2,600	2,600		
Hungary	2,052	1,952	1,936	1,690	1,599		
India	155,239	177,447	182,116	195,500	197,300	1	15.4
Indonesia	4,112	6,273	6,412	10,290	10,300		
Iran	3,388	5,239	8,000	8,000	8,000		
Iraq	1,510	2,633	2,618	1,650	1,700		
Ireland	4,211	5,926	6,936	5,637	5,899		
Italy	8,281	9,436	8,697	8,737	8,746		
Japan	2,397	3,582	4,261	4,682	4,770		
Kenya	5,859	8,433	10,652	13,247	13,793		
Korea, S.	598	1,226	1,595	2,040	2,050		
Laos	179	418	805	805	840		
Madagascar	5,709	10,073	10,100	10,243	10,250		
Mali	3,000	5,400	4,953	4,880	4,900		
Mauritania	514	2,367	1,195	1,260	1,270		
Mexico	14,489	24,668	30,933	31,156	28,200	8	2.2
Mongolia	1,988	2,026	2,452	2,541	2,600		
Morocco	2,188	3,617	3,271	3,500	3,600		
Mozambique	699	1,274	1,400	1,370	1,380		
Namibia	1,492	2,500	2,300	2,060	2,070		
Nepal	5,507	6,292	6,900	6,285	6,350		
Netherlands	2,659	4,281	4,975	4,606	4,731		
New Zealand	4,932	8,734	8,485	8,279	7,999		
Nicaragua	1,068	2,275	2,480	1,650	1,650		
Niger	2,818	4,168	3,206	3,600	3,700		
Nigeria	9,546	11,183	12,267	12,200	12,200		
Norway	1,204	949	978	932	940		
Pakistan	29,970	13,667	15,038	17,363	17,573		
Panama	567	1,201	1,522	1,500	1,500		
Paraguay	4,600	4,433	5,301	8,074	8,100		
Peru	2,830	3,999	3,913	4,003	3,800		
Philippines	710	1,701	1,872	1,682	1,629		
Poland	6,895	10,990	12,495	10,733	10,600		
Romania	4,387	4,947	6,260	7,170	7,167		
Senegal	720	2,557	2,344	2,673	2,700		
Somalia	1,430	3,750	3,900	5,200	5,400		
South Africa	11,912	11,114	12,698	11,850	11,850		
Spain	4,356	4,236	4,547	5,188	5,300		
Sri Lanka	1,131	1,601	1,637	1,800	1,800		
Sudan	3,957	12,300	18,148	20,500	21,000		
Sweden	2,595	1,934	1,927	1,688	1,678		
Switzerland	1,544	1,866	2,008	1,858	1,848		
Tanzania	6,356	11,753	12,556	14,000	14,500		
Thailand	4,608	4,470	4,971	5,120	5,367		
Turkey	10,121	13,235	15,467	11,800	11,600		
Uganda	2,547	3,987	4,933	3,912	3,950		
UK	10,277	12,632	13,375	11,902	11,933		
Uruguay	8,154	8,730	10,741	9,447	8,723		
USA	80,569	112,321	112,126	99,180	99,337	4	7.7
USSR	55,778	96,707	114,748	119,600	118,400	3	9.3
Venezuela	5,769	8,292	10,607	13,074	13,819		
Vietnam	1,541	1,750	1,685	3,026	3,100		
Yemen	972	842	900	1,179	1,200		
Yugoslavia	4,866	5,143	5,470	4,559	4,705		
Zaïre	640	968	1,183	1,450	1,500		
Zambia	881	1,580	2,152	2,770	2,800		
Zimbabwe	2,956	5,127	5,370	6,453	6,550		

Cheese

thousand tonnes

	1950	1970	1980	1989	1990	Rank	%
World	...	7,732	11,363	14,494	14,838		
Africa	...	297	364	474	479		
Asia	...	522	674	729	740		
Europe	...	3,783	5,616	6,914	7,007		
North America	...	1,574	2,474	3,435	3,622		
Oceania	...	178	237	319	185		
South America	...	372	460	567	567		
Afghanistan	...	9	10	10	10		
Albania	1	6	10	14	14		
Argentina	95	178	245	281	296		
Australia	45	76	144	191	171		
Austria	16	59	75	114	109		
Belgium-Lux.	13	39	46	67	69		
Brazil	24	49	58	60	60		
Bulgaria	38	127	162	195	192		
Burma	4	12	14	36	36		
Canada	48	121	204	279	274		
Chile	10	18	19	28	25		
China	...	171	171	138	140		
Colombia	...	39	44	51	51		
Cuba	3	1	11	16	17		
Cyprus	2	9	13	8	8		
Czechoslovakia	13	111	175	233	234		
Denmark	69	113	219	277	302	10	2
Ecuador	...	10	12	13	13		
Egypt	129	204	243	315	320	8	2.2
El Salvador	...	15	18	15	16		
Finland	14	40	72	90	93		
France	252	780	1,151	1,405	1,400	3	9.4
Germany	193	660	986	1,051	1,358	4	9.2
Greece	41	142	171	221	221		
Guatemala	1	11	14	16	16		
Honduras	8	7	8	8	8		
Hungary	5	42	68	86	88		
Iran	15	74	99	112	113		
Iraq	13	25	27	34	34		
Ireland	3	30	53	80	78		
Israel	7	32	56	73	74		
Italy	265	470	603	719	722	5	4.9
Japan	...	40	68	85	85		
Lebanon	5	7	10	11	10		
Mexico	7	77	97	140	141		
Netherlands	130	284	441	572	596	6	4
New Zealand	101	102	93	128	129		
Nicaragua	...	14	10	9	9		
Norway	24	53	70	83	77		
Peru	8	36	34	17	17		
Poland	13	249	389	447	440	7	3
Portugal	8	23	37	54	51		
Romania	13	83	132	99	101		
South Africa	9	20	31	40	40		
Spain	41	80	135	164	162		
Sudan	...	38	55	43	64		
Sweden	57	64	101	115	116		
Switzerland	53	87	123	132	130		
Syria	11	27	50	62	62		
Turkey	44	94	126	142	143		
UK	44	139	237	279	307	9	2.1
Uruguay	5	8	12	17	17		
USA	673	1,319	2,103	2,942	3,132	1	21.1
USSR	229	1,006	1,539	2,056	2,112	2	14.2
Venezuela	16	28	28	94	82		
Yemen	...	13	17	13	13		
Yugoslavia	54	99	147	136	137		

Cocoa beans

thousand tonnes

	1950	1970	1980	1989	1990	Rank	%
World	763	1,491	1,636	2,438	2,436		
Africa	500	1,085	999	1,333	1,171		
Asia	4	11	48	358	342		
North America	65	81	94	123	125		
Oceania	4	31	34	53	45		
South America	191	282	461	572	557		
Bolivia	2	1	2	3	3		
Brazil	130	183	317	392	372	2	15.3
Cameroon	52	115	118	120	109	6	4.5
Colombia	11	20	36	56	62	8	2.5
Congo	...	2	4	2	2		
Costa Rica	5	5	6	4	4		
Cuba	3	1	2	3	2		
Dominican Rep.	30	32	34	49	50	9	2.1
Ecuador	27	57	88	94	95	7	3.9
Equat. Guinea	16	25	8	8	7		
Gabon	3	5	4	2	2		
Ghana	253	430	259	295	299	3	12.3
Grenada	3	3	3	2	2		
Guatemala	1	1	3	3	3		
Guinea	...	...	4	3	4		
Haiti	2	3	3	5	5		
Indonesia	1	2	9	6	6		
Ivory Coast	53	195	403	710	740	1	30.4
Jamaica	2	2	2	2	2		
Liberia	1	2	4	1	1		
Madagascar	...	1	2	3	3		
Malaysia	—	4	32	225	250	4	10.3
Mexico	8	27	36	50	50	10	2.1
Nigeria	105	261	158	160	170	5	7
Panama	2	1	1	1	1		

Cocoa beans · continued

	1950	1970	1980	1989	1990	Rank	%
Papua New G.	...	27	30	47	40		
Peru	4	2	4	12	11		
Philippines	1	4	5	9	10		
Samoa	3	3	2	1	1		
Sao Tome	8	10	8	5	4		
Sierra Leone	2	5	8	8	9		
Sri Lanka	3	2	2	3	3		
Tanzania	...	...	2	2	2		
Togo	4	27	15	8	9		
Trinidad & T.	8	4	2	1	2		
Venezuela	18	19	14	14	14		
Zaire	2	6	4	5	5		

Coconuts
thousand tonnes

	1950	1970	1980	1988	1989	Rank	%
World	18,078	27,800	35,270	37,624	38,091		
Africa	1,171	1,459	1,491	1,929	1,943		
Asia	14,496	22,054	29,444	30,557	31,171		
North America	634	1,538	1,408	1,848	1,684		
Oceania	1,355	2,137	2,340	2,219	2,235		
South America	422	612	587	1,072	1,058		
Bangladesh	53	66	75	86	88		
Benin	38	20	20	20	20		
Brazil	241	331	254	695	686	9	1.8
Burma	17	67	97	330	330		
Cambodia	10	44	28	44	46		
China	...	53	59	83	85		
Colombia	36	35	53	78	79		
Comoros	60	69	57	53	54		
Dominican Rep.	29	57	69	24	24		
Ecuador	13	25	78	47	42		
El Salvador	...	30	55	75	75		
Fiji	215	260	215	152	152		
Fr. Polynesia	169	178	132	95	130		
Ghana	73	257	160	200	200		
Guam	...	22	30	37	38		
Guinea-Bissau	...	31	25	25	25		
Guyana	44	56	29	45	45		
Haiti	27	24	34	38	36		
India	3,656	4,472	4,444	4,600	4,739	3	12.4
Indonesia	3,614	5,892	10,800	12,100	12,300	1	32.3
Ivory Coast	10	49	155	470	470		
Jamaica	74	150	186	205	200		
Kenya	65	77	90	73	74		
Kiribati	45	62	65	110	107		
Madagascar	...	19	40	82	83		
Malaysia	1,006	1,039	1,221	1,186	1,186	6	3.1
Martinique	18	2	2	2	2		
Mexico	204	998	780	1,159	1,006	7	2.6
Mozambique	338	400	413	420	420		
New Caledonia	20	21	14	14	14		
Nigeria	150	86	90	105	105		
Pacific Is	75	101	199	195	196		
Panama	50	28	24	20	22		
Papua New G.	383	741	832	832	798	8	2.1
Philippines	4,453	7,601	9,850	8,640	8,300	2	21.8
Puerto Rico	23	9	9	6	7		
Samoa	122	193	207	185	210		
Sao Tome	41	42	35	35	35		
Seychelles	53	42	29	10	11		
Solomon Is.	77	184	217	209	203		
Sri Lanka	1,975	1,963	1,692	1,469	1,698	4	4.5
St. Lucia	21	35	38	29	29		
Tanzania	246	310	310	350	360		
Thailand	413	713	829	1,378	1,437	5	3.8
Togo	45	17	14	14	14		
Tonga	100	88	99	53	65		
Trinidad & T.	129	105	71	62	62		
Vanuatu	135	247	273	304	290		
Venezuela	74	147	152	181	180		
Vietnam	80	100	329	625	676	10	1.8

Coffee
thousand tonnes

	1950	1970	1980	1989	1990	Rank	%
World	2,222	4,264	5,173	6,021	6,040		
Africa	278	1,270	1,211	1,244	1,221		
Asia	74	349	572	1,105	1,006		
North America	366	733	916	1,058	1,180		
Oceania	2	28	50	69	65		
South America	1,502	1,885	2,425	2,545	2,568		
Angola	50	216	47	5	5		
Bolivia	2	11	21	24	25		
Brazil	1,077	1,197	1,403	1,532	1,440	1	23.8
Burundi	7	21	25	32	40		
Cameroon	9	90	103	84	100		
Central Afr. Rep.	4	10	17	26	27		
China	...	4	10	29	31		
Colombia	352	483	748	664	780	2	12.9
Costa Rica	23	82	109	147	172	8	2.8
Cuba	31	29	25	29	24		
Dominican Rep.	28	44	48	42	49		
Ecuador	21	60	82	128	134		
El Salvador	75	139	165	97	165	10	2.7
Equat. Guinea	6	7	6	7	7		
Ethiopia	27	172	191	200	195	7	3.2
Ghana	1	6	2	1	1		
Guatemala	58	125	162	220	240	5	4
Guinea	4	9	15	24	26		
Haiti	35	31	32	33	33		
Honduras	13	39	80	90	104		
India	20	82	130	215	132		
Indonesia	39	186	253	399	390	3	6.5
Ivory Coast	50	243	292	239	220	6	3.6
Kenya	11	57	84	104	96		
Liberia	...	5	11	5	5		
Madagascar	33	63	92	88	89		
Malaysia	2	6	9	6	4		
Mexico	63	182	216	326	309	4	5.1
Nicaragua	20	38	56	44	51		
Nigeria	...	4	4	5	5		
Panama	3	5	6	12	13		
Papua New G.	...	26	49	69	65		
Paraguay	...	5	9	18	18		
Peru	6	68	98	106	101		
Philippines	4	48	133	156	105		
Puerto Rico	10	12	12	15	14		
Rwanda	5	14	27	43	37		
Sierra Leone	1	6	10	9	9		
Sri Lanka	...	9	11	6	6		
Tanzania	14	49	57	52	50		
Thailand	...	...	10	46	50		
Togo	3	11	8	11	12		
Uganda	35	215	124	174	168	9	2.8
USA	3	1	1	1	1		
Venezuela	44	60	61	73	70		
Vietnam	2	5	6	14	15		
Yemen	5	4	4	5	5		
Zaire	21	69	81	103	98		
Zimbabwe	...	...	5	13	14		

Copra
thousand tonnes

	1950	1970	1980	1989	1990	Rank	%
World	2,636	3,832	4,733	4,344	5,011		
Africa	101	150	169	238	240		
Asia	2,223	3,153	4,014	3,546	4,206		
North America	73	202	183	252	237		
Oceania	219	297	335	270	288		
South America	20	29	31	38	39		
Brazil	0	2	2	3	3		
Cambodia	1	8	5	8	8		
Comoros	4	6	4	4	4		
Dominican Rep.	1	6	10	23	25		
Ecuador	3	4	14	7	7		
Fiji	35	30	23	13	13		
Fr. Polynesia	28	24	17	11	11		
Ghana	2	10	7	7	7		
Guinea-Bissau	...	6	5	5	5		
Guyana	3	6	5	5	5		
India	202	354	372	370	390	3	7.8
Indonesia	719	753	1,241	1,027	1,260	2	25.1
Ivory Coast	...	6	23	75	75	9	1.5
Jamaica	6	17	7	8	8		

	1950	1970	1980	1989	1990	Rank	%
Kenya	1	6	7	13	13		
Kiribati	8	8	8	13	14		
Madagascar	...	2	6	9	9		
Malaysia	164	186	210	88	93	8	1.9
Mexico	42	146	138	200	183	4	3.7
Mozambique	46	59	68	69	70	10	1.4
Nigeria	3	8	10	13	13		
Pacific Is.	12	12	27	18	18		
Papua New G.	64	131	151	134	116	7	2.3
Philippines	875	1,582	1,960	1,700	2,072	1	41.3
Samoa	18	15	21	15	26		
Sao Tome	4	5	4	4	4		
Seychelles	7	5	4	2	2		
Singapore	...	4	4	8	8		
Solomon Is.	12	25	30	35	45		
Sri Lanka	230	209	129	151	170	5	3.4
St. Lucia	3	6	6	5	5		
Tanzania	28	31	28	30	30		
Thailand	15	31	48	65	68		
Togo	4	3	2	2	2		
Tonga	16	10	13	2	2		
Trinidad & T.	16	13	5	5	5		
Vanuatu	23	35	39	25	40		
Venezuela	9	16	12	22	23		
Vietnam	16	20	39	124	131	6	2.6

Cotton lint
thousand tonnes

	1950	1970	1980	1989	1990	Rank	%
World	7,478	12,010	14,402	16,700	18,148		
Africa	678	1,317	1,167	1,363	1,362		
Asia	1,871	4,564	5,612	7,873	8,582		
Europe	59	190	178	327	375		
North America	3,345	2,787	3,656	2,891	3,505		
Oceania	...	27	78	286	295		
South America	554	991	1,025	1,304	1,416		
Afghanistan	7	25	31	20	20		
Angola	6	28	12	11	11		
Argentina	112	114	134	195	270	10	1.5
Australia	...	27	78	286	295	9	1.6
Benin	...	...	7	43	46		
Brazil	350	631	578	625	660	6	3.6
Burkina Faso	...	10	24	59	68		
Burma	19	14	20	20	22		
Cameroon	...	21	29	43	47		
Chad	18	40	30	54	55		
China	786	1,988	2,638	3,790	4,421	1	24.4
Colombia	8	122	99	117	122		
Egypt	396	520	511	298	341	8	1.9
El Salvador	8	49	59	10	7		
Ethiopia	2	13	20	18	19		
Greece	21	115	107	255	265		
Guatemala	2	71	148	38	48		
India	575	1,088	1,323	1,612	1,497	5	8.2
Iran	26	155	82	113	144		
Israel	...	37	82	46	45		
Ivory Coast	1	14	54	128	106		
Mali	1	22	48	97	97		
Mexico	222	354	352	162	168		
Mozambique	24	42	18	27	28		
Nicaragua	8	80	70	22	30		
Nigeria	13	62	32	38	42		
Pakistan	261	595	731	1,456	1,541	4	8.5
Paraguay	13	10	84	220	230		
Peru	76	89	92	103	93		
Senegal	...	5	10	15	15		
South Africa	3	19	53	54	56		
Spain	8	52	56	62	101		
Sudan	67	235	117	142	125		
Syria	30	150	122	155	180		
Tanzania	10	70	56	84	59		
Thailand	7	31	65	29	33		
Turkey	119	441	489	585	628	7	3.5
Uganda	66	79	8	6	6		
USA	3,105	2,225	3,011	2,653	3,245	2	17.9
USSR	971	2,132	2,685	2,656	2,613	3	14.4
Venezuela	4	15	16	18	31		
Zaire	46	21	10	26	26		
Zimbabwe	...	43	55	90	79		

Cotton seed
thousand tonnes

	1950	1970	1980	1989	1990	Rank	%
World	13,742	22,492	27,738	31,061	33,227		
Africa	1,285	2,427	2,050	2,378	2,296		
Asia	3,726	8,848	10,962	15,472	16,811		
Europe	117	365	326	632	530		
North America	5,697	4,729	6,045	4,619	5,541		
Oceania	...	45	128	449	493		
South America	1,032	1,836	1,976	2,286	2,491		
Afghanistan	14	49	59	40	40		
Angola	12	55	23	22	22		
Argentina	214	220	253	318	435		
Australia	...	45	128	449	493	10	1.5
Benin	3	18	14	60	60		
Brazil	749	1,197	1,139	1,131	1,220	6	3.7
Bulgaria	22	26	10	9	9		
Burkina Faso	7	19	43	110	110		
Burma	37	25	39	40	46		
Cameroon	1	33	49	44	48		
Central Afr. Rep.	24	33	18	15	15		
Chad	35	65	51	87	87		
China	1,572	3,977	5,276	7,580	8,842	1	26.6
Colombia	15	205	179	227	230		
Egypt	725	901	841	480	530	8	1.6
El Salvador	13	80	99	12	12		
Ethiopia	5	27	39	36	38		
Greece	43	227	211	510	515	9	1.5
Guatemala	3	138	256	62	69		
India	1,150	2,176	2,646	3,230	3,000	5	9
Iran	53	285	165	241	307		
Iraq	8	29	10	10	10		
Israel	...	58	134	76	74		
Ivory Coast	2	20	73	155	130		
Korea, N.	20	6	6	16	17		
Korea, S.	42	9	4	4	4		
Mali	3	39	84	190	135		
Mexico	384	623	558	255	263		
Morocco	...	...	11	19	21		
Mozambique	57	84	34	62	64		
Nicaragua	16	130	113	35	42		
Nigeria	26	125	63	120	108		
Pakistan	522	1,190	1,461	2,911	3,082	4	9.3
Paraguay	26	20	166	339	375		
Peru	121	148	176	205	160		
Romania	22	...	1	1	1		
Senegal	...	9	18	18	18		
South Africa	5	38	96	99	100		
Spain	16	94	86	100	160		
Sudan	121	442	221	280	280		
Syria	58	241	211	267	248		
Tanzania	20	135	108	163	114		
Thailand	13	62	129	57	65		
Turkey	227	705	782	935	1,005	7	3
Uganda	134	173	15	10	12		
USA	5,277	3,742	4,990	4,242	5,140	2	15.5
USSR	1,885	4,241	6,251	5,225	4,900	3	14.7
Venezuela	6	25	27	50	54		
Zaïre	93	42	17	50	50		
Zimbabwe	1	86	106	164	148		

Cucumbers and gherkins
thousand tonnes

	1950	1970	1980	1988	1989	Rank	%
World	...	8,417	11,190	13,108	12,774		
Africa	...	184	326	345	356		
Asia	...	3,986	5,531	7,135	7,175		
Europe	793	1,791	2,578	3,071	2,744		
North America	...	898	1,161	1,017	1,029		
Oceania	...	13	17	21	21		
South America	...	39	43	49	49		
Australia	5	10	14	18	17		
Austria	24	45	41	25	25		
Bangladesh	...	17	12	13	13		
Belgium	15	29	41	42	55		
Bulgaria	22	52	146	155	127		
Canada	35	71	82	85	85		
Chile	...	23	25	28	28		
China	...	1,987	2,662	3,814	3,930	1	30.8
Cuba	8	13	29	40	40		
Czechoslovakia	196	122	120	103	63		
Egypt	25	165	290	290	300	10	2.3
Finland	10	14	22	37	35		
France	41	81	88	104	105		
Germany	124	116	130	148	152		
Greece	26	96	135	168	156		
Hungary	62	115	135	159	130		
Indonesia	...	133	165	210	222		
Iraq	60	127	210	291	345	9	2.7
Israel	18	48	49	65	63		
Italy	37	103	121	113	96		
Japan	324	976	1,060	975	980	3	7.7
Jordan	8	7	35	67	68		
Korea, North	...	24	38	63	64		
Korea, South	33	94	119	90	90		
Kuwait	...	...	...	24	24		
Lebanon	12	28	60	77	79		
Malaysia	...	1	28	23	23		
Mexico	...	98	194	270	280		
Netherlands	72	300	372	413	415	7	3.2
Poland	111	366	415	496	352	8	2.8
Romania	...	103	284	500	486	6	3.8
Saudi Arabia	...	5	11	84	84		
South Africa	5	11	20	27	28		
Spain	49	85	277	355	300		
Sri Lanka	3	5	22	23	24		
Sweden	15	23	28	16	16		
Syria	25	61	240	253	153		
Thailand	35	62	269	205	205		
Tunisia	3	5	13	22	22		
Turkey	300	394	503	800	750	4	5.9
UK	26	34	54	87	88		
USA	392	705	834	591	593	5	4.6
USSR	...	1,506	1,533	1,470	1,400	2	10.9
Yemen	...	...	3	14	14		
Yugoslavia	27	73	131	114	106		

Dates
thousand tonnes

	1950	1970	1980	1989	1990	Rank	%
World	1,326	2,128	2,664	3,329	3,378		
Africa	453	821	1,036	1,188	1,215		
Asia	848	1,272	1,595	2,107	2,129		
Europe	7	17	10	11	11		
North America	20	18	22	22	22		
Algeria	98	145	205	210	212	6	6.2
Bahrain	...	15	39	46	46		
Chad	20	22	27	32	32		
Egypt	185	360	422	560	580	1	17.1
Iran	230	293	300	539	540	2	15.9
Iraq	313	410	397	488	490	4	14.4
Israel	—	2	5	9	9		
Libya	34	52	90	105	108	9	3.2
Mauritania	10	14	14	13	13		
Mexico	5	2	2	2	2		
Morocco	43	92	104	46	46		
Niger	3	5	6	7	7		
Oman	...	44	57	121	125	8	3.7
Pakistan	80	161	202	290	302	5	8.9
Saudi Arabia	183	237	418	500	500	3	14.7
Somalia	...	5	6	10	10		
Spain	7	16	10	11	11		
Sudan	31	91	114	130	130	7	3.8
Tunisia	34	32	51	72	73	10	2.1
UAE	...	8	48	65	68		
USA	15	16	20	20	20		
Yemen	56	101	124	28	28		

Eggs – hen's
thousand tonnes

	1950	1970	1980	1989	1990	Rank	%
World	9,364	20,922	28,316	35,509	35,883		
Africa	302	578	915	1,427	1,475		
Asia	1,132	6,184	9,298	14,096	14,505		
Europe	2,598	5,892	7,119	7,171	7,046		
North America	4,088	4,942	5,357	5,706	5,784		
Oceania	146	242	269	244	246		
South America	413	835	1,614	2,184	2,286		
Afghanistan	13	15	17	14	14		
Argentina	130	182	254	287	290		
Australia	116	185	204	187	188		
Austria	24	86	96	97	96		
Bangladesh	...	34	47	56	58		
Belgium-Lux.	109	229	192	155	158		
Benin	11	2	3	17	18		
Brazil	177	337	811	1,189	1,300	5	3.6
Bulgaria	36	89	131	152	144		
Burma	6	17	26	53	55		
Canada	216	328	323	324	303		
Chile	13	58	62	83	83		
China	566	3,309	4,632	7,392	7,700	1	21.5
Colombia	43	95	191	253	258		
Cuba	14	64	101	114	113		
Czechoslovakia	76	186	210	281	281		
Denmark	117	84	77	82	83		
Dominican Rep.	3	17	27	29	29		
Ecuador	4	13	50	42	42		
Egypt	19	55	90	180	198		
El Salvador	3	22	36	25	25		
Ethiopia	39	66	73	79	79		
Finland	21	64	76	76	77		
France	411	644	838	891	906	9	2.5
Germany	280	1,125	1,132	1,055	998	8	2.8
Ghana	18	8	13	8	8		
Greece	21	100	115	121	127		
Guatemala	13	29	40	62	63		
Hungary	51	176	246	254	236		
India	53	262	733	1,080	1,100	6	3.1
Indonesia	94	32	91	365	380		
Iran	34	56	155	260	270		
Iraq	6	10	21	60	58		
Ireland	56	40	31	33	34		
Israel	17	74	93	104	103		
Italy	274	549	632	666	682	10	1.9
Japan	120	1,735	1,998	2,423	2,415	4	6.7
Korea, N.	2	55	104	140	142		
Korea, S.	3	131	260	415	420		
Lebanon	2	27	41	53	53		
Malaysia	11	63	120	200	210		
Mexico	97	334	583	1,047	1,091	7	3
Morocco	37	44	77	86	88		
Netherlands	145	268	532	640	615		
New Zealand	30	52	57	47	48		
Nicaragua	2	13	30	26	26		
Nigeria	33	102	180	307	310		
Norway	23	38	44	54	54		
Pakistan	18	15	96	202	210		
Peru	11	28	60	96	105		
Philippines	38	117	200	270	302		
Poland	185	387	490	448	414		
Portugal	23	37	61	75	75		
Romania	50	165	330	380	380		
Saudi Arabia	...	7	21	153	153		
South Africa	33	115	160	199	205		
Spain	138	460	669	641	621		
Sudan	11	17	32	42	42		
Sweden	84	102	114	127	117		
Switzerland	27	39	44	37	38		
Syria	6	16	66	70	72		
Tanzania	8	15	35	65	65		
Thailand	66	117	104	111	111		
Tunisia	9	13	33	58	59		
Turkey	45	98	212	360	380		
UK	426	878	787	648	647		
Uruguay	13	17	17	22	22		
USA	3,717	4,057	4,116	3,970	4,014	3	11.2
USSR	684	2,250	3,745	4,680	4,540	2	12.7
Venezuela	4	76	121	136	106		
Vietnam	4	84	124	92	96		
Yugoslavia	43	137	221	231	241		
Zaïre	16	6	7	8	8		

Fish – catch by fishing area

thousand tonnes

		1970	1980	1987	1988
	World	70,000	72,008	93,415	97,985
	Marine areas	60,930	64,383	80,688	84,561
18	Arctic Sea	0	0	0	0
21	Atlantic, NW	4,239	2,867	3,093	3,044
27	Atlantic, NE	10,666	11,799	10,421	10,510
31	Atlantic, WC	1,440	1,795	2,161	1,897
34	Atlantic, EC	2,830	3,432	3,229	3,561
37	Mediterranean & Black Sea	1,120	1,649	1,949	2,012
41	Atlantic, SW	1,081	1,274	2,166	2,228
47	Atlantic, SE	2,460	2,170	2,704	2,458
48	Atlantic, Antarctic	0	453	434	443
51	Indian, W	1,620	2,098	2,654	2,881
57	Indian, E	789	1,458	2,559	2,651
58	Indian, Antarctic	0	138	39	15
61	Pacific, NW	12,997	18,759	25,871	26,703
67	Pacific, NE	2,652	1,975	3,447	3,338
71	Pacific, WC	4,176	5,487	6,299	6,537
77	Pacific, EC	902	2,417	2,477	2,438
81	Pacific, SW	221	382	907	983
87	Pacific, SE	13,732	6,228	10,278	12,863
88	Pacific, Antarctic	0	3.2	0	0
	Inland waters	9,020	7,625	12,727	13,425
1	Africa	1,308	1,404	1,753	1,802
2	N. America	136	146	572	564
3	S. America	249	280	385	368
4	Asia	6,251	4,673	8,556	9,188
5	Europe	225	373	449	484
6	Oceania	2.0	2.1	23	24
7	USSR	853	747	988	996

The figures include the produce from inland and marine fishing. In 1988 fish from inland waters accounted for 14% of the world total. The figures above refer to fish caught in the inland waters or the craft of the country. The flag of the vessel rather than the port of landing is the criteria for assigning the catch to a country. All figures are, or have been, converted to the live weight of the catch. Whales are not included.

Fish – catch by nation

thousand tonnes

	1948	1970	1980	1987	1988	Rank	%
World	19,600	70,000	72,042	93,415	97,985		
Africa	...	4,440	4,108	5,324	5,310		
Asia	...	26,300	30,934	41,891	43,601		
Europe	...	11,970	12,479	12,635	12,874		
North America	...	4,930	6,838	9,631	9,568		
Oceania	...	200	355	806	887		
South America	...	14,860	7,819	11,869	14,412		
Algeria	...	...	48	94	107		
Angola	113	368	86	83	102		
Argentina	71	215	385	559	491		
Australia	39	103	132	200	202		
Bangladesh	...	247	647	817	829		
Brazil	145	517	820	732	750		
Bulgaria	64	92	126	110	117		
Burma	...	432	580	686	704		
Cameroon	22	71	77	83	83		
Canada	1,053	1,389	1,347	1,562	1,596		
Chad	...	120	115	110	110		
Chile	65	1,181	2,817	4,815	5,210	6	5.3
China	2,500	6,868	4,235	9,346	10,359	3	10.6
Colombia	15	76	76	86	85		
Cuba	8	106	186	215	231		
Denmark	226	1,227	2,028	1,706	1,972		
Ecuador	3	92	643	680	769		
Egypt	43	72	140	250	250		
Faroe Is.	92	208	275	386	357		
Finland	46	81	143	106	121		
France	513	764	794	861	898		
Germany	414	945	542	398	389		
Ghana	24	172	224	382	361		
Greece	34	78	104	133	129		
Greenland	21	40	104	100	134		
Hong Kong	34	124	195	228	238		
Iceland	478	734	1,514	1,633	1,759		
India	...	1,756	2,442	2,908	3,146	7	3.2
Indonesia	500	1,229	1,842	2,585	2,703	9	2.8
Iran	...	...	44	150	156		

	1948	1970	1980	1987	1988	Rank	%
Ireland	25	79	149	249	253		
Italy	183	387	448	560	559		
Ivory Coast	...	58	78	102	89		
Japan	2,526	9,366	10,428	11,849	11,897	1	12.1
Kenya	30	34	48	131	137		
Korea, N.	...	800	1,400	1,700	1,700		
Korea, S.	294	934	2,091	2,876	2,727	8	2.8
Malaysia	145	368	736	612	604		
Mali	...	90	80	56	57		
Mauritania	...	...	22	99	98		
Mexico	68	387	1,222	1,419	1,363		
Morocco	69	256	330	491	551		
Netherlands	294	301	340	446	399		
New Zealand	36	59	95	431	503		
Nigeria	...	543	480	261	261		
Norway	1,422	2,980	2,409	1,949	1,826		
Oman	...	100	79	136	166		
Pakistan	...	158	279	428	445		
Panama	1	42	216	180	112		
Peru	84	12,613	2,735	4,584	6,637	4	6.8
Philippines	195	992	1,557	1,988	2,042		
Poland	47	469	640	671	655		
Portugal	292	498	271	395	347		
Romania	...	59	174	264	268		
Senegal	...	189	250	252	257		
South Africa	118	1,583	615	1,424	1,298		
Spain	547	1,539	1,265	1,393	1,430		
Sri Lanka	24	98	186	186	198		
Sweden	194	295	241	215	251		
Taiwan	122	613	936	1,236	1,361		
Tanzania	30	185	229	342	340		
Thailand	161	1,448	1,793	2,201	2,350	10	2.4
Tunisia	...	...	60	99	103		
Turkey	80	184	427	628	628		
Uganda	11	129	166	200	241		
UK	1,206	1,099	847	952	946		
Uruguay	4	13	120	138	107		
USA	2,417	2,777	3,635	5,986	5,966	5	6.1
USSR	1,485	7,252	9,476	11,160	11,332	2	11.6
Venezuela	92	126	187	309	294		
Vietnam	200	817	613	871	874		
Yugoslavia	21	45	58	81	72		
Zaire	18	137	102	166	166		

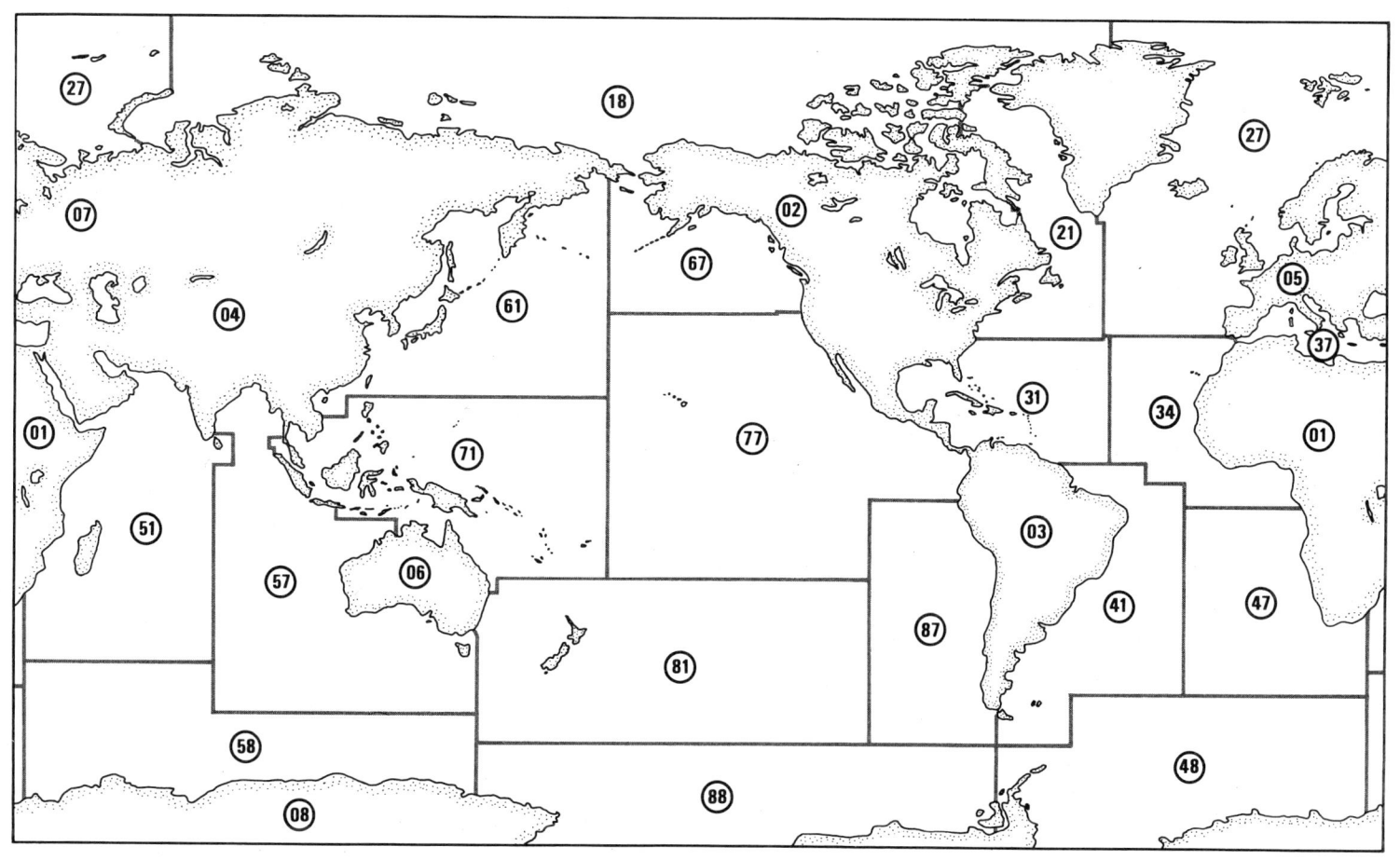

Flax
thousand tonnes

	1950	1970	1980	1988	1989	Rank	%
World	440	753	613	755	769		
Africa	4	9	25	16	16		
Asia	...	80	112	202	222		
Europe	200	184	170	211	202		
Oceania	3	1	2				
South America	2	4	3	4	4		
Belgium	31	14	10	12	12	9	1.6
China	...	77	110	201	221	2	28.7
Czechoslovakia	9	20	22	19	25	6	3.3
Egypt	4	9	25	16	16	8	2.1
France	30	49	50	91	74	3	9.6
Netherlands	...	...	6	32	31	5	4
Poland	40	55	53	18	19	7	2.5
Romania	5	18	25	36	38	4	4.9
USSR	255	476	301	323	326	1	42.4

Goats
thousand head

	1950	1970	1980	1989	1990	Rank	%
World	294,043	405,568	458,370	544,512	555,649		
Africa	84,678	134,669	146,569	173,477	175,920		
Asia	137,680	220,606	263,877	309,777	321,451		
Europe	23,955	12,378	11,580	15,831	12,995		
North America	12,395	14,018	11,145	14,889	13,795		
Oceania	170	182	351	2,065	2,409		
South America	19,897	18,360	19,100	21,973	22,598		
Afghanistan	6,000	2,662	2,962	2,100	2,200		
Albania	815	868	670	1,106	1,200		
Algeria	2,685	2,546	2,755	3,600	3,650		
Angola	420	777	935	980	985		
Argentina	5,908	5,380	3,000	3,200	3,300		
Austria	315	70	35	32	36		
Bangladesh	...	8,786	11,433	10,879	11,071		
Benin	232	578	926	994	1,000		
Bolivia	1,299	2,332	3,009	2,400	2,450		
Botswana	469	1,055	625	1,480	1,500		
Brazil	8,309	5,742	8,023	11,000	11,500	10	2.1
Bulgaria	712	354	425	434	409		
Burkina Faso	934	2,485	2,794	5,350	5,400		
Burma	179	573	607	1,033	1,050		
Burundi	466	449	650	770	790		
Cameroon	1,090	1,500	2,391	3,210	3,300		
Central Afr. Rep.	218	522	920	1,200	1,210		
Chad	2,000	2,200	2,267	2,310	2,350		
Chile	661	843	600	600	600		
China	23,885	62,501	78,928	91,152	98,332	2	17.7
Colombia	525	630	645	1,018	1,020		
Czechoslovakia	952	323	64	50	50		
Djibouti	180	540	527	501	550		
Dominican Rep.	310	310	378	543	550		
Ecuador	371	192	252	300	310		
Egypt	703	1,166	1,443	3,850	3,880		
Ethiopia	9,692	17,464	17,177	18,000	17,200	6	3.1
France	1,251	923	1,149	1,215	1,220		
Germany	90	85	72	70	71		
Ghana	386	1,408	2,067	2,000	2,000		
Greece	3,672	4,063	4,565	5,970	3,488		
Haiti	854	1,134	997	1,250	1,300		
India	45,155	66,529	71,598	107,000	110,000	1	19.8
Indonesia	6,650	6,941	7,800	10,720	10,950		
Iran	11,000	13,817	13,570	13,500	13,500	8	2.4
Iraq	2,950	3,233	3,658	1,600	1,650		
Italy	2,462	1,032	989	1,214	1,224		
Ivory Coast	279	833	1,250	1,500	1,550		
Jamaica	350	358	380	440	440		
Japan	518	204	67	40	40		
Jordan	300	415	515	500	510		
Kenya	3,791	4,237	7,761	7,500	7,600		
Korea, S.	...	...	223	139	139		
Lebanon	430	357	413	475	480		
Lesotho	623	880	777	1,040	1,050		
Libya	856	1,222	1,488	970	970		
Madagascar	292	956	1,308	1,200	1,250		
Malawi	252	618	645	1,000	1,200		
Mali	3,830	5,483	6,750	5,650	5,750		
Mauritania	1,614	3,423	2,583	3,350	3,350		
Mexico	7,773	9,120	7,491	10,241	9,086		
Mongolia	4,800	3,967	4,662	4,300	4,400		
Morocco	8,718	8,467	6,105	5,960	6,071		
Mozambique	333	522	335	380	385		
Namibia	900	1,700	2,117	2,550	2,600		
Nepal	2,260	2,267	2,502	5,210	5,210		
Niger	3,494	6,102	7,023	7,570	7,800		
Nigeria	14,700	23,367	24,567	26,000	26,000	4	4.7
Oman	...	148	240	770	770		
Pakistan	10,530	15,533	30,272	34,194	35,000	3	6.3
Peru	1,421	1,845	1,983	1,682	1,700		
Philippines	333	798	1,460	2,212	2,187		
Poland	595	129	32	10	10		
Portugal	947	668	747	745	746		
Romania	520	578	378	1,000	1,050		
Rwanda	778	478	875	1,100	1,150		
Saudi Arabia	1,901	735	1,972	2,700	3,800		
Senegal	747	1,067	1,067	1,200	1,300		
Somalia	3,595	15,167	16,267	20,300	20,600	5	3.7
South Africa	5,460	5,598	5,317	5,860	5,862		
Spain	7,233	2,656	2,079	3,640	3,200		
Sri Lanka	385	548	482	520	520		
Sudan	4,440	8,900	12,532	14,500	14,800	7	2.7
Syria	1,325	762	1,075	1,053	1,078		
Tanzania	3,372	4,441	5,673	6,650	6,650		
Togo	189	501	723	1,200	1,200		
Tunisia	1,719	540	888	1,120	1,200		
Turkey	18,517	20,129	18,755	13,100	13,100	9	2.4
Uganda	2,254	1,822	2,155	2,900	3,000		
USA	2,602	2,640	1,380	1,850	1,870		
USSR	15,632	5,355	5,747	6,500	6,480		
Venezuela	1,288	1,276	1,368	1,530	1,530		
Yemen	810	7,730	8,766	3,125	3,300		
Yugoslavia	1,097	160	125	61	65		
Zaire	1,204	2,250	2,751	3,050	3,050		
Zambia	...	...	290	519	530		
Zimbabwe	573	1,572	1,107	2,500	2,500		

Goat meat
thousand tonnes

	1970	1980	1988	1989	Rank	%
World	1,613	1,995	2,287	2,365		
Africa	439	549	617	631		
Asia	963	1,221	1,428	1,482		
Europe	83	89	99	102		
North America	23	30	47	48		
Oceania	2	2	3	6		
South America	64	64	68	69		
Afghanistan	20	27	21	22		
Albania	8	8	9	9		
Algeria	7	11	15	15		
Angola	2	3	3	3		
Argentina	11	6	6	6		
Bangladesh	31	45	54	55	8	2.3
Benin	2	3	3	3		
Bolivia	6	6	5	5		
Botswana	3	3	3	3		
Brazil	22	23	27	27		
Bulgaria	4	3	5	6		
Burkina Faso	8	6	13	13		
Burma	3	3	6	6		
Burundi	2	2	3	3		
Cameroon	6	7	10	11		
Central Afr. Rep.	2	3	4	4		
Chad	6	8	8	8		
Chile	6	5	5	5		
China	260	345	381	401	1	17
Colombia	2	2	3	3		
Cyprus	4	4	4	4		
Czechoslovakia	2	2	2	2		
Djibouti	2	2	2	2		
Egypt	16	20	24	24		
Ethiopia	57	55	66	66	6	2.8
France	10	8	8	8		
Ghana	4	6	6	6		
Greece	31	44	40	41		
Haiti	4	6	5	5		
India	254	271	378	385	2	16.3
Indonesia	39	38	53	53	9	2.2
Iran	55	45	46	46	10	1.9
Iraq	12	13	8	8		
Italy	4	5	4	4		
Ivory Coast	5	7	6	7		
Jamaica	1	1	2	2		
Jordan	4	4	2	2		
Kenya	13	17	28	29		
Korea, S.	—	2	3	3		
Lebanon	3	4	4	4		
Lesotho	1	2	3	3		
Libya	4	2	3	3		
Madagascar	2	3	6	6		
Malawi	2	2	3	4		
Mali	15	21	21	23		
Mauritania	5	5	5	5		
Mexico	16	20	37	38		
Mongolia	20	27	21	19		
Morocco	25	19	13	14		
Mozambique	2	1	2	2		
Namibia	4	5	6	6		
Nepal	10	10	28	29		
New Zealand	2	1	2	5		
Niger	19	29	30	30		
Nigeria	82	127	134	134	4	5.7
Oman	1	2	4	4		
Pakistan	86	179	258	280	3	11.8
Peru	10	10	8	8		
Philippines	3	7	22	24		
Poland	2	1	1	1		
Portugal	3	3	4	4		
Romania	3	4	7	8		
Rwanda	2	3	4	4		
Saudi Arabia	8	9	16	16		
Senegal	4	3	5	5		
Somalia	48	53	55	55	7	2.3
South Africa	24	28	33	34		
Spain	13	11	20	21		
Sudan	19	44	38	39		
Swaziland	2	3	3	3		
Syria	5	7	6	6		
Tanzania	18	15	18	18		
Togo	1	2	3	3		
Tunisia	3	3	5	5		
Turkey	101	126	75	75	5	3.2
UAE	2	3	5	6		
Uganda	7	9	12	12		
USSR	35	40	25	28		
Venezuela	6	10	11	11		
Vietnam	1	2	2	2		
Yemen	34	43	23	23		
Zaïre	6	7	8	8		
Zimbabwe	6	4	8	9		

Grapefruit
thousand tonnes

	1950	1970	1980	1989	1990	Rank	%
World	1,733	3,339	4,394	4,894	4,184		
Africa	26	198	287	320	322		
Asia	82	597	889	1,120	1,158		
Europe	1	12	19	44	41		
North America	1,600	2,318	2,850	3,055	2,289		
Oceania	7	18	34	26	34		
South America	17	197	315	329	340		
Algeria	1	4	5	2	2		
Argentina	7	129	151	150	170	6	4.1
Australia	5	13	27	20	29		
Bangladesh	1	3	4	6	7		
Belize	7	11	14	30	30		
Brazil	...	35	36	50	50		
Cambodia	11	4	1	2	2		
Central Afr. Rep.	2	3	3	3	3		
China	7	115	146	227	238	5	5.7
Cuba	7	15	81	266	280	3	6.7
Cyprus	6	50	83	68	70	8	1.7
Dominica	...	3	12	17	17		
Dominican Rep.	2	2	3	3	3		
Ecuador	2	0	57	28	29		
Egypt	...	26	25	3	3		
Greece	...	...	1	6	5		
Haiti	...	7	11	10	10		
Honduras	...	10	22	35	36		
India	6	20	20	50	50		
Israel	43	303	515	368	390	2	9.3
Italy	...	1	4	7	7		
Jamaica	19	39	27	39	40		
Jordan	...	1	1	6	6		
Kenya	...	4	7	13	14		
Laos	...	2	2	6	6		

Grapefruit continued

	1950	1970	1980	1989	1990	Rank	%
Lebanon	...	5	24	25	25		
Madagascar	...	1	4	9	9		
Mexico	2	32	119	75	100	7	2.4
Morocco	3	9	10	15	15		
Mozambique	2	14	16	17	17		
New Zealand	2	4	6	4	4		
Paraguay	6	19	56	79	69	9	1.6
Peru	...	7	6	5	5		
Philippines	8	28	35	34	34		
Portugal	...	5	5	8	8		
Puerto Rico	10	10	6	3	3		
Somalia	2	5	6	28	29		
South Africa	16	104	109	66	60		
Spain	1	6	9	23	24		
Sudan	...	44	57	62	64	10	1.5
Surinam	3	5	2	1	1		
Swaziland	...		37	36	36		
Thailand	...	9	16	250	250	4	6
Trinidad & T.	26	20	5	3	3		
Tunisia	...	2	23	50	50		
Turkey	1	7	18	28	30		
Uruguay	...	3	6	8	8		
USA	1,530	2,179	2,571	2,580	1,772	1	42.4
Venezuela		6	6	9	9		
Vietnam	...	...	...	8	8		
Zaire	...	8	9	13	13		

Grapes
thousand tonnes

	1950	1970	1980	1989	1990	Rank	%
World	34,381	54,554	65,993	59,478	60,964		
Africa	2,432	2,559	2,174	2,727	2,781		
Asia	2,378	6,138	7,122	8,613	8,498		
Europe	23,080	33,302	39,479	31,444	33,509		
North America	2,775	3,554	5,088	5,925	5,537		
Oceania	454	635	799	939	842		
South America	2,481	4,148	4,861	4,929	5,098		
Afghanistan	138	332	461	450	450		
Albania	21	62	61	88	90		
Algeria	1,738	1,163	429	360	363		
Argentina	1,586	2,636	3,071	2,971	2,950	7	4.8
Australia	450	620	761	879	782		
Austria	132	337	373	350	260		
Brazil	247	528	604	697	768		
Bulgaria	504	1,128	1,023	754	800		
Canada	36	69	72	51	56		
Chile	473	762	960	1,320	1,050		
China	48	147	185	979	1,050		
Cyprus	59	190	207	198	212		
Czechoslovakia	60	128	220	166	166		
Egypt	79	115	272	555	580		
France	7,995	9,706	10,632	7,207	7,800	2	12.8
Germany	324	1,049	885	1,893	1,902	9	3.1
Greece	1,124	1,569	1,563	1,700	1,600	10	2.6
Hungary	527	809	862	580	700		
India	...	219	196	293	304		
Iran	254	628	958	1,320	1,050		
Iraq	30	75	433	450	470		
Israel	15	69	77	92	92		
Italy	7,076	10,638	12,840	9,449	10,000	1	16.4
Japan	38	240	338	275	270		
Korea, S.	2	35	61	180	193		
Lebanon	81	98	128	210	215		
Mexico	40	160	476	489	506		
Morocco	92	213	214	180	189		
Peru	54	62	52	42	45		
Portugal	1,113	1,403	1,420	1,092	1,050		
Romania	623	1,020	1,518	2,000	2,200	8	3.6
Saudi Arabia	...	24	58	100	100		
South Africa	412	933	1,094	1,456	1,463		
Spain	2,541	4,048	6,533	4,883	5,659	3	9.3
Switzerland	102	129	123	227	228		
Syria	172	221	334	425	520		
Tunisia	105	125	135	120	130		
Turkey	1,500	3,779	3,567	3,430	3,350	6	5.5
Uruguay	114	129	110	122	124		
USA	2,670	3,325	4,540	5,380	4,970	4	8.2
USSR	753	4,220	6,471	4,900	4,700	5	7.7
Yemen		21	58	133	135		
Yugoslavia	867	1,232	1,406	1,022	1,022		

Groundnuts
thousand tonnes

	1950	1970	1980	1989	1990	Rank	%
World	9,572	18,217	18,277	22,993	22,858		
Africa	2,476	5,537	4,816	4,553	4,708		
Asia	5,875	9,959	10,717	15,936	15,774		
Europe	25	19	26	24	24		
North America	922	1,465	1,727	1,954	1,727		
Oceania	9	33	55	37	26		
South America	263	1,204	936	485	595		
Angola	10	20	20	20	20		
Argentina	100	280	401	243	370	10	1.6
Australia	9	30	47	32	21		
Bangladesh	...	45	25	45	42		
Benin	10	47	62	70	74		
Brazil	138	876	433	150	136		
Burkina Faso	51	68	76	131	150		
Burma	154	492	399	438	524	7	2.3
Burundi	2	21	38	86	87		
Cambodia	4	21	12	12	13		
Cameroon	92	178	113	140	143		
Central Afr. Rep.	10	68	123	100	100		
Chad	64	95	105	80	80		
China	2,114	2,634	3,369	5,455	6,100	2	26.7
Congo	10	17	14	25	26		
Cuba	14	15	15	15	15		
Dominican Rep.	14	76	41	34	34		
Egypt	18	40	31	25	26		
Ethiopia	21	24	20	53	54		
Gambia	62	129	103	133	130		
Ghana	43	88	99	200	180		
Greece	8	8	11	10	10		
Guinea	18	25	83	45	52		
Guinea-Bissau	50	36	32	35	30		
Haiti	2	2	33	50	50		
India	3,197	5,807	5,596	8,450	7,500	1	32.8
Indonesia	283	462	786	879	919	4	4
Israel	2	17	21	20	20		
Ivory Coast	15	42	53	137	140		
Japan	21	120	61	30	30		
Korea, S.	1	8	17	24	24		
Madagascar	12	42	36	32	32		
Malawi	...	182	177	193	193		
Malaysia	2	5	23	5	5		
Mali	88	144	166	157	162		
Mexico	55	81	81	39	40		
Morocco	—	3	33	37	37		
Mozambique	22	140	83	70	70		
Niger	61	223	94	80	60		
Nigeria	690	1,660	563	650	680	6	3
Pakistan	2	55	56	78	82		
Paraguay	11	17	24	45	45		
Philippines	19	17	50	39	40		
Rwanda	3	7	17	18	18		
Senegal	558	794	688	844	800	5	3.5
Sierra Leone	7	19	20	18	19		
South Africa	93	364	316	171	140		
Spain	12	7	6	5	5		
Sudan	19	370	830	244	400	8	1.7
Syria	1	18	20	20	20		
Taiwan	60	122	86	85	85		
Tanzania	20	32	54	55	60		
Thailand	60	128	125	172	177		
Togo	13	20	35	24	224		
Turkey	7	40	50	50	50		
Uganda	153	207	199	110	120		
USA	839	1,289	1,546	1,810	1,567	3	6.9
Venezuela	—	8	36	16	16		
Vietnam	8	78	87	204	210		
Zaire	155	258	314	400	400	9	1.7
Zambia	2	28	24	30	28		
Zimbabwe	21	111	144	95	119		

Hemp
thousand tonnes

	1950	1970	1980	1988	1989	Rank	%
World	390	363	254	211	217		
Asia	110	185	163	132	136		
Europe	190	79	53	51	53		
South America	4	4	4	4	4		
Bangladesh	...	5	2	1	1		
Bulgaria	10	9	4	1	1		

	1950	1970	1980	1988	1989	Rank	%
Chile	4	4	4	4	4	8	1.8
China	...	93	90	68	70	1	32.3
Hungary	14	18	12	8	8	6	3.7
India	80	62	48	45	46	2	21.2
Korea, N.	...	2	3	9	9	5	4.1
Korea, S.	8	6	1	1	1		
Pakistan	11	7	6	4	4	9	1.8
Poland	6	14	4	3	2	10	0.9
Romania	28	20	25	34	35	3	16.1
Turkey	10	8	12	5	6	7	2.8
USSR	89	95	34	24	24	4	11.1
Yugoslavia	42	15	6	1	2		

Honey
thousand tonnes

	1950	1970	1980	1988	1989	Rank	%
World	...	862	897	1,098	1,109		
Africa	...	75	87	105	107		
Asia	...	235	199	278	299		
Europe	...	118	133	185	183		
North America	...	174	211	202	184		
Oceania	...	25	29	32	32		
South America	...	40	52	75	74		
Afghanistan	...	3	3	3	3		
Algeria	...	1	2	2	2		
Angola	...	16	15	15	15		
Argentina	7	21	35	40	38	7	3
Australia	14	20	21	23	23	10	2
Austria	1	5	3	6	5		
Belgium	1	1	1	1	1		
Brazil	6	6	7	16	17		
Bulgaria	2	7	9	11	9		
Cameroon	...	2	2	3	3		
Canada	16	24	32	37	28	8	3
Central Afr. Rep.	2	5	6	7	8		
Chile	4	7	5	5	5		
China	...	200	102	154	177	2	16
Colombia	...	2	2	4	4		
Cuba	4	5	8	10	10		
Czechoslovakia	2	6	6	12	10		
Dominican Rep.	1	1	1	1	2		
Egypt	2	5	7	13	13		
El Salvador	...	1	2	3	3		
Ethiopia	...	17	21	23	23	10	2
France	3	11	12	24	26	9	2
Germany	17	21	126	24	23	10	2
Greece	3	7	10	12	12		
Guatemala	2	3	3	4	4		
Hungary	3	8	14	14	16		
India	...	...	43	51	50	5	5
Iran	...	4	6	6	6		
Israel	1	2	2	2	2		
Italy	7	7	4	9	8		
Japan	4	7	7	5	6		
Kenya	...	7	11	15	16		
Korea, South	1	1	5	8	9		
Madagascar	8	10	3	4	4		
Mexico	15	35	66	46	53	4	5
Morocco	2	1	5	3	3		
New Zealand	5	5	10	8	9		
Poland	3	9	12	14	15		
Portugal	2	3	3	3	3		
Romania	2	9	14	17	17		
Spain	7	8	13	21	21		
Sweden	1	2	2	3	4		
Switzerland	2	2	2	3	3		
Tanzania	...	5	10	14	14		
Turkey	5	15	27	43	40	6	4
UK	7	3	2	2	2		
Uruguay	1	1	2	8	8		
USA	109	104	94	96	80	3	7
USSR	182	194	219	222	230	1	21
Yugoslavia	4	5	5	5	5		

Hops
thousand tonnes

	1950	1970	1980	1988	1989	Rank	%
World	65	98	112	112	112		
Asia	...	2	2	7	7.3		
Europe	34	62	67	66	66		
North America	27	22	33	25	26		
Oceania	1	2	2	3	2.9		
Australia	1	2	2	3	2.4	8	2.1
Belgium-Lux.	1	2	2	1	0.6		
Bulgaria	...	1	1	1	0.6		
Canada	1	1	1	1	0.5		
China	...	...	1	3	3	7	2.7
Czechoslovakia	5	10	11	15	12	3	10.7
France	2	2	1	1	0.9		
Germany	9	29	33	33	37	1	33
Hungary	...	...	1	1	0.7		
Japan	...	2	2	2	2	10	1.8
Korea, N.	...	...	1	2	2.1	9	1.9
Korea, S.	...	...	...	..	0.2		
Poland	...	2	2	2	1.8		
Romania	...	...	1	...	0.3		
Spain	...	1	2	2	1.9		
UK	15	11	10	5	5	5	4.5
USA	26	21	32	25	25	2	22.3
USSR	3	9	7	10	10	4	8.9
Yugoslavia	1	5	4	5	4.2	6	3.8

Horses
thousand head

	1950	1970	1980	1988	1989	Rank	%
World	79,740	62,642	59,689	60,379	60,461		
Africa	3,209	3,723	3,728	4,635	4,732		
Asia	13,800	14,780	17,521	17,003	17,086		
Europe	16,551	7,714	5,392	4,360	4,253		
North America	14,243	15,732	13,979	14,092	14,105		
Oceania	1,298	583	632	512	496		
South America	18,009	12,471	12,814	13,892	13,899		
Afghanistan	195	400	403	400	400		
Algeria	209	142	173	187	190		
Argentina	7,265	3,623	3,024	2,900	2,900	6	4.8
Australia	1,055	454	486	335	317		
Belgium	257	76	38	23	23		
Bolivia	158	291	330	315	320		
Brazil	6,942	4,861	5,070	5,850	5,850	4	9.7
Bulgaria	511	183	121	123	122		
Burkina Faso	55	88	70	70	70		
Burma	12	71	112	139	140		
Canada	1,580	362	363	338	338		
Chad	120	153	166	189	195		
Chile	523	478	450	490	490		
China	5,503	7,300	11,144	10,691	10,691	1	17.7
Colombia	1,208	1,114	1,683	1,950	1,950	9	3.2
Costa Rica	76	109	113	114	114		
Cuba	410	683	810	703	703		
Czechoslovakia	613	144	47	45	44		
Dominican Rep.	137	184	204	310	310		
Ecuador	111	242	314	438	430		
El Salvador	130	79	88	93	93		
Ethiopia	1,021	1,393	1,602	2,550	2,600	7	4.3
Finland	389	88	33	36	36		
France	2,403	691	352	292	269		
Germany	2,266	411	447	469	465		
Greece	259	251	115	60	60		
Guatemala	166	149	100	112	112		
Haiti	253	335	411	430	432		
Honduras	178	173	165	170	170		
Hungary	665	233	127	88	76		
India	1,514	1,010	900	953	955		
Indonesia	560	666	616	664	725		
Iran	358	400	274	270	270		
Iraq	290	98	54	55	58		
Ireland	384	122	69	53	53		
Italy	779	292	271	250	250		
Japan	1,083	159	23	22	21		
Lesotho	101	103	102	119	120		
Mali	60	170	144	56	58		
Mexico	3,181	5,579	6,195	6,160	6,170	2	10.2
Mongolia	2,085	2,262	2,047	2,047	2,060	8	3.4
Morocco	350	380	278	190	200		
Netherlands	273	95	66	64	64		
New Zealand	204	75	71	97	98		
Nicaragua	150	212	275	250	250		
Niger	75	197	253	296	298		
Nigeria	260	325	250	250	250		
Norway	191	36	18	17	16		
Pakistan	475	385	473	455	457		
Panama	138	160	165	171	171		
Paraguay	302	334	316	328	330		
Peru	496	669	650	655	660		
Philippines	210	300	302	300	300		
Poland	2,673	2,573	1,787	1,051	1,005	10	1.7
Portugal	87	0	29	26	26		
Romania	979	686	564	693	700		
Senegal	25	196	222	208	208		
South Africa	682	258	225	230	231		
Spain	686	287	243	250	241		
Syria	101	70	52	40	43		
Thailand	219	172	28	19	18		
Tunisia	73	96	51	56	56		
Turkey	1,136	1,103	804	620	620		
UK	552	143	147	180	184		
Uruguay	667	427	493	466	470		
USA	7,757	7,667	5,053	5,203	5,203	5	8.6
USSR	12,800	7,641	5,623	5,885	5,890	3	9.7
Venezuela	344	423	478	495	495		
Vietnam	40	105	123	136	133		
Yugoslavia	1,063	1,078	630	362	340		

Jute
thousand tonnes

	1950	1970	1980	1989	1990	Rank	%
World	2,300	3,292	3,620	3,705	3,761		
Africa	23	33	20	19	16		
Asia	2,200	3,137	3,440	3,568	3,648		
North America	...	7	15	17	11		
South America	15	61	96	53	37		
Bangladesh	1,015	1,115	949	811	849	2	22.6
Bhutan	...	4	6	...	...		
Brazil	15	54	85	40	24	8	0.6
Burma	...	35	73	...	...		
Cambodia	...	7	3	8	8		
Chile	...	...	8	10	10	10	0.3
China	168	301	575	565	590	3	15.7
Cuba	...	2	14	16	10	10	0.3
Egypt	...	14	10	5	5		
India	720	1,189	1,469	1,890	1,920	1	51.1
Indonesia	1	13	15	27	30	7	0.8
Mozambique	...	6	4	4	4		
Nepal	...	52	64	18	19	9	0.5
Peru	...	7	3	3	3		
Taiwan	20	14	...	...	...		
Thailand	11	401	243	164	160	4	4.3
USSR	...	53	48	42	49	5	1.3
Vietnam	...	18	30	36	35	6	0.9

Lemons and limes
thousand tonnes

	1950	1970	1980	1989	1990	Rank	%
World	1,639	3,672	5,000	6,692	6,615		
Africa	65	217	238	454	457		
Asia	417	933	1,225	1,961	1,946		
Europe	405	1,078	1,303	1,484	1,382		
North America	555	978	1,483	1,624	1,477		
Oceania	21	33	49	41	49		
South America	176	433	702	1,128	1,305		
Algeria	7	16	8	9	6		
Argentina	54	198	332	350	500	6	7.6
Australia	16	28	43	35	43		
Bangladesh	...	4	4	7	7		
Bolivia	46	17	13	11	11		
Brazil	16	55	197	470	490	7	7.4
Cambodia	3	1	0	1	1		
Chile	41	37	74	62	60		
China	...	55	81	147	152		
Cuba	2	11	23	65	65		
Cyprus	4	28	32	42	34		
Dominica	8	8	9	6	6		
Dominican Rep.	12	12	13	9	9		
Ecuador	8	15	17	26	26		

(continuation of preceding table)

	1950	1970	1980	1989	1990	Rank	%
Egypt	37	85	79	212	215	10	3.3
El Salvador	...	15	19	23	23		
Ethiopia	...	5	5	6	6		
Ghana	4	27	30	30	30		
Greece	32	135	169	179	173		
Haiti	12	23	25	25	25		
India	285	450	485	550	563	5	8.5
Iran	53	29	34	464	465	8	7
Iraq	...	22	8	16	17		
Israel	7	41	56	43	40		
Italy	314	813	742	668	616	2	9.3
Jamaica	5	18	22	24	24		
Jordan	2	5	10	35	35		
Laos	...	2	3	9	9		
Lebanon	22	72	63	65	66		
Libya	...	2	8	24	25		
Madagascar	...	4	6	6	6		
Mexico	69	326	500	727	612	3	9.3
Morocco	4	3	7	21	20		
New Zealand	2	3	3	3	3		
Oman	...	9	16	25	25		
Pakistan	5	18	30	67	67		
Paraguay	3	15	15	16	17		
Peru	...	90	63	127	130		
Philippines	1	9	41	50	52		
Portugal	7	18	20	17	17		
Puerto Rico	...	4	3	2	2		
South Africa	5	20	35	45	45		
Spain	52	112	371	619	575	4	8.7
Sri Lanka	...	31	48	39	40		
Sudan	...	34	38	54	55		
Syria	1	5	10	31	32		
Tonga	2	2	3	3	3		
Trinidad & T.	4	2	1	1	1		
Tunisia	8	13	20	18	18		
Turkey	12	130	288	335	300	9	4.5
UAE	...	...	6	7	7		
Uruguay	10	6	18	50	55		
USA	441	572	866	738	706	1	10.7
Venezuela	...	...	11	14	15		
Zaïre	...	5	6	7	7		
Zimbabwe	...	5	7	8	8		

Linseed
thousand tonnes

	1950	1970	1980	1988	1989	Rank	%
World	3,108	3,481	2,476	1,903	2,121		
Africa	109	81	52	65	64		
Asia	483	508	524	621	605		
Europe	258	198	148	121	118		
North America	1,321	1,563	826	418	569		
Oceania	14	37	17	7	4		
South America	647	632	691	451	530		
Afghanistan	0	17	15	14	14		
Argentina	513	545	642	430	509	2	24
Australia	7	26	9	7	3		
Bangladesh	...	7	7	14	44	6	2.1
Belgium-Lux.	16	8	6	8	9		
Brazil	19	23	9	17	19		
Bulgaria	4	3	1	1	1		
Canada	238	833	586	373	531	1	25
Chile	5	1	1	1	1		
China	35	32	78	137	165	5	7.8
Czechoslovakia	10	15	12	12	16		
Denmark	13	0	0	...	...		
Egypt	5	13	35	25	23		
Ethiopia	51	62	14	36	37	8	1.7
France	15	24	30	32	25	10	1.2
Germany	30	5	0	...	...		
Hungary	7	20	10	11	10		
India	397	424	411	393	349	3	16.5
Iran	1	5	3	3	3		
Italy	12	1	2	...	...		
Mexico	54	31	5	4	4		
Morocco	35	5	0	...	...		
Nepal	...	...	...	22	22		
Netherlands	19	10	4	5	5		
New Zealand	7	11	8	...	...		
Pakistan	13	3	7	5	5		
Poland	56	66	38	12	12		
Romania	7	46	44	40	40	7	1.9
Tunisia	7	1	3	4	4		

Linseed continued

	1950	1970	1980	1988	1989	Rank	%
Turkey	31	9	3	3	3		
Uruguay	104	63	39	3	2		
USA	1,029	699	235	41	34	9	1.6
USSR	283	462	217	220	230	4	10.8

Maize
thousand tonnes

	1950	1970	1980	1989	1990	Rank	%
World	134,763	278,615	421,372	471,977	468,802		
Africa	8,884	21,762	18,169	37,508	32,519		
Asia	21,873	45,990	84,758	114,985	117,735		
Europe	12,885	38,635	54,216	55,455	45,995		
North America	74,697	278,615	212,798	211,722	223,820		
Oceania	137	261	325	370	368		
South America	10,536	25,663	32,631	36,633	32,365		
Afghanistan	350	707	780	750	750		
Albania	107	220	250	280	280		
Angola	285	467	290	204	180		
Argentina	2,839	8,717	9,533	4,260	5,049		
Austria	120	677	1,338	1,491	1,400		
Benin	181	201	335	455	460		
Bolivia	163	291	337	351	370		
Brazil	5,841	13,680	19,259	26,590	21,405	3	4.6
Bulgaria	720	2,436	2,652	2,421	1,200		
Cameroon	212	283	466	375	350		
Canada	388	2,487	5,544	6,379	7,033	10	1.5
Chile	68	217	471	938	823		
China	14,082	27,820	60,952	79,258	82,345	2	17.6
Colombia	733	856	854	1,044	1,215		
Cuba	243	85	95	95	95		
Czechoslovakia	316	511	831	1,000	508		
Egypt	1,378	2,370	2,956	3,748	3,750		
El Salvador	191	340	512	582	596		
Ethiopia	160	909	1,104	1,888	2,000		
France	452	7,394	9,624	12,926	8,996	9	1.9
Germany	20	500	748	1,574	1,440		
Ghana	168	417	397	749	493		
Greece	225	498	1,040	1,700	1,500		
Guatemala	437	751	1,011	1,232	1,277		
Haiti	203	245	179	130	130		
Honduras	205	298	343	540	580		
Hungary	2,081	4,542	6,856	6,996	4,500		
India	2,165	6,087	6,469	9,409	9,000	8	1.9
Indonesia	1,535	2,575	3,870	6,193	6,741		
Italy	2,306	4,601	6,617	6,360	5,788		
Ivory Coast	...	257	285	670	670		
Kenya	591	2,292	1,939	2,925	2,700		
Korea, N.	375	1,493	2,117	4,450	4,400		
Malawi	307	1,066	1,300	1,510	1,343		
Mexico	3,090	9,025	11,758	10,945	12,019	5	2.6
Morocco	302	380	245	403	436		
Mozambique	122	364	250	330	453		
Nepal	829	796	672	1,201	950		
Nigeria	644	1,215	1,543	2,132	1,832		
Pakistan	384	697	942	1,179	1,279		
Paraguay	112	197	578	1,000	1,139		
Peru	275	605	564	1,010	705		
Philippines	695	2,009	3,153	4,470	4,200		
Portugal	421	599	486	674	776		
Romania	2,495	7,354	11,593	6,750	9,200	7	2
South Africa	2,629	6,691	11,227	12,061	9,200	6	2
Spain	520	1,804	2,188	3,276	3,170		
Tanzania	347	817	800	3,159	2,445		
Thailand	31	1,979	3,383	4,393	4,296		
Turkey	747	1,058	1,230	2,000	2,000		
Uganda	121	420	360	300	300		
Uruguay	141	161	129	60	20		
USA	74,308	122,649	192,919	191,155	201,508	1	43
USSR	5,751	9,993	8,609	15,305	16,000	4	3.4
Venezuela	303	698	665	921	1,200		
Vietnam	180	272	478	843	850		
Yugoslavia	3,078	7,399	9,734	9,415	6,600		
Zaïre	324	426	515	790	756		
Zambia	...	786	833	1,861	1,093		
Zimbabwe	294	966	1,873	1,927	1,993		

Mangoes
thousand tonnes

	1970	1980	1989	1990	Rank	%
World	12,236	13,181	15,029	15,122		
Africa	653	846	1,009	1,013		
Asia	9,895	10,205	11,791	11,852		
North America	813	1,285	1,525	1,544		
South America	867	836	688	696		
Bangladesh	424	210	159	160		
Brazil	668	570	410	415	6	2.7
Cambodia	31	8	21	23		
Chad	24	30	32	32		
China	161	274	445	465	5	3.1
Colombia	14	20	29	30		
Cuba	18	63	81	85		
Dominican Rep.	153	177	190	193	10	1.3
Ecuador	10	26	34	35		
Egypt	61	117	100	95		
El Salvador	25	15	17	17		
Haiti	240	325	350	353	8	2.3
Honduras	12	13	15	15		
India	8,300	8,365	9,500	9,500	1	62.8
Indonesia	300	345	430	500	4	3.3
Kenya	10	13	21	22		
Madagascar	113	173	195	196	9	1.3
Malawi	19	27	32	37		
Malaysia	8	11	28	28		
Mexico	298	599	790	800	2	5.3
Mozambique	35	30	33	34		
Pakistan	486	546	740	761	3	5
Panama	23	26	4	4		
Paraguay	28	14	19	19		
Peru	62	80	59	60		
Philippines	143	371	370	375	7	2.5
Senegal	25	32	33	33		
South Africa	7	13	25	25		
Sri Lanka	39	67	60	60		
St. Lucia	38	43	46	46		
St. Vincent	12	8	2	1		
Sudan	45	68	132	134		
Tanzania	142	175	185	186		
Venezuela	78	117	127	127		
Zaïre	145	141	158	160		

Milk – cow's
thousand tonnes

	1950	1970	1980	1988	1989	Rank	%
World	233,025	363,193	426,393	468,834	474,020		
Africa	5,864	9,291	10,536	13,538	13,787		
Asia	17,045	24,083	37,121	48,301	49,882		
Europe	93,704	147,864	176,680	171,042	172,450		
North America	62,438	67,858	76,510	86,460	86,273		
Oceania	10,271	13,628	12,192	14,133	14,004		
South America	10,474	18,502	23,200	29,410	30,025		
Afghanistan	238	521	552	315	330		
Algeria	92	292	519	585	595		
Argentina	4,101	4,527	5,270	6,168	6,725		
Australia	5,538	7,468	5,574	6,423	6,462		
Austria	1,998	3,218	3,430	3,353	3,370		
Bangladesh	...	678	958	728	735		
Belgium-Lux.	3,323	3,971	4,032	3,915	3,965		
Brazil	2,581	7,317	10,423	13,609	13,609	8	2.9
Bulgaria	278	1,249	1,837	2,168	2,126		
Burma	250	400	283	585	590		
Canada	7,051	8,247	7,932	8,229	8,250		
Chile	683	1,029	1,075	1,149	1,290		
China	2,005	3,233	5,395	3,863	3,985		
Colombia	1,835	2,250	2,471	3,155	3,170		
Cuba	460	672	1,197	1,120	1,180		
Czechoslovakia	3,188	4,823	5,831	6,963	7,101		
Denmark	4,915	4,588	5,107	4,739	4,730		
Dominican Rep.	140	282	400	290	290		
Ecuador	174	705	759	1,406	1,400		
Egypt	362	575	667	980	1,000		
Ethiopia	344	516	617	602	624		
Finland	2,414	3,401	3,236	2,753	2,660		
France	14,578	27,467	32,925	26,606	27,250	4	5.7
Germany	15,997	29,904	32,741	33,178	33,500	3	7.1
Greece	147	546	704	652	715		
Hungary	1,443	1,851	2,560	2,873	2,812		
India	7,743	7,633	13,033	22,000	23,000	5	4.9
Iran	539	915	1,562	1,754	1,680		

	1950	1970	1980	1988	1989	Rank	%
Ireland	2,265	3,685	4,830	5,607	5,600		
Israel	97	447	729	964	964		
Italy	7,009	9,427	10,533	10,726	10,650	10	2.2
Japan	369	4,703	6,529	7,608	7,750		
Kenya	120	841	840	2,250	2,295		
Korea, S.	1	51	442	1,632	1,850		
Mexico	1,539	3,912	6,828	8,830	8,970		
Morocco	271	450	753	860	870		
Netherlands	5,437	8,182	11,832	11,315	11,250	9	2.4
New Zealand	4,720	6,112	6,555	7,650	7,482		
Nicaragua	174	201	235	124	121		
Nigeria	334	313	354	360	360		
Norway	1,532	1,726	1,945	1,972	1,952		
Pakistan	3,700	1,935	2,189	3,001	3,161		
Peru	237	828	798	850	802		
Poland	9,091	14,951	16,227	15,632	15,700	6	3.3
Portugal	219	607	691	1,287	1,400		
Puerto Rico	151	372	420	346	348		
Romania	1,900	2,774	4,264	4,300	4,350		
South Africa	1,740	2,816	2,500	2,600	2,600		
Spain	2,191	4,427	6,001	5,812	6,168		
Sudan	667	753	940	1,750	1,800		
Sweden	4,609	2,996	3,455	3,445	3,509		
Switzerland	2,483	3,172	3,667	3,768	3,840		
Syria	96	197	478	742	652		
Tanzania	356	692	729	448	456		
Turkey	1,315	2,565	3,469	3,000	3,000		
Uganda	...	...	300	370	370		
UK	9,967	13,007	15,921	14,981	14,750	7	3.1
Uruguay	449	744	826	990	1,020		
USA	52,349	53,173	58,146	65,840	65,432	2	13.8
USSR	33,228	81,967	90,333	105,950	107,600	1	22.7
Venezuela	203	947	1,335	1,715	1,630		
Yugoslavia	1,463	2,591	4,359	4,494	4,550		

Millet
thousand tonnes

	1950	1960	1980	1989	1990	Rank	%
World	20,000	33,180	28,476	29,820	30,193		
Africa	6,190	9,680	10,120	10,299	8,939		
Asia	7,160	20,762	16,417	15,132	15,925		
Europe	120	55	31	27	28		
North America	...	...	107	180	180		
Oceania	7	37	21	39	33		
South America	300	168	245	50	91		
Afghanistan	10	20	39	35	30		
Angola	...	78	43	63	63		
Argentina	151	168	245	50	91		
Australia	7	37	21	23	30		
Burkina Faso	195	352	387	649	597	8	2
Burma	56	65	60	143	185		
Burundi	22	26	50	10	10		
Cameroon	350	343	405	52	50		
Central Afr. Rep.	29	43	48	15	15		
Chad	600	615	577	179	172		
China	...	9,257	5,953	3,754	4,501	3	14.9
Egypt	1,000	847	638	426	215		
Ethiopia	190	117	191	155	133		
Gambia	20	40	30	56	51		
Ghana	100	120	96	180	124		
India	6,064	10,182	9,352	10,514	10,500	1	34.8
Ivory Coast	33	31	47	48	54		
Kenya	50	127	125	60	72		
Korea, N.	82	407	450	600	600	7	2
Mali	500	784	867	842	695	6	2.3
Mauritania	45	81	40	14	7		
Namibia	...	...	34	60	65		
Nepal	...	128	125	225	240		
Niger	400	974	1,245	1,335	1,133	5	3.8
Nigeria	1,000	2,792	3,200	4,594	4,000	4	13.2
Pakistan	342	339	269	204	210		
Saudi Arabia	2	125	12	5	5		
Senegal	...	539	600	639	508	9	1.7
Sudan	181	424	500	162	112		
Tanzania	250	140	155	300	200		
Togo	80	121	124	97	80		
Uganda	510	721	473	417	420	10	1.4
USSR	1,705	2,477	1,642	4,108	5,000	2	16.6
Yemen	...	70	70	61	52		
Yugoslavia	2	2	2	2	2		
Zambia	...	...	22	26	32		
Zimbabwe	50	190	167	142	143		

Mules

thousand head

	1950	1970	1980	1988	1989	Rank	%
World	11,828	11,761	13,349	14,498	14,724		
Africa	1,595	2,084	1,499	1,291	1,339		
Asia	562	2,353	4,780	5,903	6,062		
Europe	2,046	1,143	585	395	383		
North America	1,794	3,473	3,592	3,647	3,658		
South America	4,299	2,705	2,891	3,260	3,281		
Afghanistan	30	34	30	30	30		
Albania	12	22	22	22	22		
Algeria	231	188	203	113	130		
Argentina	333	200	165	165	165	9	1.1
Bolivia	54	86	86	80	80		
Brazil	3,072	1,631	1,647	1,980	2,000	3	13.6
Bulgaria	36	30	31	25	24		
Chile	22	17	10	10	10		
China	1,554	1,600	4,019	5,248	5,403	1	36.7
Colombia	525	367	576	600	600	4	4.1
Cuba	33	32	24	32	32		
Dominican Rep.	45	92	97	132	132		
Ecuador	54	92	98	116	117		
El Salvador	40	23	22	23	23		
Ethiopia	1,009	1,400	727	550	570	5	3.9
France	89	33	14	12	12		
Greece	162	185	117	82	81		
Guatemala	58	49	38	38	38		
Haiti	56	72	80	85	86		
Honduras	75	65	68	69	69		
India	57	84	128	135	138	10	0.9
Iran	126	136	131	127	126		
Iraq	120	59	27	26	26		
Italy	393	189	91	50	50		
Mexico	1,381	3,048	3,155	3,160	3,170	2	21.5
Morocco	171	378	450	500	510	6	3.5
Nicaragua	41	43	44	45	45		
Pakistan	41	55	66	65	66		
Peru	153	220	217	220	220	7	1.5
Portugal	121	92	92	80	80		
Somalia	18	20	22	23	24		
South Africa	98	20	14	14	14		
Spain	1,178	568	203	110	100		
Syria	63	64	39	27	27		
Tunisia	47	59	70	76	77		
Turkey	104	288	306	210	210	8	1.4
USA	...	...	26	28	28		
Venezuela	62	76	74	72	72		
Yugoslavia	32	18	12	10	11		

Oats

thousand tonnes

	1950	1970	1980	1989	1990	Rank	%
World	62,100	54,494	43,531	41,575	43,764		
Africa	300	166	215	207	172		
Asia	510	1,470	1,151	928	968		
Europe	19,970	17,939	14,269	11,695	12,123		
North America	25,240	18,891	10,487	9,074	8,796		
Oceania	610	1,435	1,409	1,704	1,703		
South America	880	618	766	1,139	1,203		
Algeria	136	41	97	54	40		
Argentina	743	420	490	668	670	10	1.5
Australia	560	1,378	1,353	1,638	1,615	7	3.7
Austria	275	281	298	249	262		
Belgium-Lux.	518	331	167	63	60		
Brazil	9	27	74	228	243		
Bulgaria	148	93	60	104	68		
Canada	6,220	5,508	3,192	3,546	3,507	3	8
Chile	83	106	151	165	205		
China		833	633	600	600		
Czechoslovakia	961	882	407	330	414		
Denmark	922	699	162	115	109		
Ethiopia	6	5	12	36	44		
Finland	718	1,297	1,183	1,444	1,662	6	3.8
France	3,393	2,317	1,842	1,025	875	9	2
Germany	3,711	3,567	3,315	2,010	2,123	4	4.9
Greece	119	108	80	57	40		
Hungary	213	79	95	149	150		
Ireland	617	222	95	99	136		
Italy	495	488	437	296	313		
Japan	119	63	12	10	10		
Korea, N.	4	97	135	60	60		
Mexico	47	33	72	105	105		
Mongolia	...	24	27	38	20		
Morocco	54	13	15	58	35		
Netherlands	419	243	106	32	27		
New Zealand	47	57	56	66	88		
Norway	170	215	393	421	596		
Poland	2,238	3,156	2,400	2,186	2,078	5	4.7
Portugal	124	92	79	127	62		
Romania	367	138	56	140	135		
South Africa	80	92	77	51	45		
Spain	519	508	520	494	519		
Sweden	804	1,561	1,608	1,455	1,614	8	3.7
Switzerland	68	35	52	60	55		
Turkey	326	446	342	216	274		
UK	2,853	1,300	588	530	535		
Uruguay	25	25	37	68	75		
USA	18,970	13,350	7,223	5,423	5,184	2	11.8
USSR	13,000	13,974	15,235	16,828	18,800	1	42.9
Yugoslavia	287	310	296	279	260		

Olives

thousand tonnes

	1950	1970	1980	1989	1990	Rank	%
World	4,943	7,650	9,458	10,016	9,411		
Africa	545	1,040	1,267	1,370	1,094		
Asia	424	666	1,254	780	1,649		
Europe	3,889	5,783	6,697	7,643	6,435		
North America	47	65	99	129	130		
South America	37	94	138	95	102		
Albania	12	33	51	32	25		
Algeria	151	150	114	173	179	9	1.9
Argentina	19	70	111	73	78		
Chile	15	12	10	11	11		
Cyprus	11	14	12	9	18		
Egypt	5	7	5	32	35		
France	27	16	16	11	11		
Greece	515	876	1,361	1,520	1,110	3	11.8
Iran	13	10	9	10	10		
Iraq	9	9	12	4	4		
Israel	9	13	23	19	39		
Italy	1,218	2,530	2,995	2,910	2,018	2	21.4
Jordan	28	16	20	30	50		
Lebanon	32	48	37	70	72		
Libya	27	57	142	118	118	10	1.3
Mexico	...	10	31	14	14		
Morocco	88	329	347	400	350	6	3.7
Peru	3	9	13	8	10		
Portugal	490	422	293	221	250	8	2.7
Spain	1,602	1,897	1,967	2,934	3,012	1	33
Syria	57	110	295	132	450	5	4.8
Tunisia	277	498	658	600	330	7	3.5
Turkey	268	438	810	500	1,000	4	10.6
USA	46	54	65	112	113		
Yugoslavia	28	10	13	17	18		

Olive oil

thousand tonnes

	1950	1970	1980	1989	1990	Rank	%
World	940	1,553	1,667	1,713	1,691		
Africa	96	176	156	212	235		
Asia	71	128	182	82	207		
Europe	765	1,228	1,308	1,400	1,235		
North America	3	2	3	2	2		
South America	4	18	18	10	11		
Albania	1	4	4	5	4		
Algeria	18	21	19	12	10	10	0.6
Argentina	2	17	16	9	10		
Chile	2	1	1	1	1		
Cyprus	2	2	2	1	2		
France	3	2	2	2	2		
Greece	115	178	275	300	210	3	12.4
Iran	1	1	1	1	1		
Israel	2	1	2	2	2		
Italy	241	553	591	522	382	2	22.6
Jordan	14	3	4	7	10		
Lebanon	2	11	6	5	5		
Libya	4	12	26	24	24	9	1.4
Morocco	13	41	25	70	40	8	2.4
Portugal	65	70	41	35	41	7	2.4
Spain	334	419	393	540	592	1	35
Syria	9	23	56	22	98	5	5.8
Tunisia	60	103	90	110	165	4	9.8
Turkey	47	86	113	45	90	6	5.3
USA	3	1	1	1	1		
Yugoslavia	4	2	3	3	3		

Onions

thousand tonnes

	1950	1970	1980	1988	1989	Rank	%
World	5,677	15,243	21,279	25,826	26,319		
Africa	480	1,079	1,517	1,974	2,045		
Asia	1,208	6,856	10,024	11,756	12,265		
Europe	2,252	3,657	4,073	4,663	4,664		
North America	1,053	1,519	1,811	2,411	2,416		
Oceania	60	128	121	191	197		
South America	415	1,045	1,790	2,192	2,233		
Algeria	60	43	119	200	200		
Argentina	112	215	247	414	410		
Australia	49	94	121	191	196		
Austria	14	23	35	62	56		
Bangladesh	...	182	124	141	145		
Bolivia	3	36	34	38	38		
Brazil	115	282	721	756	785	8	3
Bulgaria	15	102	78	96	110		
Burma	53	91	123	192	200		
Canada	60	108	117	112	131		
Chile	72	93	265	255	250		
China	...	2,145	2,646	3,670	3,826	1	14.5
Colombia	24	197	277	473	473		
Cuba	2	14	12	27	27		
Czechoslovakia	105	147	123	129	156		
Denmark	14	22	27	29	28		
Dominican Rep.	4	18	14	16	28		
Egypt	281	547	609	650	700	10	2.7
Ethiopia	2	26	36	41	42		
France	187	186	144	184	202		
Germany	149	79	128	168	180		
Ghana	20	17	22	28	28		
Greece	121	119	145	158	151		
Guatemala	10	15	21	30	25		
Hungary	78	149	138	252	210		
India	...	1,473	2,551	2,450	2,480	3	9.4
Indonesia	...	163	431	300	324		
Iran	...	254	607	297	500		
Iraq	30	84	119	122	130		
Israel	8	40	47	43	43		
Italy	192	472	518	470	470		
Japan	312	1,040	1,149	1,250	1,274	6	4.8
Korea, N.	...	16	25	44	46		
Korea, S.	8	91	316	527	545		
Laos	...	25	30	48	50		
Lebanon	38	34	21	29	40		
Libya	2	21	64	91	94		
Morocco	31	87	157	272	275		
Netherlands	173	336	474	452	350		
Niger	20	31	107	126	127		
Pakistan	120	235	424	633	707	9	2.7
Paraguay	6	19	19	47	55		
Peru	42	151	134	133	142		
Philippines	7	26	40	46	65		
Poland	126	339	353	515	564		
Portugal	82	70	67	58	60		
Romania	178	241	302	435	420		
Saudi Arabia	...	17	58	15	15		
Senegal	...	10	29	32	32		
South Africa	33	109	148	212	215		
Spain	463	920	950	1,020	1,008	7	3.8
Sri Lanka	17	38	65	59	60		
Sudan	5	36	33	43	45		
Syria	41	68	163	90	64		
Tanzania	8	46	40	49	50		
Thailand	4	105	124	180	180		
Tunisia	5	18	50	50	50		
Turkey	186	612	1,017	1,345	1,300	5	4.9
Uganda	5	12	13	30	32		
UK	88	164	223	311	302		
USA	943	1,340	1,622	2,189	2,168	4	8.2
USSR	350	958	1,943	2,640	2,500	2	9.5
Venezuela	15	36	75	57	58		
Vietnam	...	107	130	160	164		
Yemen	...	...	7	36	37		
Yugoslavia	125	243	315	269	339		
Zaïre	...	15	22	29	30		
Zambia	...	16	20	27	28		

Oranges

thousand tonnes

	1950	1970	1980	1989	1990	Rank	%
World	12,203	25,453	36,387	52,161	51,201		
Africa	1,084	2,688	3,301	3,823	3,746		
Asia	1,394	4,433	5,749	12,016	10,943		
Europe	1,689	3,812	3,913	5,448	5,212		
North America	5,233	9,242	11,988	10,333	10,383		
Oceania	125	291	413	416	475		
South America	2,664	4,915	10,782	20,019	20,192		
Algeria	215	351	293	177	177		
Argentina	362	892	688	580	750		
Australia	123	283	399	399	460		
Belize	...	33	34	52	52		
Bolivia	112	53	86	70	72		
Brazil	1,316	3,084	8,352	17,774	17,781	1	34.7
Cambodia	2	41	22	42	42		
Chile	63	42	51	78	72		
China	227	727	670	4,692	3,332	3	6.5
Colombia	...	91	233	80	80		
Costa Rica	13	59	75	83	86		
Cuba	46	112	281	474	520		
Cyprus	24	119	114	50	50		
Dominican Rep.	22	63	71	65	66		
Ecuador	168	160	522	85	86		
Egypt	263	760	1,239	1,370	1,420	8	2.8
El Salvador	...	38	98	96	90		
Ethiopia	42	8	9	12	13		
Ghana	17	107	20	50	50		
Greece	97	418	514	836	831		
Honduras	8	18	27	49	49		
India	226	900	1,153	1,800	1,854	6	3.6
Indonesia	...	93	191	412	412		
Iran	47	65	72	1,262	1,262	9	2.5
Iraq		28	140	175	180		
Israel	241	917	946	566	630		
Italy	548	1,403	1,604	2,067	1,824	7	3.6
Jamaica	54	70	34	60	60		
Japan	378	321	387	297	310		
Korea, S.	...	5	172	...	...		
Lebanon	47	167	231	270	280		
Libya	3	19	37	77	78		
Madagascar	10	49	85	84	85		
Mexico	480	1,377	1,649	1,166	2,200	5	4.3
Morocco	158	665	695	994	844		
Nicaragua	...	45	53	65	66		
Pakistan	135	312	513	1,100	1,139	10	2.2
Panama	33	51	67	36	37		
Paraguay	171	194	219	364	366		
Peru	368	246	149	163	165		
Portugal	68	100	107	85	75		
Puerto Rico	30	32	31	25	20		
South Africa	199	533	567	510	520		
Spain	975	1,884	1,678	2,434	2,457	4	4.8
Sudan	...	35	49	16	17		
Swaziland	...	...	41	43	43		
Syria	3	5	32	148	150		
Thailand	...	44	54	56	56		
Tunisia	25	58	113	144	150		
Turkey	66	445	707	745	750		
Uruguay	58	42	52	97	95		
USA	4,509	7,302	9,530	8,118	7,084	2	13.8
USSR	20	72	242	106	250		
Venezuela	60	182	411	426	427		
Vietnam	...	43	88	110	110		
Zaire	...	126	141	152	153		

Figures for 1950 include tangerines.

Palm kernels

thousand tonnes

	1950	1970	1980	1988	1989	Rank	%
World	889	1,179	1,742	2,869	3,014		
Africa	754	731	677	592	627		
Asia	39	178	718	1,889	1,987		
North America	12	21	15	31	31		
Oceania	...	...	26	69	71		
South America	84	248	305	288	298		
Angola	11	16	12	12	12		
Benin	42	69	29	20	25		
Brazil	76	219	266	202	205	4	6.8
Burundi	...	2	2	3	3		
Cameroon	32	40	40	27	30		
China	...	28	41	52	53	7	1.8
Colombia	...	7	17	42	48	8	1.6
Congo	12	2	1	1	1		
Costa Rica	1	3	6	13	10		
Ecuador	6	6	6	20	19		
Equatorial Guinea	7	2	3	3	3		
Ghana	6	36	30	30	30		
Guatemala	...	...	...	7	7		
Guinea	25	35	38	40	40	10	1.3
Guinea-Bissau	15	8	9	14	14		
Honduras	...	2	1	7	7		
Indonesia	29	49	130	314	298	3	9.9
Ivory Coast	8	19	34	33	40		
Liberia	17	14	8	8	8		
Madagascar	...	...	4	6	6		
Malaysia	11	99	540	1,473	1,580	1	52.4
Mexico	11	16	7	3	3		
Nigeria	372	287	342	270	300	2	10
Papua N. Guinea	...	...	24	66	67	6	2.2
Paraguay	4	16	14	19	19		
Peru	...	...	2	4	5		
Philippines	...	1	2	14	15		
Senegal	2	9	5	6	6		
Sierra Leone	74	60	31	30	30		
Solomon Is.	...	1	3	4	4		
Tanzania	...	2	5	6	6		
Thailand	...	...	4	35	41	9	1.4
Togo	9	23	11	1	1		
Zaire	8	99	68	74	74	5	2.5

Palm oil

thousand tonnes

	1950	1970	1980	1989	1990	Rank	%
World	1,008	1,971	4,979	10,787	11,346		
Africa	836	1,102	1,338	1,584	1,651		
Asia	163	791	3,405	8,435	8,860		
North America	6	31	45	141	154		
Oceania	...	...	51	168	164		
South America	2	47	140	459	518		
Angola	36	40	40	40	40		
Benin	34	28	33	40	40		
Brazil	...	7	14	63	68		
Cameroon	22	63	79	102	108		
China	...	108	188	210	215	6	1.9
Colombia	...	27	73	224	260	4	2.3
Congo	4	8	9	16	16		
Costa Rica	...	13	23	50	65		
Ecuador	...	6	37	126	138	10	1.2
Equat. Guinea	4	4	5	5	5		
Gambia	2	2	3	3	3		
Ghana	30	19	21	78	85		
Guinea	10	44	42	45	45		
Guinea-Bissau	8	4	5	3	3		
Honduras	...	6	12	77	75		
Indonesia	114	223	668	1,948	2,200	2	19.4
Ivory Coast	6	46	164	189	220	5	1.9
Liberia	40	14	24	35	35		
Malaysia	49	457	2,528	6,057	6,200	1	55
Nigeria	422	528	667	770	790	3	7
Papua New G.	...	...	35	146	143	9	1.3
Paraguay	1	7	5	7	7		
Peru	...	...	6	30	30		
Philippines	...	2	12	51	65		
Senegal	...	5	6	6	6		

	1950	1970	1980	1989	1990	Rank	%
Sierra Leone	35	49	48	44	44		
Solomon Is.	...	...	16	22	21		
Thailand	...	...	10	170	180	8	1.6
Togo	3	17	20	14	14		
Zaire	172	223	164	178	180	7	1.6

Paper and paper board

million tonnes

	1961 –65 av.	1970	1980	1987	1988	Rank	%
World	86.7	128	170	213	224		
Africa	0.5	0.9	1.7	2.4	2.5		
Asia	10.4	19.1	29.6	43.5	47.4		
Europe	26.2	38.8	50.1	60.2	64.3		
North America	43.3	58.4	72.4	86.4	89.8		
Oceania	0.9	1.5	2.1	2.2	2.4		
South America	1.5	2.8	5.6	7.7	7.7		
Argentina	0.4	0.6	0.7	1	1		
Australia	0.6	1.1	1.4	1.5	1.7		
Austria	0.6	1	1.6	2.4	2.7		
Belgium	0.5	0.8	0.9	0.8	0.8		
Brazil	0.7	1.2	3.4	4.7	4.6	10	2.1
Bulgaria	0.1	0.2	0.4	0.4	0.5		
Canada	8.6	11.3	13.4	16.1	16.6	3	7.4
Chile	0.1	0.3	0.4	0.4	0.4		
China	2.9	4.2	6.8	13	14.1	4	6.3
Colombia	0.1	0.2	0.4	0.5	0.5		
Czechoslovakia	0.7	0.8	1.2	1.3	1.3		
Denmark	0.2	0.3	0.2	0.3	0.3		
Finland	2.8	4.3	5.9	8	8.7	7	3.9
France	3	4.1	5.2	5.6	6.3	9	2.8
Germany	4.7	6.6	8.8	11.3	11.9	5	5.3
Greece	0.1	0.2	0.3	0.3	0.3		
Hungary	0.2	0.3	0.4	0.5	0.5		
India	0.5	0.9	1	1.9	1.9		
Indonesia	...	...	0.2	0.8	1		
Italy	1.9	3.5	4.9	4.9	4.5		
Japan	6.4	12.9	18.1	22.5	24.6	2	10.9
Korea, S.	0.1	0.3	1.7	3.2	3.7		
Mexico	0.5	0.9	2	2.6	3.4		
Netherlands	1.2	1.6	1.7	2.2	2.5		
New Zealand	0.3	0.5	0.7	0.6	0.7		
Norway	0.9	1.4	1.4	1.6	1.7		
Peru	0.1	0.1	0.2	0.2	0.3		
Philippines	0.1	0.1	0.3	0.3	0.3		
Poland	0.8	0.9	1.3	1.4	1.5		
Portugal	0.1	0.2	0.5	0.6	0.6		
Romania	0.3	0.5	0.8	0.8	0.8		
South Africa	0.3	0.6	1.2	1.6	1.6		
Spain	0.5	1.3	2.6	3.3	3.4		
Sweden	2.7	4.4	6.2	7.8	8.2	8	3.7
Switzerland	0.6	0.7	0.9	1.1	1.2		
Taiwan	0.1	0.2	0.5	0.8	0.9		
Thailand	...	0.1	0.3	0.5	0.5		
Turkey	0.1	0.2	0.5	0.4	0.4		
UK	4.2	4.9	3.8	4.2	4.3		
USA	34.1	46.1	56.8	67.5	69.5	1	31
USSR	4	6.7	8.7	10.2	10.2	6	4.6
Venezuela	0.1	0.3	0.5	0.7	0.7		
Yugoslavia	0.3	0.6	1.1	1.4	1.4		

Peaches and nectarines

thousand tonnes

	1950	1970	1980	1988	1989	Rank	%
World	2,443	6,269	7,128	8,211	8,586		
Africa	44	196	240	281	313		
Asia	68	809	1,170	1,588	1,604		
Europe	1,434	2,530	3,073	3,523	4,001		
North America	1,434	1,716	1,719	1,678	1,517		
Oceania	64	144	100	102	88		
South America	191	545	575	588	604		
Algeria	6	8	12	34	35		
Argentina	110	254	247	254	250	10	2.9
Australia	55	117	77	74	60		
Austria	5	7	10	7	12		
Bolivia	...	22	32	30	30		
Brazil	29	111	111	100	105		
Bulgaria	6	148	97	63	97		

Peaches and nectarines *continued*

	1950	1970	1980	1988	1989	Rank	%
Canada	44	48	35	44	47		
Chile	22	91	117	151	163		
China	1	289	384	641	664	4	7.7
Czechoslovakia	2	11	14	15	27		
Egypt	3	13	10	32	35		
France	111	575	461	457	551	6	6.4
Germany	47	44	23	25	27		
Greece	12	184	376	591	582	5	6.8
Hungary	33	119	81	59	80		
India	...	10	15	19	20		
Iraq	...	3	28	28	28		
Israel	...	19	26	36	36		
Italy	291	1,086	1,436	1,442	1,650	1	19.2
Japan	35	273	269	203	205		
Korea, N.	...	29	63	98	100		
Korea, S.	16	71	90	135	135		
Lebanon	5	11	22	24	24		
Libya	2	2	3	10	10		
Madagascar	4	7	6	7	7		
Mexico	48	161	185	265	265	9	3.1
Morocco	1	14	20	28	29		
New Zealand	9	27	22	29	28		
Pakistan	...	9	10	18	20		
Peru	...	31	26	21	25		
Portugal	19	36	33	45	51		
Romania	8	28	51	86	82		
South Africa	29	147	170	147	170		
Spain	94	233	412	655	751	3	8.7
Syria	1	6	23	51	47		
Tunisia	2	7	20	21	25		
Turkey	11	87	237	328	318	8	3.7
Uruguay	26	31	28	17	15		
USA	1,341	1,507	1,498	1,370	1,205	2	14
USSR	...	329	252	450	460	7	5.4
Venezuela	...	2	8	7	7		
Yugoslavia	16	59	79	77	91		

Pears

thousand tonnes

	1950	1970	1980	1988	1989	Rank	%
World	3,926	7,936	8,616	9,884	9,675		
Africa	26	130	219	351	367		
Asia	597	1,757	2,759	3,955	4,129		
Europe	2,405	4,451	3,693	3,715	3,304		
North America	683	682	879	859	839		
Oceania	75	181	150	174	154		
South America	140	196	254	350	382		
Algeria	5	10	17	44	45		
Argentina	101	90	148	201	217	10	2.2
Australia	66	162	134	162	142		
Austria	210	158	120	189	133		
Belgium-Lux.	167	73	66	80	74		
Brazil	27	54	35	23	23		
Bulgaria	54	130	96	73	73		
Canada	24	35	36	23	25		
Chile	8	31	40	99	115		
China	394	910	1,621	2,542	2,705	1	28
Czechoslovakia	80	50	26	49	48		
Denmark	26	12	14	4	4		
Ecuador	1	3	9	5	6		
Egypt	1	17	51	45	60		
France	317	551	445	345	344	9	3.6
Germany	483	536	411	575	406	8	4.2
Greece	30	112	115	91	93		
Hungary	53	87	85	87	88		
India	...	52	64	83	85		
Iran	...	29	45	71	71		
Israel	...	26	15	21	17		
Italy	328	1,749	1,178	997	797	2	8.2
Japan	79	465	511	454	472	6	4.9
Korea, N.	...	22	64	108	112		
Korea, S.	31	49	66	192	193		
Lebanon	6	8	15	14	14		
Mexico	15	34	44	55	50		
Morocco	2	7	13	40	41		
Netherlands	132	120	103	84	94		
New Zealand	9	20	16	12	11		
Norway	6	10	7	8	2		
Pakistan	11	16	29	34	35		
Peru	...	6	8	9	9		

	1950	1970	1980	1988	1989	Rank	%
Poland	46	110	96	56	59		
Portugal	26	43	51	41	55		
Romania	15	63	90	130	128		
South Africa	17	92	126	196	195		
Spain	70	295	478	459	550	4	5.7
Sweden	30	16	17	11	11		
Switzerland	252	155	135	229	120		
Syria	2	6	10	21	13		
Tunisia	1	4	10	23	23		
Turkey	75	172	316	410	407	7	4.2
UK	38	70	54	34	47		
Uruguay	4	7	10	9	9		
USA	644	612	800	781	764	3	7.9
USSR	...	538	660	480	500	5	5.2
Yugoslavia	69	112	107	173	177		

Pigs

thousand head

	1950	1970	1980	1989	1990	Rank	%
World	300,801	625,150	779,897	846,855	860,861		
Africa	4,464	6,868	9,505	13,224	13,588		
Asia	94,502	292,872	371,947	419,888	432,822		
Europe	69,169	132,398	175,070	186,662	185,961		
North America	75,489	85,970	92,972	89,373	88,686		
Oceania	1,872	4,327	4,746	5,346	5,344		
South America	35,591	45,186	53,866	54,262	55,560		
Argentina	3,250	4,466	3,751	4,200	4,400		
Australia	1,146	2,414	2,415	2,671	2,610		
Austria	2,048	3,245	3,906	3,874	3,766		
Belgium	1,142	3,263	5,083	6,233	6,350		
Bolivia	509	900	1,454	2,127	2,200		
Brazil	24,879	30,851	35,732	31,700	33,200	5	3.9
Bulgaria	998	2,159	3,800	4,119	4,352		
Burma	418	1,562	1,456	2,451	2,600		
Cambodia	370	1,167	192	1,550	1,585		
Cameroon	220	430	1,190	1,299	1,300		
Canada	4,792	6,624	9,116	10,763	10,532		
Chile	710	1,025	1,085	1,400	1,450		
China	73,759	246,320	313,660	349,172	360,594	1	41.9
Colombia	2,368	1,470	2,080	2,600	2,630		
Cuba	1,315	1,483	1,950	1,850	1,850		
Czechoslovakia	3,612	5,234	7,694	7,384	7,400		
Denmark	2,829	8,336	9,718	9,120	9,300		
Dominican Rep.	739	714	300	429	435		
Ecuador	547	1,918	3,552	4,160	4,160		
Finland	403	963	1,430	1,327	1,340		
France	6,582	10,516	11,470	12,410	12,200		
Germany	13,853	29,156	34,769	35,053	34,178	4	4
Greece	549	407	947	1,226	1,300		
Guatemala	408	810	791	800	800		
Haiti	1,137	1,525	1,060	950	960		
Honduras	445	803	550	600	600		
Hungary	4,006	6,271	8,232	8,327	7,660		
India	3,910	6,207	10,033	10,300	10,400		
Indonesia	1,161	3,143	3,184	6,590	6,600		
Ireland	615	1,210	1,120	961	995		
Italy	4,026	8,501	8,886	9,359	9,254		
Japan	513	6,343	9,851	11,866	11,880		
Korea, N.	281	1,333	2,100	3,145	3,200		
Korea, S.	439	1,287	2,248	4,852	4,900		
Laos	149	1,100	1,117	1,300	1,350		
Madagascar	381	581	671	1,400	1,425		
Malaysia	389	1,055	1,711	2,350	2,400		
Mexico	6,340	10,273	12,900	16,157	17,300	7	2
Netherlands	1,561	5,521	10,058	13,820	14,200	10	1.6
New Zealand	556	571	530	411	410		
Nicaragua	454	407	578	680	680		
Nigeria	420	844	1,100	1,100	1,100		
Papua New G.	...	1,065	1,400	1,789	1,800		
Paraguay	340	601	1,294	2,305	2,305		
Peru	960	1,894	2,070	2,289	2,300		
Philippines	3,827	6,619	7,790	7,909	7,990		
Poland	7,439	14,348	20,346	18,835	19,800	6	2.3
Portugal	1,330	1,835	3,243	2,326	2,300		
Romania	2,110	6,061	10,926	15,400	15,510	9	1.8
Singapore	...	995	1,203	321	320		
South Africa	1,372	1,283	1,330	1,470	1,480		
Spain	6,041	6,657	10,392	16,715	16,910	8	2
Sweden	1,279	2,167	2,736	2,264	2,175		
Switzerland	892	1,796	2,113	1,869	1,857		
Taiwan	2,483	2,900	4,820	6,900	7,000		
Thailand	2,427	4,814	5,300	4,679	4,650		

	1950	1970	1980	1989	1990	Rank	%
UK	3,363	8,205	7,844	7,626	7,383		
USA	58,895	61,720	64,070	55,469	53,852	3	6.3
USSR	19,714	57,528	51,588	78,100	78,900	2	9.2
Venezuela	1,454	1,563	2,249	3,053	2,326		
Vietnam	4,515	9,467	10,000	11,643	11,700		
Yugoslavia	3,955	4,327	7,707	7,396	7,231		

Pig meat

thousand tonnes

	1950	1970	1980	1989	1990	Rank	%
World	16,389	38,619	54,581	67,258	67,115		
Africa	132	269	361	490	497		
Asia	3,743	11,818	19,699	26,882	27,267		
Europe	5,176	13,109	18,547	21,855	21,745		
North America	5,490	7,233	8,757	9,327	9,080		
Oceania	128	236	285	379	395		
South America	719	1,316	1,721	1,917	1,994		
Argentina	151	221	260	200	215		
Australia	89	173	218	297	312		
Austria	125	263	332	462	465		
Belgium-Lux.	177	455	669	811	815		
Brazil	351	763	960	1,000	1,050		
Bulgaria	69	160	310	394	413		
Burma	8	60	79	77	80		
Cambodia	12	33	8	25	25		
Canada	408	611	829	1,184	1,140		
Chile	42	44	50	113	120		
China	3,298	9,639	16,189	22,148	22,245	1	33.1
Colombia	51	69	101	140	144		
Cuba	36	34	59	89	89		
Czechoslovakia	300	566	819	915	937		
Denmark	305	722	955	1,167	1,168		
Ecuador	7	28	63	67	67		
Finland	57	109	171	169	171		
France	751	1,388	1,866	1,852	1,840	5	2.7
Germany	1,189	3,061	3,930	4,702	4,668	4	7
Greece	19	54	144	160	160		
Hong Kong	5	120	206	194	196		
Hungary	194	598	898	1,022	1,010		
India	24	52	70	96	88		
Indonesia	35	72	90	145	155		
Ireland	65	147	152	142	140		
Italy	233	590	1,074	1,269	1,278	10	1.9
Japan	46	722	1,434	1,594	1,635	8	2.4
Korea, N.	12	52	113	168	176		
Korea, S.	7	107	236	480	522		
Malaysia	31	51	68	148	148		
Mexico	58	240	467	727	732		
Netherlands	206	704	1,142	1,632	1,632	9	2.4
New Zealand	39	40	35	46	45		
Norway	36	66	84	95	95		
Paraguay	42	42	80	120	120		
Peru	14	64	66	92	94		
Philippines	89	304	406	575	665		
Poland	718	1,312	1,618	1,828	1,753	6	2.6
Portugal	78	98	169	183	207		
Romania	28	447	961	840	920		
Singapore	9	25	55	75	78		
South Africa	49	80	85	124	125		
Spain	147	468	985	1,722	1,720	7	2.6
Sweden	165	239	317	299	304		
Switzerland	83	201	277	285	283		
Thailand	60	210	239	334	335		
UK	296	934	936	1,017	947		
Uruguay	17	21	16	21	21		
USA	4,904	6,227	7,255	7,173	6,961	2	10.4
USSR	1,002	4,638	5,210	6,595	6,750	3	10.1
Venezuela	16	46	90	117	110		
Vietnam	41	345	423	494	505		
Yugoslavia	151	515	728	871	800		

Pineapples
thousand tonnes

	1950	1970	1980	1989	1990	Rank	%
World	1,448	4,911	9,287	9,820	9,659		
Africa	70	731	1,196	1,291	1,303		
Asia	241	1,990	5,763	5,764	5,652		
Europe	2	2	2	1	1		
North America	908	1,315	1,362	1,301	1,326		
Oceania	47	152	140	157	167		
South America	181	721	823	1,306	1,209		
Angola	...	19	35	35	35		
Australia	46	139	124	126	135		
Bangladesh	15	97	150	135	135		
Bolivia	1	6	8	8	8		
Brazil	134	437	392	828	720	4	7.5
Cambodia	8	23	5	11	12		
Cameroon	...	5	33	34	34		
China	49	781	299	742	845	3	8.7
Colombia	...	85	119	230	240		
Congo	...	90	102	115	100		
Costa Rica	...	5	10	82	82		
Cuba	120	15	16	21	22		
Dominican Rep.	5	13	21	44	44		
Ecuador	30	57	133	55	55		
El Salvador	...	33	19	15	15		
Ghana	7	28	6	10	10		
Guatemala	...	17	32	47	47		
Guinea	3	14	17	36	36		
Honduras	2	6	33	35	35		
India	...	98	548	598	602	5	6.2
Indonesia	...	107	207	350	356	8	3.7
Ivory Coast	7	127	295	213	189		
Japan	...	80	56	37	37		
Kenya	6	28	177	212	226		
Laos	...	25	31	50	52		
Liberia	...	6	7	7	7		
Madagascar	...	39	49	50	50		
Malaysia	34	270	185	207	207		
Martinique	4	25	18	15	16		
Mexico	125	262	530	324	324	9	3.4
Mozambique	...	18	13	15	16		
Nicaragua	...	25	34	41	42		
Panama	...	5	6	16	16		
Papua New G.	...	7	9	12	12		
Paraguay	11	32	28	44	45		
Peru	...	63	49	52	52		
Philippines	55	251	1,059	1,179	1,200	2	12.4
Portugal	2	2	2	1	1		
Puerto Rico	32	50	37	49	52		
South Africa	35	158	217	252	265	10	2.7
Sri Lanka	10	38	61	40	40		
Swaziland	...	10	29	42	42		
Taiwan	60	338	228	230	230		
Tanzania	12	37	47	68	70		
Thailand	56	187	2,857	2,005	1,745	1	18.1
USA	617	850	597	526	545	6	5.6
Venezuela	5	37	89	77	78		
Vietnam	15	34	303	485	490	7	5.1
Zaire	...	144	154	185	190		

Plantains
thousand tonnes

	1970	1980	1988	1989	Rank	%
World	16,378	21,826	24,766	24,972		
Africa	10,723	13,317	17,817	17,834		
Asia	1,187	2,372	800	840		
North America	1,244	1,593	1,694	1,840		
South America	3,210	4,540	4,454	4,458		
Bolivia	100	148	118	110		
Burma	406	413	220	220		
Cameroon	968	988	1,100	1,150	7	4.6
Central Afr. Rep.	53	61	66	67		
Colombia	1,619	2,328	2,278	2,463	2	9.8
Congo	29	34	65	66		
Costa Rica	56	84	105	110		
Cuba	60	89	165	165		
Dominican Rep.	529	592	651	809		
Ecuador	445	730	959	959	10	3.8
El Salvador	22	15	12	12		
Gabon	79	63	135	132		
Ghana	807	872	1,135	1,036	8	4.1
Guatemala	40	52	55	55		

	1970	1980	1988	1989	Rank	%
Guinea	177	225	350	350		
Guinea-Bissau	23	25	25	25		
Guyana	23	21	25	25		
Haiti	189	297	275	250		
Honduras	117	165	160	180		
Ivory Coast	653	808	1,076	1,030	9	4.1
Jamaica	15	18	26	26		
Kenya	168	233	270	272		
Liberia	24	31	33	33		
Malawi	100	106	114	115		
Nicaragua	62	84	97	98		
Nigeria	1,635	2,217	1,800	1,800	4	7.2
Panama	101	102	29	30		
Peru	654	800	594	420		
Philippines	402	280	200	200		
Puerto Rico	48	86	94	80		
Rwanda	1,656	2,058	2,140	2,150	3	8.6
Sierra Leone	15	21	28	28		
Sri Lanka	389	1,679	580	620		
Tanzania	558	773	1,300	1,350	6	5.4
Uganda	2,650	3,450	6,630	6,700	1	26.8
Venezuela	363	511	475	475		
Zaire	1,217	1,442	1,550	1,530	5	6.1

Plums
thousand tonnes

	1950	1970	1980	1988	1989	Rank	%
World	3,020	5,707	5,505	6,729	6,518		
Africa	35	38	78	128	138		
Asia	112	602	843	1,326	1,264		
Europe	2,224	3,697	2,869	3,342	3,019		
North America	539	617	719	769	877		
Oceania	26	31	26	22	24		
South America	51	98	97	142	147		
Afghanistan	...	24	36	37	37		
Albania	2	10	14	16	13		
Algeria	13	10	14	30	31		
Argentina	32	64	71	59	50		
Australia	24	27	22	18	20		
Austria	55	82	73	86	82		
Belgium	18	12	5	7	5		
Bulgaria	118	325	164	139	139		
Canada	17	9	8	4	6		
Chile	13	21	17	72	86		
China	...	339	436	819	770	4	11.8
Czechoslovakia	145	81	48	55	49		
Denmark	15	4	2	1	1		
Egypt	1	1	6	38	40		
France	136	184	151	229	146	9	2.2
Germany	481	589	429	574	382	6	5.9
Greece	10	21	13	8	8		
Hungary	144	216	163	182	180	7	2.8
India	...	19	28	36	38		
Iraq	...	4	29	33	33		
Israel	3	10	18	27	15		
Italy	88	140	166	154	133		
Japan	42	52	58	68	70		
Korea, South	...	2	14	32	40		
Lebanon	...	7	11	11	12		
Mexico	33	75	68	84	86		
Morocco	4	5	28	36	40		
Netherlands	30	10	6	5	6		
Norway	10	14	14	13	13		
Pakistan	...	29	36	48	50		
Poland	35	125	143	98	75		
Portugal	20	15	6	8	10		
Romania	190	706	615	800	765	5	11.7
South Africa	14	13	16	17	18		
Spain	42	61	95	141	145	10	2.2
Switzerland	25	44	52	33	31		
Syria	2	4	24	40	31		
Tunisia	1	6	8	4	6		
Turkey	49	110	152	175	166	8	2.5
UK	115	52	37	24	15		
USA	522	532	642	680	786	3	12.1
USSR	...	625	873	1,000	1,050	1	16.1
Yugoslavia	524	1,002	666	765	819	2	12.6

Potatoes
thousand tonnes

	1950	1970	1980	1989	1990	Rank	%
World	247,443	275,758	258,465	278,568	268,390		
Africa	1,273	2,931	5,056	6,742	6,970		
Asia	17,998	25,234	37,716	62,759	60,039		
Europe	130,155	126,640	109,583	103,026	105,274		
North America	12,739	17,482	18,598	21,081	21,891		
Oceania	609	1,025	1,130	1,329	1,393		
South America	4,430	8,707	9,722	11,427	10,822		
Afghanistan	...	160	267	300	300		
Algeria	212	253	570	1,030	1,107		
Argentina	1,169	2,212	1,836	2,600	2,500		
Australia	477	782	849	1,049	1,107		
Austria	2,270	2,787	1,356	845	850		
Bangladesh	...	842	942	1,089	1,100		
Belgium-Lux.	2,127	1,631	1,445	1,800	1,750		
Bolivia	176	660	822	659	534		
Brazil	699	1,557	2,004	2,129	2,239		
Bulgaria	240	378	377	538	475		
Canada	1,808	2,312	2,608	2,842	2,870		
Chile	461	707	893	882	829		
China	12,090	11,029	15,204	31,260	28,050	3	10.5
Colombia	506	871	1,931	2,697	2,649		
Czechoslovakia	7,055	4,864	3,307	3,167	2,800		
Denmark	2,170	815	865	1,242	1,614		
Ecuador	98	560	309	413	398		
Egypt	187	496	1,118	1,005	921		
Finland	1,442	906	629	981	820		
France	13,734	8,569	6,749	5,750	6,000	10	2.2
Germany	37,430	16,236	19,465	17,115	17,520	5	6.5
Greece	385	700	984	1,107	1,100		
Hungary	1,976	1,874	1,501	1,332	1,200		
India	1,547	4,482	9,353	14,893	15,000	6	5.6
Iran	37	417	694	1,295	1,450		
Ireland	2,903	1,450	1,122	668	687		
Italy	2,732	3,632	2,959	2,458	2,454		
Japan	2,451	3,490	3,350	3,587	3,500		
Kenya	190	206	358	300	400		
Korea, N.	575	960	1,550	2,050	2,100		
Korea, S.	302	598	452	450	500		
Mexico	134	489	829	966	810		
Morocco	56	283	502	800	851		
Nepal	...	265	277	641	658		
Netherlands	4,679	5,367	6,330	6,856	6,500	8	2.4
Norway	1,174	776	488	499	499		
Pakistan	139	213	412	645	670		
Peru	1,239	1,877	1,567	1,690	1,300		
Poland	29,641	45,012	39,521	34,390	36,300	2	13.5
Portugal	1,081	1,222	1,067	1,077	999		
Romania	1,661	2,671	4,399	7,200	7,600	7	2.8
Rwanda	67	134	221	184	190		
South Africa	230	596	694	1,245	1,250		
Spain	3,346	4,985	5,648	5,323	5,444		
Sweden	1,814	1,221	1,215	1,179	1,233		
Switzerland	1,021	1,016	924	795	800		
Syria	34	62	266	337	29		
Turkey	605	1,984	2,923	4,060	4,100		
Uganda	25	147	337	150	150		
UK	9,443	7,359	6,567	6,215	6,300	9	2.3
USA	10,681	14,483	14,802	16,803	17,835	4	6.6
USSR	80,239	93,739	76,660	72,205	62,000	1	23.1
Yugoslavia	1,486	3,020	2,555	2,359	2,200		

Poultry meat
thousand tonnes

	1950	1970	1980	1989	1990	Rank	%
World	...	17,760	27,245	37,968	39,378		
Africa	...	568	1,168	1,565	1,600		
Asia	...	4,106	6,171	8,962	9,314		
Europe	...	4,176	7,087	8,247	8,335		
North America	...	6,898	8,049	12,005	12,876		
Oceania	...	148	333	467	482		
South America	...	825	2,297	3,506	3,671		
Algeria	14	28	46	64	64		
Argentina	33	182	362	345	358		
Australia	45	128	295	396	410		
Austria	3	47	76	83	85		
Bangladesh	...	60	82	163	173		
Belgium-Lux.	29	116	128	181	185		
Brazil	87	360	1,264	2,137	2,321	4	5.8
Bulgaria	24	94	154	198	202		

Poultry meat (continued)

	1950	1970	1980	1989	1990	Rank	%
Burma	...	16	70	119	119		
Canada	142	542	526	693	690		
Chile	12	52	101	240	245		
China	...	2,583	2,959	3,152	3,281	2	8.3
Colombia	27	46	115	252	260		
Cuba	5	40	71	98	102		
Czechoslovakia	24	109	181	238	247		
Denmark	23	76	99	127	122		
Dominican Rep.	...	28	65	112	112		
Ecuador	1	9	21	62	62		
Egypt	24	67	106	108	114		
Ethiopia	50	54	59	73	74		
France	248	631	1,133	1,425	1,465	6	3.7
Germany	67	344	535	590	597		
Greece	11	71	142	162	156		
Guatemala	3	8	49	59	59		
Hong Kong	1	21	35	53	53		
Hungary	87	215	350	460	490		
India	45	83	111	289	319		
Indonesia	37	53	106	447	483		
Iran	18	53	214	260	265		
Iraq	1	16	56	214	220		
Ireland	28	32	47	66	67		
Israel	6	96	195	175	172		
Italy	57	619	1,002	1,111	1,113	7	2.8
Ivory Coast	4	9	22	31	31		
Jamaica	...	16	28	23	23		
Japan	10	494	1,125	1,482	1,487	5	3.8
Jordan	...	4	26	64	64		
Kenya	...	16	32	47	46		
Korea, N.	...	20	32	46	46		
Korea, S.	2	46	95	158	158		
Lebanon	2	21	46	62	62		
Libya	...	2	26	54	55		
Madagascar	12	35	44	84	87		
Malaysia	11	53	142	298	300		
Mexico	55	225	469	640	783	10	2
Morocco	16	30	103	104	104		
Nepal	9	18	6	6	6		
Netherlands	9	297	340	442	448		
New Zealand	3	17	32	59	60		
Nigeria	...	59	212	204	200		
Pakistan	...	17	46	174	182		
Peru	4	58	145	205	223		
Philippines	35	103	184	255	277		
Poland	33	130	439	332	320		
Portugal	7	54	141	127	144		
Puerto Rico	5	13	28	40	40		
Romania	45	128	415	380	390		
Saudi Arabia	...	8	30	217	220		
Singapore	2	27	53	66	66		
South Africa	21	107	229	374	384		
Spain	66	454	779	831	823	9	2.1
Sweden	13	27	43	47	47		
Syria	2	14	43	60	60		
Thailand	28	64	171	587	607		
Tunisia	3	10	40	49	49		
Turkey	18	104	243	296	302		
UK	95	574	750	1,011	997	8	2.5
USA	2,160	5,979	6,715	10,153	10,876	1	27.6
USSR	278	1,040	2,140	3,216	3,100	3	7.8
Venezuela	3	85	229	301	231		
Vietnam	...	70	90	162	161		
Yugoslavia	36	127	277	334	334		

Rapeseed

thousand tonnes

	1950	1970	1980	1989	1990	Rank	%
World	2,820	6,617	11,093	22,487	24,298		
Africa	20	20	22	118	128		
Asia	2,001	3,076	5,139	20,318	11,287		
Europe	718	1,889	3,246	8,363	8,873		
North America	16	1,523	2,566	3,155	3,393		
Oceania	...	31	28	84	122		
South America	...	74	76	125	65		
Argentina	...	2	16	2	2		
Australia	...	31	28	74	114		
Austria	5	8	7	96	117		
Bangladesh	...	132	126	207	220		
Belgium	5	1	1	15	18		
Canada	9	1,517	2,563	3,096	3,325	3	13.7
Chile	...	72	59	113	53		
China	782	996	2,864	5,436	6,600	1	27.2
Czechoslovakia	25	70	161	387	385		
Denmark	9	30	219	655	904	8	3.7
Ethiopia	20	20	22	30	40		
Finland	6	9	68	125	100		
France	154	603	871	1,803	2,011	5	8.3
Germany	193	370	634	1,869	2,153	4	8.9
Hungary	4	46	81	98	97		
India	815	1,629	1,845	4,412	4,220	2	17.4
Italy	14	5	1	40	33		
Japan	129	34	4	2	2		
Korea, S.	...	31	26	9	9		
Mexico	6	6	3	3	3		
Netherlands	33	22	28	23	21		
Norway	...	6	10	9	10		
Pakistan	270	249	249	249	233		
Poland	100	455	431	1,586	1,200	7	4.9
Romania	4	3	18	40	40		
Spain	...	...	11	10	10		
Sweden	146	218	324	422	401	10	1.7
Switzerland	4	19	33	54	55		
Turkey	4	5	23	3	3		
UK	...	10	274	953	1,231	6	5.1
USA	...	1	1	56	65		
USSR	63	4	16	424	430	9	1.8
Yugoslavia	5	13	75	64	52		

Raspberries

thousand tonnes

	1950	1970	1980	1988	1989	Rank	%
World	91	196	243	370	376		
Europe	58	106	121	141	196		
North America	29	21	21	34	32		
Oceania	5	3	2	2	2		
Australia	4	1	1	1	1		
Bulgaria	...	11	7	4	4	10	1.1
Canada	7	7	9	21	19	7	5.1
Czechoslovakia	1	1	1	1	1		
France	2	5	7	7	7	9	1.9
Germany	19	24	20	28	27	4	7.2
Hungary	1	12	18	24	22	6	5.9
Ireland	1	1	1	1	1		
Italy	1	1	1	2	2		
New Zealand	1	2	1	2	2		
Norway	2	2	3	3	3		
Poland	4	11	23	43	44	3	11.7
UK	18	19	21	23	23	5	6.1
USA	21	15	13	13	13	8	3.5
USSR	...	65	99	142	145	1	38.6
Yugoslavia	4	15	16	53	61	2	16.2

Rice

thousand tonnes

	1950	1970	1980	1989	1990	Rank	%
World	169,575	311,469	396,259	506,934	520,513		
Africa	3,462	7,315	8,414	11,123	11,216		
Asia	158,005	285,927	360,233	463,630	480,616		
Europe	1,319	1,820	1,833	2,242	2,478		
North America	2,533	5,345	911	9,448	9,007		
Oceania	84	289	716	838	960		
South America	3,970	9,500	13,374	17,093	13,436		
Afghanistan	331	374	458	450	430		
Argentina	137	347	288	469	467		
Australia	63	267	689	805	924		
Bangladesh	...	16,540	20,140	27,691	29,400	4	5.6
Brazil	2,921	6,847	8,535	11,030	7,457		
Burma	5,481	8,107	12,730	13,515	13,623	7	2.6
Cambodia	1,635	3,016	1,003	2,500	2,400		
China	58,188	111,599	145,128	171,438	188,403	1	36.2
Colombia	248	756	1,843	1,884	2,050		
Cuba	164	309	474	536	500		
Dominican Rep.	65	210	392	467	450		
Ecuador	135	165	700	806	760		
Egypt	971	2,567	2,456	2,680	2,700		
Guinea	208	364	343	426	500		
India	33,383	62,861	75,135	111,147	112,500	2	21.6
Indonesia	9,441	19,136	29,686	44,726	44,490	3	8.5
Iran	424	1,041	1,344	1,852	1,800		
Italy	723	858	996	1,246	1,226		
Ivory Coast	104	335	532	650	650		
Japan	12,736	16,281	13,320	12,934	12,500	8	2.4
Korea, N.	1,158	2,392	4,850	5,500	5,500		
Korea, S.	3,385	5,573	6,741	8,192	7,786	10	1.5
Laos	530	870	1,025	1,404	1,500		
Madagascar	829	1,865	2,051	2,380	2,400		
Malaysia	631	1,689	2,138	1,744	1,800		
Mexico	173	390	533	637	291		
Nepal	2,460	2,297	2,310	3,390	3,300		
Nigeria	250	352	1,027	2,000	2,000		
Pakistan	12,399	3,431	4,865	4,830	5,250		
Peru	191	539	555	1,091	1,037		
Philippines	2,767	5,225	7,688	9,657	9,600	9	1.8
Sierra Leone	274	474	471	430	450		
Sri Lanka	479	1,463	2,023	2,063	2,200		
Taiwan	1,771	2,663	2,354	1,800	1,700		
Tanzania	62	172	212	570	660		
Thailand	6,846	13,475	17,375	18,500	18,500	6	3.6
USA	1,925	3,953	7,007	7,007	7,027		
USSR	202	1,272	2,528	2,560	2,800		
Venezuela	41	208	691	313	459		
Vietnam	3,815	9,752	11,669	18,990	19,150	5	3.7

Rubber – natural

thousand tonnes

	1950	1070	1980	1989	1990	Rank	%
World	1740	3,004	3,823	4,865	4,943		
Africa	60	215	200	312	280		
Asia	1,650	2,754	3,584	4,489	4,593		
North America	2	5	5	18	16		
Oceania	2	6	4	4	4		
South America	28	30	36	43	50		
Bolivia	1	3	5	5	5		
Brazil	22	24	27	33	40		
Burma	10	11	16	14	14		
Cambodia	16	20	10	28	30		
Cameroon	2	13	17	36	38		
China	5	5	103	240	245	5	5
Ghana	0.4	7	9	3	3		
Guatemala	8	10	11	18	16		
India	17	90	151	259	289	4	5.8
Indonesia	635	838	934	1,260	1,300	2	26.3
Ivory Coast	10	11	21	58	68	10	1.4
Liberia	31	78	76	118	70	9	1.4
Malaysia	722	1,285	1,602	1,419	1,420	1	28.7
Nigeria	14	63	48	78	80	8	1.6
Papua N. Guinea	2	6	4	5	5		
Philippines	1	19	63	172	180	6	3.6
Sri Lanka	102	150	140	111	120	7	2.4
Thailand	104	295	520	936	930	3	18.8
Vietnam	35	36	45	50	65		
Zaïre	10	43	28	18	20		

Rye

thousand tonnes

	1950	1970	1980	1989	1990	Rank	%
World	38,330	30,782	24,684	36,102	36,619		
Africa	15	10	8	6	6		
Asia	1,000	2,066	1,807	1,192	1,286		
Europe	17,790	14,689	12,573	13,521	13,050		
North America	990	1,458	1,132	1,220	1,196		
Oceania	15	19	12	20	20		
South America	560	305	211	86	61		
Argentina	526	271	186	71	48		
Austria	343	417	327	381	375	8	1
Belgium-Lux.	232	82	43	12	12		
Bulgaria	240	27	29	51	40		
Canada	469	474	647	873	939	5	2.6
China	...	1,233	1,200	1,000	1,000	4	2.7
Czechoslovakia	1,110	586	519	708	530	7	1.4
Denmark	365	136	219	487	565	6	1.5
Finland	201	130	88	196	244		
France	573	297	368	262	248		
Germany	5,582	4,456	3,829	3,900	4,025	3	11
Greece	47	9	8	33	25		

Rye *(continued)*

thousand tonnes

	1950	1970	1980	1989	1990	Rank	%
Hungary	732	193	118	267	240		
Italy	123	65	34	21	20		
Korea, N.	...	40	53	46	50		
Netherlands	455	196	39	33	36		
Poland	6,374	7,143	6,185	6,216	5,800	2	15.8
Portugal	162	164	128	106	77		
Romania	162	52	40	55	60		
Spain	482	292	243	337	274		
Sweden	258	239	205	319	340	9	0.9
Turkey	500	781	548	191	285	10	0.8
UK	52	14	24	31	32		
USA	524	984	487	347	257		
USSR	17,960	12,235	8,941	20,057	21,000	1	57.3
Yugoslavia	248	132	78	75	72		

Seed cotton

thousand tonnes

	1970	1980	1988	1989	Rank	%
World	34,857	43,076	53,393	49,085		
Africa	3,790	3,225	3,812	3,867		
Asia	13,413	16,537	24,365	23,257		
Europe	564	566	1,101	976		
North America	7,540	9,890	9,970	7,838		
South America	2,902	3,053	4,694	3,734		
Oceania	82	220	762	813		
Afghanistan	75	87	120	60		
Argentina	370	447	849	580		
Australia	82	220	762	813	9	1.7
Benin	28	19	106	108		
Brazil	1,828	1,681	2,535	1,885	6	3.8
Burkina Faso	31	66	179	179		
Burma	39	59	82	82		
Cameroon	58	81	112	119		
Chad	107	83	137	125		
China	5,965	7,882	12,600	11,757	1	24
Colombia	352	291	380	320		
Egypt	1,434	1,341	854	821	8	1.7
El Salvador	135	169	28	22		
Ethiopia	40	87	58	55		
Greece	342	356	720	752	10	1.5
Guatemala	190	434	133	130		
India	3,264	3,952	4,430	4,430	4	9
Iran	441	258	305	394		
Israel	95	219	167	106		
Ivory Coast	37	131	256	290		
Mali	66	129	228	250		
Mexico	988	984	830	600		
Mozambique	127	51	94	94		
Nicaragua	236	220	100	78		
Nigeria	185	96	100	120		
Pakistan	1,786	2,191	4,278	4,331	5	8.8
Paraguay	31	268	537	515		
Peru	245	262	282	321		
South Africa	57	151	187	153		
Spain	155	174	351	194		
Sudan	684	342	460	480		
Syria	391	342	473	431		
Tanzania	210	163	254	275		
Thailand	93	170	106	102		
Togo	7	19	67	87		
Turkey	1,146	1,269	1,690	1,443	7	2.9
Uganda	258	17	18	15		
USA	5,967	8,038	8,854	6,986	3	14.2
USSR	6,566	9,586	8,689	8,600	2	17.5
Venezuela	41	41	73	83		
Zaire	63	23	77	77		
Zambia	8	16	59	53		
Zimbabwe	129	158	279	276		

This is raw cotton before it is separated into fibre, oil etc.

Sesame seed

thousand tonnes

	1950	1970	1980	1988	1989	Rank	%
World	1,777	2,014	29,000	2,354	2,352		
Africa	218	511	492	591	592		
Asia	1,440	1,174	1,257	1,605	1,600		
North America	98	195	183	77	86		
South America	16	127	66	80	73		
Afghanistan	...	33	36	26	27		
Angola	1	2	2	2	2		
Bangladesh	23	29	51	55	57	6	2.4
Benin	...	2	1	4	5		
Brazil	4	3	3	4	12		
Burkina Faso	4	5	7	9	11		
Burma	44	116	159	190	200	4	8.5
Cambodia	...	10	3	7	8		
Cameroon	2	6	8	13	14		
Central Afr. Rep.	5	13	11	18	20		
Chad	5	14	10	8	12		
China	...	368	396	450	500	2	21.3
Colombia	7	21	13	7	9		
Egypt	12	19	16	15	15		
El Salvador	4	1	10	9	9		
Ethiopia	31	81	37	37	38		
Guatemala	1	7	13	21	15		
Haiti	...	1	4	4	4		
Honduras	1	1	3	5	5		
India	429	487	461	667	600	1	25.5
Indonesia	2	4	9	7	6		
Iran	9	6	3	3	3		
Iraq	9	13	6	11	13		
Ivory Coast	1	2	3	3	3		
Kenya	1	3	8	8	8		
Korea, S.	2	10	14	52	55	7	2.3
Mexico	80	178	145	34	50	9	2.1
Mozambique	2	3	3	3	3		
Nicaragua	...	6	7	4	2		
Nigeria	13	59	73	75	70	5	2.9
Pakistan	10	11	18	10	10		
Somalia	2	7	44	46	47	10	2
Sri Lanka	4	10	11	7	9		
Sudan	99	256	224	278	268	3	11.4
Syria	8	4	19	11	9		
Tanzania	5	11	17	21	22		
Thailand	8	20	26	27	29		
Togo	...	1	2	2	2		
Turkey	32	40	26	45	39		
Uganda	31	21	20	36	40		
Venezuela	7	101	49	68	52	8	2.2
Vietnam	2	3	7	25	26		
Yemen	...	6	9	8	8		
Zaïre	...	5	4	11	11		

Sheep

thousand head

	1950	1970	1980	1989	1990	Rank	%
World	782,744	1,072,109	1,112,226	1,180,321	1,194,028		
Africa	105,108	160,311	181,493	201,294	205,189		
Asia	171,941	264,780	323,323	333,688	337,902		
Europe	119,826	127,204	134,440	151,721	155,571		
North America	38,988	31,022	22,348	19,042	19,236		
Oceania	145,370	237,208	202,373	222,179	226,123		
South America	124,632	115,151	105,657	112,896	113,008		
Afghanistan	14,000	18,900	19,467	12,500	13,000		
Albania	1,633	1,249	1,169	1,555	1,560		
Algeria	3,990	7,940	13,064	12,500	12,550		
Argentina	52,940	42,773	32,406	29,345	28,571		
Australia	111,485	177,491	134,534	161,603	167,781	1	14
Bangladesh	300	547	1,060	1,150	1,160		
Bolivia	7,224	6,825	8,791	12,300	12,400		
Brazil	14,427	17,768	17,935	20,500	21,000		
Bulgaria	8,377	9,518	10,358	8,609	7,988		
Burkina Faso	590	1,657	1,852	3,050	3,050		
Cameroon	638	2,000	2,168	3,170	3,200		
Canada	1,176	635	466	729	754		
Chad	2,000	2,200	2,300	2,310	2,350		
Chile	5,789	6,179	6,059	6,600	6,300		
China	31,070	78,000	101,388	110,571	113,508	3	10
Colombia	1,153	1,935	2,399	2,650	2,650		
Ecuador	1,724	1,897	2,777	1,883	1,900		
Egypt	1,254	2,030	1,624	3,750	3,800		

	1950	1970	1980	1989	1990	Rank	%
Ethiopia	9,559	24,077	23,250	24,000	22,960		
Finland	1,102	963	110	59	59		
France	7,499	10,023	12,178	11,943	11,900		
Germany	3,231	2,534	3,147	4,098	4,136		
Ghana	443	1,324	1,683	2,200	2,250		
Greece	6,978	7,646	7,974	10,376	10,400		
Hungary	1,003	2,986	2,960	2,215	2,069		
India	36,824	40,657	41,267	53,486	54,588	5	5
Indonesia	2,038	3,169	4,155	5,530	5,700		
Iran	11,727	32,000	33,959	34,000	34,000	6	3
Iraq	10,000	12,000	11,633	9,500	9,500		
Ireland	2,419	4,092	3,343	4,991	5,782		
Italy	10,187	8,097	9,120	11,623	11,569		
Ivory Coast	234	833	1,200	1,500	1,500		
Kenya	7,240	3,935	4,333	6,325	6,350		
Lesotho	1,561	1,610	1,131	1,450	1,450		
Libya	1,390	2,125	5,901	5,800	5,850		
Mali	4,080	5,700	6,200	5,650	5,700		
Mauritania	1,065	4,427	5,100	4,200	4,300		
Mexico	5,041	8,687	7,719	5,863	5,500		
Mongolia	12,575	12,678	14,261	13,451	13,500		
Morocco	11,249	17,087	14,180	17,500	17,500		
Namibia	2,887	4,000	5,433	6,500	6,600		
Nepal	2,437	2,156	2,372	880	880		
New Zealand	33,871	59,708	67,832	60,569	58,334	4	5
Niger	1,324	2,632	2,805	3,500	3,500		
Nigeria	5,200	8,083	11,683	13,200	13,200		
Norway	1,819	1,769	2,308	2,248	2,248		
Pakistan	8,000	14,400	26,297	28,345	29,239	10	2
Peru	17,515	16,698	14,539	12,903	12,750		
Poland	2,205	3,236	4,176	4,409	4,300		
Portugal	4,307	4,405	5,187	5,354	5,380		
Romania	10,519	13,984	15,766	18,800	19,000		
Saudi Arabia	3,572	1,950	4,100	7,698	7,700		
Senegal	506	1,533	2,005	3,886	3,900		
Somalia	2,261	8,967	10,100	13,800	13,900		
South Africa	33,237	35,585	31,625	30,935	32,605	7	3
Spain	24,489	18,712	14,598	24,252	27,400		
Sudan	5,560	10,300	17,708	20,000	20,300		
Syria	2,968	5,843	9,723	13,903	14,125		
Tanzania	2,333	2,823	3,782	5,000	5,200		
Tunisia	2,462	3,200	4,728	5,000	5,000		
Turkey	24,282	36,470	46,199	34,850	31,500	8	3
Uganda	1,063	799	1,070	1,780	1,800		
UK	19,945	26,332	31,196	29,045	29,521	9	3
Uruguay	23,149	19,906	19,232	25,560	26,000		
USA	31,565	20,501	12,663	10,858	11,368		
USSR	76,879	136,434	142,591	139,500	137,000	2	12
Yemen	10,000	2,800	3,002	3,681	3,700		
Yugoslavia	10,493	9,136	7,360	7,564	7,596		

Sheep meat

thousand tonnes

	1950	1970	1980	1988	1989	Rank	%
World	4,255	5,708	5,821	6,445	6,473		
Africa	581	634	709	829	850		
Asia	1,143	1,225	1,631	1,814	1,866		
Europe	681	940	1,115	1,293	1,316		
North America	331	280	170	194	200		
Oceania	656	1,317	1,102	1,207	1,114		
South America	369	367	263	268	277		
Afghanistan	80	86	113	105	110		
Albania	18	16	17	19	19		
Algeria	34	40	62	75	70		
Argentina	194	181	121	90	90		
Australia	332	753	543	586	542	3	8.4
Belgium	2	7	5	7	7		
Bolivia	12	14	20	28	28		
Brazil	32	36	29	35	37		
Bulgaria	43	71	59	65	72		
Cameroon	4	9	8	12	13		
Canada	16	8	5	8	8		
Chad	9	7	10	9	10		
Chile	47	24	16	14	16		
China	220	350	399	400	420	4	6.5
Colombia	1	2	9	11	11		
Cyprus	3	4	6	5	5		
Czechoslovakia	7	5	6	10	9		
Ecuador	3	11	8	7	7		
Egypt	36	29	24	32	35		
Ethiopia	75	80	77	79	82		
Finland	7	1	1	1	1		

Sheep meat — continued

	1950	1970	1980	1988	1989	Rank	%
France	82	113	169	156	150		
Germany	55	21	46	51	52		
Ghana	3	4	5	6	6		
Greece	38	61	76	85	85		
Hungary	8	6	6	5	4		
Iceland	6	12	15	10	10		
India	272	112	120	148	158	10	2.4
Indonesia	26	13	22	45	45		
Iran	90	145	234	234	241	7	3.7
Iraq	25	39	38	20	21		
Ireland	21	44	41	50	58		
Italy	47	49	65	69	69		
Ivory Coast	1	4	6	7	7		
Jordan	5	5	5	5	5		
Kenya	1	13	21	23	25		
Kuwait	...	8	17	33	33		
Lebanon	2	11	9	9	10		
Libya	8	19	52	57	59		
Mali	25	18	24	19	19		
Mauritania	9	6	7	6	6		
Mexico	10	16	16	27	27		
Mongolia	150	69	89	100	102		
Morocco	60	47	44	50	53		
Namibia	2	12	17	22	23		
Nepal	16	7	8	3	3		
Netherlands	6	10	19	12	13		
New Zealand	324	563	560	621	572	2	8.8
Niger	18	4	11	16	16		
Nigeria	72	28	40	45	45		
Norway	15	17	19	24	25		
Pakistan	66	67	135	187	193	9	3
Peru	14	23	22	19	20		
Poland	7	22	18	26	25		
Portugal	16	19	23	24	25		
Romania	14	62	77	60	65		
Saudi Arabia	...	14	20	54	54		
Senegal	4	6	8	12	13		
Somalia	15	11	12	35	36		
South Africa	79	168	133	130	133		
Spain	78	123	126	211	200	8	3.1
Sudan	22	64	82	103	105		
Syria	43	40	81	92	95		
Tanzania	30	10	10	13	14		
Tunisia	19	27	23	35	36		
Turkey	52	226	298	305	304	6	4.7
Uganda	12	4	5	9	11		
UK	143	221	255	322	347	5	5.4
Uruguay	63	72	33	58	61		
USA	283	250	144	152	157		
USSR	494	945	830	840	850	1	13.1
Venezuela	1	1	12	3	3		
Yemen	...	15	19	37	37		
Yugoslavia	34	50	60	70	66		

Silk

tonnes

	1950	1970	1980	1988	1989	Rank	%
World	20,860	42,146	65,068	73,866	73,589		
Asia	17,174	37,997	59,649	67,212	66,824		
Europe	2,015	846	508	429	436		
South America	87	270	1,033	1,900	1,900		
Afghanistan	...	45	50	60	60		
Brazil	87	270	1,033	1,900	1,900	6	2.6
Bulgaria	...	253	201	165	160		
China	3,600	10,258	34,050	42,041	42,044	1	57.1
Greece	90	103	89	11	11		
India	927	2,245	2,658	10,255	10,500	2	14.3
Iran	176	145	267	900	850	9	1.2
Italy	1,452	321	75	20	20		
Japan	11,624	20,562	15,868	6,862	7,000	3	9.5
Korea, N.	...	1,400	2,700	3,800	3,000	5	4.1
Korea, S.	455	2,876	3,681	1,355	1,400	7	1.9
Lebanon	31	12	16	5	6		
Romania	...	106	120	170	185		
Spain	36	29	15	15	15		
Syria	8	54	34	...	...		
Thailand	...	...	...	1,250	1,250	8	1.7
Turkey	...	146	70	251	250		
USSR	1,855	3,009	3,873	4,300	4,400	4	6
Vietnam	...	233	246	420	450	10	0.6

Sisal

thousand tonnes

	1970	1980	1988	1989	Rank	%
World	645	508	377	430		
Africa	373	163	108	113		
Asia	14	6	17	16		
North America	16	98	54	66		
South America	242	240	197	235		
Angola	65	7	1	1		
Brazil	229	234	190	227	1	52.8
China	9	6	17	16	6	3.7
Ethiopia	2	1	1	1		
Haiti	16	12	10	10	7	2.3
Indonesia	5	1	1	1		
Kenya	46	42	43	43	3	10
Madagascar	26	15	20	20	5	4.7
Mexico	...	86	44	55	2	12.8
Morocco	1	1	2	2	10	0.5
Mozambique	28	11	1	1		
South Africa	8	6	8	8	8	1.9
Tanzania	198	80	33	38	4	8.8
Venezuela	12	6	8	8	9	1.9

Sorghum

thousand tonnes

	1950	1970	1980	1989	1990	Rank	%
World	27,000	55,629	64,331	59,730	59,034		
Africa	6,000	9,107	10,543	14,037	12,812		
Asia	6,400	18,781	20,244	18,563	19,830		
Europe	35	446	699	633	556		
North America	3,900	22,419	24,744	21,983	21,013		
Oceania	78	761	1,052	1,284	934		
South America	300	4,092	6,945	3,071	3,728		
Argentina	100	3,823	5,570	1,360	2,016	6	3.4
Australia	75	759	1,046	1,283	933	9	1.6
Benin	50	52	74	110	110		
Brazil	...	...	172	236	229		
Burkina Faso	370	528	640	991	917	10	1.6
Burundi	60	105	95	72	87		
China	...	8,607	7,525	4,535	5,770	4	9.8
Colombia	...	153	488	695	809		
El Salvador	115	144	146	156	166		
Ethiopia	150	827	685	974	984	8	1.7
France	...	203	332	300	271		
Ghana	79	147	135	245	156		
Guatemala	12	46	70	107	18		
Haiti	174	210	115	61	65		
India	5,981	8,516	11,217	12,915	13,000	2	22
Italy	5	15	64	139	98		
Kenya	50	215	209				
Korea, N.	10	115	135	15	15		
Lesotho	23	57	68	24	28		
Malawi	38	78	123	20	16		
Mexico	200	2,573	4,800	5,754	5,900	3	10
Morocco	81	70	24	14	14		
Mozambique	...	202	150	145	135		
Nicaragua	36	57	70	72	89		
Niger	215	262	333	421	415		
Nigeria	1,500	3,632	3,800	4,831	4,000	5	6.8
Pakistan	239	308	238	262	255		
Rwanda	80	141	170	164	155		
Saudi Arabia	14	185	100	40	40		
Somalia	30	132	140	291	250		
South Africa	180	376	535	466	400		
Spain	4	168	200	85	79		
Sudan	601	1,525	2,470	1,536	1,503	7	2.5
Tanzania	250	156	220	503	368		
Thailand	...	123	330	231	233		
Uganda	240	337	422	260	300		
Uruguay	10	53	110	79	69		
USA	3,896	19,314	19,206	15,632	14,516	1	24.6
USSR	10	62	103	159	160		
Venezuela	...	5	530	595	500		
Yemen	...	878	634	547	500		
Zambia	122	49	35	37	20		
Zimbabwe	100	106	91	76	90		

Soybeans

thousand tonnes

	1950	1970	1980	1989	1990	Rank	%
World	16,000	43,487	85,940	106,941	108,588		
Africa	15	80	318	480	478		
Asia	8,300	9,329	9,956	15,284	16,990		
Europe	30	117	629	2,783	2,491		
North America	7,400	31,684	56,506	54,596	54,134		
Oceania	—	5	84	135	90		
South America	950	1,751	17,950	32,706	33,485		
Argentina	4	39	3,657	6,400	11,000	4	10.1
Australia	—	5	84	135	90		
Bolivia	...	2	49	226	225		
Brazil	83	1,547	13,457	24,052	19,981	2	18.4
Bulgaria	...	8	126	23	21		
Canada	86	257	672	1,219	1,327	9	1.2
China	7,200	8,131	7,801	10,238	11,508	3	10.6
Colombia	...	114	130	177	235		
Ecuador	...	2	32	125	133		
Egypt	...	...	100	136	140		
India	...	2	450	1,700	2,000	5	1.8
Indonesia	270	468	658	1,301	1,412	8	1.3
Iran	...	15	65	90	90		
Italy	...	...	...	1,624	1,491	7	1.4
Japan	376	128	192	272	270		
Korea, N.	...	255	340	453	455		
Korea, S.	136	228	241	252	260		
Mexico	...	252	581	992	474		
Nigeria	3	61	75	75	78		
Paraguay	...	46	575	1,615	1,826	6	1.7
Romania	5	102	366	450	420		
Taiwan	15	65	26	15	15		
Thailand	14	51	131	672	686		
Turkey	2	11	3	161	165		
USA	7,312	31,174	55,251	52,354	52,303	1	48.2
USSR	166	400	494	956	920	10	0.8
Vietnam	10	26	39	82	86		
Yugoslavia	4	5	64	209	145		
Zimbabwe	...	...	...	121	110		

Strawberries

thousand tonnes

	1950	1970	1980	1988	1989	Rank	%
World	...	1,212	1,768	2,286	2,362		
Africa	...	...	4	25	31		
Asia	22	158	305	311	335		
Europe	183	583	901	1,178	1,185		
North America	193	357	424	623	660		
Oceania	2	8	8	8	8		
South America	1	6	14	14	13		
Argentina	1	2	5	6	5		
Australia	1	4	3	5	5		
Austria	...	8	10	17	15		
Belgium-Lux.	15	34	23	28	22		
Brazil	...	1	2	2	2		
Bulgaria	5	28	21	15	17		
Canada	18	18	26	31	25		
China	...	...	...	8	9		
Czechoslovakia	4	33	17	25	33		
Denmark	8	11	8	9	9		
Egypt	...	...	1	20	25		
Finland	3	3	8	9	9		
France	17	65	81	95	88	8	3.7
Germany	11	41	72	83	88	9	3.7
Greece	2	5	14	6	7		
Hungary	1	20	18	14	15		
Ireland	...	...	5	5	5		
Israel	0	4	8	12	12		
Italy	20	89	201	189	201	5	8.5
Japan	22	138	194	219	220	4	9.3
Korea, South	...	4	79	28	41		
Mexico	2	112	94	79	96	7	4.1
Netherlands	22	32	20	26	25		
New Zealand	1	4	5	3	3		
Norway	5	13	18	18	18		
Peru	...	2	3	3	3		
Poland	31	87	188	249	269	2	11.4
Portugal	...	2	2	3	3		
Romania	...	29	31	39	36		
South Africa	...	...	3	5	6		
Spain	3	10	92	243	227	3	9.6
Switzerland	...	3	3	6	6		

Strawberries *continued*

	1950	1970	1980	1988	1989	Rank	%
Turkey	...	12	23	42	52		
UK	35	50	53	55	55	10	2.3
USA	173	227	314	514	539	1	22.8
USSR	...	101	101	127	130	6	5.5
Venezuela	...	1	4	2	2		
Yugoslavia	4	22	16	43	47		

Sugar

thousand tonnes

	1950	1970	1980	1988	1989	Rank	%
World	32,296	72,176	88,623	104,102	105,639		
Africa	1,560	4,863	6,441	8,347	7,901		
Asia	3,757	13,852	18,206	27,068	28,878		
Europe	7,810	15,240	21,187	20,558	21,406		
North America	12,118	17,499	19,200	21,339	21,617		
Oceania	1,035	2,840	3,689	4,096	4,097		
South America	3,315	9,161	12,883	13,495	12,175		
Angola	47	74	31	21	29		
Argentina	638	985	1,584	1,132	915		
Australia	913	2,511	3,243	3,679	3,679	10	3.5
Austria	104	315	450	357	446		
Bangladesh	...	90	134	194	119		
Barbados	168	146	113	80	66		
Belgium-Lux.	330	711	994	1,005	1,038		
Belize	2	62	101	83	96		
Bolivia	1	120	250	182	175		
Brazil	1,649	5,161	7,991	8,582	7,409	4	7
Bulgaria	64	218	165	75	82		
Canada	122	130	118	108	135		
Chile	...	217	131	418	448		
China	986	3,878	3,809	5,966	5,634	6	5.3
Colombia	165	710	1,192	1,364	1,425		
Congo	...	78	13	31	35		
Costa Rica	26	160	199	206	210		
Cuba	5,786	6,388	7,510	8,119	8,188	3	7.8
Czechoslovakia	697	731	808	614	755		
Denmark	319	307	493	549	530		
Dominican Rep.	539	1,010	1,142	777	810		
Ecuador	56	258	349	292	325		
Egypt	196	564	666	1,029	977		
El Salvador	28	128	210	174	237		
Ethiopia	...	98	168	195	190		
Fiji	122	329	446	363	385		
Finland	20	60	104	145	160		
France	1,085	2,872	4,720	4,424	4,130	7	3.9
Germany	1,523	2,645	3,936	3,553	3,777	9	3.6
Greece	...	163	286	235	421		
Guadeloupe	73	154	86	76	83		
Guatemala	31	191	409	679	687		
Guyana	218	354	291	169	166		
Haiti	53	64	55	35	45		
Honduras	5	60	182	172	197		
Hungary	232	321	511	480	576		
India	1,316	4,188	5,380	9,447	10,200	1	9.7
Indonesia	287	760	1,286	1,964	1,917		
Iran	54	571	630	634	603		
Ireland	93	164	178	212	228		
Italy	598	1,271	1,956	1,607	1,880		
Ivory Coast	...	...	97	140	145		
Jamaica	279	383	238	222	192		
Japan	21	628	789	984	945		
Kenya	15	132	386	458	470		
Madagascar	16	104	115	122	120		
Malawi	...	32	140	175	175		
Martinique	40	29	6	7	7		
Mauritius	443	659	615	634	568		
Mexico	733	2,495	2,796	3,806	3,678		
Morocco	...	168	364	690	610		
Mozambique	85	278	189	19	24		
Netherlands	364	769	1,000	1,075	1,240		
Nicaragua	27	147	194	209	205		
Pakistan	55	557	738	1,936	2,011		
Panama	16	78	183	105	110		
Paraguay	...	50	85	103	105		
Peru	496	782	571	592	580		
Philippines	827	1,859	2,289	1,335	1,590		
Poland	871	1,581	1,530	1,825	1,850		
Puerto Rico	1,157	383	156	93	83		
Réunion	116	220	247	251	183		
Romania	116	434	595	450	400		
South Africa	555	1,629	2,011	2,469	2,276		
Spain	315	882	934	1,306	1,038		
St. Christ.-N.	45	30	37	26	23		
Sudan	...	89	166	445	378		
Swaziland	...	167	317	465	497		
Sweden	285	233	350	395	422		
Switzerland	27	65	119	150	152		
Taiwan	500	588	830	600	600		
Tanzania	10	100	119	102	110		
Thailand	34	436	1,534	2,704	4,052	8	3.8
Trinidad & T.	151	229	117	91	97		
Turkey	161	700	1,178	1,414	1,314		
Uganda	56	154	4	15	20		
UK	626	1,033	1,215	1,418	1,304		
USA	2,785	5,215	5,345	6,269	6,464	5	6.1
USSR	2,700	8,722	7,017	9,200	9,565	2	9.1
Venezuela	60	448	344	583	556		
Vietnam	1	7	196	366	390		
Yugoslavia	132	440	798	636	930		
Zambia	...	37	105	136	132		
Zimbabwe	1	155	336	453	460		

These figures refer to raw suger manufactured from cane and beet

Sugar beet

thousand tonnes

	1950	1970	1980	1989	1990	Rank	%
World	90,827	227,615	269,332	312,261	307,654		
Africa	100	1,359	2,385	3,876	4,084		
Asia	1,909	17,576	21,672	28,230	30,847		
Europe	59,117	107,209	148,892	156,161	154,258		
North America	10,598	25,604	22,994	23,625	25,974		
South America	53	1,772	1,264	2,956	2,491		
Albania	6	114	265	290	300		
Austria	726	1,847	2,580	2,641	2,760		
Belgium-Lux.	2,194	4,683	6,796	5,732	6,200		
Bulgaria	522	1,664	1,568	912	550		
Canada	835	971	989	805	800		
Chile	...	1,371	858	2,810	2,800		
China	355	6,387	5,337	9,240	10,000	8	3.3
Czechoslovakia	4,967	6,098	7,558	6,390	6,000		
Denmark	2,193	1,968	3,066	3,302	3,300		
Egypt	...	...	...	650	700		
Finland	197	410	760	933	950		
France	10,790	18,469	28,069	27,694	29,925	2	9.7
Germany	11,659	19,218	27,752	26,987	28,900	3	9.4
Greece	...	1,258	2,377	2,500	2,500		
Hungary	1,733	2,500	4,089	5,301	5,000		
Iran	454	3,750	2,906	3,535	3,535		
Ireland	607	1,039	1,184	1,350	1,350		
Italy	4,671	9,622	14,505	16,891	13,855	6	4.5
Japan	166	2,204	3,417	3,664	3,835		
Morocco	100	1,199	2,239	2,876	2,969		
Netherlands	2,598	4,922	6,161	7,679	7,650	10	2.5
Pakistan	...	208	376	342	350		
Poland	5,383	12,207	13,364	14,374	15,200	5	4.9
Romania	888	3,560	5,693	6,650	7,000		
Spain	2,291	5,602	6,641	7,273	7,286		
Sweden	1,777	1,579	2,313	2,654	255		
Switzerland	207	419	786	950	950		
Syria	20	216	466	116	120		
Turkey	966	4,522	8,842	10,929	12,500	7	4.1
UK	4,525	6,772	7,547	8,113	8,000	9	2.6
Uruguay	53	399	405	142	140		
USA	9,762	24,557	22,005	22,797	25,032	4	8.1
USSR	19,050	74,095	72,124	97,414	90,000	1	29.3
Yugoslavia	1,180	3,182	5,759	6,797	5,200		

Sugar cane

thousand tonnes

	1950	1970	1980	1989	1990	Rank	%
World	275,757	586,432	749,874	1,009,498	1,044,268		
Africa	15,309	46,418	61,025	71,563	74,797		
Asia	88,930	237,001	276,676	398,689	412,495		
Europe	322	486	376	256	256		
North America	104,214	153,847	171,283	186,797	182,174		
Oceania	7,594	20,557	27,175	31,427	30,559		
South America	59,388	128,123	213,338	320,766	343,986		
Angola	444	776	377	330	330		
Argentina	7,756	10,213	15,527	14,000	14,500		
Australia	6,686	17,607	23,420	27,146	26,226	10	2.5
Bangladesh	...	7,551	6,737	6,707	6,700		
Barbados	1,464	1,325	1,040	675	675		
Belize	...	622	990	864	864		
Bolivia	290	1,327	3,153	1,906	1,900		
Brazil	32,837	78,460	146,274	252,290	272,540	1	26.1
Burma	944	1,458	1,602	3,100	3,100		
Cameroon	...	123	697	1,300	1,300		
China	5,656	38,529	31,974	55,567	61,767	4	5.9
Colombia	8,406	13,167	25,567	25,920	28,000	8	2.7
Congo	135	672	184	400	400		
Costa Rica	827	1,837	2,551	2,600	2,600		
Cuba	50,466	60,467	68,895	81,003	85,900	3	8.2
Dominican Rep.	4,485	8,986	10,320	8,718	9,000		
Ecuador	1,900	6,500	6,621	5,700	5,700		
Egypt	2,910	7,096	8,886	10,765	10,850		
El Salvador	763	1,662	2,446	2,407	2,500		
Ethiopia	50	1,092	1,369	1,750	1,750		
Fiji	908	2,603	3,743	3,958	3,900		
Guadeloupe	919	1,785	1,003	971	970		
Guatemala	588	2,692	5,396	7,897	7,400		
Guyana	2,552	4,081	3,993	2,541	2,700		
Haiti	4,324	2,702	2,967	3,000	3,000		
Honduras	560	1,368	2,807	2,700	2,700		
India	53,865	128,689	143,670	204,630	210,000	2	20.1
Indonesia	3,538	10,352	16,880	26,713	26,800	9	2.6
Iran	120	556	1,433	1,630	1,850		
Ivory Coast	...	...	1,440	1,750	1,750		
Jamaica	2,663	4,098	2,745	2,293	2,548		
Japan	95	2,407	2,285	2,594	2,435		
Kenya	152	1,645	4,272	4,500	4,500		
Madagascar	348	1,153	1,407	2,000	2,000		
Malawi	...	296	1,427	1,750	1,750		
Malaysia	...	203	797	1,400	1,400		
Martinique	491	499	261	207	200		
Mauritius	3,737	5,400	5,459	5,436	6,800		
Mexico	10,419	33,269	35,509	4,000	34,893	6	3.3
Morocco	...	...	430	1,095	1,100		
Mozambique	838	2,767	2,050	270	280		
Nicaragua	629	1,866	2,643	2,300	2,300		
Nigeria	...	556	895	1,300	1,300		
Pakistan	10,063	23,836	29,061	35,493	38,000	5	3.6
Panama	305	1,242	2,522	1,600	1,600		
Paraguay	343	1,215	1,403	2,909	2,900		
Peru	4,257	7,915	5,597	7,085	6,965		
Philippines	7,396	16,271	20,101	17,591	19,600		
Puerto Rico	9,947	4,952	2,001	1,216	1,260		
Réunion	1,127	2,026	2,338	1,756	2,200		
South Africa	4,789	14,561	16,382	18,500	19,000		
Spain	...	437	300	250	250		
Sudan	80	825	1,867	4,000	4,000		
Swaziland	300	1,548	2,771	3,800	3,800		
Taiwan	4,801	5,991	8,851	7,000	7,000		
Tanzania	182	1,141	1,302	1,320	1,320		
Thailand	990	5,856	17,101	36,668	33,561	7	3.2
Trinidad & T.	1,441	2,481	1,400	1,250	1,250		
Uganda	496	1,667	543	650	650		
USA	13,186	21,404	25,202	26,695	22,407		
Venezuela	1,064	4,883	4,633	7,809	8,000		
Vietnam	400	333	3,911	5,850	5,850		
Zaire	50	626	649	1,150	1,150		
Zambia	...	303	936	1,350	1,350		
Zimbabwe	17	1,363	2,871	3,622	3,575		

Sunflower seed

thousand tonnes

	1950	1970	1980	1989	1990	Rank	%
World	3,888	9,876	14,197	21,113	22,242		
Africa	83	156	540	697	933		
Asia	131	512	1,437	2,978	2,693		
Europe	737	1,930	2,986	5,997	6,458		
North America	11	169	2,632	869	1,158		
Oceania	1	26	155	144	92		
South America	1,018	1,029	1,557	3,358	4,109		
Albania	...	18	32	24	24		
Angola	...	10	10	10	10		
Argentina	889	949	1,447	3,100	3,850	2	17.3
Australia	1	26	155	144	92		
Austria	1	1	2	67	70		
Brazil	...	...	37	45	45		
Bulgaria	180	471	417	447	450		

Sunflower seed continued

	1950	1970	1980	1989	1990	Rank	%
Burma	...	...	16	129	115		
Canada	6	39	186	69	115		
Chile	56	26	27	32	27		
China	40	76	747	1,064	1,200	5	5.4
Czechoslovakia	9	4	26	70	75		
France	8	55	278	2,130	2,314	3	10.4
Greece	1	2	4	50	40		
Hungary	250	122	499	699	650	9	2.9
India	...	...	86	600	600	10	2.7
Iran	...	39	7	24	24		
Italy	6	9	66	340	303		
Kenya	3	3	15	5	5		
Mexico	...	9	28	2	11		
Morocco	6	12	25	96	160		
Mozambique	1	4	21	20	20		
Portugal	...	1	15	45	45		
Romania	200	769	843	750	770	8	3.5
South Africa	37	106	381	431	585		
Spain	1	146	515	906	1,265	4	5.7
Tanzania	6	13	37	22	30		
Turkey	89	383	638	1,250	850	7	3.8
Uruguay	73	55	47	48	45		
USA	6	121	2,418	798	1,032	6	4.6
USSR	1,906	6,055	4,889	7,070	6,800	1	30.6
Yugoslavia	93	334	382	420	402		
Zambia	...	1	18	18	22		
Zimbabwe	1	3	11	66	66		

Sweet potatoes
thousand tonnes

	1970	1980	1988	1989	Rank	%
World	111,469	134,243	127,860	133,234		
Africa	4,000	5,502	6,459	6,105		
Asia	102,446	125,374	117,917	123,600		
Europe	140	130	86	86		
North America	1,386	1,313	1,294	1,324		
Oceania	489	530	568	571		
South America	3,008	1,393	1,536	1,549		
Angola	147	180	180	170		
Argentina	457	290	450	460		
Bangladesh	830	764	558	545		
Brazil	2,155	769	760	750	8	0.6
Burundi	831	476	619	426		
Cameroon	134	133	152	154		
China	89,845	114,257	108,390	114,000	1	85.6
Cuba	237	320	260	260		
Ethiopia	...	123	143	144		
Haiti	280	276	350	330		
India	2,260	1,479	1,347	1,350	5	1
Indonesia	2,215	2,122	2,159	2,106	2	1.6
Japan	2,590	1,378	1,326	1,330	6	1
Kenya	230	351	542	550		
Korea, North	273	374	497	500		
Korea, South	2,053	1,199	562	600	10	0.5
Laos	15	84	187	185		
Madagascar	348	379	472	475		
Mexico	142	47	52	52		
Nigeria	139	237	260	260		
Papua N. Guinea	346	442	473	475		
Peru	167	149	121	142		
Philippines	680	1,060	695	661	9	0.5
Rwanda	379	899	1,032	810	7	0.6
Tanzania	234	514	337	340		
Thailand	280	223	106	104		
Uganda	693	1,284	1,709	1,800	4	1.4
USA	586	561	537	542		
Vietnam	1,108	2,207	1,875	2,000	3	1.5
Zaïre	264	313	372	372		

Tangerines
thousand tonnes

	1970	1980	1989	1990	Rank	%
World	5,093	7,747	9,183	8,784		
Africa	388	526	726	552		
Asia	2,810	3,970	4,434	4,434		
Europe	672	1,238	2,084	2,091		
North America	616	805	731	557		
Oceania	25	36	54	52		
South America	582	904	1,155	1,099		
Algeria	132	130	80	95		
Argentina	237	225	290	240		
Australia	21	28	36	41		
Bolivia	12	24	40	41		
Brazil	238	517	650	640	3	7.3
China	150	265	381	385	7	4.4
Cuba	10	26	17	30		
Cyprus	2	3	10	9		
Ecuador	11	31	25	25		
Egypt	88	81	155	160		
Ethiopia	7	7	8	8		
France	7	32	29	30		
Greece	32	42	73	73		
Haiti	8	9	9	9		
Iran	40	51	376	376	8	4.3
Iraq	12	41	75	79		
Israel	57	90	120	120		
Italy	284	333	476	481	5	5.5
Jamaica	12	9	20	20		
Japan	2,359	3,110	2,015	2,100	1	23.9
Jordan	10	13	30	30		
Korea, S.	150	172	579	500	4	5.7
Laos	6	8	22	22		
Lebanon	15	32	50	55		
Mexico	105	126	157	169		
Morocco	140	265	420	223		
Pakistan	116	256	410	425	6	4.8
Paraguay	30	59	60	60		
Peru	14	20	19	19		
Philippines	22	34	24	25		
Portugal	16	21	20	20		
Spain	332	811	1,484	1,484	2	16.9
Tunisia	18	32	50	53		
Turkey	66	165	336	300	10	3.4
Uruguay	35	30	37	40		
USA	480	633	525	326	9	3.7
Venezuela	...	25	34	34		
Zimbabwe	5	7	8	8		

This list includes tangerines, mandarins, clementines and satsumas. 1950 figures in oranges list

Tea
thousand tonnes

	1950	1970	1980	1988	1989	Rank	%
World	639	1,341	1,841	2,466	2,537		
Africa	19	117	199	301	323		
Asia	598	1,123	1,462	1,963	2,002		
Oceania	—	1	8	8	8		
South America	1	34	44	50	59		
Argentina	...	26	29	34	43	10	1.7
Bangladesh	22	25	39	44	38		
Bolivia	...	...	2	2	2		
Brazil	1	6	10	10	10		
Burundi	...	...	2	5	5		
Cameroon	...	1	2	3	3		
China	70	246	329	557	521	2	20.5
Ecuador	—	—	1	1	1		
India	273	416	565	684	735	1	28.9
Indonesia	39	65	93	156	165	5	6.5
Iran	5	18	23	46	46	9	1.8
Japan	41	91	101	90	90	8	3.5
Kenya	6	38	93	181	193	4	7.6
Malawi	7	18	32	38	42		
Malaysia	2	3	3	5	5		
Mauritius	...	4	5	6	6		
Mozambique	3	17	19	2	2		
Papua New G.	—	1	8	8	8		
Peru	—	2	3	3	4		
Rwanda	—	1	6	11	12		
South Africa	—	1	5	10	10		
Sri Lanka	140	215	203	207	225	3	8.9
Taiwan	11	28	24	23	23		
Tanzania	1	9	17	16	18		
Thailand	...	...	2	5	5		
Turkey	...	34	83	136	140	7	5.5
Uganda	2	18	2	5	6		
USSR	21	65	128	143	144	6	5.7
Vietnam	6	10	22	32	34		
Zaïre	...	8	7	6	6		
Zimbabwe	...	3	10	17	17		

Tobacco
thousand tonnes

	1950	1970	1980	1989	1990	Rank	%
World	2,851	4,614	5,321	7,151	7,032		
Africa	135	203	296	356	375		
Asia	849	2,099	2,348	4,388	4,215		
Europe	385	569	712	689	722		
North America	1,129	1,071	1,086	876	890		
Oceania	4	22	19	13	15		
South America	192	393	565	596	616		
Albania	2	12	14	20	20		
Algeria	20	5	3	4	4		
Argentina	30	60	61	80	68		
Australia	2	18	15	11	14		
Bangladesh	...	43	44	39	45		
Brazil	113	246	397	444	468	4	6.7
Bulgaria	45	113	138	83	110		
Burma	47	53	53	50	47		
Cambodia	5	12	5	12	13		
Canada	62	105	101	76	66		
Chile	7	7	6	11	14		
China	132	801	1,134	2,851	2,711	1	38.6
Colombia	21	42	52	34	35		
Cuba	32	31	25	42	44		
Czechoslovakia	8	6	5	5	5		
Dominican Rep.	18	22	47	29	19		
France	48	45	50	29	26		
Germany	24	8	12	12	12		
Greece	49	88	123	126	132	10	1.9
Guatemala	2	5	9	8	12		
Honduras	4	6	8	4	5		
Hungary	23	20	18	17	15		
India	247	353	450	491	490	3	7
Indonesia	64	73	84	150	158	8	2.2
Iran	12	18	21	19	21		
Iraq	3	15	12	3	4		
Italy	76	79	129	197	205	6	2.9
Japan	90	163	146	74	74		
Kenya	1	2	3	7	9		
Korea, N.	12	40	45	64	65		
Korea, S.	15	60	97	78	73		
Madagascar	4	5	5	4	4		
Malawi	13	21	55	86	91		
Malaysia	2	3	9	13	11		
Mexico	35	72	72	67	52		
Morocco	2	3	8	8	8		
Nepal	6	7	6	5	5		
Nigeria	11	13	13	10	9		
Pakistan	68	118	71	74	63		
Paraguay	8	20	19	2	5		
Philippines	21	58	47	80	71		
Poland	25	82	70	64	66		
Portugal	7	10	11	4	4		
Romania	17	26	38	31	32		
South Africa	23	35	40	34	34		
Spain	19	24	37	37	34		
Sri Lanka	3	9	8	9	9		
Syria	6	8	14	18	20		
Tanzania	2	12	19	15	15		
Thailand	27	77	86	59	74		
Turkey	91	157	217	253	210	5	3
Uganda	2	5	3	4	4		
USA	959	819	810	642	683	2	9.7
USSR	157	256	297	233	200	7	2.8
Venezuela	10	12	20	15	16		
Vietnam	5	12	23	24	30		
Yemen	1	2	7	7	7		
Yugoslavia	24	46	64	57	54		
Zaïre	3	5	6	8	8		
Zambia	3	6	5	5	5		
Zimbabwe	42	63	103	130	142	9	2

Tomatoes
thousand tonnes

	1950	1970	1980	1989	1990	Rank	%
World	11,420	34,823	50,343	68,791	69,902		
Africa	708	3,126	5,060	8,313	8,619		
Asia	451	7,578	12,144	16,910	17,226		
Europe	3,886	10,870	14,572	18,195	18,607		
North America	4,386	7,074	9,292	13,500	13,430		
Oceania	122	236	272	368	370		
South America	505	1,769	2,829	4,404	4,350		
Albania	...	...	216	49	50		
Algeria	120	110	187	500	537		
Argentina	219	373	500	750	760		
Australia	100	177	213	319	320		
Austria	12	22	32	23	23		
Bangladesh	...	75	65	84	85		
Belgium-Lux.	60	90	98	215	230		
Bolivia	...	50	38	40	42		
Brazil	136	762	1,507	2,173	2,132	9	3
Bulgaria	140	711	855	850	755		
Canada	321	350	458	626	625		
Chile	38	199	153	426	426		
China	...	2,584	4,158	5,430	5,599	5	8
Colombia	28	146	244	480	480		
Cuba	40	64	194	260	260		
Czechoslovakia	70	105	90	122	125		
Denmark	16	21	16	18	18		
Dominican Rep.	...	62	144	170	170		
Ecuador	19	25	35	40	40		
Egypt	396	1,580	2,541	4,800	5,000	6	7.2
Ethiopia	6	37	46	52	52		
France	354	514	813	756	770		
Germany	78	49	62	99	100		
Ghana	15	73	132	96	100		
Greece	289	976	1,694	2,150	2,100	10	3
Guatemala	...	71	86	126	130		
Hungary	161	359	443	418	400		
India	...	620	743	800	806		
Indonesia	...	310	91	200	214		
Iran	...	180	329	690	800		
Iraq	66	305	465	710	650		
Israel	41	165	249	425	425		
Italy	1,128	3,571	4,593	5,730	5,796	3	8.3
Japan	165	811	1,017	773	770		
Jordan	17	142	166	210	240		
Kenya	...	41	53	25	25		
Korea, S.	5	54	52	30	30		
Lebanon	26	68	75	125	120		
Libya	13	134	221	215	218		
Mexico	337	901	1,315	1,889	1,746		
Morocco	70	246	423	215	218		
Netherlands	75	365	400	580	595		
New Zealand	22	56	53	44	44		
Nigeria	...	220	403	635	650		
Paraguay	...	47	63	77	52		
Peru	20	60	65	70	71		
Philippines	29	97	138	179	180		
Poland	126	339	334	451	380		
Portugal	73	764	624	865	1,005		
Romania	189	780	1,463	2,200	2,350	8	3.4
Saudi Arabia	30	87	169	385	390		
South Africa	83	225	275	490	500		
Spain	787	1,687	2,142	2,876	3,114	7	4.5
Sudan	...	109	148	168	150		
Syria	60	211	541	500	580		
Tunisia	20	160	285	450	460		
Turkey	...	1,756	3,650	5,750	5,700	4	8.2
UK	197	108	141	144	146		
Uruguay	14	30	46	66	66		
USA	3,671	5,516	6,944	10,233	10,300	1	14.7
USSR	...	4,171	6,184	7,100	7,300	2	10.4
Venezuela	30	76	166	195	195		
Yugoslavia	138	332	450	434	440		
Zaire	...	30	31	38	40		

Watermelons
thousand tonnes

	1950	1970	1980	1988	1989	Rank	%
World	...	19,720	25,458	28,881	28,423		
Africa	...	1,478	1,995	2,374	2,449		
Asia	...	10,594	14,158	15,911	15,529		
Europe	...	2,336	3,049	2,672	2,498		
North America	...	1,509	1,604	1,700	1,675		
Oceania	...	36	50	59	61		
South America	...	967	899	922	912		
Afghanistan	...	54	34	31	32		
Algeria	...	194	170	320	330		
Argentina	45	182	152	124	120		
Australia	...	36	44	51	52		
Brazil	216	466	288	430	420		
Bulgaria	390	289	327	318	79		
Chile	...	159	245	86	88		
China	...	3,127	4,229	5,875	6,082	1	21.4
Cyprus	6	21	21	28	29		
Ecuador	7	6	28	69	70		
Egypt	264	919	1,197	1,390	1,400	4	4.9
El Salvador	...	41	41	87	86		
Greece	305	555	644	550	600	9	2.1
Hungary	189	102	182	125	124		
Iran	...	758	937	960	925	6	3.3
Iraq	200	527	465	483	585	10	2.1
Israel	...	97	82	117	118		
Italy	370	748	719	651	708	8	2.5
Japan	232	1,048	1,008	790	790	7	2.8
Jordan	...	34	25	67	67		
Korea, North	...	20	55	90	92		
Korea, South	...	130	311	465	420		
Lebanon	24	21	31	32	32		
Libya	...	30	159	147	150		
Malaysia	...	10	100	32	32		
Mexico	...	246	442	470	475		
Morocco	75	140	134	146	147		
Paraguay	50	62	72	73	73		
Peru	40	48	38	40	40		
Philippines	22	63	245	105	100		
Saudi Arabia	...	505	247	320	320		
South Africa	...	41	25	40	40		
Spain	225	217	545	627	581		
Sudan	...	75	101	125	135		
Syria	132	256	770	477	105		
Taiwan	18	225	345	375	380		
Thailand	...	350	660	380	380		
Tunisia	19	77	205	200	240		
Turkey	1,738	3,486	4,723	5,250	5,000	3	17.6
USA	1,146	1,216	1,102	1,130	1,100	5	3.9
USSR	...	2,800	3,702	5,243	5,300	2	18.6
Venezuela	25	32	54	62	63		
Vietnam	...	60	94	145	150		
Yemen	...	26	93	237	236		
Yugoslavia	196	417	627	391	396		

Whaling
number

	1970	1980	1986	1987	1988
Minke whale					
World	...	11,828	5,875	5,736	421
Japan	...	3,658	1,941	1,941	273
Norway	...	2,002	379	373	29
USSR	...	3,879	3,028	3,028	0
Greenland	...	...	147	90	119
Brazil	...	902	0	0	0
Korea, S.	...	925	69	0	0
Bryde's whale					
World	910	522	317	317	0
Japan	...	307	317	317	0
Sei whale					
World	10,466	102	40	20	10
Iceland	...	100	40	20	10
Fin whale					
World	4,547	472	85	89	77
Iceland	...	236	76	80	68
Denmark	...	...	9	9	9
Spain	...	219	0	0	0
Humpback whale					
Denmark	24	15	0	0	0

	1970	1980	1986	1987	1988
Baleen whale					
World	...	217	191	180	173
USA	...	...	22	22	23
USSR	...	178	169	158	150
Sperm whale					
World	22,904	2,210	200	191	8
Japan	...	1,192	200	188	0
Portugal	...	330	0	3	
Longfin Pilot whale					
World	...	2,820	1,709	1,422	1,705
Faroe Is.	...	2,773	1,709	1,422	1,690
Killer whale					
World	...	971	24	3	0
Japan	8	3	0		
Beluga whale					
World	...	1,909	1,254	716	13
Greenland	...	889	375	682	0
Canada	...	774	700	0	5
USSR	...	236	178	34	8
Narwhal					
Greenland	...	520	263	0	0
Toothed whale					
World	...	381	344	115	0
Canada	...	339	344	330	0

Wheat
thousand tonnes

	1950	1970	1980	1989	1990	Rank	%
World	190,000	329,033	443,858	541,920	595,652		
Africa	4,576	7,945	8,740	12,502	13,646		
Asia	37,292	80,243	135,100	192,142	199,009		
Europe	41,160	72,910	91,555	127,501	131,962		
North America	45,062	56,112	89,350	84,429	110,014		
Oceania	5,300	9,371	14,810	14,256	15,920		
South America	7,420	9,648	12,173	18,782	17,101		
Afghanistan	1,700	2,150	2,788	1,925	1,925		
Albania	89	242	500	633	633		
Algeria	996	1,359	1,330	850	750		
Argentina	5,175	5,873	7,993	10,100	11,000		
Australia	5,161	9,014	14,486	14,121	15,700	9	2.6
Austria	348	912	1,025	1,363	1,370		
Bangladesh	55	103	803	1,022	1,100		
Belgium-Lux.	555	848	955	1,478	1,527		
Brazil	498	1,743	2,614	5,556	3,200		
Bulgaria	1,776	2,899	3,877	5,402	5,225		
Canada	13,443	13,901	20,287	24,578	31,798	6	5.3
Chile	942	1,296	882	1,766	1,718		
China	16,000	31,005	57,965	90,810	96,004	2	16.1
Czechoslovakia	1,493	3,436	4,507	6,356	6,712		
Denmark	285	509	678	3,224	4,101		
Egypt	1,111	1,509	1,834	3,183	4,000		
Ethiopia	460	643	470	833	816		
Finland	263	445	267	507	627		
France	7,791	14,112	21,996	31,817	33,363	5	5.6
Germany	3,912	8,471	11,248	14,509	15,793	8	2.7
Greece	894	1,867	2,703	2,005	1,656		
Hungary	1,909	3,410	4,862	6,540	6,159		
India	6,087	20,859	34,599	54,110	49,652	4	8.3
Iran	1,863	3,946	5,700	5,525	7,000		
Iraq	448	1,080	1,297	491	805		
Ireland	327	375	240	477	603		
Italy	7,170	9,756	9,017	7,413	8,245		
Japan	1,375	557	579	1,025	1,000		
Korea, N.	84	250	380	900	900		
Mexico	534	2,141	2,771	4,374	3,630		
Mongolia	...	227	240	687	500		
Morocco	786	1,819	1,500	3,927	3,614		
Nepal	...	230	444	830	850		
Netherlands	324	675	867	1,047	1,076		
New Zealand	139	357	325	135	220		
Pakistan	3,630	6,796	10,698	14,419	14,300	10	2.4
Poland	1,833	4,925	4,197	8,462	8,635		
Portugal	499	614	331	615	268		
Romania	2,486	4,433	5,634	7,880	8,000		
Saudi Arabia	16	111	150	3,100	3,100		
South Africa	555	1,461	1,882	2,003	1,800		
Spain	3,626	4,734	4,435	5,465	4,759		
Sweden	677	958	1,086	1,751	2,173		
Switzerland	260	378	403	620	565		
Syria	761	763	1,877	1,020	2,069		

Wheat — *continued*

	1950	1970	1980	1989	1990	Rank	%
Tunisia	452	520	837	420	1,122		
Turkey	4,770	11,423	17,054	16,221	20,000	7	3.4
UK	2,397	4,140	8,034	14,200	13,900		
Uruguay	469	379	379	542	375		
USA	31,065	40,034	66,242	55,429	74,534	3	12.5
USSR	35,759	92,804	92,131	92,307	108,000	1	18.1
Yugoslavia	2,171	4,760	4,624	5,599	6,348		

Wine

thousand tonnes

	1950	1970	1980	1988	1989	Rank	%
World	18,901	28,854	34,626	27,510	29,055		
Africa	1,726	1,565	1,071	1,060	1,117		
Asia	42	144	230	250	253		
Europe	14,125	20,340	24,815	19,103	20,333		
North America	927	1,239	1,690	2,100	2,049		
Oceania	148	278	413	456	543		
South America	1,644	2,658	3,289	2,760	2,860		
Algeria	1,348	888	280	62	100		
Argentina	1,136	1,935	2,392	1,910	2,000	4	6.9
Australia	145	259	366	408	494		
Austria	92	239	262	350	350		
Brazil	102	179	252	376	376		
Bulgaria	248	436	466	340	340		
Canada	22	39	45	54	57		
Chile	313	443	572	380	390		
China	...	...	38	80	85		
Cyprus	13	45	56	63	61		
Czechoslovakia	31	100	148	142	139		
France	5,245	6,298	7,050	5,782	5,891	2	20.3
Germany	232	754	701	998	1,165	7	4
Greece	370	454	535	473	497		
Hungary	316	476	547	471	450		
Israel	6	33	31	16	15		
Italy	4,342	6,825	8,213	6,186	5,980	1	20.6
Japan	4	16	29	61	61		
Mexico	1	17	50	200	216		
Morocco	64	100	89	50	38		
New Zealand	3	18	46	48	49		
Portugal	817	959	1,017	360	845	10	2.9
Romania	402	584	889	1,000	1,000	8	3.4
South Africa	224	489	623	916	944	9	3.2
Spain	1,539	2,485	4,132	2,266	3,030	3	10.4
Switzerland	70	91	89	111	111		
Tunisia	74	79	66	20	23		
Turkey	15	44	39	23	23		
USA	904	1,183	1,594	1,845	1,775	6	6.1
USSR	277	2,630	3,120	1,780	1,900	5	6.5
Uruguay	79	86	55	74	74		
Yugoslavia	414	603	720	576	486		

Wood – coniferous roundwood

million cubic metres

	1961 -65 av.	1970	1980	1987	1988	Rank	%
World	959	1,081	1,195	1,355	1,369		
Africa	6.8	9.9	13.6	14.8	15.4		
Asia	122	135	172	189	187		
Europe	185	204	220	230	238		
North America	316	371	432	526	530		
Oceania	7.4	11	14.2	16	17		
South America	19.6	30.5	47.1	53.7	54.1		
Afghanistan	1.7	2.3	2.4	2.2	2.3		
Algeria	...	...	1.1	13.7	14.1		
Argentina	0.5	0.5	0.8	2	2		
Australia	1.8	2.5	4.3	6.8	7		
Austria	9.4	10	12.1	11.1	12.2		
Belgium	1.4	1.5	1.5	2.4	2.6		
Brazil	16.9	26.1	38	41.5	41.8	6	3.1
Bulgaria	1.5	1.5	1.2	1	1.1		
Canada	89.9	111	145	162	164	3	12
Chile	2.2	3.7	8	9.9	10.1		
China	65.2	77.1	113	134	134	4	9.8
Czechoslovakia	10.2	10.6	14.3	14.6	14.4		
Denmark	0.9	1.3	1.2	1.4	1.4		
Ethiopia	1.8	2.4	2.3	2.5	2.6		
Finland	34.1	32.6	39.1	34.6	39.2	7	2.9
France	13.6	15	18.1	20.7	21.8	9	1.6
Germany	23.3	25.1	30.9	31.6	32.8	8	2.4
Guatemala	3.1	2.8	4.9	5.9	6.1		
Honduras	2.3	2.4	2.3	2.5	2.6		
India	3.5	4.3	8	9.5	9.6		
Ireland	0.2	0.2	0.3	1.2	1.3		
Italy	1.5	1.4	1.7	1.5	1.9		
Japan	34.7	26.8	21.4	20.6	19.1	10	1.4
Kenya	0.8	...	1.4	1.7	1.9		
Korea, N.	2.7	3.2	2.3	2.7	2.8		
Korea, S.	6.5	7.3	5.6	4.7	4.6		
Mexico	5.9	7.5	9.2	10.8	10.9		
Mongolia	1.1	1.9	2.2	2.2	2.2		
New Zealand	5.5	8.4	9.7	9.1	9.8		
Norway	8.4	7.7	8.3	9.3	9.9		
Poland	14.1	15.3	16.8	18.4	17.7		
Portugal	4	4.5	5.5	5.4	5.4		
Romania	6.4	7.3	7.0	6.7	6.4		
South Africa	2.5	4	6.3	6.1	6.5		
Spain	5.7	5.6	7.3	10.1	10.1		
Swaziland	0.4	0.8	1.4	1.6	1.6		
Sweden	40.5	53.3	41.7	44.8	44.2	5	3.2
Switzerland	2.8	3	3.1	3.5	3.4		
Turkey	5.8	10.4	14.2	9.9	9.9		
UK	1.4	1.8	2.6	4.7	5.4		
USA	214	246	269	343	345	1	25.2
USSR	301	321	298	325	327	2	23.9
Yugoslavia	4.3	3.8	4.3	4.7	4.3		

Roundwood is timber as it is felled

Wood – coniferous sawnwood

million cubic metres

	1961 -65 av.	1970	1980	1987	1988	Rank	%
World	277	311	334	377	379		
Africa	1	1.6	1.9	2.2	2.3		
Asia	33.6	46	51.1	53.1	53.1		
Europe	55.6	62.1	70.4	66.9	68.2		
North America	88.6	91.9	112	152	150		
Oceania	2.3	2.6	3.2	3.1	3.4		
South America	3.5	5.3	9.2	10.9	10.9		
Afghanistan	0.2	0.3	0.4	0.4	0.4		
Australia	0.7	0.8	1.1	1.2	1.4		
Austria	4.4	5.1	6.3	5.7	6.2		
Belgium	0.3	0.4	0.3	0.7	0.7		
Brazil	2.9	4.5	7.1	8.4	8.4	8	2.2
Bulgaria	0.7	0.9	1	1.1	1		
Canada	21.3	25.4	42.9	61.1	59.6	3	15.7
Chile	0.5	0.7	1.8	2.3	2.3		
China	7	8.9	13.2	17.1	17	5	4.5
Czechoslovakia	3.3	2.9	4	4.3	4.2		
Denmark	0.3	0.4	0.4	0.5	0.5		
Finland	6.9	7.1	10.2	7.5	7.7	9	2
France	5.1	5.7	5.7	5.9	6.5	10	1.7
Germany	8.8	9	10.2	10	10.6	7	2.8
Honduras	0.5	0.4	0.5	0.4	0.4		
Hungary	0.3	0.4	0.5	0.4	0.4		
India	0.5	0.7	1.5	2.4	2.4		
Italy	0.8	0.6	1.1	0.7	0.9		
Japan	23.8	32.8	29.9	26.2	26.2	4	6.9
Korea, S.	0.5	0.6	1.6	2.9	2.9		
Mexico	1	1.4	1.5	1.9	1.9		
Mongolia	0.2	0.4	0.5	0.5	0.5		
New Zealand	1.6	1.8	2	1.8	1.9		
Norway	1.6	1.9	2.5	2.4	2.4		
Poland	5.9	5.9	5.9	5.4	4.9		
Portugal	1.1	1.6	2	1.7	1.7		
Romania	2.8	2.8	2.1	1.3	1.3		
South Africa	0.7	1.1	1.4	1.5	1.6		
Spain	1.1	1.6	1.6	1.9	1.9		
Sweden	8.7	12	11.1	11.3	11	6	2.9
Switzerland	1.1	1.3	1.5	1.4	1.5		
Taiwan	...	0.7	0.4	0.3	0.1		
Turkey	0.9	1.8	3.6	3.3	3.3		
UK	0.3	0.3	1.1	1.6	1.7		
USA	65.5	64.6	66.6	88.3	88.3	2	23.2
USSR	92.4	102	86	89.4	90.3	1	23.8
Yugoslavia	1.6	1.7	2.3	2.4	2.3		

Sawnwood is timber in its first stage of processing – sawn into rectangular forms

Wood – nonconiferous roundwood

million cubic metres

	1961 -65 av.	1970	1980	1987	1988	Rank	%
World	1,162	1,315	1,634	1,905	1,934		
Africa	240	284	333	405	416		
Asia	419	501	719	832	841		
Europe	122	128	112	121	124		
North America	122	121	192	235	238		
Oceania	16.5	16.5	19.5	22.6	23		
South America	178	199	197	224	228		
Afghanistan	3.5	4.1	3.9	3.6	3.6		
Albania	1.1	1.6	1.5	1.5	1.5		
Angola	5.9	7.3	4.3	5.1	5.3		
Argentina	10.4	11.2	7.3	7.4	7.4		
Australia	12.1	11	12.6	13.1	13.5		
Austria	1.6	1.8	2.3	2.6	2.7		
Bangladesh	7.8	9.5	23.8	28.6	29.4		
Benin	1.7	2.1	3.6	4.5	4.6		
Bhutan	...	...	3.2	3.2	3.2		
Bolivia	4.4	4.2	1.4	1.3	1.3		
Brazil	117	133	145	167	170	4	8.8
Bulgaria	3.9	3.6	2.9	3.4	3.3		
Burkina Faso	3.5	4.1	6.6	7.9	8.1		
Burma	13.4	15.6	17.4	20.2	21		
Burundi	0.7	0.8	3.1	3.8	3.9		
Cambodia	3.6	4	4.7	5.5	5.7		
Cameroon	6.4	7.6	10.2	12.4	12.6		
Canada	9.5	10.7	13.6	15.2	15.5		
Central Afr. Rep.	1.9	2.3	3	3.4	3.4		
Chad	2.9	3.3	1.2	1.4	1.4		
Chile	4.4	3.9	5.6	6.3	6.4		
China	81.4	93.8	121	142	142	5	7.3
Colombia	24.9	26.6	13.4	14.8	15.1		
Congo	1.7	2.2	2.2	2.8	3.3		
Costa Rica	2	2.8	3.4	3.8	3.9		
Cuba	2.2	1.8	2.8	2.8	2.8		
Czechoslovakia	2.7	3.9	4.5	4.1	4.1		
Dominican Rep.	1.7	1.8	0.5	0.6	0.6		
Ecuador	2.4	2.8	5.6	6.9	6.9		
Egypt	0.1	0.2	1.8	2.2	2.2		
El Salvador	2.7	2.4	3.7	4.1	4.1		
Ethiopia	18.6	20.8	30.1	34.3	35	10	1.8
Finland	12.1	12.5	8	8.2	9.4		
France	19.7	19.8	20.7	19.8	20.2		
Gabon	2.6	2.9	3.1	3.5	3.6		
Germany	10.2	10.4	12.2	13	13		
Ghana	9.6	10.1	10.6	13.7	13.9		
Greece	2.5	2.3	2	2.4	2.4		
Guatemala	0.6	4.3	1.1	1.3	1.3		
Guinea	2.3	2.8	3.8	4.5	4.6		
Haiti	3	3.4	4	4.5	4.6		
Honduras	1.3	1.7	2.6	3.3	3.4		
Hungary	3.7	4.8	5.8	6.2	6		
India	91.2	107	204	239	243	1	12.6
Indonesia	86	107	145	170	172	3	8.8
Iran	6.5	6.9	6.4	6.4	6.4		
Italy	9.8	13.7	7.3	7.5	7.8		
Ivory Coast	7.2	8.9	11.2	11.2	11.6		
Japan	26	22.9	13	11.5	9		
Kenya	8.8	10.3	15.8	20.8	21.7		
Korea, N.	1.4	1.6	2	2	2		
Korea, S.	2.9	3.3	3	2.1	2.1		
Laos	2.4	2.8	3	3.2	3.2		
Liberia	1.2	1.6	3.3	3.7	3.9		
Madagascar	3.7	5.3	6.1	7.4	7.6		
Malawi	3.5	4.2	5.6	7.1	7.3		
Malaysia	12.9	24.1	33.9	42.1	42.2	8	2.2
Mali	2.3	2.7	4.2	5.2	5.3		
Mexico	6.7	6.9	8.7	10.4	10.7		
Morocco	1.8	2.2	0.9	1.2	1.2		
Mozambique	7.6	8.5	12.8	15.3	15.4		
Nepal	7.7	8.8	14	16.7	17.1		
Nicaragua	1.9	1.9	2.6	3.1	3.2		
Niger	1.9	2.3	3.4	4.2	4.3		
Nigeria	46.6	56.9	74.8	93.2	96.2	6	5
Pakistan	6.2	8.1	16.3	21.7	22.5		
Panama	1.6	1.4	2	2	2		
Papua New G.	3.8	4.7	7	8.2	8.2		
Paraguay	2.4	3.7	5.8	7.3	7.2		
Peru	4.7	5.7	8.1	8.6	8.7		
Philippines	24.9	33.7	34.8	37.4	38	9	2
Poland	2.6	3.2	4	4.7	5		
Portugal	1.9	1.9	3	4.7	4.7		
Romania	14.8	15	10.7	13.3	13.2		

Wood – nonconiferous roundwood

continued

	1961 –65 av.	1970	1980	1987	1988	Rank	%
Rwanda	3	3.7	4.8	5.8	5.8		
Senegal	2.1	2.5	3.1	3.7	3.8		
Sierra Leone	2.7	2.7	2.5	2.8	2.9		
Somalia	2.4	2.9	4.6	5.9	6.1		
South Africa	4.1	5.6	12.5	12.8	12.8		
Spain	9.2	8.1	5.1	5.6	5.6		
Sri Lanka	4	4.7	7.9	8.7	8.8		
Sudan	19.7	20.9	6.3	7.8	8		
Sweden	6	6.7	7.5	8.5	8.5		
Tanzania	25.9	31.4	22.6	29.5	30.6		
Thailand	16	18.7	30.6	34.2	34.6		
Tunisia	1.2	1.4	1.7	2	2		
Turkey	5.2	6.4	8.4	7	7		
Uganda	11.5	13.9	10	12.6	13		
UK	1.1	1.5	1.3	0.8	1		
Uruguay	1	1	2.2	2.7	2.7		
USA	88.6	82	147	184	185	2	9.6
USSR	65.9	64.4	58.9	64.4	64.6	7	3.3
Venezuela	5.6	7	1.2	1.3	1.5		
Vietnam	14.9	17.6	22.3	25.8	26.4		
Yugoslavia	13.3	12.5	8.8	10.4	10.8		
Zaïre	12	14	26.8	33.2	34.2		
Zambia	3.8	4.7	5	6.2	6.3		
Zimbabwe	4.5	5.4	6	7.1	7.3		

Roundwood is timber as it is felled

Wood – nonconiferous sawnwood

million cubic metres

	1961 –65 av.	1970	1980	1987	1988	Rank	%
World	77.7	93.1	113	124	125		
Africa	1.9	2.8	5.4	6.1	6.2		
Asia	20.1	29	42.6	50.8	50.8		
Europe	14.7	17.5	18.5	16.1	16.4		
North America	17.7	18.6	19.7	21.1	21.2		
Oceania	2.6	2.7	2.3	2	2.1		
South America	5.9	7.9	12.6	15.3	15.3		
Angola	0.1	0.2	–	–	–		
Argentina	0.7	0.6	0.7	1.2	1.2		
Australia	2.4	2.5	2	1.8	1.9		
Austria	0.2	0.2	0.4	0.2	0.2		
Bangladesh	0.3	0.4	0.2	0.1	0.1		
Belgium	0.3	0.3	0.3	0.3	0.3		
Bolivia	...	0.1	0.2	0.1	0.1		
Brazil	2.8	3.5	7.7	9.7	9.8	5	7.8
Bulgaria	0.9	0.7	0.4	0.4	0.3		
Burma	0.5	0.6	0.4	0.5	0.5		
Cameroon	0.1	0.1	0.4	0.6	0.6		
Canada	1.1	1.3	1.4	1	1		
Chile	0.5	0.3	0.4	0.4	0.4		
China	4.2	5.6	7.8	9.2	9.2	6	7.4
Colombia	0.9	1.8	0.9	0.7	0.7		
Costa Rica	0.3	0.4	0.5	0.5	0.5		
Czechoslovakia	0.5	0.7	0.8	0.8	0.8		
Denmark	0.3	0.5	0.4	0.4	0.4		
Ecuador	0.3	0.7	0.9	1.3	1.3		
Finland	0.1	0.2	0.1	0.1	0.1		
France	2.8	3.6	3.7	3.4	3.7	9	3
Germany	2.1	2.3	2.4	2	2		
Ghana	0.4	0.4	0.2	0.4	0.4		
Greece	0.1	0.1	0.2	0.2	0.2		
Hong Kong	0.2	0.2	0.2	0.2	0.2		
Hungary	0.5	0.5	0.8	0.8	0.8		
India	1.4	1.9	9.2	14.8	14.8	2	11.8
Indonesia	1.7	1.7	4.8	9.8	10.2	4	8.2
Italy	1.1	1.7	1.5	1.1	1.1		
Ivory Coast	0.2	0.3	0.7	0.8	0.8		
Japan	6.1	9.8	7	3.9	3.9	8	3.1
Korea, S.	0.1	0.5	1.4	0.1	0.1		
Liberia	—	—	—	0.3	0.3		
Madagascar	0.1	0.1	0.2	0.1	0.2		
Malaysia	1.7	3	6.2	6.2	6.2	7	5
Mexico	0.1	0.1	0.1	0.2	0.2		
Mozambique	0.1	0.1	0.1	0.1	0.1		
Nepal	0.2	0.2	0.2	0.2	0.2		
Netherlands	0.2	0.2	0.2	0.2	0.2		
Nicaragua	0.1	0.1	0.2	0.1	0.1		
Nigeria	0.3	0.6	2.8	2.7	2.7	10	2.2

	1961 –65 av.	1970	1980	1987	1988	Rank	%
Papua New G.	2.4	0.1	0.1	0.1	0.1		
Paraguay	0.1	0.2	0.6	0.9	0.9		
Peru	0.1	0.2	0.6	0.6	0.5		
Philippines	1.1	1.3	1.5	1.2	1		
Poland	0.9	0.9	1.1	0.8	0.8		
Portugal	0.1	0.2	0.2	0.3	0.3		
Romania	1.8	2.5	2.5	1.5	1.5		
Singapore	0.2	0.8	0.4	1.2	1		
South Africa	0.1	0.1	0.2	0.2	0.2		
Spain	0.6	0.7	0.5	0.6	0.6		
Sri Lanka	...	0.2	–	–	–		
Sweden	0.2	0.2	–	0.2	0.2		
Switzerland	0.1	0.2	0.2	0.2	0.2		
Taiwan	0.3	0.3	0.2	0.2	0.2		
Tanzania	0.1	0.1	0.1	0.1	0.1		
Thailand	1.1	1.2	1.5	1.1	1		
Turkey	0.3	0.5	1.0	1.5	1.5		
UK	0.7	0.5	0.7	0.3	0.3		
USA	15.9	16.6	17.2	19.2	19.2	1	15.4
USSR	15	14.6	12.2	12.6	12.7	3	10.2
Venezuela	0.2	0.3	0.3	0.3	0.3		
Vietnam	0.4	0.4	0.4	0.3	0.3		
Yugoslavia	1	1.3	1.9	2.2	2.3		
Zaïre	0.2	0.2	0.1	0.1	0.1		

Sawnwood is timber in its first stage of processing – sawn into rectangular forms

Wood pulp

million tonnes

	1961 –65 av.	1970	1980	1987	1988	Rank	%
World	69	105	126	146	151		
Africa	0.5	0.8	1.4	1.7	1.7		
Asia	5.6	10.4	12.3	13.6	14.3		
Europe	19.8	27.7	30.5	35.5	36.7		
North America	37.8	56.4	66.6	77.3	79.7		
Oceania	0.6	1.3	1.8	2	2.2		
South America	0.7	1.4	4.3	5.9	6.1		
Argentina	0.1	0.2	0.3	0.6	0.6		
Australia	0.3	0.7	0.7	0.9	1		
Austria	0.7	0.9	1.3	1.4	1.5		
Belgium	0.2	0.3	0.4	0.3	0.3		
Brazil	0.4	0.8	3.1	4.2	4.4	7	2.9
Bulgaria	0.1	0.1	0.3	0.2	0.2		
Canada	11.7	16.6	19.9	22.7	23.6	2	15.6
Chile	0.2	0.4	0.8	0.9	0.9		
China	0.8	1.2	1.3	1.7	1.8		
Colombia	0.1	0.1	0.1	0.2	0.2		
Czechoslovakia	0.6	0.6	0.9	1.3	1.3		
Finland	4.7	6.5	7.2	8.5	9	6	6
France	1.3	1.8	1.8	2.1	2.2	9	1.5
Germany	2.0	2.5	2.7	2.9	3	8	2
India	...	0.1	0.5	1	1		
Italy	0.6	0.9	0.7	0.6	0.7		
Japan	4.6	8.8	9.8	9.7	10.4	3	6.9
Korea, S.	...	0.1	0.2	0.3	0.3		
Mexico	0.2	0.3	0.4	0.6	0.7		
Netherlands	0.1	0.2	0.2	0.1	0.2		
New Zealand	0.4	0.6	1.1	1.1	1.2		
Norway	1.6	2.2	1.5	2	2	10	1.3
Philippines	...	...	0.1	0.2	0.2		
Poland	0.5	0.6	0.7	0.8	0.8		
Portugal	0.1	0.4	0.6	1.4	1.4		
Romania	0.2	0.4	0.7	0.7	0.7		
South Africa	0.4	0.6	1.1	1.3	1.3		
Spain	0.2	0.6	1.3	1.6	1.6		
Swaziland	0.1	0.1	0.2	0.2	0.2		
Sweden	5.8	8.1	8.7	10	10.1	5	6.7
Switzerland	0.2	0.3	0.3	0.3	0.3		
Turkey	0.1	0.1	0.3	0.4	0.4		
UK	0.3	0.4	0.4	0.4	0.4		
USA	25.8	39.5	46.2	54.1	55.5	1	36.8
USSR	3.8	6.3	8.8	10.4	10.4	4	6.9
Yugoslavia	0.3	0.5	0.7	0.8	0.8		

Wool – greasy

thousand tonnes

	1950	1970	1980	1989	1990	Rank	%
World	1,835	2,806	2,797	3,197	3,375		
Africa	139	211	203	246	253		
Asia	160	273	425	515	516		
Europe	211	255	274	322	329		
North America	139	94	59	47	49		
Oceania	694	1,231	1,060	1,264	1,418		
South America	320	330	314	329	344		
Afghanistan	...	26	23	15	17		
Albania	2	2	3	4	4		
Algeria	4	13	20	46	49	10	1.5
Argentina	186	169	167	164	161	5	4.8
Australia	515	899	705	959	1,100	1	32.6
Bolivia	4	7	9	12	13		
Brazil	20	32	31	32	33		
Bulgaria	14	29	35	27	29		
Canada	4	2	1	1	1		
Chile	20	23	20	20	20		
China	45	60	170	237	238	4	7.1
Czechoslovakia	1	4	4	6	6		
Ecuador	...	2	3	2	2		
Egypt	2	3	3	2	2		
Ethiopia	...	12	12	13	13		
Falkland Is.	2	2	2	2	2		
France	20	20	22	23	22		
Germany	11	10	16	25	23		
Greece	8	8	8	9	9		
Hungary	4	10	12	9	10		
India	26	37	35	30	32		
Iran	...	19	16	32	32		
Iraq	14	16	18	18	17		
Ireland	6	10	9	15	17		
Italy	15	12	13	14	14		
Jordan	...	3	3	2	3		
Lesotho	4	4	3	3	3		
Libya	...	4	11	9	9		
Mexico	...	8	9	6	5		
Mongolia	...	19	20	16	17		
Morocco	11	16	13	34	35		
Namibia	7	4	5	2	2		
Nepal	2	4	4	1	1		
New Zealand	179	332	355	305	318	3	9.4
Norway	3	5	5	5	5		
Pakistan	11	20	43	58	61	9	1.8
Peru	9	13	13	11	12		
Poland	4	9	13	14	14		
Portugal	10	9	9	9	9		
Romania	16	30	38	44	45		
Saudi Arabia	...	4	5	4	4		
South Africa	105	137	110	95	97	7	2.9
Spain	34	33	21	30	33		
Sudan	...	10	15	17	17		
Syria	8	14	18	27	27		
Tunisia	3	5	9	12	12		
Turkey	33	47	61	49	43		
UK	40	47	51	73	74	8	2.2
Uruguay	79	81	67	83	98	6	2.9
USA	129	85	49	40	43		
USSR	694	412	463	474	465	2	13.8
Yemen	...	...	6	4	4		
Yugoslavia	15	12	10	10	10		
Zimbabwe	...	...	9	11	12		

Wool – scoured

thousand tonnes

	1950	1970	1980	1989	1990	Rank	%
World	1,050	1,620	1,662	2,017	2,090		
Africa	68	101	107	123	130		
Asia	95	155	241	270	270		
Europe	112	140	160	189	199		
North America	62	45	30	25	27		
Oceania	420	748	667	925	983		
South America	190	184	180	200	202		
Afghanistan	...	14	13	9	9		
Albania	2	1	2	2	2		
Algeria	2	6	10	25	26		
Argentina	108	91	90	95	93	5	4.4
Australia	295	511	421	622	711	1	34
Bolivia	2	4	5	7	7		
Brazil	12	20	19	10	10		

Wool – scoured continued

	1950	1970	1980	1989	1990	Rank	%
Bulgaria	9	15	18	13	15		
Canada	2	1	1	1	1		
Chile	10	12	10	20	20		
China	27	36	102	120	119	4	5.7
Czechoslovakia	1	2	3	3	3		
Egypt	1	2	2	3	3		
Ethiopia	...	6	6	7	7		
Falkland Is.	2	1	2	1	2		
France	8	9	12	13	12		
Germany	5	6	8	12	13		
Greece	4	4	5	5	5		
Hungary	2	5	5	4	4		
India	18	23	23	20	22		
Iran	...	11	9	18	18		
Iraq	8	7	8	10	9		
Ireland	4	8	7	8	10		
Italy	7	6	6	6	6		
Lesotho	2	2	1	2	2		
Libya	...	1	3	3	3		
Mexico	...	4	4	3	3		
Mongolia	...	11	12	10	11		
Morocco	5	6	6	16	17		
Namibia	4	2	3	1	1		
Nepal	1	2	2	1	1		
New Zealand	124	237	246	303	272	3	15.2
Norway	2	3	3	3	3		
Pakistan	7	12	25	34	35	9	1.7
Peru	4	7	6	6	6		
Poland	2	5	8	8	8		
Portugal	5	3	4	4	4		
Romania	10	18	23	28	29	10	1.4
Saudi Arabia	...	2	3	2	2		
South Africa	52	68	63	47	51	8	2.4
Spain	14	13	11	17	19		
Sudan	...	4	6	7	7		
Syria	4	7	9	15	15		
Tunisia	1	2	4	6	6		
Turkey	18	26	33	27	24		
UK	26	31	38	53	58	7	2.8
Uruguay	51	49	46	59	62	6	3
USA	57	40	26	22	23		
USSR	103	247	277	284	279	2	13.3
Yemen	...	...	2	2	2		
Yugoslavia	9	7	6	5	5		
Zimbabwe	...	...	...	6	6		

Yams

thousand tonnes

	1980	1988	1989	Rank	%
World	23,384	23,498	23,459		
Africa	22,291	22,319	22,266		
Asia	165	193	195		
North America	338	369	350		
Oceania	231	247	249		
South America	355	368	397		
Benin	687	922	1,049	3	4.5
Brazil	179	200	210		
Burkina Faso	71	78	51		
Cameroon	203	230	230	9	0.9
Central Afr. Rep.	191	230	230	10	0.9
Chad	163	240	240	8	1.0
Colombia	143	125	142		
Ethiopia	276	240	250	7	1.1
Gabon	80	100	100		
Ghana	614	902	782	4	3.3
Guinea	64	90	100		
Haiti	112	120	130		
Ivory Coast	2,079	2,452	2,370	2	10.1
Jamaica	142	167	133		
Japan	144	168	169		
Nigeria	17,000	16,000	16,000	1	68.2
Papua N. Guinea	164	176	177		
Philippines	21	25	26		
Sudan	114	120	125		
Togo	498	378	400	5	1.7
Tonga	34	35	35		
Venezuela	33	42	45		
Zaire	186	264	265	6	1.1

■ *Mineral and manufactured products*

Aluminium
thousand tonnes

	1948	1970	1980	1988	1989	Rank	%
World	1,608	12,427	19,931	22,993	23,602		
Argentina	0	0	140	164	169		
Australia	0	224	342	1,149	1,318	5	5.6
Austria	13.4	97.2	109	125	127		
Bahrain	0	0	126	183	187		
Brazil	0	62.1	311	941	946	7	4
Cameroon	0	52.4	43.1	80	87		
Canada	333	994	1,133	1,621	1,654	3	7
China	.	135	350	713	744	9	3.2
Czechoslovakia	0	40	38.3	67	69		
Egypt	0	0	120	181	180		
Finland	...	...	9	30	30		
France	90	469	602	552	568		
Germany	110	648	1,196	1,336	1,333	4	5.6
Ghana	0	113	188	164	169		
Greece	0	87.5	146	155	154		
Hungary	9.1	66	73.5	75	75		
Iceland	0	38.7	74.8	83	89		
India	3.5	161	185	335	423		
Indonesia	0	0	0	185	197		
Iran	0	0	17.9	57	65		
Italy	45.8	388	537	604	610	10	2.6
Japan	9.9	1,050	1,880	1,105	1,135	6	4.8
Korea, S.	0	16.8	24.9	41	72		
Mexico	0	41.5	59.7	387	404		
Netherlands	0	82	312	387	404		
New Zealand	0	0	158	270	263		
Norway	31	527	671	836	867	8	3.7
Poland	0	98.8	95.1	48	48		
Romania	0	101	241	265	269		
South Africa	0	0	114	199	195		
Spain	1.0	147	325	379	431		
Surinam	0	54.9	54.9	10	28		
Sweden	3.3	86.2	106	131	130		
Switzerland	19	106	106	100	103		
Taiwan	2.5	27	82.7	68	67		
Turkey	0	0	33.6	57	62		
UAE	0	0	38	163	168		
UK	112	289	565	535	552		
USA	825	4,544	6,231	6,066	6,084	1	25.8
USSR	140	1,700	2,420	2,440	2,380	2	10.1
Venezuela	0	22.4	338	471	576		
Yugoslavia	1.9	47.7	185	394	386		

Includes recovery from scrap in western countries

Aluminium ore (Bauxite)
thousand tonnes

	1948	1970	1980	1988	1989	Rank	%
World	7,810	52,650	93,300	101,070	106,560		
Australia	6	3,294	27,179	36,370	38,583	1	36.2
Brazil	15	400	4,632	7,728	7,894	4	7.4
China	...	500	1,700	3,500	3,650	7	3.4
Dominican Rep.	0	0	511	168	165		
France	804	2,992	1,925	978	720		
Ghana	133	342	225	285	348		
Greece	44	2,283	3,259	2,533	2,576	10	2.4
Guinea	0	2,642	13,911	16,800	17,500	2	16.4
Guyana	1,996	4,103	3,052	1,774	1,340		
Haiti	...	...	461	0	0		
Hungary	478	2,022	2,950	2,906	2,352		
India	21	1,370	1,785	4,013	4,345	6	4.1
Indonesia	0	1,229	1,249	513	862		
Iran	0	0	0	93	100		
Italy	153	225	23	17	12		
Jamaica	...	12,106	11,978	7,409	9,395	3	8.8
Malaysia	0	1,139	920	361	355		
Romania	0	792	450	435	345		
Sierra Leone	0	443	674	1,403	1,548		
Surinam	1,983	6,011	4,893	3,434	3,530	8	3.3
Turkey	0	51	547	269	562		
USA	1,752	2,562	1,559	588	670		
USSR	...	4,300	6,400	5,900	5,750	5	5.4
Venezuela	0	0	0	550	702		
Yugoslavia	144	2,099	3,138	3,034	3,252	9	3.1

Antimony
thousand tonnes

metal content

	1948	1970	1980	1988	1989	Rank	%
World	48.2	68.6	63.3	64.1	62.9		
Australia	0.2	1.0	1.4	1.4	1.4	8	2.2
Austria	0.3	0.6	0.7	0.3	0.4		
Bolivia	12.3	11.6	15.5	9.9	8.5	2	13.5
Burma	0.1	0.1	0.4	—	—		
Canada	0.1	0.3	2.4	3.4	2.9	5	4.6
China	3.3	12	10	28	29	1	46.1
Czechoslovakia	4.1	0.8	0.5	0.5	0.3		
Guatemala	—	0.3	0.6	1.4	1.2	9	1.9
Italy	0.6	0.8	0.7	0.1	—		
Mexico	7.4	4.5	2.2	2.2	1.9	7	3
Morocco	0.8	2.0	0.5	0.2	0.1		
Peru	1.6	1.2	0.7	0.4	0.3		
Romania	—	—	—	0.6	0.5		
South Africa	4.0	17.3	13	6.3	5.2	4	8.3
Thailand	0.1	2.5	3.6	0.7	0.7		
Turkey	0.5	3.4	1	1.7	1	10	1.6
USA	5.9	1.0	0.3	0.2	2.5	6	4
USSR	3.0	6.7	7	6	5.8	3	9.2
Yugoslavia	2.3	2.0	1.7	0.7	0.8		
Zimbabwe	—	—	0.2	0.1	0.1		

Asbestos
thousand tonnes

	1948	1970	1980	1989	1990	Rank	%
World	870	4,682	4,700	4,300	4,100		
Australia	1.3	1	92.4	—	—		
Brazil	1.5	340	170	178	200	3	4.9
Canada	650	1,508	1,323	700	650	2	15.9
China	...	170	132	150	150	5	3.7
Cyprus	8.1	29	35.5	—	—		
Egypt	1.6	—	0.3	...	...		
Finland	10.8	14.0	—	—	—		
Greece	–	–	–	73	70	7	1.7
India	0.1	10	33.7	35	35	8	0.8
Italy	13	119	158	95	...		
Japan	4.8	21	3.9	–	–		
Korea, S.	–	2	9.9				
South Africa	41.5	287	268	155	145	6	3.5
Swaziland	29.4	33	32.8	27	25	9	0.6
Turkey	0.2	3	8.9	25	20	10	0.5
USA	33.6	114	80.1	17	20		
USSR	...	1,066	2,070	2,600	2,600	1	63.4
Yugoslavia	2.3	12	10.5	9	...		
Zimbabwe	62.3	80	251	187	180	4	4.4

Cars
thousands

	1948	1970	1980	1989	1990	Rank	%
World	4,640	22,550	28,999	34,000	...		
Argentina	0	169	218	114	...		
Australia	29	391	316	332	361		
Austria	...	1.2	8	7	14		
Belgium	...	734	891	1,120	...	9	3.4
Brazil	0	255	652	313	268		
Bulgaria	0	8	0	0	0		
Canada	167	923	847	984	941	10	2.9
Chile	0	21	25	4	4		
Colombia	0	8	32	40	...		
Czechoslovakia	18	143	186	188	191		
Egypt	0	4.0	16	10	8		
France	100	2,458	3,487	3,414	3,215	4	10
Germany	33	3,655	3,688	4,762	4,618	3	14
Hungary	...	10	0	0	0		
India	0	45	47	178	180		
Indonesia	...	...	43	33	57		
Iran	0	33	...	...	...		
Ireland	14	47	45	...	...		
Israel	0	4	...	...	...		
Italy	44	1,720	1,445	1,970	...	5	5.8
Japan	0.4	3,179	7,038	8,370	9,948	1	24.6
Korea, S.	0	13	58	846	958		
Malaysia	0	21	79	95	103		
Mexico	...	137	312	455	614		
Morocco	0	20	15	48	...		
Netherlands	2	85	80	134	...		
New Zealand	10	55	73	60	60		
Pakistan	0	0	2	20	26		
Peru	0	10	0	1	1		
Philippines	...	8	2	27	...		
Poland	0	65	352	286	266		
Portugal	0	55	22	75	...		
Romania	0	23	89	120	...		
South Africa	...	195	268	238	...		
Spain	0.2	455	1,048	1,696	...	6	5
Sweden		272	257	400	...		
Switzerland	...	19	0	0	0		
Taiwan	0	9	133	275	...		
Thailand	0	0	24	70	...		
Trinidad & T.	0	0	10	2	...		
Tunisia	0	0.5	1	1	...		
Turkey	0	5	52	116	...		
UK	335	1,641	924	1,300	1,296	7	3.8
USA	3,909	8,505	6,376	6,808	6,052	2	20
USSR	20	344	1,327	1,217	1,259	8	3.6
Venezuela	0	46	...	...	...		
Yugoslavia	0	109	187	312	292		

Rank for 1989

Cement
thousand tonnes

	1948	1970	1980	1989	1990	Rank	%
World	103,000	582,982	869,000	1,110,000	1,140,000		
Albania	...	340	1,000	1,000	1,000		
Algeria	130	928	4,156	8,000	8,000		
Argentina	1,265	4,770	7,289	6,000	6,000		
Australia	1,029	5,100	5,201	6,888	7,068		
Austria	721	4,806	5,455	4,752	4,908		
Belgium	3,451	6,729	7,482	6,720	6,924		
Brazil	1,265	9,002	25,880	26,508	25,848		
Bulgaria	378	3,668	5,359	4,968	4,680		
Canada	2,221	7,283	10,340	12,588	11,808		
Chile	540	1,349	1,583	1,800	1,800		
China	660	25,720	79,857	203,844	204,000	1	17.8
Colombia	364	2,757	4,356	6,648	6,360		
Congo	127	92	34	50	50		
Cuba	285	742	2,831	3,756	3,600		
Cyprus	...	266	1,233	1,044	960		
Czechoslovakia	1,658	7,402	10,546	10,884	10,368		
Denmark	769	2,604	1,917	2,004	1,656		
Dominican Rep.	43	493	928	1,300	1,300		
Ecuador	40	458	1,389	3,000	3,000		
Egypt	768	3,684	3,638	10,056	9,900		
Finland	563	1,838	1,787	1,608	1,668		
France	5,830	29,009	29,104	25,992	26,508	10	2.3
Germany	...	46,309	46,991	40,764	40,000	6	3.5
Greece	288	4,848	11,591	12,528	13,944		
Hong Kong	53	430	1,489	2,136	2,160		
Hungary	552	2,771	4,660	3,852	3,936		
India	1,578	13,956	17,803	44,568	43,200	5	3.8
Indonesia	38	515	5,289	15,660	15,972		
Iran	53	2,575	8,114	15,000	15,000		
Iraq	11	1,542	5,500	10,000	10,000		
Ireland	398	868	1,812	1,320	1,320		
Israel	241	1,384	2,092	2,292	2,868		
Italy	3,211	33,076	41,862	39,708	39,600	7	3.5
Japan	1,859	57,189	87,957	80,316	84,444	3	7.4
Jordan	...	378	913	1,932	1,800		
Kenya	18	792	1,280	1,500	1,500		
Korea, N.	292	4,000	8,000	10,000	10,000		
Korea, S.	17	5,782	15,612	30,120	33,912	8	3
Kuwait	...	...	1,307	1,000	1,000		
Lebanon	209	1,339	2,200	900	900		
Libya	...	98	1,787	3,000	3,000		
Malaysia	0	1,030	2,349	4,800	5,880		
Mexico	833	7,267	16,398	23,760	24,504		

Cement continued

	1948	1970	1980	1989	1990	Rank	%
Morocco	262	1,421	3,561	4,644	4,500		
Netherlands	589	3,830	3,745	3,540	3,708		
New Zealand	238	829	720	720	804		
Nigeria	...	664	1,714	3,000	3,000		
Norway	526	2,680	2,206	1,380	1,260		
Pakistan	329	2,656	3,343	6,936	7,488		
Peru	282	1,144	2,758	3,300	2,184		
Philippines	120	2,447	4,516	3,624	6,348		
Poland	1,824	12,180	18,428	17,112	12,564		
Portugal	498	2,347	5,914	6,000	6,000		
Puerto Rico	413	1,509	1,282	1,140	1,332		
Romania	657	8,127	14,607	13,260	13,200		
Saudi Arabia	...	667	2,888	13,000	13,000		
Singapore	...	...	1,952	1,704	2,000		
South Africa	1,308	5,752	7,125	6,936	7,000		
Spain	1,803	16,702	28,752	27,372	28,092	9	2.5
Sweden	1,486	4,061	2,445	2,500	2,500		
Switzerland	1,022	4,797	4,252	5,436	5,500		
Syria	54	964	1,995	3,972	4,000		
Taiwan	236	4,541	14,062	17,000	17,000		
Thailand	83	2,627	5,359	15,024	15,000		
Tunisia	162	547	1,780	3,780	4,140		
Turkey	336	6,373	14,802	23,808	24,000		
UK	8,657	17,171	14,805	16,548	15,600		
USA	35,210	67,682	69,589	71,388	70,944	4	6.2
USSR	6,455	95,248	125,049	140,436	137,328	2	12
Venezuela	215	2,318	4,842	5,268	6,072		
Vietnam	98	790	850	2,000	2,000		
Yugoslavia	1,169	4,399	9,716	8,556	7,956		

Chromium ore
thousand tonnes

	1948	1970	1980	1989	1990	Rank	%
World	650	2,730	9,749	12,500	12,000		
Albania	–	200	760	700	700	4	5.8
Brazil	0.8	10.5	313	450	400	8	3.3
Cuba	16.8	...	28.5	50	55		
Cyprus	5.2	16	16.6	–	–		
Finland	0	50.5	186	500	450	7	3.8
Greece	0.6	10.2	33.6	75	65		
India	11.2	135	321	800	750	3	6.3
Iran	...	145	80	60	60		
Japan	3.9	10.9	13.6	10	10		
Madagascar	...	43.4	147	60	70	10	0.1
New Caledonia	36.8	...	2.2	60	60		
Pakistan	8.9	14	3.1	5	5		
Philippines	88.7	197	496	180	170	9	1.4
South Africa	186	643	3,414	4,200	4,200	1	35
Sudan	...	14	26	10	10		
Swaziland	...	13.9	25.6	...	...		
Turkey	140	295	383	800	600	5	5
USSR	...	735	2,450	3,800	3,800	2	31.7
Vietnam	...	...	15.5	5	5		
Yugoslavia	17	14.8	–	–	–		
Zimbabwe	125	181	553	600	600	6	5

Coal – bituminous
million tonnes

	1948	1970	1980	1989	1990	Rank	%
World	1,412	2,131	2,740	3,474	3,562		
Australia	15.0	44.3	78.4	149	162	6	4.5
Belgium	26.7	11.4	5.3	2.9	2.3		
Brazil	2.0	2.4	5.1	6.5	6.5		
Canada	15.3	11.6	30.6	38.8	37.7		
Chile	2.0	1.4	1.0	1.9	1.9		
China	32.4	360	590	1,021	1,051	1	29.5
Colombia	...	...	10	19	20		
Czechoslovakia	16.7	28.1	28.2	25.1	22.9		
France	43.3	37.4	18.1	12.2	11.3		
Germany	103	112	87.1	77.4	76.4	9	2.1
Hungary	1.2	4.2	3.1	2.2	1.7		
India	30.6	73.7	109	199	198	4	5.6
Japan	33.7	39.7	18.0	10.2	8.6		
Korea, N.	2.1	24.0	18.5	55	55	10	1.5
Korea, S.	0.8	12.4	18.6	18.9	20.8		
Mexico	...	...	5	10	10		
Netherlands	11.0	4.3	0	0	0		

	1948	1970	1980	1989	1990	Rank	%
Poland	70.3	140	193	170	148	7	4.2
Romania	2.0	6.4	9.7	8.2	3.8		
South Africa	24.0	54.6	116	174	180	5	5.1
Spain	10.4	10.8	12.7	14.5	15.5		
Taiwan	1.6	4.5	2.6	1.2	1		
Turkey	2.7	5.0	6.6	6.3	5.6		
UK	213	145	130	100	92.9	8	2.6
USA	593	550	715	810	861	2	24.2
USSR	150	433	493	576	472	3	13.3
Vietnam	0.4	3.0	5.3	5.4	5.5		
Zimbabwe	1.7	3.4	3.1	5.1	5.2		

Coal – lignite and brown
million tonnes

	1948	1970	1980	1988	1989	Rank	%
World	316	793	996	1,257	1,176		
Australia	6.8	24.3	32.9	48.3	47.7	9	4.1
Austria	3.3	3.7	2.9	2.2	2.5		
Bulgaria	4.1	28.9	29.9	34.1	31.5		
Canada	1.4	3.5	6.0	31.7	30.7		
China	...	...	9.3	81.2	81.5	5	6.9
Czechoslovakia	23.6	81.8	95.7	93.9	85.5	3	7.3
Denmark	2.4	0.1	0	0	0		
France	1.8	2.8	2.6	2.2	2.1		
Germany	177	369	388	411	357	1	30.4
Greece	0.1	7.7	23.5	51.9	51.7	8	4.4
Hungary	9.4	23.7	22.6	17.9	15.8		
India	0	3.5	4.5	11.2	11.2		
Italy	0.9	1.4	1.9	1.5	1.5		
Japan	2.6	0.2	0	0	0		
New Zealand	1.9	1.9	0.2	0.2	0.2		
Poland	5.0	32.8	36.9	71.8	67.6	6	5.7
Romania	0.9	14.1	28.1	52.5	33.6	10	2.9
Spain	1.4	2.8	15.4	21.5	21.3		
Thailand	0	0.4	1.4	9	10		
Turkey	0.7	4.5	17.0	36	32.4		
USA	2.8	5.4	43.4	78.6	82.6	4	7
USSR	58.2	145	163	164	157	2	13.4
Yugoslavia	9.7	27.8	46.6	67.5	64.8	7	5.5

Cobalt ore
tonnes

	1948	1970	1980	1987	1988	Rank	%
World	6,400	24,500	33,700	45,700	43,900		
Albania	500	500	600	600	600	9	1.4
Australia	10	20	3,704	1,900	2,000	6	4.6
Botswana	0	0	226	200	300	10	0.7
Brazil	...	...	...	100	100		
Canada	701	2,372	2,118	2,500	2,500	4	5.7
China	...	...	200	250	250		
Cuba	0	1,542	1,500	1,500	2,000	7	4.6
Finland	1,128	1,650	1,360	200	200		
Morocco	267	595	838	–	200		
New Caledonia	0	0	2,598	1,700	2,100	5	4.8
Norway	0	735	0	–	–		
Philippines	0	0	1,331	–	–		
South Africa	...	...	750	750	750	8	1.7
USA	312	0	0	0	0		
USSR	...	1,500	2,150	2,800	2,900	3	6.6
Zaïre	3,600	13,598	14,482	29,000	25,400	1	57.9
Zambia	395	2,052	3,310	5,900	6,700	2	15.3
Zimbabwe	...	...	115	–	–		

Commercial vehicles
thousands

	1948	1970	1980	1989	1990	Rank	%
World	2,030	6,780	9,870	14,000	...		
Algeria	0	3	7	6	...		
Argentina	...	50	61	19	...		
Australia	19	88	46	28	23		
Austria	1	7	7	5	5		
Belgium	...	61	43	76			
Brazil	0	161	516	725	672	5	5.2
Bulgaria	0	3	...	...	...		
Canada	97	236	528	949	808	3	6.8
Chile	0	4	4	6	...		

	1948	1970	1980	1989	1990	Rank	%
Colombia	...	10	11	14	...		
Cuba	...	2					
Czechoslovakia	17	28	88	91	58		
Denmark	2	1	...	3	...		
Egypt	0	2	4	5	4		
France	98	292	505	509	553	6	3.6
Germany	31	345	420	247	349	9	1.8
Greece	0	3	...	...	...		
Hungary	...	10	14	13	8		
India	0	41	66	115	...		
Indonesia	...	...	131	141	207		
Iran	0	14	0	0	0		
Ireland	3	4	...	...	...		
Israel	...	6	2	1	...		
Italy	15	135	167	246	...	10	1.8
Japan	37	2,125	4,006	4,010	3,550	2	28.6
Korea, S.	0	15	64	244	322		
Malaysia	0	8	22	39	50		
Mexico	...	53	132	190	124		
Morocco	0	5	2	10	...		
Netherlands	2	16	17	28	...		
New Zealand	6	17	17	13	...		
Nigeria	0	7	...	...	...		
Pakistan	0	0	12	16	16		
Peru	0	4	...	3	3		
Philippines	0	6	18	...	...		
Poland	...	53	60	53	43		
Portugal	0	13	69	57	...		
Romania	0	52	41	5	...		
South Africa	...	76	108	120	...		
Spain	0.4	77	146	310	...	8	2.2
Sweden	...	32	60	70	...		
Switzerland	...	19	...	0	...		
Thailand	0	0	51	155	...		
Trinidad & T.	0	0	2	1	...		
Tunisia	0	1	8	1	...		
Turkey	0	12	19	31	...		
UK	173	...	389	338	274	7	2.4
USA	1,376	1,692	1,667	4,056	3,720	1	39
USSR	177	815	874	896	860	4	6.4
Venezuela	0	13	...	...	...		
Yugoslavia	0.5	46	64	71	49		

Rank for 1989

Copper
thousand tonnes

	1948	1970	1980	1988	1989	Rank	%
World	3,875	7,578	7,300	15,074	15,328		
Albania	–	4	9	16	15		
Argentina	...	...	10	12.5	11		
Australia	...	146	182	263	296		
Austria	...	22	43.3	38.4	46.3		
Belgium-Lux.	132	338	424	364	334	10	2.2
Brazil	...	18.6	38.9	168	173		
Bulgaria	...	38.3	63	58	55.8		
Canada	201	493	505	563	547	6	3.6
Chile	374	465	811	1,013	1,071	4	7
China	...	130	295	460	470	7	3.1
Czechoslovakia	...	16.7	25.6	27.1	26.9		
Finland	20.7	34	40.5	53.9	55.7		
France	...	33.6	46.5	197	193		
Germany	84.9	456	425	697	749	5	4.9
Hungary	...	12.2	27.8	15.3	13.1		
India	6	9.3	23.2	40.1	41.8		
Iran	...	6	1.0	32	40		
Italy	13.5	15.5	12.2	280	291		
Japan	54.3	705	1,014	1,653	1,626	2	10.6
Korea, N.	1.9	13	23	40	40		
Korea, S.	0.5	5.1	72.9	216	223		
Mexico	9.4	53.7	103	137	144		
Norway	4.5	25.8	25.8	51.7	45.2		
Oman	0	0	0	16.3	15		
Peru	0.2	36.2	224	175	224		
Philippines	0	0	0	132	132		
Poland	1.4	72.2	357	401	390	9	2.5
Romania	...	16	65	42	48		
South Africa	10.7	75.3	148	139	144		
Spain	14.3	79.7	154	197	205		
Sweden	24.1	51.2	55.7	101	115		
Taiwan	0.4	3.8	19.5	68.3	74.5		
Turkey	...	14.1	18.8	68.4	75.6		
UK	196	206	161	256	248		

Copper   *continued*

	1948	1970	1980	1988	1989	Rank	%
USA	1,241	2,035	1,686	2,803	2,868	1	18.7
USSR	220	1,075	1,300	1,380	1,355	3	8.8
Yugoslavia	14.4	89.3	131	199	203		
Zaïre	83	190	144	288	255		
Zambia	62.4	581	607	448	466	8	3
Zimbabwe	...	20.7	7.1	27.5	24		

Includes recovery from scrap in western countries

Copper ore
thousand tonnes

metal content

	1948	1970	1980	1988	1989	Rank	%
World	2,120	6,320	7,900	8,811	9,129		
Albania	–	7	12	17	17		
Australia	13	142	244	249	296	10	3.2
Bolivia	7	9	2	–	–		
Botswana	–	–	16	24	22		
Brazil	–	4	...	44	47		
Bulgaria	1	43	62	50	50		
Canada	223	610	716	777	732	4	8
Chile	445	711	1,068	1,451	1,609	1	17.6
China	...	100	165	370	380	8	4.2
Cuba	15	7	3	3	3		
Finland	20	34	53	20	15		
Germany	6	10	12	7	5		
India	8	10	27	52	53		
Indonesia	–	–	59	126	149		
Iran	–	–	1	51	68		
Japan	26	120	53	15	13		
Korea, N.	2	13	12	12	12		
Malaysia	–	–	27	22	24		
Mexico	59	61	175	242	291		
Mongolia	–	–	44	...	...		
Morocco	–	3	8	130	135		
Namibia	8	23	39	40	32		
Norway	15	20	29	17	20		
Oman	–	–	–	18	17		
Papua New G.	...	...	147	214	205		
Peru	18	206	367	298	364	9	4
Philippines	2	160	305	218	193		
Poland	–	83	343	441	385	7	4.2
Portugal	...	...	1	5	103		
Romania	–	–	28	40	43		
South Africa	30	148	201	192	197		
Spain	4	10	48	18	28		
Sweden	15	26	43	75	70		
Turkey	11	22	21	38	45		
USA	757	1,560	1,181	1,419	1,498	2	17.3
USSR	...	925	1,130	990	950	3	10.4
Yugoslavia	41	91	115	104	119		
Zaïre	156	387	462	465	441	6	4.8
Zambia	217	684	610	476	500	5	5.5
Zimbabwe	...	...	27	16	16		

Diamonds
thousand carats

	1948	1970	1980	1989	1990	Rank	%
World	10,270	47,634	42,000	99,000	108,000		
Angola	796	2,396	1,500	1,200	1,300	6	1.2
Australia	...	...	...	37,000	36,000	1	33.3
Botswana	...	464	5,146	15,200	17,300	3	16
Brazil	...	47	667	500	500	10	0.5
Central Afr. Rep.	119	494	342	600	500		
China	...	...	1,000	1,000	1,000	7	0.9
Ghana	786	2,550	1,149	200	200		
Guinea	80	74	38	200	100		
Guyana	37	61	10	5	5		
India	2	20	14	15	15		
Indonesia	...	20	15	30	30		
Liberia	...	826	298	300	300		
Namibia	201	1,865	1,560	900	800	8	0.7
Sierra Leone	466	2,050	592	600	700	9	0.6
South Africa	1,382	8,112	8,520	9,000	8,500	5	7.9
Tanzania	150	708	270	60	60		
USSR	...	14,057	10,850	12,000	15,000	4	13.9
Venezuela	76	509	666	160	160		
Zaïre	5,825	14,057	10,334	20,000	24,000	2	22.2

This table does not show synthetic diamond production

Fertilizers – nitrogenous
thousand tonnes

	1970	1980	1988	1989	Rank	%
World	37,825	62,766	82,290	85,151		
Africa	552	1,066	1,902	2,149		
Asia	7,407	18,614	28,244	30,209		
Europe	13,101	17,414	18,153	18,011		
North America	9,713	14,524	16,731	17,483		
Oceania	182	214	288	300		
South America	336	778	1,452	1,395		
Albania	36	73	66	69		
Algeria	50	24	117	106		
Argentina	38	30	43	45		
Australia	182	214	220	230		
Austria	230	300	250	222		
Bangladesh	92	161	593	667		
Belgium	641	743	760	700		
Brazil	78	384	746	705		
Bulgaria	523	730	808	824		
Burma	...	...	150	99		
Canada	772	1,755	2,466	2,762	5	3.2
Chile	108	101	125	130		
China	2,245	10,286	13,685	13,954	2	16.4
Colombia	77	42	91	98		
Cuba	2	112	157	142		
Czechoslovakia	410	705	625	585		
Denmark	77	147	176	175		
Egypt	152	400	657	677		
Finland	243	266	281	293		
France	1,476	1,640	1,435	1,450		
Germany	1,899	2,379	2,350	2,301	6	2.7
Greece	240	311	416	419		
Hungary	374	651	692	591		
India	1,054	2,164	5,466	6,712	4	7.9
Indonesia	60	958	1,979	2,033	7	2.4
Iran	143	71	87	114		
Iraq	26	355	77	330		
Ireland	85	190	286	268		
Israel	24	62	76	73		
Italy	1,046	1,318	1,279	1,298		
Japan	2,199	1,202	986	977		
Jordan	...	...	109	111		
Korea, N.	230	553	650	660		
Korea, S.	418	688	594	679		
Kuwait	270	218	400	385		
Libya	0	0	302	144		
Malaysia	...	...	282	247		
Mexico	356	739	...	...		
Morocco	...	...	83	233		
Netherlands	1,217	1,624	1,742	1,757	9	2.1
Nigeria	...	...	73	243		
Norway	396	428	384	426		
Pakistan	274	580	1,097	1,112		
Peru	22	74	86	66		
Philippines	55	34	120	128		
Poland	1,147	1,290	1,543	1,622	10	1.9
Portugal	150	172	180	170		
Qatar	...	286	320	359		
Romania	874	1,707	1,916	1,920	8	2.3
Saudi Arabia	69	152	449	417		
South Africa	248	444	380	440		
Spain	724	960	975	955		
Sweden	169	180	164	168		
Syria	16	13	81	81		
Trinidad & T.	114	15	246	248		
Tunisia	0	47	173	197		
Turkey	145	672	730	725		
UAE	...	...	244	265		
UK	751	1,167	1,105	1,100		
USA	8,433	11,825	12,162	12,691	3	14.9
USSR	6,533	10,155	15,539	15,604	1	18.3
Venezuela	10	145	379	344		
Yugoslavia	360	391	615	630		
Zimbabwe	58	72	705	723		

Fertilizers – phosphate
thousand tonnes

	1970	1980	1988	1989	Rank	%
World	23,673	34,506	39,173	41,532		
Africa	925	1,286	1,924	2,479		
Asia	2,724	5,649	8,556	9,749		
Europe	8,776	8,208	7,180	7,233		
North America	6,762	10,436	10,135	10,380		
Oceania	1,321	1,210	933	1,040		
South America	382	1,693	1,605	1,495		
Algeria	38	31	55	71		
Australia	900	829	703	810	9	2
Austria	139	105	105	110		
Bangladesh	...	...	54	66		
Belgium	780	590	368	325		
Brazil	278	1,582	1,472	1,395	5	3.4
Bulgaria	126	217	128	179		
Canada	738	724	436	464		
China	1,031	2,368	3,341	3,766	3	9.1
Christmas I.	–	35	20	10		
Colombia	40	46	31	21		
Czechoslovakia	334	361	307	313		
Denmark	97	135	150	130		
Egypt	116	106	189	202		
Finland	213	159	231	216		
France	1,611	1,351	955	940	8	2.3
Germany	1,395	1,057	653	639		
Greece	153	170	204	257		
Hungary	161	216	244	220		
India	330	859	1,703	2,290	4	5.5
Indonesia	–	220	554	552		
Iran	62	30	–	–		
Iraq	–	–	220	414		
Ireland	136	44	–	–		
Israel	–	41	160	145		
Italy	500	424	450	412		
Japan	729	648	531	490		
Jordan	–	13	278	283		
Korea, N.	105	127	137	137		
Korea, S.	163	494	515	488		
Lebanon	42	103	20	20		
Malaysia	26	27	...	...		
Mexico	217	201	399	375		
Morocco	154	177	537	970	6	2.3
Netherlands	351	328	334	367		
New Zealand	421	346	210	220		
Norway	130	143	168	196		
Pakistan	8	58	96	102		
Philippines	42	37	193	199		
Poland	763	843	942	962	7	2.3
Portugal	89	102	80	75		
Romania	313	687	690	700		
Senegal	...	...	35	35		
South Africa	340	380	33	38		
Spain	576	497	451	441		
Sweden	181	132	123	110		
Syria	1	–	78	46		
Tunisia	194	408	726	747	10	1.8
Turkey	125	564	610	634		
UK	444	326	200	248		
USA	5,795	9,500	9,276	9,520	1	22.9
USSR	2,784	6,023	8,840	9,155	2	22
Venezuela	10	23	70	50		
Vietnam	42	23	62	72		
Yugoslavia	231	302	365	365		
Zimbabwe	38	42	44	41		

Fertilizers – potash
thousand tonnes

	1970	1980	1988	1989	Rank	%
World	20,185	27,445	30,819	31,151		
Africa	284	...	...	...		
Asia	922	832	2,012	2,052		
Europe	7,269	9,150	8,750	8,541		
North America	6,261	9,389	9,132	9,202		
Canada	3,820	7,337	7,841	8,089	2	26
China	300	20	40	54		
Congo	284	...	...	...		
France	1,664	1,933	1,564	1,411	4	4.5
Germany	4,956	6,123	5,830	5,778	3	18.5
Israel	622	812	1,252	1,218	5	3.9
Italy	131	95	128	159	10	0.5
Jordan	...	...	720	780	7	2.5
Spain	514	691	754	748	8	2.4
Taiwan	62	88	...	...		
UK	...	307	474	445	9	1.4
USA	2,441	2,052	1,291	1,113	6	3.6
USSR	5,433	8,064	10,888	11,300	1	36.3

Gases, other than natural gas†
*thousand terajoules **

	1970	1980	1988	1989	Rank	%
World	11,287	14,717	16,174	16,313		
Africa	73	212	442	456		
Asia	...	2,649	3,261	3,435		
Europe	...	3,665	3,454	3,508		
North America	...	5,137	5,690	5,650		
Oceania	...	269	264	264		
South America	...	424	691	698		
Algeria	5.8	32.0	219	224		
Argentina	45.6	60.8	90.3	90.1		
Australia	164	267	252	253		
Austria	48.7	55.9	48.4	46.5		
Belgium	159	127	139	134		
Brazil	52.7	184	347	348	9	2.1
Bulgaria	17.4	20.9	28.6	29.6		
Canada	304	602	685	690	5	4.2
Chile	39.3	38.3	36	36.4		
China	...	...	107	109		
Colombia	22.1	22.5	41	41.4		
Czechoslovakia	182	225	174	177		
Denmark	14.6	9.9	19.9	20		
Egypt	...	10.1	35.2	39.5		
Finland	10.4	19.5	32.6	30.8		
France	658	508	314	341	10	2.1
Germany	1,012	1,019	768	747	4	4.6
Greece	7.5	20.9	30.8	33.3		
Hungary	43.8	54.2	45.3	41.4		
India	38.6	48.8	106	115		
Indonesia	8.3	10.4	64.2	87.1		
Iran	25.0	50.6	41	49.9		
Iraq	1.0	11.8	46	65		
Ireland	6.4	6.2	4.2	4.4		
Israel	4.2	8.2	9	9.2		
Italy	318	342	313	340		
Japan	...	1,832	1,798	1,843	3	11.3
Korea, S.	2.1	19.3	57.2	68.7		
Kuwait	50.1	109	128	138		
Libya	0	18.3	16	18.2		
Luxembourg	55.6	22.3	15.4	16.2		
Malaysia	2.6	5.4	27.4	29.4		
Mexico	97.4	265	347	357	8	2.2
Netherlands	76.5	223	289	306		
Neths. Antilles	3.9	5.9	3	3		
New Zealand	2.9	1.3	12.3	11.3		
Norway	14.9	*20.9	66.4	64.2		
Philippines	5.9	9.0	9.5	9.1		
Poland	210	241	227	221		
Portugal	11.8	17.8	28.1	33.5		
Puerto Rico	26.3	29.3	2.5	2.5		
Qatar	0	5.2	29.1	59		
Romania	96.2	143	203	201		
Saudi Arabia	66.5	380	460	474	6	2.9
Singapore	1.9	9.5	18.4	17.2		
South Africa	58.9	133	144	145		
Spain	159	206	210	220		
Sweden	31.1	30.4	28.9	29.6		
Switzerland	9.9	11.7	13.5	10.4		
Trinidad & T.	36.8	20.9	9	8.7		
Turkey	42.6	58.2	109	106		
UAE	0	35.6	110	117		
UK	625	274	380	384	7	2.4
USA	4,140	4,194	4,626	4,572	1	28
USSR	1,826	2,360	2,373	2,299	2	14.1
Venezuela	81.1	98.1	150	156		
Yugoslavia	39.7	65.8	76.3	79		

* A terajoule is a measure of energy equivalent to 23.46 tonnes of oil or 34.12 tonnes of coal
† These include liquefied petroleum gas, refinery gas, gasworks gas, coke oven gas and blast furnace gas

Gold
tonnes

metal content

	1948	1970	1980	1889	1990	Rank	%
World	698	1,483	1,220	1,700	1,750		
Australia	27.5	19.4	17.0	205.0	241.0	4	13.8
Brazil	4.1	5.6	13.8	100.0	75.0	7	4.3
Canada	110.0	74.9	50.6	160.0	165.0	5	9.4
Chile	5.1	2.0	6.8	27.1	31.6		
China	...	...	45.0	80.0	80.0	6	4.6
Colombia	10.4	6.3	15.9	31.7	32.5	10	1.9
Dominican Rep.	0.0	0.0	11.5	6.5	6.5		
Ecuador	2.5	0.3	0.1	9.5	9.5		
Fiji	2.9	3.2	0.8	4.5	4.5		
Ghana	20.9	21.9	11.0	15.3	17.3		
India	5.6	3.2	2.5	1.9	1.9		
Japan	3.1	3.0	3.2	7.5	7.5		
Korea, N.	...	...	5.0	5.0	5.0		
Korea, S.	0.1	1.6	1.3	12.0	12.0		
Mexico	11.4	6.2	6.1	9.2	9.2		
New Zealand	2.9	0.4	0.2	2.5	2.4		
Nicaragua	6.9	3.5	1.9	0.8	0.8		
Papua New G.	2.7	0.7	14.7	33.8	33.6	9	1.9
Peru	3.5	3.0	3.9	5.2	5.2		
Philippines	6.5	18.7	20.0	38.0	37.5	8	2.1
South Africa	360.0	1,002.0	673.0	610.0	610.0	1	34.9
Sweden	2.2	1.4	2.0	3.9	3.9		
Taiwan	0.6	...	0.4	0.2	0.2		
Tanzania	1.8	0.2	0.3	0.2	0.2		
USA	63.0	56.4	30.2	260.0	295.0	2	16.9
USSR	...	195.0	260.0	280.0	280.0	3	16
Yugoslavia	0.8	1.7	3.3	4.6	4.6		
Zaire	9.3	5.5	2.2	3.7	3.7		
Zimbabwe	16.0	15.6	11.4	15.0	15.0		

Iron and ferro-alloys
thousand tonnes

	1948	1970	1980	1989	1990	Rank	%
World	113,000	436,416	520,000	506,000	506,000		
Algeria	...	409	669	1,500	1,500		
Argentina	17	847	1,095	2,100	2,100		
Australia	1,149	5,769	7,276	5,880	6,192		
Austria	613	2,970	3,493	3,828	3,444		
Belgium	3,929	10,950	9,845	8,868	9,420		
Brazil	552	4,334	14,774	24,384	21,084	6	4.2
Bulgaria	...	1,251	1,523	1,488	1,140		
Canada	2,140	8,424	11,185	10,200	11,000		
Chile	14	481	707	684	680		
China	...	17,060	39,014	58,956	64,548	3	12.7
Czechoslovakia	1,645	7,653	9,992	10,056	9,840		
Egypt	...	300	650	140	140		
Finland	90	1,164	2,072	2,316	2,280		
France	6,628	19,575	19,595	15,072	14,412	8	2.8
Germany	5,857	35,985	36,513	35,784	32,724	5	6.5
Hungary	384	1,837	2,227	1,956	1,692		
India	1,487	7,338	8,864	12,204	12,000	10	2.4
Italy	525	8,529	12,411	12,012	12,000		
Japan	836	69,728	88,907	81,360	81,360	2	16.1
Korea, N.	105	2,400	2,446	2,500	2,500		
Korea, S.	...	35	5,686	15,144	15,528	7	3.1
Luxembourg	2,624	4,814	3,568	2,688	2,724		
Mexico	176	2,353	5,330	5,280	5,280		
Netherlands	442	3,594	4,328	5,160	4,956		
Norway	202	1,251	1,991	1,140	900		
Peru	...	86	851	250	250		
Poland	1,134	7,111	11,682	9,492	8,640		
Portugal	...	314	577	372	360		
Romania	186	4,210	9,013	9,252	9,200		
South Africa	651	4,328	8,542	6,516	6,500		
Spain	537	4,278	6,725	5,904	6,000		
Sweden	804	2,842	2,612	2,640	2,736		
Trinidad & Tob.	...	...	200	550	550		
Turkey	100	1,176	2,048	3,576	3,800		
UK	9,453	19,023	6,413	11,832	12,492	9	2.5
USA	56,166	85,303	63,748	50,676	51,000	4	10
USSR	13,742	85,933	102,960	113,928	110,184	1	21.8
Venezuela	...	510	2,450	3,580	3,580		
Yugoslavia	183	1,377	2,668	3,228	2,628		
Zimbabwe	17	250	862	800	800		

Iron ore
million tonnes

metal content

	1948	1970	1980	1989	1990	Rank	%
World	104	417	520	983	962		
Algeria	1.0	1.5	1.9	2.8	2.9		
Angola	–	3.8	–	–	–		
Australia	1.4	28.7	62.1	62.5	70.3	4	7.3
Austria	0.4	1.3	1.0	0.7	0.7		
Brazil	1.1	20.4	41.2	99.2	104.5	2	10.9
Bulgaria	–	0.8	0.5	0.5	0.3		
Canada	1.5	29.2	29.9	24.1	21.7	7	2.3
Chile	1.7	6.9	5.4	5.3	...		
China	...	24.2	32.5	81.1	84.7	3	8.8
Colombia	...	0.5	0.2	0.3	...		
Czechoslovakia	0.4	0.4	0.5	0.5	0.5		
Egypt	...	0.2	0.8	1.2	...		
Finland	...	...	0.6	–	–		
France	7.6	17.8	9.0	2.8	2.6		
Germany	1.8	1.8	0.5	–	–		
India	1.5	19.7	25.8	33.6	34	6	3.5
Italy	0.3	0.4	–	–	–		
Japan	0.3	0.9	0.3	...	...		
Korea, N.	...	4.0	3.7	0.4	0.4		
Liberia	...	15.4	12.5	7.9	8		
Luxembourg	1.0	1.6	0.3	–	–		
Malaysia	...	2.5	0.3	0.3	...		
Mauritania	...	5.9	5.6	6.5	...		
Mexico	0.2	2.6	5.2	3.4	3.9		
Morocco	0.7	0.5	0.1	0.1	...		
Norway	0.1	2.6	2.5	1.5	1.3		
Peru	...	6.1	3.4	2.6	1.9		
Poland	0.2	0.7	0.1	...	...		
Romania	0.1	0.9	0.6	0.1	...		
Sierra Leone	0.6	1.4	0.2	...	...		
South Africa	0.7	5.9	16.4	19.4	19.9	8	2.1
Spain	0.8	3.5	4.4	2.3	...		
Swaziland	...	1.5	–	–	–		
Sweden	8.2	19.8	17.0	14	12.9	9	1.3
Tunisia	0.4	0.4	0.2	0.1	0.1		
Turkey	0.1	1.7	1.4	2.2	2.9		
UK	4.0	3.4	0.3	0.008	0.01		
USA	50.9	53.3	43.9	37.2	34.9	5	3.6
USSR	16.2	106	148	145	142	1	14.8
Venezuela	...	14.1	10.4	11.5	11	10	1.1
Yugoslavia	0.3	1.3	1.6	1.8	1.4		
Zimbabwe	...	0.3	0.9	0.6	0.6		

Lead
thousand tonnes

	1948	1970	1980	1988	1989	Rank	%
World	490	4,002	5,403	5,792	5,828		
Argentina	23.7	38.1	41.7	28.9	24.5		
Australia	162	213	234	180	204	9	3.5
Austria	9.3	13.5	18.1	22.4	22		
Belgium-Lux.	66	94.3	106	105	93		
Brazil	1	19.5	85	98	86		
Bulgaria	2.1	98.6	118	105	99		
Burma	7.6	7.6	5.9	4.3	3.8		
Canada	145	186	235	268	245	8	4.2
China	0.8	110	175	241	302	6	5.2
Czechoslovakia	5.8	17.6	20	26.1	26		
Denmark	...	16	24.5	–	–		
Finland	1.1	2.4	3.7	–	–		
France	34.7	170	219	256	267	7	4.6

Lead

continued

	1948	1970	1980	1988	1989	Rank	%
Germany	10	335	398	387	500	3	8.6
Greece	0.8	16.1	21.1	15.1	7		
India	0.6	1.9	25.5	32.8	37.3		
Indonesia	...	–	2	7	10		
Iran	0	0	0	10	9		
Ireland	...	...	7	11.7	12		
Italy	42.7	79.3	134	178	181	10	3.1
Japan	10.7	209	317	340	332	5	5.7
Korea, N.	...	55	65	70	75		
Korea, S.	0	3.6	15	90	106		
Malaysia	0	0	2.5	15	16		
Mexico	...	180	160	179	174		
Morocco	...	24.9	42.4	71	66.2		
Namibia	...	67.9	42.7	44.4	44.2		
Netherlands	...	17.6	32.2	39	41.5		
New Zealand	...	...	7	5	5		
Peru	34.3	72	87	53.6	74.4		
Philippines	0	0	4.8	7	7		
Poland	19	54.5	82.0	90.7	78.2		
Portugal	2.5	0.6	2	7	7		
Romania	4	40	41	44	45		
South Africa	...	...	35.4	36.2	36.9		
Spain	22.1	75.5	121	111	114		
Sweden	14.1	57.1	42.3	85.4	71.4		
Taiwan	...	...	16.8	67.3	58.2		
Thailand	0	0	5	15.6	18.7		
Tunisia	18	22.5	19.2	–	–		
Turkey	...	3.5	3	11	9		
UK	...	287	325	374	350	4	6
USA	863	748	1,151	1,091	1,169	1	20.1
USSR	75	540	780	795	750	2	12.9
Venezuela	0	...	10.0	13.1	14.1		
Yugoslavia	49.2	97.4	102	131	117		
Zambia	13.2	27.6	10	7.7	4.6		

Lead ore

thousand tonnes

metal content

	1948	1970	1980	1988	1989	Rank	%
World	1,370	3,400	3,600	3,403	3,341		
Argentina	18	36	33	29	27		
Australia	220	459	397	462	495	2	14.8
Austria	4	6	5	2	2		
Bolivia	26	26	17	13	16		
Brazil	...	18	22	17	15		
Bulgaria	13	96	100	85	85	8	2.5
Burma	...	4	6	6	5		
Canada	172	358	297	367	275	5	8.2
Chile	6	1	0	1	1		
China	...	100	160	312	341	4	10.2
Czechoslovakia	...	7	3	3	3		
France	8	29	29	2	1		
Germany	22	41	32	18	9		
Greece	1	10	22	26	25		
Greenland	...	...	30	24	24		
Honduras	...	15	14	11	6		
India	...	3	14	23	25		
Iran	...	23	12	17	11		
Ireland	...	63	58	33	32		
Italy	30	35	24	17	15		
Japan	7	64	43	23	19		
Korea, N.	...	70	100	90	80	10	2.4
Korea, S.	...	11	11	15	17		
Mexico	193	177	145	178	163	7	4.9
Morocco	29	85	115	70	65		
Namibia	34	71	48	23	24		
Norway	...	3	2	3	3		
Peru	49	157	189	149	192	6	5.7
Poland	15	67	48	50	47		
Romania	4	42	23	33	38		
South Africa	...	...	86	90	78		
Spain	31	60	87	75	63		
Sweden	24	78	72	85	82	9	2.5
Thailand	–	1	11	30	24		
Tunisia	13	22	9	2	2		
Turkey	3	5	8	10	14		
UK	2	4	4	1	1		
USA	354	519	550	394	419	3	12.5
USSR	...	450	580	520	500	1	15
Yugoslavia	63	127	121	78	79		
Zambia	13	33	14	14	12		

Magnesite

thousand tonnes

metal content

	1948	1970	1980	1989	1990	Rank	%
World	900	8,836	11,900	20,000	20,000		
Australia	33.0	23.5	32.2	65	65		
Austria	412	1,609	1,318	1,200	1,100	6	5.5
Brazil	...	236	317	700	700	8	3.5
China	...	1,000	2,000	3,500	3,500	2	17.5
Colombia	0	0.3	1.6	15	15		
Czechoslovakia	...	631	666	2,700	2,800	4	14
Greece	11.2	718	1,501	850	800	7	4
India	49.1	354	380	500	500	9	2.5
Iran	0	4.4	4.0	5	5		
Korea, N.	...	1,633	1,850	3,000	3,000	3	15
Nepal	...	...	40	28	25		
Pakistan	...	...	...	4	4		
Poland	...	38.7	19.6	25	25		
South Africa	10.7	84.3	78.8	75	110		
Spain	9.9	222	506	450	450	10	2.3
Turkey	3.6	300	826	1,200	1,200	5	6
USA	...	...	100	100	100		
USSR	...	1,423	2,000	5,000	5,000	1	25
Yugoslavia	51.7	512	262	350	375		
Zimbabwe	5.7	...	78.2	35	35		

Manganese ore

tonnes

	1948	1970	1980	1989	1990	Rank	%
World	2,100	7,400	12,786	24,100	24,000		
Angola	0.2	9.4	–	–	–		
Argentina	0.5	10.2	1.9	2	2		
Australia	1.6	397	969	1,990	2,300	4	9.6
Botswana	–	16.3	–	–	–		
Brazil	78.7	830	1,027	1,900	1,850	6	7.7
Chile	10.6	11.1	10.5	30	30		
China	...	300	480	2,700	2,700	3	11.3
Czechoslovakia	...	14.1	0.2	0.1	0.1		
Gabon	...	729	1,073	2,390	2,200	5	9.2
Ghana	314	191	100	280	260	9	1.1
Hungary	20.0	34.8	24.8	80	80	10	0.3
India	255	616	592	1,300	1,300	7	5.4
Italy	7.4	14.9	2.8	9	9		
Japan	19.0	79.0	27.9	...	...		
Mexico	24.0	98.6	161	450	450	8	1.9
Morocco	85.2	59.6	65.3	30	30		
Philippines	11.5	2.5	1.1	1	1		
Romania	17.7	26.7	18.0	65	65		
South Africa	116	1,182	4,743	3,620	3,795	2	15.8
Spain	7.0	3.6	–	–	–		
Thailand	...	8.3	24.4	7	8		
Turkey	4.1	5.5	16.6	7	7		
USA	68.6	47.4	20.5	4	4		
USSR	...	...	6,000	8,000	9,000	1	37.5
Vanuatu	–	11.7	–	–	–		
Yugoslavia	4.3	4.6	10.3	40	40		
Zaire	6.4	156	3.0	...	...		

Mercury

tonnes

	1948	1970	1980	1989	1990	Rank	%
World	3,250	8,810	6,818	6,287	5,541		
Algeria	...	...	841	700	700	4	12.6
China	...	690	800	700	1,042	2	18.8
Czechoslovakia	–	17	159	165	49.9	10	0.9
Finland	...	...	75.0	125	100	7	1.8
Italy	1,318	1,530	3	–	–		
Japan	58	222	–	–	–		
Mexico	165	1,043	145	300	195	5	3.5
Spain	782	1,570	1,721	1,380	715	3	12.9
Turkey	1.0	324	154	100	100	8	1.8
USA	496	941	1,058	414	104	6	1.9
USSR	...	1,655	1,800	2,300	2,500	1	45.1
Yugoslavia	377	533	–	69.9	69.9	9	1.3

Molybdenum

tonnes

metal content

	1948	1970	1980	1987	1988	Rank	%
World	12,830	82,340	110,000	89,100	94,700		
Bulgaria	–	–	150	200	200		
Canada	83	15,319	11,889	14,800	12,400	3	13.1
Chile	532	5,701	13,668	16,900	17,000	2	18
China	...	1,500	2,000	2,000	2,000	7	2.1
Iran	...	...	500	500	500	9	0.5
Japan	0	265	56	100	100		
Korea, S.	19	251	302	300	270	10	0.3
Mexico	...	85	74	4,400	4,300	5	4.5
Mongolia	...	...	487	1,100	1,100	8	1.2
Niger	...	...	122	8	15		
Norway	79	303	9	...	...		
Peru	0	607	2,407	3,300	2,400	6	2.5
Turkey	0	0	100	100	100		
USA	12,114	50,508	68,350	34,100	43,000	1	45.4
USSR	...	7,700	10,400	11,500	11,500	4	12.1

Natural gas

*thousand terajoules**

	1970	1980	1989	1990	Rank	%
World	38,455	53,938	70,497	70,546		
Afghanistan	94.7	102	115	...		
Albania	3.8	15.0	16	...		
Algeria	63.9	759	1,489	1,575	6	2.2
Argentina	209	326	785	785		
Australia	55.9	359	621	702		
Austria	77.1	78.0	49	46		
Bahrain	13.7	114	182	182		
Bangladesh	0	45.0	155	165		
Bolivia	1.3	79.8	114	119		
Brazil	3.3	41.2	133	136		
Brunei	7.9	399	376	350		
Bulgaria	16.7	6.7	0	...		
Burma	2.2	13.3	42	...		
Canada	2,113	2,790	3,981	4,088	3	5.8
Chile	48.1	27.1	66	593		
China	144	555	586	600		
Colombia	55.2	151	174	170		
Czechoslovakia	37.5	21.2	27	23		
Egypt	3.3	40.5	251	250		
France	271	289	165	165		
Germany	479	774	537	530		
Hungary	114	237	227	183		
India	19.0	50.6	405	400		
Indonesia	46.8	607	1,215	1,239	7	1.8
Iran	452	278	866	955		
Iraq	30.5	50.0	191	190		
Ireland	...	34.3	91	92		
Israel	4.8	5.6	2	...		
Italy	498	477	658	671		
Japan	10.3	85.1	82	84		
Kuwait	79.6	133	212	143		
Libya	0	134	181	201		
Malaysia	2.9	10.3	559	575		
Mexico	451	1,035	984	1,005		
Netherlands	1,116	3,205	2,270	2,286	4	3.2
New Zealand	4.2	38.9	187	192		
Nigeria	4.3	51.7	169	164		
Norway	0	1,065	1,288	1,121	9	1.6
Pakistan	95.8	246	449	490		
Peru	15.9	43.7	20	...		
Poland	186	193	169	121		
Qatar	39.1	176	214	200		
Romania	990	1,455	1,108	1,157	8	1.6
Saudi Arabia	...	48.7	1,042	1,069	10	1.5
Thailand	0	0	234	...		
Trinidad & T.	73	146	147	140		
Tunisia	0	16.7	16	...		
UAE	8.9	471	190	187		
UK	434	1,464	1,724	1,842	5	2.6
USA	22,860	19,256	17,177	17,480	2	24.8
USSR	6,670	15,148	25,843	27,580	1	39.5
Venezuela	382	643	792	919		
Yugoslavia	39.3	75.4	101	89		

* A terajoule is a measure of energy equivalent to 23.46 tonnes of oil or 34.12 tonnes of coal

Nickel
thousand tonnes

	1970	1980	1988	1989	Rank	%
World	607	733	837	865		
Albania	–	4.5	4.5	5		
Australia	1	35.3	42	42.9	5	5
Austria	...	0.5	2	3		
Brazil	2.5	2.5	13.1	13.7		
Canada	207	134	135	129	2	14.9
China	...	11	25.5	26.3	10	3
Colombia	...	...	16.9	16.9		
Cuba	2.5	20	24.1	26.5	9	3.1
Czechoslovakia	1.5	2.2	3.5	3.5		
Dominican Rep.	–	18	29.3	31.3	7	3.6
Finland	4	12.8	15.7	13.4		
France	11	9.8	9	9.9		
Germany	1.8	3	3.3	2.7		
Greece	8.6	13.9	13.1	16.1		
Indonesia	0	4.4	4.9	5		
Japan	89.8	109	101	107	3	12.4
Korea, S.	–	–	–	5		
New Caledonia	28	32.6	37.4	36.3	6	4.2
Norway	38.5	37.1	52.5	54.9	4	6.3
Philippines	0	22.7	–	–		
Poland	18.3	1.8				
South Africa	9	18.1	27.2	30	8	3.5
Taiwan	...	...	10.5	10		
UK	36.7	19.3	28	26.1		
USA	14.1	40.1		0.3		
USSR	124	165	215	225	1	26
Yugoslavia	0	0	5.6	6.3		
Zimbabwe	5	15.1	18.3	18.6		

Nickel ore
thousand tonnes

metal content

	1948	1970	1980	1988	1989	Rank	%
World	144	651	738	837	865		
Albania	...	...	4.8	8	8		
Australia	–	18	74.3	62.4	65	4	7.5
Botswana	–	–	15.4	22.5	19.8	10	2.3
Brazil	–	2.9	4.3	13.1	13.7		
Canada	120	278	185	217	203	2	23.5
China	–	–	11	26	27.5	9	3.2
Colombia	–	–	–	16.9	16.9		
Cuba	–	35.2	38.2	43.9	46.5	6	5.4
Dominican Rep.	–	–	18	29.3	31.3	8	3.6
Finland	–	6.7	6.4	11.7	10.5		
France	–	–	6.6	–	–		
Germany	...	...	2.7	1.5	1.5		
Greece	–	–	13.9	13.1	16.1		
Indonesia	–	18	40.5	59.8	59.6	5	6.9
New Caledonia	3.6	138	86.6	71.2	96.2	3	11.1
Norway	–	–	0.6	0.3	1.3		
Philippines	...	...	34.9	10.3	15.4		
Poland	–	2	1.1	–			
South Africa	0.5	11.6	25.7	34.8	34	7	3.9
USA	0.8	17	10.2	–	–		
USSR	17.5	109	143	205	210	1	24.3
Yugoslavia	–	–	0.5	5.6	6.3		
Zimbabwe	0	10.9	14.3	12.1	12.7		

Oil – crude
million tonnes

	1948	1970	1980	1989	1990	Rank	%
World	469	2,244	2,829	3,106	3,150		
Algeria	–	51.2	50.6	51.5	54.8		
Angola	–	5.4	7.5	22.6	23.6		
Argentina	3.3	19.1	24.2	23.5	23.9		
Australia	–	8.7	18.9	22.7	26.8		
Austria	1	2.7	1.4	1.2	1.2		
Bahrain	1.5	3.8	2.4	2.1	2.1		
Brazil	–	8	9.1	29.8	31.6		
Brunei	2.7	7.4	11.5	7	7.1		
Cameroon	–	0	2.9	8.6	8.3		
Canada	1.7	62.9	70.3	76.4	76.3		
China	0.1	24.9	106	138	138	5	4.4
Colombia	3.3	10.7	6.2	20.4	21.9		

	1948	1970	1980	1989	1990	Rank	%
Congo	–	–	2.8	7.4	7.6		
Denmark	–	–	0.3	5.5	5.9		
Ecuador	0.3	0.2	11.1	14.6	14.7		
Egypt	2.1	16.3	29.7	43	45.8		
France	0.1	2.3	1.3	3.2	3.1		
Gabon	–	5.4	9	10.7	13.8		
Germany	0.6	7.3	4.5	3.8	3.7		
Hungary	–	–	–	1.9	2		
India	0.3	6.9	9.1	33.7	33.3		
Indonesia	4.3	42.5	78.5	69.3	74.5		
Iran	25.3	166	73	147	157	4	5
Iraq	3.4	77.1	131	142	101	9	3.2
Italy	–	–	1.8	4.6	4.6		
Kuwait	6.4	136	68.8	91.2	61.4		
Libya	–	165	88.9	54.2	65.2		
Malaysia	–	–	14.3	27.9	29.7		
Mexico	8.4	21.4	96.4	131	133	6	4.2
Netherlands	0.5	1.8	1.2	3.4	3.5		
New Zealand	–	–	0.5	1.7	1.7		
Nigeria	–	53.9	102	85.2	86.5		
Norway	–	–	26.3	74.9	81.9		
Oman	–	16.6	14.1	30.8	34.5		
Peru	1.9	3.6	9.5	7.5	6.9		
Qatar	–	18	23.5	18.4	17.1		
Romania	4.1	13.4	11.5	9.2	7.9		
Saudi Arabia	19.1	177	480	260	323	3	10.3
Spain	–	0.2	1.6	1.1	1.1		
Syria	–	4.1	8.2	18.3	19.2		
Trinidad & T.	2.9	7	10.5	7.7	7.8		
Tunisia	–	4.4	5.5	4.9	4.8		
Turkey	–	3.4	2.2	2.8	3.8		
UAE	–	96.7	85.1	89.1	102	8	3.2
UK	0.2	0.1	80.6	87.4	87.9	10	2.8
USA	273	480	427	382	366	2	11.6
USSR	29.2	353	603	607	570	1	18.1
Venezuela	71.7	185	108	100	118	7	3.7
Yemen	–	–	4.2	9.4	9.9		
Yugoslavia	0	2.9	4.2	3.4	3.1		

Petroleum products†
million tonnes

	1970	1980	1988	1989	Rank	%
World	2,032	2,762	2,822	2,857		
Africa	...	69.9	110	113		
Asia	...	523	627	657		
Europe	...	672	599	599		
North America	...	923	868	875		
Oceania	...	32.5	32.7	33.5		
South America	...	145	146	146		
Algeria	2.3	13	39.1	37.9		
Argentina	19.4	23.7	19.9	20		
Australia	22.1	28.2	28.7	29.2		
Austria	5.5	8.8	7.3	7.2		
Bahamas	2.3	8.3	0	0		
Bahrain	11.2	10.6	10.2	10.4		
Belgium	26.0	29.5	23.5	23.9		
Brazil	23.5	47.7	51.4	51.4		
Bulgaria	5.8	11.7	9.8	10.2		
Cameroon	0	0.8	1.8	1.8		
Canada	57.4	98.3	82.8	85.2	5	3
Chile	3.6	5	5	5.9		
China	21.8	64.2	81.5	84.7	6	3
Colombia	6.2	7.7	10.4	10.6		
Cuba	4.2	5.5	7	7.3		
Czechoslovakia	8.7	16.1	12.9	13		
Denmark	9.4	6.2	7.6	8		
Ecuador	1.1	4.5	5.7	5.3		
Egypt	3.2	13.4	21.7	22.4		
Finland	7.0	10.9	9.1	7.8		
France	88.5	103	68.7	67.5		
Germany	105.1	115.8	90.4	96	4	3.4
Greece	4.6	13.1	14.5	15.1		
Hungary	5.0	8.9	7.7	7.5		
India	14.3	19.4	36.6	39.6		
Indonesia	10.4	23.8	28.2	28.1		
Iran	24.2	32.9	29.5	33.1		
Iraq	3.3	8.2	17.7	19.9		
Ireland	2.6	2.0	1.3	1.5		
Israel	5	5.6	6.7	6.7		
Italy	104	85.4	77.1	77.2	9	2.7
Japan	...	185	138	144	3	5
Kenya	2.1	2.7	1.9	2		

† The principal products are aviation gasolene, fuel oils, diesel oil, jet fuel, kerosene liquefied petroleum gas, motor gasolene and residual fuel oil

	1970	1980	1988	1989	Rank	%
Korea, S.	8.8	22	29.6	34.4		
Kuwait	17	17.8	31.9	33.7		
Libya	0.4	5.3	8.4	11.2		
Malaysia	4	5.6	7.7	8		
Mexico	21.1	56.1	67.2	68.6	10	2.4
Morocco	1.4	3.8	4.3	4.7		
Netherlands	51.6	51.9	56.2	54.6		
Neths. Antilles	45.6	26.1	9.4	9.7		
New Zealand	2.7	2.7	4	4.2		
Nigeria	0.9	6.2	7.2	7.5		
Norway	5.5	7.5	8.8	9.8		
Pakistan	4	3.9	5.2	5.3		
Panama	3.6	1.9	1.2	1.1		
Peru	4.1	7.2	7.9	7.5		
Philippines	8.1	8.8	8.6	8.9		
Poland	6.1	12.8	11.9	12.1		
Portugal	3.3	7.1	7.5	9.4		
Puerto Rico	9.4	10.2	4.4	4.2		
Romania	14.0	24.6	28.9	28.1		
Saudi Arabia	25.5	36.9	76.1	77.2	8	2.7
Singapore	10.7	25.1	28.6	29.8		
South Africa	7.7	12.7	13.7	13.8		
Spain	29.3	44.2	43.7	45.7		
Sweden	10.4	16.7	13.6	15.5		
Switzerland	4.9	4.4	3.8	2.9		
Syria	1.7	5.1	10.5	10.5		
Taiwan	5.9	20.3	20	20		
Thailand	3.2	7.2	9.9	11.7		
Trinidad & T.	21.7	11.5	5	4.6		
Turkey	6.8	11.4	20.5	19.3		
UAE	0	1.5	12.4	12.4		
UK	90.6	77.6	79.5	80.9	7	2.8
US Virgin Is.	13.1	22.5	10.4	10.9		
USA	496	677	684	688	1	24.1
USSR	246	396	440	433	2	15.2
Venezuela	60.6	46.3	43.6	43.3		
Yemen	6.3	1.9	3.6	3.5		
Yugoslavia	6.4	13.7	13.9	13.7		

Phosphates
thousand tonnes

	1948	1970	1980	1989	1990	Rank	%
World	17,100	61,077	138,000	155,000	153,000		
Algeria	671	492	1,025	1,224	1,102		
Brazil	...	700	2,612	3,655	3,700	7	2.4
China	...	1,200	10,726	13,000	13,000	4	8.5
Christmas I.	178	1,064	1,638	0	0		
Egypt	300	584	658	1,397	1,280		
Finland	...	...	400	580	580		
India	...	172	543	704	800		
Iraq	–	–	–	1,143	850		
Israel	...	1,162	2,307	3,922	3,516	8	2.3
Jordan	...	913	3,911	6,657	5,925	6	3.9
Korea, N.	...	300	500	560	550		
Mexico	–	46	387	657	650		
Morocco	3,226	11,424	18,824	18,000	21,000	3	13.7
Nauru	268	2,200	2,087	1,181	926		
Senegal	...	998	1,290	2,383	2,289		
South Africa	40	1,685	3,282	2,963	3,086	9	2
Syria	...	...	2,250	1,670			
Togo	...	1,508	2,933	3,356	2,314	10	1.5
Tunisia	1,864	3,021	4,502	6,621	6,566	5	4.3
Turkey	...	...	100	85	200		
USA	9,539	10,884	54,415	49,000	38,000	1	24.8
USSR	...	20,400	24,700	35,000	35,000	2	22.9
Venezuela	...	...	3	195	200		
Vietnam	...	...	90	500	500		
Zimbabwe	...	...	130	155	125		

Potash

thousand tonnes

potash content

	1948	1970	1980	1989	1990	Rank	%
World	3,450	18,388	27,900	29,000	27,000		
Canada	...	3,103	7,303	7,300	7,000	2	25.9
Chile	1	22	25	10	10		
China	...	...	12	35	55		
France	769	1,904	2,039	1,195	1,300	5	4.8
Germany	936	5,064	6,159	5,386	4,850	3	18
Israel	...	530	797	1,271	1,260	6	4.7
Italy	...	241	130	152	100	10	0.4
Jordan	...	...	...	792	842	7	3.1
Spain	152	598	791	742	690	8	2.6
UK	0	...	321	463	455	9	1.7
USA	1,034	2,476	2,239	1,595	1,635	4	6.1
USSR	...	...	8,064	10,238	9,100	1	33.7

Radio receivers

thousands

	1948	1970	1980	1986	1987	Rank	%
World	...	110,000	165,000	128,213	105,687		
Algeria	...	76	112	204	160		
Australia	335	729	199	238	...		
Austria	96	223	...	...	...		
Belgium	...	1,943	2,246	1,233	1,165		
Brazil	...	809	6,769	8,519	8,676	5	8.2
Bulgaria	...	145	51	38	56		
Canada	640	2,004	...	...	...		
China	...	...	30,038	15,895	17,638	2	16.7
Cuba	...	19	200	237	227		
Czechoslovakia	268	416	250	263	200		
Denmark	143	147	88	42	47		
Egypt	...	148	209	180	19		
Finland	56	174	219	0	0		
France	956	2,921	2,141	2,161	2,080	10	2
Germany	...	7,536	4,622	5,104	6,381	7	6
Hong Kong	...	22,096	48,262	29,618	...		
Hungary	...	215	271	93	100		
India	25	1,771	1,918	1,084	1,178		
Indonesia	...	393	1,530	966	997		
Iran	...	165	57	245	...		
Italy	...	3,300	...	...	...		
Japan	801	32,618	15,343	13,204	10,496	4	9.9
Korea, S.	...	1,088	3,972	1,493	1,425		
Mexico	...	1,015	1,029	276	207		
New Zealand	...	119	...	...	...		
Nigeria	...	215	...	...	...		
Norway	155	144	...	...	...		
Philippines	...	129	...	...	...		
Poland	34	987	2,695	2,729	2,833	9	2.7
Portugal	...	517	593	960	1,188		
Romania	0	455	863	580	618		
Singapore	...	...	593	16,583	19,868	1	18.8
South Africa	...	313	861	813	776		
Sweden	70	170	26	-	-		
Taiwan	...	6,247	9,490	10,139	11,599	3	11
Thailand	...	...	1,113	...	...		
Turkey	...	198	52	159	441		
UK	1,630	1,313	439	408	...		
USA	16,708	13,628	7,672	4,833	4,315	8	4.1
USSR	536	7,815	8,478	8,924	8,143	6	7.7
Yugoslavia	23	277	125	177	198		

Rubber – synthetic

thousand tonnes

	1960	1970	1980	1989	1990
World	2,439	5,055	6,515	...	...
Argentina	...	39	33.4	48	...
Australia	11.4	41.9	46.2	42	...
Belgium	0	50	115	131	120
Brazil	10.1	96.4	287	257	...
Bulgaria	0	6	30	...	...
Canada	168	218	256	197	...
China	...	...	250	289	315
Czechoslovakia	0	64.2	79.3	70	70
France	17.5	343	528	591	522
Finland	...	...	...	17	29
Germany	218	442	540	666	680
India	0	30.3	22.5	51	...
Italy	67	155	250	295	...
Japan	18.6	760	1,161	1,352	1,426
Korea, S.	0	5.1	75.5	139	183
Mexico	43.7	90.8	132	134	119
Netherlands	12	200	212	212	236
Poland	20.2	61.7	118	125	235
Romania	...	61.2	150	149	102
South Africa	0	28.6	38.7	62	...
Spain	...	38.6	81.1	82	73
Sweden	...	7.8	13.2	...	...
Turkey	0	0	23	37	39
UK	138	345	227	311	297
USA	1,757	2,436	2,302	2,261	2,114
Yugoslavia	0	6.2	7.4	...	...

Salt

thousand tonnes

	1948	1970	1980	1989	1990	Rank	%
World	...	143,200	169,800	184,000	191,000		
Algeria	...	...	...	240	240		
Argentina	288	958	1,004	800	800		
Australia	249	2,053	5,665	7,100	74,000	1	38.7
Austria	200	563	672	650	650		
Bahamas	32	756	684	900	900		
Bangladesh	...	...	463	400	400		
Brazil	781	1,826	3,932	4,300	4,300		
Burma	43	165	268	200	200		
Canada	672	4,862	7,423	11,400	11,100	6	5.8
Chile	54	517	441	-	-		
China	...	16,000	17,280	22,700	28,600	3	15
Colombia	124	763	838	700	700		
Cuba	...	...	200	230	230		
Czechoslovakia	140	213	371	350	350		
Denmark	...	436	347	550	550		
Egypt	126	440	636	900	900		
France	5,289	11,004	6,543	6,000	8,000	8	4.2
Germany	...	23,202	15,783	15,000	16,300	4	8.5
India	4,602	5,588	8,009	11,000	9,000	7	4.7
Indonesia	...	63	690	600	600		
Iran	...	390	600	700	700		
Israel	...	100	130	400	400		
Italy	3,222	8,596	5,267	4,500	4,400		
Japan	293	951	1,112	1,400	1,400		
Korea, N.	...	544	570	550	550		
Korea, S.	90	405	455	1,000	1,000		
Mexico	...	4,063	6,575	7,700	7,900	9	4.1
Namibia	15	110	230	150	150		
Netherlands	249	2,871	3,464	3,700	3,700		
Neths. Antilles	-	...	400	225	225		
Pakistan	281	1,524	687	700	700		
Peru	63	191	457	350	350		
Philippines	118	210	346	500	500		
Poland	420	5,808	4,533	4,700	5,600		
Portugal	...	401	620	700	700		
Romania	352	2,862	5,056	5,500	5,400		
Senegal	...	90	140	25	25		
South Africa	153	420	573	700	700		
Spain	1,980	4,160	3,508	3,100	3,200		
Switzerland	112	333	360	260	260		
Taiwan	376	...	722	110	110		
Thailand	...	200	177	170	170		
Tunisia	105	300	460	450	450		
Turkey	266	649	1,179	1,300	1,300		
UK	7,698	18,217	7,154	5,800	5,700	10	3
USA	29,672	83,271	36,606	35,600	35,700	2	18.7
USSR	...	12,428	14,500	14,800	14,900	5	7.8
Venezuela	36	266	243	500	500		
Vietnam	65	270	435	300	300		
Yemen	...	...	210	200	200		
Yugoslavia	102	126	379	370	370		

Ships

Merchant Vessels launched
thousand gross registered tons

	1948	1970	1980	1989	1990	Rank	%
World	2,308	21,690	15,397	12,721	14,680		
Argentina	0	18	60	37	10		
Australia	9	54	8	15	25		
Belgium	52	155	99	46	57		
Brazil	0	100	615	261	105		
Bulgaria	...	53	152	72	81		
Canada	102	33	80	32	3		
China	...	...	222	453	5	3.1	
Denmark	99	514	227	313	405	6	2.8
Egypt	...	9	6	7	20		
Finland	7	222	198	207	161	10	1.1
France	138	960	328	78	108		
Germany	...	2,021	802	595	863	3	5.9
Greece	...	73	22	1	3		
India	...	29	74	54	1		
Indonesia	...	...	...	5	12		
Ireland	0	28	7	...	...		
Italy	112	598	168	203	352	8	2.4
Japan	...	10,476	7,288	6,030	6,531	1	44.5
Korea, S.	...	...	629	2,618	3,295	2	22.4
Netherlands	142	461	125	95	163	9	1.1
Norway	47	639	319	31	83		
Peru	...	35	17	-	-		
Poland	5	463	395	94	136		
Portugal	4	16	167	61	24		
Romania	...	...	31	59	5		
Singapore	...	6	27	36	42		
Spain	22	926	509	275	383	7	2.6
Sweden	246	1,711	338	4	27		
Taiwan	0	217	572	...	...		
Turkey	...	11	14	29	8		
UK	1,176	1,237	244	101	78		
USA	126	338	558	3	12		
Yugoslavia	6	393	123	482	462	4	3.1

Silver

tonnes

metal content

	1948	1970	1980	1988	1989	Rank	%
World	5,000	9,393	11,101	14,435	14,654		
Argentina	212	64	73	79	83		
Australia	313	855	767	1,114	1,075	7	7.3
Bolivia	235	186	190	232	295	10	2
Brazil	...	11	45	90	64		
Bulgaria	...	...	24	25	20		
Burma	14	17	30	10	7		
Canada	529	1,376	1,070	1,484	1,306	5	8.9
Chile	27	74	299	507	536	8	3.7
China	...	25	60	145	165		
Czechoslovakia	...	...	35				
Dominican Rep.	...	...	48	40	38		
Ecuador	6	2	1	-	-		
El Salvador	7	...	5	1	1		
Finland	5	23	48	24	31		
France	15	150	74	22	19		
Germany	94	205	85	61	69		
Greece	...	...	60	61	62		
Honduras	97	119	54	50	37		
India	...	2	11	41	36		
Indonesia	...	9	22	62	68		
Iran	...	...	...	25	25		
Italy	21	33	43	16	14		
Japan	94	343	268	252	156		
Korea, N.	...	22	290	300	300	9	2.1
Korea, S.	1	53	63	70	70		
Malaysia	...	...	22	20	13		
Mexico	1,789	1,332	1,557	2,360	2,306	1	15.7
Morocco	15	21	98	225	195		
Namibia	12	38	105	110	109		
New Zealand	7	...	23	-	-		
Papua New G.	...	1	37	70	92		
Peru	289	1,239	1,315	1,552	1,840	3	12.6
Philippines	6	53	61	55	51		
Poland	...	...	766	1,063	1,083	6	7.4
Portugal	...	...	2	4	19		
Romania	...	25	28	20	20		
South Africa	36	110	232	200	178		
Spain	11	51	178	221	250		

Silver
continued

	1948	1970	1980	1988	1989	Rank	%
Sweden	35	123	166	225	200		
USA	1,220	1,400	1,006	1,661	2,007	2	13.7
USSR	...	1,182	1,550	1,500	1,500	4	10.2
Yugoslavia	47	106	149	139	133		
Zaïre	118	53	79	74	60		
Zambia	5	48	24	29	20		
Zimbabwe	3	2	30	22	22		

Steel
thousand tonnes

	1948	1970	1980	1989	1990	Rank	%
World	155,800	593,791	690,000	676,000	685,000		
Argentina	122	1,859	2,556	3,852	4,000		
Australia	1,245	6,874	7,895	6,708	6,684		
Austria	648	4,079	4,624	5,004	4,548		
Belgium	6,373	12,609	12,321	10,956	11,424		
Brazil	483	5,390	7,940	26,112	20,580	8	3
Bulgaria	0	1,800	2,565	2,904	2,184		
Canada	2,903	11,198	15,901	15,456	15,500		
Chile	30	547	695	816	850		
China	800	17,790	37,120	64,212	64,656	4	9.4
Czechoslovakia	2,621	11,480	15,225	15,468	14,868		
Egypt	0	304	500	750	750		
Finland	103	1,167	2,509	2,916	2,856		
France	7,236	23,773	23,176	19,308	16,800	10	2.5
Germany	5,866	5,093	51,146	48,900	45,600	5	6.7
Greece	...	435	935	960	960		
Hungary	770	3,108	3,766	3,312	2,808		
India	1,277	6,286	9,427	12,876	13,000		
Italy	2,125	17,277	26,501	25,176	25,000	6	3.6
Japan	1,715	93,322	111,395	107,904	110,328	2	16.1
Korea, N.	115	2,200	3,500	5,000	5,000		
Korea, S.	8	481	5,760	23,700	24,468	7	3.6
Luxembourg	2,624	5,462	4,620	3,720	3,480		
Mexico	292	3,846	7,003	7,392	8,220		
Netherlands	334	5,042	4,959	5,676	5,412		
Norway	74	869	853	672	396		
Poland	1,955	11,750	18,648	15,096	13,644		
Romania	353	6,517	13,175	13,500	13,000		
South Africa	596	...	8,976	9,456	9,500		
Spain	624	7,394	12,553	12,564	12,000		
Sweden	1,276	5,494	4,231	4,692	4,452		
Switzerland	120	524	929	1,000	1,000		
Taiwan	18	607	3,651	8,000	8,000		
Turkey	102	1,312	1,700	7,848	7,500		
UK	15,116	28,316	11,277	18,744	18,000	9	2.6
USA	80,413	119,309	101,456	88,392	87,000	3	12.7
USSR	18,639	115,889	147,941	160,092	154,416	1	22.5
Venezuela	...	927	1,784	3,408	3,600		
Yugoslavia	368	2,228	2,306	4,500	4,500		
Zimbabwe	9	250	804	1,000	1,000		

Television receivers
thousands

	1960	1970	1980	1986	1987	Rank	%
World	...	44,000	71,000	93,650	100,561		
Algeria	0	30	94	259	318		
Argentina	125	194	454	801	695		
Australia	435	320	332	238	211		
Austria	112	369	...	569	...		
Belgium	243	505	746	930	917		
Brazil	183	726	3,254	3,034	2,902	9	2.9
Bulgaria	0	193	91	154	199		
Canada	342	543	444	...	...		
Chile	0	80	69	34	35		
China	0	...	2,492	14,594	19,344	1	19.2
Colombia	0	45	103	82	112		
Cuba	0	...	40	56	...		
Czechoslovakia	263	383	389	434	507		
Denmark	233	70	77	134	110		
Egypt	0	64	318	443	333		
Finland	46	146	341	602	627		
France	655	1,511	1,928	1,875	2,184		
Germany	2,580	3,316	5,003	4,607	4,206	7	4.2
Greece	0	134	137	...	...		
Hong Kong	0	27	108	...	...		
Hungary	139	364	417	417	446		
India	...	...	88	1,088	972		
Indonesia	0	5	607	539	575		
Iran	0	125	246	411	...		
Ireland	38	87	...	...	...		
Israel	0	68	24	49	35		
Italy	728	2,030	1,984	1,847	2,233	10	2.2
Japan	3,578	12,488	15,205	13,862	14,777	3	14.7
Korea, S.	0	114	6,819	11,799	14,922	2	14.8
Malaysia	0	44	157	863	1,240		
Mexico	80	431	964	630	609		
New Zealand	8	45	126	90	90		
Nigeria	0	6	208	84	34		
Norway	64	103	...	...	...		
Pakistan	...	...	74	189	199		
Peru	0	35	75	116	...		
Philippines	2	44	206	166	300		
Poland	171	616	900	631	647		
Portugal	0	64	467	492	485		
Romania	0	280	541	530	484		
Singapore	0	...	1,889	1,707	2,123		
South Africa	...	...	338	236	231		
Spain	39	618	763	1,233	...		
Sweden	256	178	308	358	354		
Taiwan	...	1,254	7,041	6,216	6,442	6	6.4
Thailand	...	...	248	500	544		
Tunisia	...	...	88	84	58		
Turkey	...	4	327	886	694		
UK	2,141	2,214	2,364	2,926	3,022	8	3
USA	5,611	8,298	10,320	12,862	12,871	4	12.8
USSR	1,726	6,682	7,528	9,436	9,081	5	9
Venezuela	0	98	...	...	...		
Yugoslavia	14	320	543	631	591		

Tin
thousand tonnes

	1948	1970	1980	1988	1989	Rank	%
World	167	272	283	253	267		
Argentina	0	0.1	0.5	0.5	0.4		
Australia	1.9	5.7	4.8	0.7	0.7		
Belgium-Lux.	10.6	5.7	2.8	5	6		
Bolivia	0.1	0.7	17.5	5.5	9.7	9	3.6
Brazil	0.2	3.5	9	42	44.4	2	16.6
Burma	0	0	0	0.1	0.5		
China	4.9	22	15	24	28.3	4	10.6
Czechoslovakia	0	0.1	0.2	0.5	0.6		
France	...	1.5	0.2	0.3	0.3		
Germany	0.2	12.2	5.7	5.2	6.1	10	2.3
Indonesia	0.5	5.2	30.5	28.4	29.9	3	11.2
Italy	...	3	3.5	0.5	0.6		
Japan	0.1	6.1	1.3	5.4	4.7		
Korea, S.	–	0.3	0.4	2.5	2.4		
Malaysia	50.5	90.3	71.3	47.4	51.9	1	19.4
Mexico	0.2	1.4	1.4	1.5	3		
Netherlands	16.7	8.5	1.1	3.9	4.8		
Nigeria	...	8.1	2.7	0.6	0.3		
Norway	...	0.4	0.2	...	...		
Portugal	0.3	0.4	0.4	0.1	0.1		
Singapore	...	...	4	0.9	0.6		
South Africa	0.6	1.5	2.2	2.4	2.6		
Spain	0.5	4.5	4.8	1	2.2		
Thailand	...	22	34.7	14.7	14.6	8	5.5
UK	31.5	31.5	18.4	24.8	16.8	5	6.3
USA	37.3	24.8	6.4	16.7	16.2	7	6.1
USSR	...	10	17	17	16.5	6	6.2
Vietnam	0	0	0	0.5	0.6		
Zaïre	3.9	1.4	0.5	0.1	–		
Zimbabwe	0	0	0.9	0.9	0.8		

Includes recovery from scrap in western countries

Tin ore
thousand tonnes

metal content

	1948	1970	1980	1988	1989	Rank	%
World	154	186	236	204	223		
Argentina	0.3	2.3	0.4	0.4	0.4		
Australia	1.9	8.9	11.6	7	7.8	8	3.5
Bolivia	38	30.1	27.5	10.5	15.8	5	7.1
Brazil	0.2	4.3	6.9	44	50.2	1	22.5
Burma	1.2	0.3	1.1	0.5	0.6		
Canada	0.3	0.1	6.9	3.6	2.8		
China	4.9	...	16	30	33	2	14.8
Czechoslovakia	...	0.2	0.2	0.6	0.3		
Germany	0.2	...	1.8	3	2		
Indonesia	31.1	19.1	32.5	29.6	31.3	4	14
Japan	0.1	0.8	0.5	–	–		
Laos	0	1.4	0.6	0.2	0.3		
Malaysia	45.5	73.8	61.4	28.9	32	3	14.3
Mexico	0.2	0.5	0.1	0.3	0.2		
Mongolia	0	0	0	1.2	1.3		
Namibia	0.1	1	1	1.2	0.5		
Nigeria	9.4	8	2.7	0.9	1.2		
Peru	0.1	0	1.1	4.4	5.1	9	2.3
Portugal	0.7	0.4	0.3	0.1	0.1		
Rwanda	1.4	1.3	1.6	...	0.8		
South Africa	0.5	2.0	2.9	1.4	1.3		
Spain	0.3	0.4	0.4	0.8	2		
Thailand	4.3	21.8	33.7	14	14.7	6	6.6
Uganda	0.2	0.1	0.1	–	–		
UK	0.9	1.7	3.3	3.4	4	10	1.8
USSR	...	...	16	15	14	7	6.3
Vietnam	...	...	0.4	0.7	0.8		
Zaïre	11.7	6.5	3.2	1.9	1.6		
Zimbabwe	0.1	0.6	0.9	0.9	0.8		

Tungsten
tonnes

metal content

	1948	1970	1980	1989	1990	Rank	%
World	14,700	43,990	52,000	50,000	45,000		
Argentina	101	181	35	20	20		
Australia	749	1,760	3,575	1,200	1,300	6	2.9
Austria	0	165	2,150	1,000	1,400	4	3.1
Bolivia	1,491	2,411	2,664	900	950	9	2.1
Brazil	792	1,463	1,104	700	500		
Burma	947	278	848	400	450		
Canada	475	1,690	3,178	–	–		
China	...	10,100	15,000	20,000	20,000	1	44.4
Czechoslovakia	0	0	80	50	50		
France	340	88	577	–	–		
Japan	0	854	700	200	250		
Korea, N.	...	2,700	2,200	500	500		
Korea, S.	871	2,822	2,737	1,500	1,250	7	2.8
Malaysia	53	92	35	–	–		
Mexico	80	288	266	200	100		
Mongolia	...	...	1,500	1,500	1,500	3	3.3
New Zealand	16	2	4	–	–		
Peru	212	1,014	549	500	1,200	8	2.7
Portugal	1,702	1,860	1,580	1,300	1,400	5	3.1
Rwanda	...	410	323	40	200		
South Africa	142	5	–	–	–		
Spain	526	514	448	80	80		
Sweden	190	0	364	300	250		
Thailand	297	895	1,615	600	650	10	1.4
Turkey	–	–	400	300	300		
Uganda	117	153	20	50	50		
UK	27	0	40	–	–		
USA	2,307	5,326	2,754	200	400		
USSR	3,000	8,500	2,754	8,000	7,000	2	15.6
Zaïre	221	237	134	20	20		
Zimbabwe	47	0	90	–	–		

Tyres
thousands

	1948	1970	1980	1989	1990
No world totals					
Algeria	0	391	7	...	...
Angola	0	111	...	...	...
Argentina	862	3,530	5,297	4,200	...
Australia	2,721	7,227	7,674	...	...
Brazil	995	7,847	22,102	30,012	29,172
Bulgaria	14	546	1,532	1,884	1,296
Canada	4,868	19,109	22,335	...	...
Chile	0	676	878	1,200	...
China	...	4,250	11,460	31,356	...
Colombia	0	875	1,899	...	...
Cuba	0	202	387	228	...
Czechoslovakia	691	2,513	4,969	5,268	5,316

Tyres continued

	1948	1970	1980	1989	1990
Ecuador	0	120	...	...	...
Egypt	0	807	1,113	1,008	...
Finland	52	1,037	1,127	20,000	...
France	4,055	39,415	50,601	61,368	...
Germany	5,628	42,240	44,866	58,380	57,120
Ghana	0	122	...	...	...
Hungary	71	586	655	972	852
India	800	3,058	4,720	8,064	...
Indonesia	...	358	3,821	8,028	...
Israel	0	1,392	1,242	756	...
Italy	...	22,383	29,681	44,928	...
Jamaica	0	163	178	...	...
Japan	1,000	66,556	116,384	155,724	150,012
Korea, S.	0	899	12,327	24,540	27,900
Luxembourg	...	2,974	2,138	...	...
Malaysia	...	3,067	4,634	6,156	6,744
Mexico	668	3,369	8,749	10,152	11,772
Morocco	0	411	782	...	...
New Zealand	31	1,433	1,663	1,200	...
Nigeria	0	266	...	...	...
Pakistan	...	109	204	912	912
Panama	...	35	84	...	...
Peru	0	609	717	...	...
Philippines	...	598	1,637	2,016	...
Poland	120	2,995	6,533	6,024	4,704
Portugal	60	1,400	1,833	3,120	...
Romania	58	2,457	5,003	...	...
South Africa	994	3,600	5,254	4,824	...
Spain	...	8,909	15,626	...	...
Sri Lanka	0	76	241	108	...
Sweden	800	5,379	2,691	...	...
Thailand	...	714	1,762	4,296	...
Tunisia	0	104	88	504	...
Turkey	...	1,139	2,929	6,636	4,596
UK	8,835	30,545	26,346	31,080	29,376
USA	81,314	190,251	158,435	212,868	210,660
USSR	9,000	22,351	60,066	51,864	51,396
Venezuela	39	1,577	3,483	4,176	...
Yugoslavia	36	2,763	9,584	13,200	12,636

Uranium

tonnes

metal content

	1970	1980	1989	1990	Rank	%
World	24,000	52,000	70,000	65,000		
Argentina	50	242	100	100		
Australia	299	1,561	3,634	3,520	3	9.5
Brazil	0	0	50	50		
Canada	3,639	6,739	11,330	8,700	2	23.5
France	1,476	2,845	2,800	2,800	7	7.6
Gabon	472	1,034	870	710	10	1.9
Germany	0	41	3,000	3,000	6	8.1
India	0	150	100	100		
Namibia	0	4,036	3,000	3,200	5	8.6
Niger	54	4,132	2,960	2,800	8	7.6
Portugal	95	68	100	100		
South Africa	3,737	6,186	3,000	2,500	9	6.8
Spain	60	167	150	150		
USA	11,538	16,800	5,100	3,500	4	9.5
USSR	...	...	36,194	31,230	1	84.4

Vanadium

tonnes

metal content

	1948	1970	1980	1987	1988	Rank	%
World	1,480	12,050	34,800	30,000	32,000		
Chile	–	–	270	–	–		
China	–	–	4,500	4,500	4,500	3	14
Finland	–	1,196	3,009	–	–		
Namibia	187	600	–				
Norway	0	1,080	490	...	...		
South Africa	–	4,257	12,700	14,400	16,400	1	51.3
USA	608	4,825	4,360	1,500	1,500	4	4.7
USSR	–	–	9,500	9,600	9,600	2	30

Zinc

thousand tonnes

	1948	1970	1980	1988	1989	Rank	%
World	2,068	6,126	7,455	8,577	8,572		
Algeria	–	–	25	38	28		
Argentina	1.6	28.7	38.7	35	34		
Australia	83.1	264	361	307	296	8	3.5
Austria	...	17.5	22.7	26	27		
Belgium	154	233	258	318	307	7	3.5
Brazil	–	12.5	96.5	144	162		
Bulgaria	...	76.1	91	90	87		
Canada	178	413	592	703	670	3	7.7
China	...	100	155	425	451	6	5.2
Finland	...	55.8	147	156	163		
France	64.4	258	264	298	294	9	3.4
Germany	113	372	485	471	481	5	5.5
India	...	23.4	43.6	69	71		
Italy	25.6	157	217	248	252		
Japan	21.2	763	977	856	843	2	9.6
Korea, N.	5.4	90	105	223	240		
Korea, S.	–	2.3	75.7	225	215		
Mexico	48.3	84.6	144	191	195		
Netherlands	13.6	46.2	170	210	203		
Norway	42	61.7	90.6	122	121		
Peru	1	68.7	63.8	123	138		
Poland	87	209	218	177	164		
Portugal	–	–	2	6	5		
Romania	...	60	45.9	50	50		
South Africa	...	27.1	81.4	84	86		
Spain	21.2	88.2	162	257	258	10	3
Sweden	0	0	0	0	8		
Thailand	0	0	0	68	69		
Turkey	–	–	12.6	23	24		
UK	73.2	239	149	128	124		
USA	715	1,174	674	582	606	4	6.8
USSR	110	725	1,060	1,035	1,020	1	11.7
Vietnam	–	–	10	10	10		
Yugoslavia	7.2	65.8	84.5	130	123		
Zaire		63.7	43.8	61	54		
Zambia	22.5	53.5	32.8	20	13		

Includes recovery from scrap in western countries

Zinc ore

thousand tonnes

metal content

	1948	1970	1980	1988	1989	Rank	%
World	1,790	5,520	6,192	7,067	7,138		
Algeria	6	17	8	10	10		
Argentina	13	39	34	37	43		
Australia	194	496	495	759	803	3	11.2
Austria	3	16	22	17	15		
Bolivia	21	47	50	57	75		
Brazil	...	12	70	103	106		
Bulgaria	10	76	70	47	57		
Burma	2	4	4	2	1		
Canada	252	1,239	1,059	1,347	1,215	1	17
Chile	–	1	1	19	18		
China	...	106	180	527	620	4	8.7
Czechoslovakia	...	...	7	7	7		
Finland	8	69	58	64	58		
France	6	19	37	21	27		
Germany	29	127	121	76	64		
Greece	3	9	27	21	25		
Greenland	–	–	86	77	72		
Honduras	–	5	16	14	33		
India	–	8	32	59	65		
Iran	–	58	30	25	25		
Ireland	...	97	229	173	169		
Italy	74	110	57	38	44		
Japan	33	280	238	147	132		
Korea, N.	13	130	130	220	200	9	2.8
Korea, S.	–	23	56	23	32		
Mexico	179	266	243	288	284	7	4
Morocco	2	16	6	11	17		
Namibia	12	46	32	34	39		
Norway	6	10	28	18	15		
Peru	59	321	468	485	597	5	8.4
Poland	96	242	217	183	170	10	2.4
Romania	...	...	40	42	45		
South Africa	...	...	82	90	77		
Spain	47	98	179	277	265	8	3.7
Sweden	36	93	167	193	168		
Thailand	–	–	1	81	87		
Tunisia	2	12	9	9	10		
Turkey	2	24	23	41	37		
UK	–	–	4	6	6		
USA	572	485	349	256	288	6	4
USSR	...	610	1,000	960	940	2	13.2
Vietnam	...	...	10	10	10		
Yugoslavia	37	101	95	71	75		
Zaire	47	104	67	76	73		
Zambia	23	66	43	23	24		

Recent significant political changes

Cambodia As of May 1990, 'State of Cambodia' is its official state title, whilst 'Cambodia' remains the short-form country name. The national flag and the national emblem have also changed slightly.

China/USSR Agreement was reached in May 1991 to transfer almost all the disputed islands in the rivers Amur, Argun and Ussuri to China. The islands transferred lie between the meeting point of the Soviet, Chinese, Mongolian borders, and the junction of the Soviet, Chinese and North Korean borders.

COMECON The Council for Mutual Economic Assistance (COMECON) has been wound up in Budapest, to be replaced by a new economic organization – probably to be called the Organization for International Economic Co-operation (OIEC).

The Warsaw Pact was officially abolished on 1 July 1991, though final details will take until the end of the year to be ratified.

Czechoslovakia The new official name of Czechoslovakia is the 'Czech and Slovak Federal Republic'. It seems likely, however, that it will continue to be called Czechoslovakia in general usage. The two National Republics are Ceska Republika (Czech Rep.) and Slovenska Republika (Slovak Rep.).

EEC The countries of EFTA are willing to join. Austria, Norway, Iceland, Finland and Switzerland have signalled that they are willing to join but, at present, Sweden is reluctant. The EEC has said that no new states will be considered for membership until after the economic and monetary union of 1992.

Czechoslovakia signed a ten-year trade and co-operation agreement with the EEC as its Prime Minister announced that the country was to seek full EEC membership within the next decade. Similar agreements have already been concluded with Poland, Hungary, the Soviet Union and Yugoslavia.

Germany On 3 October 1990 East Germany (DDR) ceased to exist when the five Länder (see below) became part of the Federal Republic of Germany.

The government of Germany is to move from Bonn back to Berlin. The Bundestag voted narrowly for the move, which means they will return to the restored Reichstag building, which was gutted by fire in the 1930s. However, the complete move will take several years.

The five Länder of East Germany have resumed their identity. They are Brandenburg (capital Potsdam), Mecklenburg-Vorpommern (capital Schwerin), Sachsen (capital Dresden), Sachsen-Anhalt (capital Halle) and Thüringen (capital Erfurt).

Poland and Germany have signed a treaty recognizing Poland's present frontiers and its possession of pre-war German territories.

Ivory Coast The capital of the Ivory Coast is now Yamoussoukro (population 120,000 in 1984). It is the birthplace of the country's present leader.

From 1 January 1986, the French version of the name of the country, Côte d'Ivoire, became the official title.

Japan Japan will agree to the initial return of only two of the four Kuril Islands seized by Soviet troops at the end of World War II (Kunashiri and Etorofu). Japan had previously insisted all four islands be returned together.

Korea The Security Council approved without a vote a resolution recommending UN membership for North and South Korea. Both are to be admitted by the General Assembly in New York on 17 September 1991.

Namibia At midnight on 20 March 1990, Namibia became an independent country. The new green, red and blue flag of the new republic has replaced that of South Africa. Namibia becomes the 50th member of the Commonwealth and will become the 160th member of the UN.

New Zealand Two new regions have been formed in North Island – Waikato and Wanganui-Monawatu. In South Island, Nelson and Marlborough have amalgamated to form Nelson-Marlborough. Westland is now known as West Coast. The name for Gisborne district of North Island is East Cape.

Palau Voters in Palau once again rejected a plan to bring the island into a compact of free association with the United States.

Romania On 28 December 1989 the provisional government in Romania announced that henceforth the official name of the country is to be Romania rather than 'The Socialist Republic of Romania'. It also announced that the country's flag is now a simple tri-colour of blue, yellow and red.

UK Humberside was recommended for break-up by the Local Government Commission on 3 July 1991. The Commission proposed that the Humberside districts of Cleethorpes, Glanford, Great Grimsby and Scunthorpe would return to Lincolnshire. The districts of Boothferry, East Yorks., Holderness, Hull and Beverley (the former East Riding) would remain in a reduced county of Humberside. The Commission recognized that the Humber Bridge had failed to unite the two banks of the river, socially or economically. It is likely that Humberside will change its name to East Yorkshire.

USSR The committee working on the draft of the new Union Treaty, which is chaired by Mikhail Gorbachev, has agreed to rename the future Soviet Federation 'The Union of Soviet Sovereign Republics'.

Lithuania declared itself independent of the USSR on 11 March 1990 and has changed the name of the country to 'The Sovereign Republic of Lithuania'. The declaration transfers all power of legislation to Lithuania's capital Vilnius.

Latvia is to become the second of the Soviet Union's constituent states to secede. A draft proposal being put to the Latvian parliament renounces the 1940 annexation by the USSR. However, unlike Lithuania, it does not advocate direct immediate independence, but rather a transitory period to self-rule.

Estonia has followed Latvia down the road to independence. As in Latvia, the new laws passed call for the reinstitution of the pre-Soviet constitution, but does not go as far as cancelling Soviet authority. The name of the country has been changed to 'The Estonian Republic' as opposed to 'Soviet Socialist Republic'.

Following the lead of the Russian Federation and the Baltic Republics, Ukraine has voted to restore independent sovereignty. The parliament in Kiev said that Ukraine intended to become a neutral state, with its own army and currency.

Armenia declared itself independent on 23 August 1990, thus going further towards secession than any of the other Soviet Republics apart from Lithuania. The vote, passed by the Armenian parliament, grants Armenia an army, currency and control of foreign affairs.

Georgia declared formal and complete independence from the Soviet Union on 9 April 1991 in an effort to force the central government's hand over the secession issue.

Moldavia is the latest republic to challenge the authority of Moscow. The parliament in Kishinev voted overwhelmingly to declare sovereignty which, in effect, is a move to make the Moldavian constitution supreme over that of the Soviet Union.

Yemen The two Arab states of North and South Yemen merged to become one country, the Yemen Republic, on 22 May 1990. The withdrawal of subsidies from the Soviet Union to Communist South Yemen had left it little choice but to combine with the North.

The Republic of Yemen was welcomed into the United Nations on 23 May 1990. The republic has Sana (or Sana'a) as its political capital, with Aden as its economic and commercial capital. The new flag is that of North Yemen but without the green star in the centre.

Yugoslavia Croatia and Slovenia unilaterally declared their independence from Yugoslavia on 25 June 1991.

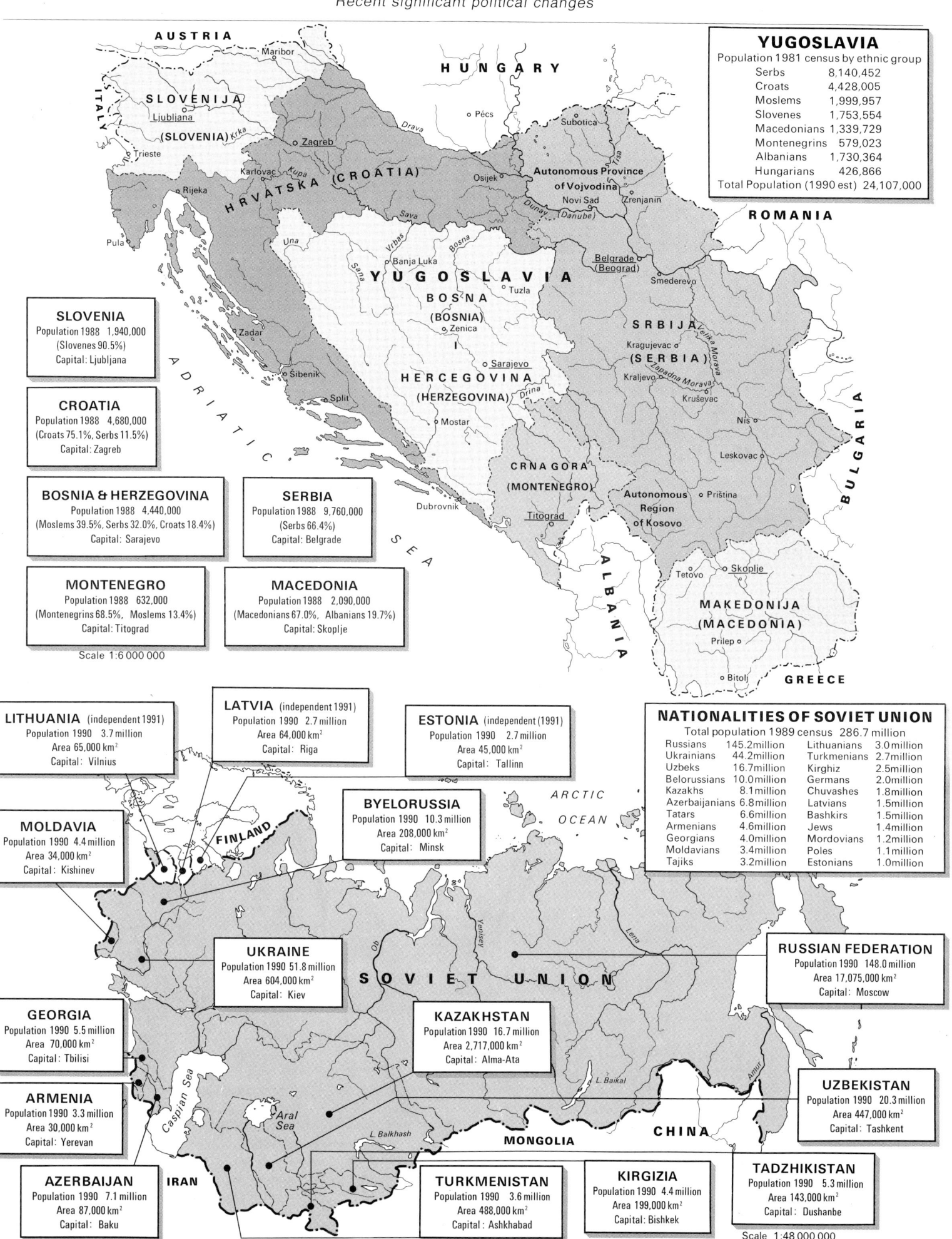

Recent placename changes

Bulgaria

Tolbukhin	Dobrich

Burma

Irrawaddy (R.)	Ayeyarwady
Salween (R.)	Thanlwin
Bassein	Pathen
Pegu	Bago
Moulmein	Mawlamyine
Tavoy	Dawei
Rangoon	Yangon
Burma	Myanmar

Czechoslovakia

Gottwaldov	Zlin

France

Département des Côtes-du-Nord	Département des Côtes d'Armor

Germany

Karl-Marx-Stadt	Chemnitz

Turkey

Urfa	Sanliurfa

US outlying areas

Truk	Chuuk
Moen	Weno

USSR

Andropov	Rybinsk	Russia
Brezhnev	Naberezhnyye Chelny	Russia
Brezhnev (region)	Zavodskoy	Russia
Chernenko	Sharypovo	Russia
Chernenko (region)	Khorgos	Russia
Frunze	Bishkek	Kirgizia
Georgiu-Dezh	Liski	Russia
Gor'kiy	Nizhniy Novgorod	Russia
Gor'kovskaya Oblast	Nizhegorodskaya Oblast	Russia
Gotval'd	Zmiyev	Russia
Kalinin	Tver	Russia
Kalininskaya Oblast	Tverskaya Oblast	Russia
Kapsukas	Mariyampole	Lithuania
Kingissep	Kuressaare	Estonia
Kirovabad	Gyandzha	Azerbaijan
Kuybyshev	Samara	Russia
Kuybyshevskaya Oblast	Samarskaya Oblast	Russia
Lazovsk (region)	Synzherey	Russia
Leninabad	Khodzhent	Tadzhikstan
Leningrad	Sankt Peterburg	Russia
Ordzhonikidze	Vladikavkaz	Russia
Rybach'ye	Issyk-Kul	Kirgizia
Shaumyan (region)	Khataya	Russia
Ustinov	Izhevsk	Russia
Voroshilov Oblast	Lugansk Oblast	Ukraine
Voroshilovgrad	Lugansk	Ukraine
Zhdanov	Mariupol	Ukraine
Zhdanov (region)	Beylagan	Ukraine

USSR names likely to change, but not confirmed as yet:

Mayakovsky	Bagdati
Sverdlovsk	Yekaterinburg
Lazo	Synzherey

Raw materials, engineering and energy projects

Coal

China Although it cannot be confirmed, it is believed that work has begun on China's largest coal project in Inner Mongolia. The open-cast mine at Heidaigou is anticipated to have a productive capacity of 12 million tonnes per year. A power station with a capacity of 200,000 kilowatts and a 215-kilometre railway linking the mine with the Beijing-Baotou and Datang-Qinhuanhdao railways are to be built. The project, which western experts believe will be completed by 1993, is reported to cover 1,300 square kilometres.

UK Kent coalfield is now no longer in production. In Scotland, only Longannet colliery is still in production following the placement of Monktonhall and Frances on care and maintenance.

British Coal announced in December 1990 that the Lea Hall Colliery at Rugeley, Staffordshire, would close at the end of the year with the loss of 800 jobs.

Copper

Chile The Zalvidar (Antofagasta) copper mine, 150 kilometres south-east of Antofagasta, is expected to start production in early 1993. Estimates suggest reserves of 60 million tonnes of average grade 1.6% copper.

France The first copper mine to open in France for more than a century is due to start up in 1992 at Chessy, near Lyons.

Zaire The Kambove (Shaba) copper mine ceased production in January 1989. The closure was a result of depleted reserves and marks the end of a century of mining in the area.

Gold

South Africa The Weltevreden (Orange Free State) gold mine is to be opened 8 kilometres south-west of Orkney. Production of the first phase should begin in late 1992.

Lead/zinc

Spain A substantial lead/zinc deposit was discovered in Los Frailes in 1989. It has the potential to be one of the largest base-metal mines in Europe. Provisional estimates put the reserves at 50 million tonnes containing 14% lead/zinc. If the go-ahead is given, production could begin in mid-1992.

Nickel

Australia Plans were announced in November 1989 to produce 20,000 tonnes of nickel per annum from a new open-pit mine at the Mount Keith (Western Australia) deposit. Reserves are estimated at 270 million tonnes containing 0.6% nickel.

Canada The McCreedy East (Ontario) mine is to be developed as the first large nickel project in the Sudbury belt for over 20 years. Production is expected to begin in 1993.

Tin

UK In March 1990 it was announced that Wheal Jane mine at Baldhu near Truro was to close over a period of nine months. This decision means that there is now only one remaining tin mine in Cornwall, that at South Crofty, which is being run by volunteers who work at an average of 30-40% below their previous pay levels.

Other minerals

Angola Development of a mineral sands project is going ahead near Angoche (Nampula), with commercial production expected to begin in 1992. Output is estimated at about 422,300 tonnes of ilmenite, 37,500 tonnes of zircon, 8,060 tonnes of rutile and 1,000 tonnes of monazite per annum. Life expectancy of the mine is at least ten years.

Ireland A coastal quarry is to be developed in Ireland to supply high-quality aggregates to Europe. The quarry will be at Bantry Bay, and will be one of only two major deep-water coastal operations in the British Isles. It will be operational from early 1992.

Tanzania In May 1991 it was announced that graphite mining is to be started at a site in the Morogoro region, west of Dar es Salaam, and there are reserves in other areas of Tanzania.

Oil and gas production

Chad Chad is to become an oil producer, it was announced in April 1991. The small Sedigi field near Lake Chad, which was discovered in the 1970s, will be developed with the oil piped to a 3,000-barrels-a-day refinery.

China China has discovered a large oilfield in the Xinjiang region, which it says will produce 2 million tonnes of crude annually within the next few years. Output at the field, which is located 170 kilometres north-west of the regional capital of Urumqi, is expected to rise to 4 million tonnes a year by 1995.

Colombia A significant oil discovery was reported in July 1991 in the new Cusiana field, about 150 kilometres north-east of Bogota, in the eastern foothills of the Andes.

Czechoslovakia In March 1991 it was announced that the Transalpine Oil Pipeline (TAL) which connects Trieste, Italy, to Karlsruhe, Germany, is to be extended to Czechoslovakia. It will run from Ingolstadt to the Kralupy and Litvinov refineries north of Prague.

Egypt Egypt's Suez-Mediterranean pipeline (SUMED) has, after 14 years of use, replaced the Suez Canal as the main crude oil transit link from the Gulf to the Mediterranean. Approximately twice the amount of oil went through the pipeline in 1990 compared to that passing through the Suez Canal.

Iran A giant oilfield, 80 kilometres long and 12 kilometres wide, was discovered in 1989 at Lamard in Fars province. It was described by the country's oil ministry as the third largest in the country.

Mexico In May 1991 it was reported that a big oilfield had been found in the Gulf of Mexico. At a depth of 1,000 metres it is one of the deepest commercially-viable fields. BP say it is potentially the biggest prospect since the discovery of the North Sea fields. Production could start within five years.

Tunisia Two new oil wells are to be drilled over the next four years at Grombalia, at the base of the Cape Bon Peninsula. A new 120-centimetre gasline is to be constructed from the Algerian border to the wells. Due for completion in 1994, it will allow for doubling Algerian gas exports to Italy.

UK Occidental Petroleum announced that it is to install another production platform in the Saltire field (formally East Piper). Production is due to start in 1992. The field itself is estimated to contain reserves of about 100 million barrels of oil.

Occidental Petroleum announced in March 1990 that it had found oil 6 kilometres west of its Claymore field in the North Sea. Further drilling is required before a decision is made as to whether to develop the find.

In March 1990 BP began seeking permission to create a 6-hectare island just outside the entrance to Poole Harbour. The area chosen is amongst the most environmentally sensitive in the country and the plan is likely to meet large-scale protest. BP plan to use the island as a site for an oil well to tap the Sherwood oil reservoir.

A gas pipeline between 266 and 290 kilometres long is to be built between Bacton, Norfolk, and the Thames Estuary. It will be the first large-scale privately built pipeline in the UK and is expected to go into service by December 1992.

Construction was due to begin in early 1990 on a 338-kilometre gas pipeline between the Beryl field (North Sea) and a planned processing plant at St Fergus, Grampian.

East Midland Oil and Gas was given approval in 1990 to develop its Whisby onshore field near Lincoln. Production is likely to be 300-400 barrels per day initially.

In 1990 an American exploration company made a large gas discovery in block 110/13 near the Morecambe field. Analysts say that the find may hold as much as 34 billion cubic metres of recoverable gas and, being close to the Mersey Estuary, would be relatively cheap to develop.

British and Italian engineers are to begin construction of the Tiffany field in the North Sea. A medium-sized field by North Sea standards, Tiffany's reserves, along with neighbouring Toni, are stated as being in the region of 175 million barrels of oil and 3 billion cubic metres of natural gas. Tiffany, Toni, together with Thelma and south-east Thelma, should all be on stream by 1996.

Production from the Wareham onshore oilfield in Dorset started in late February 1991. Initial output of 1,000 barrels a day was expected to build up to 35,000 by the end of March. Seven production wells and four water-injection wells have been drilled to tap the 5-million-barrel reservoir. Oil is piped to the Wytch Farm gathering station.

USSR In February 1990 it was announced that a new 2,750-metre well had been bored on the Sakhalin shelf. The new well will provide access to oil and silicon deposits as well as increasing knowledge of the shelf's geology.

Oilfield statistics

Offshore fields under production

Field name	Discovery date	Year of first production	Estimated recoverable reserves originally present in the field (million tonnes)	First year of peak production	1990 production (million tonnes)	Cumulative total production to end 1990 (million tonnes)
Alwyn North	1975	1987	26.4	1989	4.1	11.3
Arbroath	1969	1990	9.9	1991	1.1	1.1
Argyll	1971	1975	9.5	1977	0.2	9.5
Auk	1971	1975	12.3	1977	0.4	11.4
Balmoral	1975	1986	8.9	1987	1.7	7.0
Beatrice	1976	1981	17.0	1985	1.2	15.6
Beryl 'A'	1972	1976	68.0	1980 ⎫		
Beryl 'B'	1975	1984	32.2	1990 ⎬ 4.6		61.4
Brae North	1975	1988	26.4	1989	3.3	7.7
Brae South	1977	1983	40.0	1985	1.5	27.0
Brent	1971	1976	241.1	1984	3.5	171.9
Buchan	1974	1981	10.9	1983	0.9	10.5
Claymore	1974	1977	70.3	1984	2.0	47.0
Clyde	1978	1987	20.5	1988	1.8	7.7
Cormorant North	1974	1982	52.6	1986	3.0	31.8
Cormorant South	1972	1979	24.7	1986	1.2	15.7
Cyrus	1979	1990	1.7	1990	0.3	0.3
Deveron	1972	1984	2.4	1986	0.2	1.7
Duncan	1981	1983	2.5	1985	0.1	2.2
Dunlin	1973	1978	49.1	1979	1.7	39.7
Eider	1976	1988	11.5	1990	2.0	4.0
Forties	1970	1975	329.3	1978	9.2	274.2
Fulmar	1975	1982	60.6	1986	6.9	53.7
Glamis	1982	1989	2.3	1990	0.8	4.3
Heather	1973	1978	13.5	1982	0.6	12.0
Highlander	1976	1985	7.4	1986	0.9	6.3
Hutton	1973	1984	26.3	1986	1.4	17.5
Hutton N.W.	1975	1983	19.3	1984	0.8	13.3
Innes	1983	1985	0.8	1985	0.1	0.8
Ivanhoe/Rob Roy	1975	1989	14.0	1989	1.1	1.6
Kittiwake	1981	1991	9.2	1992	0.2	1.9
Magnus	1974	1983	89.0	1985	6.3	45.1
Maureen	1973	1983	28.0	1984	2.2	23.2
Moira	1988	1990	0.8	1991	0.1	0.6
Montrose	1969	1976	13.1	1976	0.1	10.6
Murchison	1975	1980	45.6	1983	1.4	29.3
Ness	1986	1987	3.9	1988	0.5	2.3
Ninian	1974	1978	141.0	1982	5.7	120.7
Petronella	1975	1986	1.3	1987	0.5	1.8
Piper	1973	1976	126.9	1979	0.0	111.2
Scapa	1975	1985	8.6	1988	1.2	4.2
Statfjord (UK)	1974	1979	480.0	1988	5.2	41.8
Tartan	1975	1981	15.6	1986	0.7	9.4
Thistle	1973	1978	53.0	1982	1.3	47.2

Onshore fields under production

Field name	Discovery date	Year of first production	Estimated recoverable reserves (million tonnes)	First year of peak production	1990 production (million tonnes)	Cumulative total production to end 1990 (million tonnes)
Beckingham West	1985	1987	0.01	1988 ⎫		
Crosby Warren	1986	1987	0.45	1987		
Farley's Wood	1983	1985	0.02	1986		
Horndean	1983	1988	0.04	1989		
Humbly Grove	1980	1986	0.5	1987 ⎬ 1.8		6.6
Nettleham	1983	1985	0.09	1987		
Stainton	1984	1987	0.03	1988		
Welton	1981	1985	3.6	1988		
Wytch Farm	1973	1979	37.6	1990 ⎭		

Offshore fields under development

Field name	Discovery date	Year of first production	Estimated recoverable reserves (million tonnes)	First year of peak production
Emerald	1981	1990	9.6	1991
Gannet	1973	1992	22.4	
Miller	1983	1992	40.2	1993
Nelson	1988	1992	21.5	
Osprey	1974	1991	8.3	1992

Gasfield statistics

Offshore fields under production

Field name	Discovery date	Year of first production	Estimated recoverable reserves originally present in the field (thousand million m³)	First year of peak production	1990 production (thousand million m³)	Cumulative total production to end 1990 (thousand million m³)
Alwyn North	1975	1987	27.0	1989	2.9	8.7
Amethyst	1972	1990	23.9	1992	0.7	0.7
Audrey	1976	1988	22.0	1990	2.4	4.9
Barque	1983	1991	8.9	1992	2.4	4.9
Bure	1983	1987 ⎫		1988	0.3	1.1
Thames	1969	1986 ⎬ 8.4		1987	1.0	4.2
Yare	1969	1987 ⎭		1987	0.2	1.0
Camelot	1987	1990	6.0	1990	0.7	1.0
Cleeton	1983	1988	6.0	1989	1.3	2.7
Clipper	1983	1990	15.8	1992	0.3	0.3
Esmond	1982	1985 ⎫		1986	1.4	6.5
Forbes	1970	1985 ⎬ 15.1		1986	0.1	1.4
Gordon	1969	1985 ⎭		1986	0.6	2.9
Frigg	1972	1977	69.0	1980	1.6	67.4
Hewett & Della	1966	1969	109.5	1977	2.3	99.2
Indefatigable	1966	1971	131.5	1986	4.1	105.8
Leman	1966	1968	315.9	1976	8.5	253.7
Morecambe South	1974	1985	114.0	1989	4.8	9.1
Ravenspurn	1983	1990	55.4	1992	1.6	1.7
Rough	1968	1985	10.2	1988	0.0	4.4
Sean, N. & S.	1969	1986	15.3	1987	0.6	1.9
Valiant North	1970	1988 ⎫		1991	0.7	1.3
Valiant South	1970	1988		1991	0.9	1.5
Vanguard	1982	1988 ⎬ 48.0		1991	0.3	0.8
Vulcan	1983	1988 ⎭		1991	1.7	4.0
Victor	1972	1984	26.0	1984	1.5	9.5
Viking	1965	1972	79.3	1974	1.6	72.9
Welland, N.W. & S.	1984	1990	8.5	1991	0.3	0.3
West Sole	1965	1967	53.6	1971	1.5	38.6

Onshore fields under production

Field name	Discovery date	Year of first production	Estimated recoverable reserves (thousand million m³)	First year of peak production	1990 production (thousand million m³)	Cumulative total production to end 1990 (thousand million m³)
Hatfield Moors	1981	1986	0.14	1986	0.1	0.2

Chemicals

Austria By early 1991, two plants in Austria will be making diesel fuel from rapeseed. They will use home-produced rapeseed and thus reduce Austria's energy bill.

USSR Three petrochemical plants are to be built at Tobolsk (Siberia) between 1990 and 2003. Combined capacity will be 130,000 tonnes per annum of propylene and polypropylene.

Agriculture, irrigation and water supply

Turkey Turkey announced that it would hold back the flow of the Euphrates River for one month, from January 1990, to start filling the reservoir behind its newly built Ataturk Dam. When full, the dam will hold four times the annual flow of the Euphrates. It is one element of Turkey's Great Anatolia Project (GAP), a huge investment project for the irrigation of the south-east of Turkey, which will involve 21 dams on the Tigris and Euphrates by the end of the century.

On 13 January 1990, President Ozal of Turkey performed the official ceremony that cut the flow of water along the Euphrates in order to fill the Ataturk Dam.

Turkey reopened the flow of water along the Euphrates following the diverting of the river to fill the Ataturk Dam. The closure was one day short of the month that Turkey had planned and does not seem to have had any adverse effects in the area. Syria and Iraq, who both use the river for hydroelectric power and irrigation, had claimed that the Turkish move would lead to a massive reduction in power output and an 'agricultural disaster'.

UK Thames Water are digging a tunnel 80 kilometres long to carry the new ring main for London's water. It is half finished and will be completed in 1996. It is the largest project of its kind in the world, and is 40 metres deep to avoid foundations of office blocks, London Underground and all the service lines which are underground.

USSR Uzbekistan is pressurizing the Kremlin to revive the scheme for diverting water from the Siberian rivers to Soviet Central Asia, in particular to revive the Aral Sea. In May 1991 Uzbekistan hinted that it would not sign a treaty establishing a new, looser Soviet Union without a commitment from Moscow to save the Aral Sea. The scheme is not likely to receive much support elsewhere because of the enormous costs and environmental risks.

Zimbabwe A new dam is to be built in the Ngezi communal lands, about 215 kilometres south-west of Harare. The dam was expected to be completed by July 1990.

HEP

Central African Republic A new hydroelectric project is to include the construction of a dam to harness the potential of the M'Bali River. The dam will be 780 metres long and should be completed by 1991.

China In March 1990 it was reported that the Dongjiang Dam in Hunan province had recently been completed. It has a generating capacity of 500 megawatts and is 157 metres high.

The Chinese government is to proceed with the Three Gorges Dam on the Yangtze River. It will take 18 years to build and when completed 104 towns with 14 million people will be displaced. There has been considerable opposition to the scheme from many sources.

Colombia The 245-metre-high Guavio rockfill dam in central Colombia was opened on 30 March 1990. The dam is the highest in South America and the second highest in the Americas after Chicasen in Mexico (261 metres). The dam is designed to generate 1,600 megawatts.

France In late 1990 construction began on the Puylaurent Dam in France, a 73-metre-high concrete arch dam with a crest length of 220 metres.

Indonesia In August 1990 it was reported that the Bakaru hydro plant in the Pinrang district of southern Sulawesi was almost ready to come on line, nearly a year ahead of schedule. The plant has a design capacity of 126 megawatts.

Iran/USSR Iran and the USSR agreed in early 1990 to co-operate in the construction of two multi-purpose dams on their common border along the Aras River. The Khoda-Afarin and Qeis-Qal'ehsi Dams will generate 280 megawatts of electricity and provide irrigation for 130,000 hectares of farmland in Iran.

North Korea Work restarted in early 1990 on the controversial 800-megawatt Kumgang (Diamond Mountain) Dam in North Korea. The dam, which was begun in 1986, will cause a 25% reduction in the flow of water in the Han River which flows into South Korea and through Seoul. South Korean officials claim that an accidental or deliberate breaching of the dam would cause widespread destruction in the south.

Four hydro stations were constructed in 1990 on Mount Taesong in the suburbs of Pyongyang, the capital.

Thailand/Burma In January 1990 it was announced that the Thai and Burmese governments had agreed to co-operate in the construction of seven new dams along their common border. The projects are the Kraburi water resource development project, the Kok River project, the four Mae Moei River projects and the Salween River project.

Nuclear energy

Argentina Atucha 1, Argentina's first nuclear plant, reopened on 10 January 1990 having been closed since 1988 due to technical failure.

Canada The state-run Atomic Energy of Canada Ltd is to build ten new reactors over the next 25 years. This follows a period of uncertainty about the future of Canada's nuclear industry.

Czechoslovakia In July 1990 Austria announced plans to help Czechoslovakia deactivate two potentially dangerous nuclear reactors at Bohunice. Having experienced a fall in supplies of Soviet energy and having embarked upon a programme of closing coal-fired stations for ecological reasons, Czechoslovakia is to push ahead with a modernization of its nuclear plants.

Germany The German government is to close five Soviet-built nuclear power reactors in what was East Germany because officials think they are very unsafe and too costly to adapt.

Indonesia Indonesia is to embark upon a programme to build a 600-megawatt nuclear reactor. Work is set to start in the middle of this decade and should be ready for production by the start of the next century.

Pakistan France is to help Pakistan build its first nuclear power station, it was announced on 21 February 1990.

UK Plans for a controversial nuclear-waste reprocessing plant at Dounreay, Highlands, were approved on 25 October 1989. The plant would be able to recycle fast-reactor fuel by recovering plutonium and uranium. Details of construction dates have not been finalized.

Calder Hall, Cumbria, the world's first nuclear power station, is to continue operating long after its original life expectancy. Both Calder Hall and Chapelcross in Dumfries and Galloway are likely to continue in service until after their 40th year in 1996 and 1999 respectively. Both had an original design life of 20 years.

Winfrith nuclear power station in Dorset is to be closed with the loss of 450 jobs. The 23-year-old station will shut in 1992. The area will continue to be used by BNFL as a research centre.

A one-year trial designed to measure the severity of incidents at nuclear power stations began in November 1990. The scale, which has been agreed internationally, runs from one, a minor anomaly, through to seven, a serious accident involving large releases of radioactivity and widespread health and environmental effects.

USSR The Ukrainian Supreme Soviet announced in March 1990 that it would phase out operations of the Chernobyl nuclear reactor over the next five years. Final ratification of the announcement will have to come from Moscow.

Solar power

USSR There is a new solar energy plant in the small Crimean town of Shchelkino. This replaces a nearby nuclear plant which was shut down through public pressure.

Tidal power

UK A prototype 'wave power' electricity generating station started operating in August 1991 near Portnahaven on the Rinns of Islay, Strathclyde. It is expected that wave power can generate the same amount of electricity as HEP.

The project should provide enough electricity for 4,000 people. It uses an air-driven turbine to turn energy from waves entering a narrow rock gully into electricity.

In April 1990 the Government announced plans to back a 1,800-metre barrage across the Mersey Estuary in order to generate electricity from the tides. The barrage, if built, will cost £880 million and generate about 700 megawatts of electricity.

Wind power

Denmark The world's first offshore wind farm, situated off Vindeby, on Louand Island in the Baltic Sea, was due to come on stream early in 1991. Denmark expects to have 10% of its energy produced by wind power by the end of 1991.

UK In August 1990 plans were put forward for the establishment of a wind power farm between Thornton and Oxenhope in the Pennines. The scheme, if approved, could provide enough energy for 4,500 homes. Meanwhile, a meeting of the British Association heard that 30% of Britain's energy could be produced by wind power in the future.

A public inquiry into disputed plans for Britain's first large-scale wind power scheme, on the edge of the Snowdonia National Park, was seen as a test case in April 1991 to decide the pace and scale of commercially exploited wind energy.

Manufacturing

Algeria A 220,000-tonnes-per-annum aluminium smelter complex was given the go-ahead in November 1989 for construction on the western Algerian coast. Completion is expected in 1993.

Italy On 14 November 1989 the decision was reached to close the Bagnoli (Veneto) steel plant by the end of 1990.

UK The Ravenscraig steel strip mill near Motherwell is to close with the loss of 700 jobs. The first phase of shutdown was due to take place from 10 June 1990.

Transport and communications

Air

Hong Kong China has agreed to allow Hong Kong to go ahead with the building of a new airport on the island of Lantau.

Indonesia The Indonesian airline, Garuda, in co-operation with Cathay Pacific, is to open a Surabaya-Hong Kong (and return) direct flight. Juanda Airport was classified international in January 1991, and Surabaya-Singapore (and return service) has been operating from there since then.

Ireland Knock in County Mayo was granted permission to receive transatlantic flights, thus becoming the second airport to be granted permission after Shannon. Although commonly referred to as Knock, the official name is Horan International (after the late monsignor who was responsible for promoting the site).

UK Proposals to build a fifth terminal at Heathrow Airport may be scrapped following the Department of Transport's refusal to meet all but a small proportion of the cost.

A new £400 million terminal was opened at Stansted Airport on 15 March 1991 by the Queen.

British Aerospace bought a majority stake in Liverpool Airport on 31 May 1991 and is now considering developing the facilities into an international airport.

An estimated 150,000 extra passengers have passed through Glasgow International Airport as a result of direct transatlantic flights introduced in the early part of 1990.

Manchester Airport is aiming to become one of the 12 biggest in the world, with plans to build a second runway in 1998 and treble capacity to almost 30 million passengers a year. It would increase traffic movement from 30 to 42 an hour, the same as Heathrow. However, there is fierce local opposition. At present Manchester is 17th largest in the world, and the fastest growing airport in Europe.

A feasibility study for an international airport, to be sited in the Thames Estuary, was due to be submitted to the Government at the end of March 1990. The proposed airport would be built on reclaimed land on West Shingle Bank, 8 miles north-east of the Isle of Sheppey. It would have four runways and handle up to 45 million passengers a year. This compares with Heathrow's 39 million passengers a year.

USSR Leningrad's Pulkovov Airport looks set to build a new runway and a complex of luxury hotels.

British Airways have linked up with the Soviet and Russian aviation ministries to set up a new commercial airline operating from Domodedovo, Moscow's third airport. Hard currency profits will be used to reconstruct the air terminal at Domodedovo which is currently used for internal flights.

An airport with a handling capacity of 2,500 passengers per hour is under construction at the resort of Sochi on the Black Sea. It is scheduled to open in 1992.

Zimbabwe Harare International Airport is to have a new terminal. Construction began in March 1991. Improved terminal facilities will lure more airlines to Zimbabwe and provide indirect revenue for the tourist industry.

Canals

Germany Since 11 August 1989, large draught barges have been able to transit from Dortmund in the Ruhr to Rotterdam. This follows the enlargement of the Wesel-Datteln Canal which connects Dortmund to the Rhine.

UK A new aqueduct costing £4 million was opened on 30 July 1991. It carries the Grand Union Canal over the Grafton Street dual carriageway on the north side of Milton Keynes.

Ferries

Japan/USSR The Japanese city of Wakkanai on Hokkaido and the Soviet port of Korsakov on Sakhalin have been linked again by a ferry service for the first time since the end of World War II.

UK A new ship has been built for the Isles of Scilly to operate inter-island commuter services, special trips and charters. The *Firethorn* carries 100 passengers. It will operate daily services linking Bryher, Tresco and St Mary's.

On 28 June 1991 the Sealink Stena Line ferry left Southampton for Cherbourg, thus reinstating Southampton as a continental ferry port for the first time since the mid-1980s. It will run twice a day in the peak season, once daily at other times. The service is part of an attempt to draw motorists away from the Channel Tunnel.

The Liverpool to Belfast ferry closed in October 1990.

Sea Cat, a catamaran capable of carrying 450 passengers and 80 cars, has begun operating between Portsmouth and Cherbourg. The journey time has been effectively halved from 5 hours to 2 hours 40 minutes.

There are plans to upgrade the existing ferry services along the River Thames in London.

Ports

UK Dover is building a new ship berth to cater for larger ferries. It should be completed by the end of 1992. Ferry operators at Dover have ordered seven new ships to enter service over the next 18 months.

The small Devon port of Exmouth closed permanently to commercial vessels in December 1989. The decision to close was taken three months ahead of schedule.

Traffic lights for river traffic will be installed on 19 London bridges from Putney Bridge to Tower Bridge.

An £8 million extension to Teesport Container Terminal has been opened, which will double the port's capacity to 110,000 containers a year.

Railways

Canada Canada's railway passenger system was effectively halved on 15 January 1990 when 18 of the 38 routes owned by VIA Rail, the government-owned passenger service, were eliminated. Among them was the famous Canadian Pacific line, along which trains operated daily between Montréal and Vancouver. New trains will take a more northerly route and only run three times a week.

Chile/Argentina Construction of a new tunnel through the Andes has been proposed, linking Juncal in Chile with Puente de Inca in Argentina. The tunnel would be 2,720 metres above sea level and 15 kilometres long. It would supplement the Cristo Redentor tunnel which is difficult in winter, being 3,151 metres above sea level.

China/USSR The first freight trains have started running from Xinjiang-Uigur across the Alatan Pass to Kazakhstan. This bridges the gap in a line which runs from the Atlantic to the Pacific, and will be serious competition for the Trans-Siberian Railway.

China A 173-kilometre line running between Shangqiu in Henan province and Fuyang in Anhui has been opened. It forms part of a 1,000-kilometre 'coal corridor' paralleling the Beijing-Shanghai line.

Chinese Railways have started on the construction of a 4.6-kilometre bridge over the Yangtze River at Jinjiang. It will be the third railway bridge across the river; completion is due by the end of 1992. The bridge will carry the Beijing-Jinjiang section of the planned high-speed line from Beijing to Kowloon.

Chinese Railways have announced plans for a 200-kilometre line linking Nanping and Hengfeng. To be built in 1992-96, it will duplicate the busy artery between Fujian and Jiangxi provinces.

Construction of the 113-kilometre Nanchang-Xinggan section of the Nanchang-Xi'an line was formally inaugurated on 25 October 1989.

On 30 September 1990 Chinese Railways opened the Wachang-Jinjiang line, cutting 361 kilometres from the Beijing-Nanchang journey.

Denmark Construction work on the rail tunnel under the Great Belt is 13 months behind schedule, constant delays being due to the unreliability of tunnel-boring machines. Only 840 metres have been tunnelled so far, out of 2,400 metres which should have already been completed.

France The 'TGV Atlantique' line was inaugurated on 24 September 1989 between Paris and Nantes, reducing the journey time on the Paris-Le Mans section from 100 to 55 minutes. Speeds of up to 300 kilometres per hour can be attained between Paris and Le Mans, and 220 kilometres per hour on the remaining line on to Rennes and Nantes. A branch line to Tours opened in 1990.

Two bypass lines are being worked on in France. One, in Paris, is set to connect the existing TGV lines of Sud Est and Atlantique with the Paris-Lille/Channel Tunnel/Brussels line, avoiding the city centre. The second line, an extension to the Sud Est TGV, will be a bypass of Lyons.

A new automated 'people-mover' is expected to open in Paris in 1991. It will connect Orly Airport to the RER line at Antony. It will have four stations and the journey time should be 5 minutes 45 seconds.

Germany The former major east-west line from the Rhine/Ruhr to eastern Germany, which was abandoned in 1945, was restored on 27 May 1990. The 3.5-kilometre line between Eichenberg (west) and Arenshausen (east) was reopened, once again allowing direct service.

A further 10 kilometres of the Hannover-Würzburg high-speed line opened for traffic at the end of May 1991 between Nörten-Hardenburg and Göttingen.

The station at Bayerisch-Eisenstein, on the border between Germany and Czechoslovakia, was reopened in June 1991.

The route for the new express rail line between Cologne and Frankfurt has been confirmed. It will follow the shortest and cheapest route and will serve neither Bonn nor Cologne-Bonn Airport.

The first of Germany's express rail services left Hamburg station on 2 June 1991. Preparations for express services were started in 1973. Some new lines have been built and others strengthened to take trains which will run at more than 280 kilometres per hour.

Greece Construction of two metro lines is to go ahead in Athens. The two lines will total 18 kilometres with 21 stations and 90% of the route will be underground.

India Work has started on the Roha-Dasgaon section of the Konkan line. It is expected that the 900-kilometre route will be completed in four years.

The north-south route from Delhi to Madras has been electrified. The Delhi to Bombay and Calcutta to Bombay routes will be completely electrified shortly.

A 57-kilometre line, running from Ernakulam to Alleppey (south-west India), was opened on 15 October 1989.

Iraq The Baghdad-Basra railway was reopened on 22 May 1991, following repairs to sections damaged during the Gulf War.

Israel Israel's Ports and Railway Authority announced in August 1990 that it is to proceed with plans to construct a link between Amozorov and Mikven (South) stations, across the eastern part of Tel Aviv. Construction of the 5-kilometre double track will take two years.

Malawi/Mozambique The 815-kilometre railway linking Malawi with the Indian Ocean port of Nacala (Mozambique) has reopened after being closed for five years due to sabotage.

Malaysia Johor State Government has plans for a light rail link across the causeway linking Malaysia to Singapore, joining the new MRT loop line at Woodlands.

Mongolia A 60-kilometre branch line has been opened to link a fluorspar processing plant at Bornuur to the Mongolian Railways main line south-east of Ulan Bator.

Norway Norwegian State Railways are to apply in July 1992 for powers to build the North Norway railway. The long planned link of 390 kilometres would go from Fauske to Tromso and its 81-kilometre branch from Bierkvik to Harstad.

Pakistan/Iran For the first time in 20 years, trains have started to run between Quetta in Pakistan and Zahidan in Iran. The trains from Quetta will mostly carry goods imported via Pakistan ports. On the reverse journey exports of fresh and dried fruit, scrap metal and luxury goods will be carried from Iran, for sale in Pakistan and for export elsewhere.

Senegal A 310-kilometre railway is to be built from Koudékourou to Tambacounda in order to facilitate transportation of iron ore from mines at Koudékourou. Work on the line is due to start in 1992 and should be completed by 1995.

Singapore Work is due to start on an extension of Singapore's Mass Rapid Transit network, to serve the new town of Woodlands in the north-west. The extension will open in 1996-98. A north-east line from Outram Park to Punggol on the coast has been approved, and also a branch from the East Line to Changi Airport. After 2000, when the island has reached capacity, lines will be extended to the islands of Pulau Ubin and Pulau Tekong, off the north-east coast.

Spain Work on the high-speed line from Madrid to Barcelona will start in 1993. A second high-speed line, from Barcelona to the French frontier with one stop at Gerona, could also start in 1993 if authority is given by the central government.

Switzerland Construction of the Rhaetian Railway's Vereina Tunnel will start later this year. The tunnel link will be 19.1 kilometres long.

Taiwan An 85-kilometre 'rapid transit' rail system is to be built in Taipei within the next ten years. Two rapid transit lines are planned, totalling about 60 kilometres plus branch lines.

UK On 9 November 1989 the 8-kilometre landward service tunnel between Dover and the Channel Tunnel terminal at Folkestone was completed. It is a landmark in the construction of the Channel Tunnel.

British Rail announced modifications to the proposed Channel Tunnel rail link between Folkestone and London in May 1991. It involves a 1-kilometre viaduct to avoid sensitive areas between Sandling and Upper Halling, north of Maidstone.

The route for the new fast rail link from the Channel Tunnel to London will terminate at King's Cross, not Waterloo, Malcolm Rifkind, Secretary of State for Transport, announced on 9 October 1991. Publishing his decision in a letter to the Chairman of British Rail, Sir Bob Reid, Mr Rifkind said the Government had decided that a route which approached London from the east, via Stratford, was to be preferred. Mr Rifkind said the preferred route satisfied transport objectives, minimized the impact on the environment and residential property, maximized development potential in East London and the Lower Thames area, removed blight in South London and West Kent, and by terminating at King's Cross should bring economic benefit to the north of England and Scotland as well as the South-east.

A new station has been opened at Worle, Avon, between Weston Milton and Yatton. It will be served by local trains on Bristol/Weston-super-Mare/Taunton lines.

British Rail has stated its intention to close the stations at Greatham and Grangetown on the Middlesbrough-Hartlepool-Newcastle line on 25 November 1991.

Work will commence in December 1991 on the construction of a new railway line in the Glasgow suburbs. This will connect the main Glasgow-Edinburgh line with the route from Cowlairs West junction towards Springburn.

The new bridge over the River Ness has officially opened. The old bridge was destroyed by floods on 7 February 1989.

The Solent Link and its new station at Hedge End on the line from Eastleigh to Fareham were officially opened on 9 May 1991.

Clearance has been given for an express rail route between Paddington and Heathrow Airport. The link is due to open in 1995 and the journey should take 16 minutes as against 50 minutes on London Underground. The 7-kilometre branch line will leave the main line just west of Hayes and Harlington station, and

400 metres beyond will go underground. The railway will serve all four airport terminals.

Plans were approved on 16 November 1989 for an extension to the Jubilee line on the London Underground rail network. The 16-kilometre line will link Stratford and Green Park via Waterloo and London Bridge. It will be the first major extension of the network for 25 years. The decision has meant the postponement of plans for other rail developments in London. The line will run from Green Park to Westminster, across the Thames to Waterloo, Southwark, London Bridge, Bermondsey and Canada Water before crossing back to Canary Wharf. The line then crosses once more to Greenwich Point, before going back to the north bank and Canning Town, West Ham and Stratford.

Services have begun on the Docklands Light Railway on the 1.6-kilometre tunnelled extension to an interchange with London Underground at Bank. There will be a 10-minute interval service on the single track spur from 1 July 1991. Full operation cannot begin until the second bore is completed in 1992.

A 29-kilometre light rail network is proposed in Croydon. A street-running loop would link with converted rail routes to Wimbledon, Elmers End

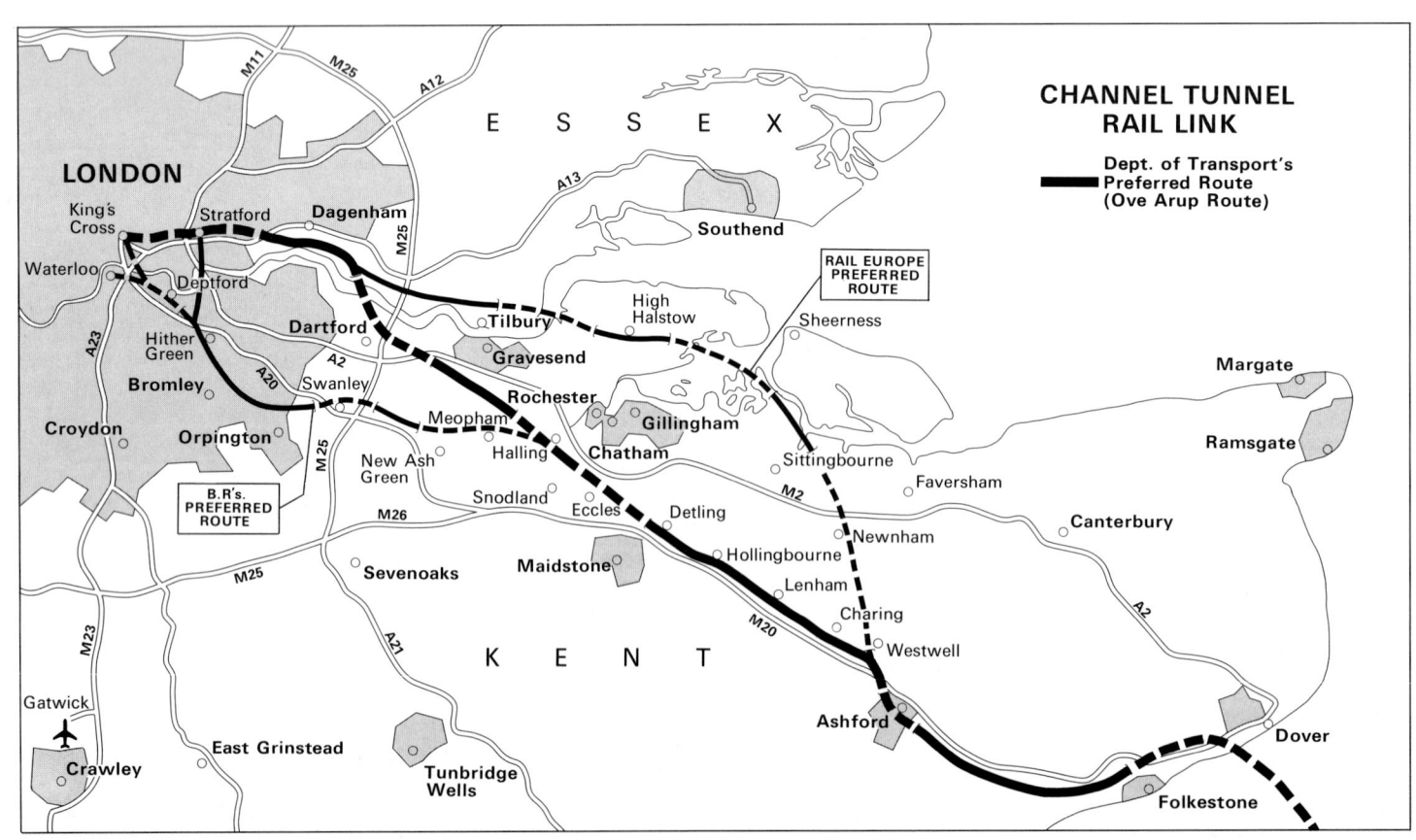

and Beckenham Junction. A new line would go to Addington.

British Rail is planning to close Primrose Hill station and withdraw the twice-daily Watford to Liverpool Street service in an effort to improve commuter services. The Watford-Liverpool Street service would be diverted to run between Gospel Oak and Stratford. If approved, the closure will be the fourth station or track closure during the last seven years – the others being Broad Street station, North Pole junction and Westbourne Park station.

London Underground has approved, in principle, the extension of its East London line to Dalston and East Dulwich. A parliamentary bill is expected to be submitted in November 1991 so the link can be open by 1994.

Holborn Viaduct tube station has closed. It will be replaced by St Paul's Thameslink station. The site of Holborn Viaduct station is being sold off for property development.

A light rail service is due to start in November 1991 between Bury and Victoria, Manchester, to be extended to the city centre a month later. In January 1992 the line to Altrincham should be open and the branch to Manchester Piccadilly should be in service by June 1992.

A new station is due to open in late 1991 at Whiston, near Liverpool. The station is situated between Huyton and Rainhill on the Liverpool to Manchester Victoria line.

Bure Valley Railway opened on 10 July 1990. It is a 15-inch narrow gauge line which runs from Aylsham to Wroxham, with halts at Brampton, Buxton and Coltishall. The railway connects to the main line at Wroxham.

Nottingham is considering a proposed tram system. As yet it is only under discussion, but if all goes well, the first line will open in 1996.

Bicester North railway station in Oxfordshire was officially opened on 10 December 1990.

Electrification of the East Coast railway line between Edinburgh and the Borders has been completed.

The Government has announced approval for Sheffield's proposed 30-kilometre Supertram system.

The four-platform interchange station at Meadowhall has been opened. Space has been reserved for a two-platform Sheffield Supertram terminus.

Swinton station opened on 14 May 1990. The station is situated on a new line which connects the Sheffield-Dearne Valley line with the Sheffield-Doncaster line.

Berry Brow station reopened on 9 October 1989 between Huddersfield and Barnsley after 23 years of closure.

A 5.6-kilometre railway line has officially opened, connecting Stansted Airport in Essex with the main line between London's Liverpool Street and Cambridge. The link has been provided to meet the expected rise in passenger traffic at Stansted after the opening of a new £400 million terminal building in March 1991.

The Alloa-Stirling railway line is to reopen from 1994. The 10-kilometre line was closed in 1968 as part of the Beeching cuts.

The 'Paisley Canal' railway line in Strathclyde has been formally reopened to passengers. The stations on the line, at Dumbreck, Corkerhill, Mosspark and Crookston, are on the sites of previous stations but have all been rebuilt. Paisley Canal station is also new, but to the east of the old line.

Three new stations have opened in the Glasgow suburban area. The stations are 'Priesthill and Darnley' between Kennishead and Nitshill on the Barrhead line; 'Shieldmuir', south of Motherwell; and 'Whinhill' on the Wemyss Bay line.

The Tanfield Railway, at Marley Hill, Tyne and Wear, is to be extended 1.6 kilometres to Causey.

British Rail has announced a £750 million investment programme designed to update the 'West Coast' line between London and Glasgow. The improvements are to allow the introduction of high-speed intercity 250 trains, capable of travelling up to 250 kilometres per hour, from 1994. The trains will reduce journey times along the line considerably. For example, journey times between London and Manchester should be reduced by 20 minutes.

Tame Bridge station between Hampstead and Bescot was officially opened on 4 June 1990, serving passengers for Walsall and Birmingham.

A new station has been opened at Bloxwich North on the Walsall to Hednesford line.

Steeton and Silsden station on the Bradford-Skipton line was reopened on 14 May 1990 after being closed since 1965.

Walsden station has reopened. It is near Todmorden on the Caldervale route. Leeds-Bradford-Manchester trains will stop there for the first time since the old station closed in 1961.

The rail connection to the Rowntree's factory off the York-Scarborough line, which closed in 1986, was finally lifted by contractors during early October 1990.

USA A 14.5-kilometre line running south-west from central Chicago to Midway Airport is due to open in 1993.

USSR Work is due to start on the construction of the first express metro lines in Moscow. Eventually five new routes will be constructed. The express metros will have fewer stations, and thus will be cheaper to build than conventional metro lines.

Yugoslavia A high-speed line is to be built from Subotica on the Hungarian frontier to Dimitrovgrad on the Bulgarian border. It forms the central part of the Budapest to Sofia route. Completion is due in 1997.

Roads

Argentina/Paraguay The 2.7-kilometre Paraná River road/rail bridge between Posadas, Argentina, and Encarnacion, Paraguay, was completed in 1987.

Australia The ring road round Australia's coast was completed in 1989 with the building of the final link between Perth and Port Hedland.

Austria The 7.9-kilometre Karawanken Tunnel connecting Austria to Yugoslavia is due to open in mid-1991. The tunnel will link the Tauren and Karawanken motorways in Austria with the Trans-Yugoslavian motorway, now under construction.

Brazil/Peru Brazil and Peru have agreed on the route for a trans-Andean highway link. This would allow Brazilian grain cheaper access to Asian markets and open new markets in Brazil for Peruvian fertilizers. The road will go from Rio Branco in Acre State, to Assis Brasil on the national border, then to Puerto Maldonado, Cuzco and Arequipa in Peru, and eventually to the Peruvian port of Matarani.

China China's longest super highway, the Shenyang-Dalian Expressway, was completed in September 1990. It is four lanes wide and 375 kilometres long. It is the fifth road of its kind to be built in China, and was started in 1984. The 142-kilometre Beijing-Tianjing-Tanggu highway should be completed in 1992.

France/Spain France and Spain have agreed to dig an 8.5-kilometre road tunnel under the Western Pyrenees. The £98 million project is scheduled for completion in 1995.

France The Calais port and Hoverport link road to the A26 autoroute is now open, cutting the time to cross Calais at peak times from 30 minutes to 5 minutes.

Germany Berlin is to reopen the Brandenburg Gate to traffic for the first time since 1961, when the Wall was built. Pedestrians have been using it since the Wall came down.

All road users in Germany, including foreign tourists, are likely to have to pay an annual toll of DM100 (£35) from early 1991 to raise money needed to maintain roads and railways. Lorries and buses face a levy of up to DM1,000.

Greece The Greek Public Works Department is considering proposals for the construction of a 2.5-kilometre suspension bridge over the Gulf of Corinth, between Rion and Antirrion. The estimated cost is £224 million, and an early start on the construction is considered desirable.

Hawaii Hawaii's H3 freeway, the largest construction project in the history of the islands, is well on the way to completion. The H3 is a four-lane highway, 17 kilometres long, and when complete will connect both sides of the island of Oahu. It crosses the largely undeveloped, rugged Koolau Mountains of the interior.

Hong Kong Hong Kong's longest road tunnel, the Tate's Cairn Tunnel, was opened on 1 July 1991.

Indonesia Construction of a 2.7-kilometre bridge to link Surabaja (East Java) and the island of Madura is scheduled to begin in the 1990s with completion due by 2000.

Japan A road tunnel, bridge and tunnel are under construction in Tokyo harbour to connect one side with the other.

Jordan What is believed to be the longest Portland cement concrete road in the Middle East, a 180-kilometre highway from Azraq to Jafr in Jordan, will become part of a 400-kilometre trade route linking Aqaba to the Iraq frontier in the north-east. The project began in 1987 and was due to be completed early in 1991.

Pakistan A highway 165 kilometres long is to be built between Karachi and Peshawar.

Sweden Sweden's parliament voted to build a bridge across the 17-kilometre-wide Oresund to Denmark, thus linking the Scandinavian peninsula to mainland Europe. The Danish parliament was expected to vote in favour in August 1991.

UK Bypasses are to be built for Christleton and Waverton, Cheshire; Turvey, Bedfordshire; Kilkhampton, Cornwall; and Bishop Burton, Humberside. Improvements are scheduled for Ilminster, Somerset; Guestling Thorn and Icklesham in East Sussex; Wooferton, Salop; and Skeffington, Leicestershire.

The 420-kilometre stretch of the Great North Road from London to Newcastle is to become a six-lane motorway by 2000. Initially, the road will be known as the A1(M), although eventually it will probably have to be renumbered. When complete, it will be Britain's longest motorway.

Work on Belfast's cross-harbour road and rail bridges is to start in the autumn of 1991. The bridges will cross the River Lagan uniting Ulster's rail network and joining main road routes. The rail and road viaducts should be completed by 1994 and the whole project by 1997.

The first privately funded turnpike road to be built since the 18th century was announced on 12 August 1991. It will be the 50-kilometre Birmingham northern relief road. Tolls will be the 1991 equivalent of £1.50 for cars and £3 for lorries. The road is not expected to be finished before 1997. Other likely candidates for private finance are a toll motorway to the west of Birmingham; a link between the A1 and M1 at Scratchwood in London; a link to connect the M25 and Chelmsford in Essex; and a new route between Birmingham and Manchester to compete with the M6.

Plans have been announced to widen the busiest section of the M25 to eight lanes. A 7-kilometre section in the Colne Valley, Buckinghamshire, is to be the first stage of plans to widen all of the M25.

A 100-metre bridge is to be built across River Tees to link new Teesdale development with Stockton town centre.

A £1.5 million bridge and road link has opened in Kendal, Cumbria.

The Dumfries bypass has opened. It will remove through-traffic from Dumfries. The road is 8.9 kilometres long and of single-carriageway construction.

The 1-kilometre-long immersed tube tunnel, crossing the River Conwy, will be open for traffic in October 1991. The immersed tube tunnel is the first of its kind to be built in the UK. The tunnel is part of the stretch between Glan Conwy and Conwy Morfa with a total length of 6 kilometres.

Salisbury's long-awaited southern bypass route has been selected. Detailed proposals will be ready within the next two years.

Approval has been given for the digging of a cutting 122 metres long and 30 metres deep through Twyford Down near Winchester. The cutting, intended to carry the final stretch of the M3 from Bar End to Campton, will destroy an area of historic natural beauty.

The M20 link between Maidstone and Ashford opened on 22 May 1991.

The M20 in Kent is to be widened to dual three-lane carriageways between junctions 5 and 6, and 7 and 8; between junctions 6 and 7 it will be dual four-lane. Six bridges are to be demolished and eight built, plus three railway bridges.

Work on the first stage of the new A20 between Folkestone and Dover will begin soon. This covers the 6 kilometres between the Round Hill tunnels at Hawkinge and the junction with the existing A20 at Court Wood.

The Dartford toll bridge across the Thames opened on 30 October 1991.

A new section of the M74 is to be built between Millbank and Nether Abington, Lanarkshire – a distance of just over 11 kilometres.

The 12-kilometre A17 Long Sutton-Sutton Bridge bypass opened in September 1989 between Gedney and the Lincolnshire/Norfolk border.

The Department of Transport has ordered a study into cross-river links east of Tower Bridge. There is likely to be a third Blackwall Tunnel, and a second one at Rotherhithe.

A public inquiry is due to open in late 1991 to discuss the future of a new road bridge across the Thames. The proposal is for a box girder bridge at Gallions Reach to connect the A13 at Beckton to the A2 at Falconwood. Tied up with this proposal is a plan to extend the City Airport in order to accommodate jet aircraft.

The first part of London's 'red route network' (priority roads kept clear of illegal parking to keep traffic moving) came into effect on 7 January 1991 on the A1 trunk road from Highgate to the Angel, Islington.

A contract has been awarded for widening the M5 between junctions 6 and 8 – between Worcester and the M50 junction. This is the final section to be upgraded from two to three lanes. All bridges have to be replaced.

The £300 million extension of the M40 between Oxford and Birmingham was opened on 16 January 1991. The 93-kilometre motorway is one of the longest stretches of carriageway to be opened since the London-to-Birmingham section of the M1 in 1959.

The go-ahead has been given for a motorway service area on the M40 at Ardley, Oxfordshire, adjacent to the Ardley interchange. It is hoped that the service area will open in 1992.

Proposals to up-grade the A1 to motorway between Dishforth and Scotch Corner were announced on 9 July 1991.

Plans were announced on 24 November 1989 for the 16-kilometre M1-A1 link road between the M1 north of the M62 Lofthouse interchange and the A1 at Hook Moor. Plans also include the upgrading of 7 kilometres of the A1 between Hook Moor and Bramham. Construction should start in the mid-1990s with completion unlikely before 2000.

After a 17-year campaign to prevent the building of a link road between the A1 and the M1 across the battlefield of Naseby, Northamptonshire, protestors have been refused a judicial review of the decision to build an access road and two bridges over the proposed link. Work has been completed on the eastern section, and work on the remaining section will start in November 1991 and finish by the summer of 1993.

The preferred route for the extension of the M6 across the English-Scottish border was announced on 13 June 1991. It will connect the M6 at junction 44 to the M74 north of the border and will complete the final link of three-lane motorway between England and Scotland.

Work is expected to begin on the second Severn Bridge early in 1992 downstream of the existing Severn and Wye bridges. The total length of the bridge with its approach viaducts will be just over 4.8 kilometres.

Construction of the road bridge to Skye is due to start in April 1992, for completion by April 1995. It will be a privately financed toll bridge, of box girder construction, with a central span of 250 metres.

General communications

Sweden/Denmark Support is growing in Sweden and Denmark for a bridge connecting Malmö to Copenhagen. At the same time, there is renewed interest in connecting the Danish island of Lolland to Germany which, with the link to Malmö, would mean there would be a direct German-Scandinavian link. The main obstacle to such links was Denmark's objection to a fixed link across the Great Belt. However, now with the road and rail tunnel/bridge planned to open in 1997, these objections have been suppressed.

UK A proposal has been put forward for a bridge supporting a cable car between Woolwich Arsenal and the Royal Docks. The project, which would cost in the region of £20 million, has the support of the London Docklands Development Corporation.

Conservation and environment

Australia The Great Barrier Reef has been designated the world's first 'particularly sensitive area'. Status was granted in February 1991 by the UN International Maritime Organization, and means that the Australian government can insist that ships passing through the reef must carry a pilot.

Brazil A NASA-funded study claimed in March 1990 that the Amazonian rain forest would be destroyed completely within the next 50 to 100 years if the present rate of destruction is not abated.

In June 1990 it was announced that the destruction of the Amazonian rain forest was more extensive than previously admitted. According to the Brazilian Satellite Research Institute, 8% of the rain forest in the area had been destroyed, rather than the 5.6% announced the previous April. This means that 404,000 square kilometres have been lost (a land area almost the size of Sweden). This is equivalent to destroying the area of a football pitch every two seconds for a year. The figures, though catastrophic, are still less than claimed by some environmental groups.

Brazil's government approved a plan in December 1990 to replant 1 million hectares of Amazon forest at a cost of $3 billion (£1.6 billion) over ten years.

Brunei In November 1990 it was reported that a new field study centre was being built in the tropical forest of Brunei, two hours by road from the town of Bangar.

Czechoslovakia From July 1991, factories in Czechoslovakia have five years in which to reduce emissions into the air, which in the past have caused incalculable damage across much of Europe. Crippling fines will be imposed if the factories fail to reach their targets. About 80% of Czechoslovakia's forests are dying from acid rain, and in the industrial areas of Northern Bohemia life expectancy is ten years below the EEC average.

Europe The ozone layer over large parts of Europe has been depleted at a worrying rate. In July 1991 it was reported that the rate of depletion had increased by 8% over an area stretching from northern Britain to southern Spain and had doubled during the last ten years. Demands for a more rapid phase-out of chlorofluorocarbons (CFCs) will be reinforced.

The European Community has agreed to ban the use of CFCs by June 1997 – three years ahead of the deadline laid down in the Montréal Protocol on the protection of the ozone layer.

Hungary The Hungarian foreign ministry announced in May 1991 that it was to abandon work on the Nagymáros Dam which was part of the Gabcikovo-Nagymáros hydroelectric scheme on the Danube.

India India's forests are being destroyed at the rate of 47,500 hectares per annum. Total coverage in February 1991 was down to 64 million hectares.

Japan Japanese plans to build an airport on a coral reef, the Shiraho Reef, have been abandoned. However, it is now planned to build the airport on the next reef along. Conservationists say it is just as likely to damage the Shiraho Reef, which is the most diverse remaining ecosystem in Japan.

Pacific In July 1991 a beach on an isolated and uninhabited atoll in the Pacific Ocean was found to be carpeted with litter and detritus. It has no fresh water to attract passing yachts, so presumably the rubbish was dumped by ships, and much more must still be floating in the ocean.

UK The Secretary of State for the Environment announced in March 1990 that Britain is to spend £1.5 billion to suspend the discharge of British raw sewage into the North Sea.

In January 1990 Shell announced plans, given approval in 1987, to build a pipeline construction yard at Morrich More, north of Inverness. Morrich More is a grade one site of special scientific interest and one of the last areas of natural salt marsh and mud flats in the UK.

The Scottish Office announced in January 1990 the award of a three-year contract to the Macauley Land Use Research Institute to produce maps of areas in Scotland sensitive to acid deposition.

The Government announced plans in February 1991 for nine new 'community forests' to be developed on existing farmland and derelict ground. These nine are in addition to three announced in 1989.

The proposal to build a new village of 3,200 houses in Great Lea, near Reading, Berkshire, was rejected by the Environment Secretary in February 1990. The plan, which was to have included a regional shopping centre and railway station, was rejected because it would have meant a spread of urban development into a mainly rural area.

Publication of maps to show which power stations most need equipment to curb acid emissions have been delayed, although they were promised by the Government for the end of 1990. Friends of the Earth have made their own maps and conclude that eight power stations need restriction to save areas acutely susceptible to acid rain. The areas affected are the Pennines, parts of Scotland, Wales and the West Country. Power stations needing modification are at South Glamorgan, Shropshire, Nottinghamshire, Cheshire, Kent and Fife.

Public access to woodlands could be limited as the Forestry Commission sells off hundreds of hectares to private buyers from early 1991. Details of Forestry Commission lands open to the public are being deleted from the latest editions of the 1:50,000 and 1:25,000 Ordnance Survey maps.

A study published by the Forestry Commission in July 1991 recommends that Britain should greatly diversify its forest planting, so that it has many different types of forests in different locations. This marks a distinct break with the previous policies of commercially viable coniferous forests in environmentally sensitive upland areas.

Britain's largest accessible cavern, the Battlefield Chamber at Ingleton, North Yorkshire, was opened to the public on 1 May 1991. Part of the White Scar cave system, the cavern was first discovered in 1974.

In May 1991 it was reported that high levels of radon gas had been found in 21 counties in England and Wales. Radon is an odourless gas produced by the decay of uranium in the ground, which then seeps into nearby houses. It is the largest cause of radiation to which people are exposed and is commonest in Cornwall. An estimated 2,500 cancer deaths per year in Britain are caused by it.

The National Trust acquired a stretch of the Cleveland coast in May 1991. It comprises 62 hectares along a 1.2-kilometre stretch of coast at Huntcliff near Saltburn. The purchase includes the Guibal Fan House, which was built in 1872 as part of the Huntcliff ironstone mine development. It housed a 4.5-metre-diameter wooden wheel and other mine ventilating equipment. It closed in 1906 and only two other examples of fan houses survive.

Acid rain in Britain will take a long time to eradicate. By 2005, 8% of UK surface area and 4% of fresh water in Scotland will still be sustaining acid rain. Soil will still be affected in the Lake District, Pennines, Peak District, Snowdonia, South Wales, Southern Uplands and south-east Highlands.

By mid-1991 sulphur dioxide emissions had been reduced by nearly 40% since the late 1960s and will be cut further.

Britain's first inner-city butterfly sanctuary opened at Stave Hill Ecological Park, Surrey Docks, south-east London, in the summer of 1991.

A wildlife reserve is to be created on the site of the Barn Elms Reservoirs, which will become redundant in 1992 when the new London water ring main comes into operation. The project aims to convert the site into a complex of lagoons, marshes and water meadows. It will be a sanctuary for wild birds and mammals.

Sellafield in Cumbria was named on 23 July 1991 as the preferred site for Britain's first nuclear depository for radioactive wastes, after an investigation costing £50 million. The alternative site was Dounreay in Caithness. Ninex is to spend another year refining the design and environmental impact, and intends to make a planning application late in 1992.

USA/USSR In January 1990 the two superpowers agreed in Washington, DC, to establish an international park spanning the Bering Strait. The US government proposed 1.1 million hectares on their side of the strait. The USSR is expected to match this in what is described as a 'joint centre for research and information'. Legislation is required in both countries, but it is hoped that the park will be open by 1992. Amongst the benefits will be the ability of Eskimo families, previously divided by political boundaries, to meet again.

USSR From 1 July 1990, the Valdai National Park, situated between St Petersburg and Moscow, will be open to the public. The park, 162,000 hectares in size, is one of the biggest in the USSR.

The Aral Sea has retreated 70-80 kilometres from its former shores. The level has fallen by 14 metres in 30 years and its area has dwindled from 63,800 square kilometres in 1989 to approximately 36,000 square kilometres in January 1991.

Plans to build a new city in Armenia were unveiled in Yerevan on 7 December 1990 at a ceremony to commemorate the anniversary of the Armenian earthquake. The city, to be named Europolis, after the Dutch company which initiated the project, is to have enough jobs and amenities for 250,000 people.

World Twenty countries, including the USA, USSR, Japan and the UK, signed an agreement to freeze 'as soon as possible' the emission of polluting gases responsible for global warming. This is regarded as a significant step towards an international climate convention.

A report from Friends of the Earth claims that tropical rain forest destruction has nearly doubled during the 1980s. The total deforestation rate is now 142,000 square kilometres per year – the equivalent of six football pitches per minute.

A study by British, Norwegian and Swedish scientists has found that the effects of acid rain are worse than at first envisaged. The report in March 1990 found that for many lakes even a 60% reduction in sulphur dioxide, proposed by the EEC for the year 2003, would not be enough to renew fish stocks.

In June 1990 it was reported that the hole in the ozone layer over Antarctica will not return to normal for at least 60 years, even if the world as a whole were to phase out CFCs immediately. At present, 56 countries have ratified an accord to phase out CFCs by the year 2000. The worry is that two of the potentially largest users, China and India, will not sign.

It was later reported that the thinning of the ozone layer over Antarctica had been far worse than expected during 1990. In October, scientists from the US National Aeronautics and Space Administration and the British National Oceanic and Atmospheric Administration concurred that levels had reached the lowest point for the year. It was feared that 1990 might prove to be the worst on record.

The World Ocean Circulation Experiment began in early 1991. It is hoped that a study of the temperature, salinity and circulation of the oceans will give scientists the key to forecast climate accurately for the first time. Meanwhile, accurate measurements will take place at 24,000 locations on the oceans. These will take several years to complete.

In June 1991, scientists at a meeting of the American Geophysical Union asserted that 75% of the drop in ozone levels could be due to the influence of the sun. Electrons and neutrons from the sun could break down ozone, the number of particles varying in each solar cycle. Thus, mankind may not be totally to blame for the rapid decline in ozone levels in the atmosphere.

The Arctic ice-cap has shrunk significantly over the last ten years, according to satellite measurements of global ice coverage in 1991. There has been no change in Antarctic sea ice over the same period.

Major industrial accidents and pollution

Australia A Greek oil tanker broke up and spilled 10,000 to 20,000 tonnes of light crude oil off the west coast of Australia on 21 July 1991. The slick was estimated to be 13 kilometres long and 900 metres wide. Conservation zones and five islands supporting seabirds and colonies of sea lions were threatened.

Brazil A fire that started on 3 February 1990 destroyed more than a third of the 5,200-hectare Polo das Antas Nature Reserve in Brazil. As much as 2,000 hectares of the reserve, 130 kilometres north-east of Rio de Janeiro and the last surviving habitat of the golden lion tamarin, were feared lost.

Canada In March 1990 firemen in Hagersville, Ontario, 145 kilometres south-west of Toronto, managed to extinguish a fire in a tyre dump that had lasted for 17 days. It was feared that toxic by-products of the fire may have found their way into the water table.

Italy An oil tanker, the *Haven*, sank on 14 April 1991 after catching fire. It was a sister ship of the *Amoco Cadiz*, which caused one of the worst oil pollution disasters in 1978. Pollution experts have worked to avert an ecological disaster and also to prevent damage to the Riviera beaches.

Kuwait The most serious environmental consequence of the 1991 Gulf War has been the firing of some 650 Kuwaiti oil wells by retreating Iraqi forces. The resulting fires led to widespread air pollution. Side-effects included acid rain, photochemical smog, reduced solar radiation and the long-term contamination of soil and groundwater. After a concerted effort, the last of the fires was extinguished in November 1991.

The vast oil slick in the Gulf did not cause as much damage as was first feared, due to hot sunshine and slow-moving currents. Environmentalists think the Gulf's high salinity made the oil float on the surface. It then evaporated under the blazing sun. As much as 50-70% of the oil would have disappeared in 24 hours as the lighter constituents evaporated.

Madeira An oil spill, 19 kilometres long and 2 kilometres wide, covered the beaches of Porto Santo Island in January 1990. The origin of the spill was unknown but large numbers of turtles and seagulls were killed.

Morocco An oil slick, 280 kilometres long, from the damaged Iranian tanker *Khark 5* threatened the coast of Morocco in 1990. Up to 70,000 tonnes of oil were believed to have escaped since a series of explosions wrecked the

ship on 19 December. MAP, the Moroccan news agency, estimated that the potential damage to fish stocks could cost 100,000 people their jobs.

Portugal A 24-kilometre oil slick threatened to pollute the coast of southern Portugal after crude oil leaked from a tanker at Cabo de Sines in September 1990.

UK A collision between a tanker and a tug on 22 November 1989 caused a 50-tonne oil spill off the coast at Milford Haven, Dyfed. A 1.7-kilometre-long slick was created.

Up to 24 kilometres of beaches along the Sussex coast were closed after six containers of potassium cyanide were washed ashore in February 1990. The beaches affected were at Brighton, Newhaven, Ovingdon and Cuckmere Haven. Police put a cordon around the area as the fumes from potassium cyanide are fatal in seconds.

Oil sludge from a slick stretching 1 kilometre wide came ashore on a headland near Bigbury Bay, east of Plymouth, on 15 May 1990. As much as 500 tonnes of sludge were expected to wash ashore overnight.

More than 2 tonnes of CFC gas leaked into the atmosphere from the Atomic Research Station near Harwell, Oxfordshire, in July 1990. The leak was equivalent to the amount used in 1 million cans of aerosol spray.

USA Up to 1,800,000 litres of oil washed ashore along the beaches of Orange County, California, from the BP-owned tanker *American Trader* (82,000 DWT). The vessel was holed as it was mooring alongside an offshore pipeline terminal in February 1990. Amongst the beaches affected along a 24-kilometre stretch of coast was the prestigious 'Huntingdon Beach'.

In February 1990, 29 kilometres of Californian beaches were closed after 7.6 million tonnes of part-treated sewage were released into a creek flowing into the Pacific Ocean. The pollution was released into the creek to avoid flooding streets following heavy rain.

At least 450,000 litres of fuel oil and crude were discharged into the Gulf of Mexico following a series of explosions that swept through the Norwegian supertanker *Mega Borg* on 11 June 1990. A further 170 million litres of light crude, which it had been feared would also be discharged, were prevented from escaping. Nevertheless, this still presents an ecological disaster of large proportions.

Two cargo ships collided in Chesapeake Bay during a severe thunderstorm in July 1990, causing an oil spill of at least 160,000 litres.

USSR The Soviet Union has admitted that 300 people died following the disaster at the Chernobyl nuclear power station. This figure is ten times greater than the figure originally released. The cost of the damage caused has officially been declared as 82 billion roubles.

Space research

Australia/USA In July 1990 Australian and US astronomers announced plans to build a large telescope near the South Pole.

French Guiana *Ariane 40* was launched from Kourou, French Guiana, on 22 January 1990. It carried one French satellite and six micro-satellites, including two from the University of Surrey. It will monitor pollution, temperature of the sea surface, destruction of rain forests and the size of polar ice-caps. It can peer through cloud and measure wave heights, current circulation and windspeed over the oceans.

Hawaii On 11 July 1991, a solar eclipse took place. It started at 7.30 am in the Hawaiian Islands, where some of the world's biggest telescopes were following it. The longest blackout, nearly seven minutes, occurred over Mexico and the eclipse ended over Brazil.

Japan Japan's first spacecraft, *Muses A*, was launched on 24 January 1990. The spacecraft will place two satellites in orbit around the Moon, making Japan the third nation to do this (after the US and USSR).

UK In early 1991 astronomers at Jodrell Bank identified a planet outside the Solar System, ten times bigger than the Earth and in orbit round a neutron star. It is believed chances of life on the planet are not high. It is bathed in radiation, mostly in the form of gamma rays.

USA Astronomers have discovered another moon circling Saturn, bringing the number to 18. The discovery was made after analysis of *Voyager 2* data in July 1990.

Astronomers have succeeded in seeing what lies at the centre of our Galaxy, it was reported in November 1990. They have identified two previously unknown objects, one of which appears to lie precisely in the centre of the Galaxy. Shining with an intensity several million times stronger than the Sun, it could be a compact cluster of hot stars or it could be associated with a black hole which many astronomers believe lies at the galactic centre.

Geographical catastrophes

Algeria Two earthquakes, one of them at Richter 6, struck northern Algeria on 29 October 1989, leaving at least 24 people dead. Most damage was in the Tipasa region, 64 kilometres west of Algiers. A week later two more tremors (Richter 4 and 3.5) threatened buildings already weakened.

Australia An earthquake measuring 5.5 on the Richter Scale killed 11 people and caused widespread damage in December 1989 in Newcastle, New South Wales. Although not the greatest magnitude of earthquake suffered in Australia, this is the first to have caused fatalities.

At least four people died and 20 were missing following huge floods in April 1990. The floods, the worst in over a hundred years, affected over 800,000 hectares of Queensland, New South Wales and Victoria.

Bangladesh In early April 1990, 166 people were reported to have lost their lives and a further 1,600 were injured when tropical storms battered Bangladesh for nine days.

The devastating cyclone at the end of April 1990 was estimated to have killed 140,000 people and affected the lives of some 12 million. The cost of rehabilitation and reconstruction in the devastated areas of Bangladesh was put at $1.78 billion. The cyclone, which threatened thousands with famine and disease, also wreaked economic havoc on one of the poorest nations in the world. The salt-producing industry was destroyed, shrimp cultivation washed away and fishing boats swamped.

At least 25 people were killed and a further 20,000 left homeless when the Brahmaputra burst its banks and flooded 300 villages in Bangladesh in July 1990.

Further flooding occurred in Bangladesh in July 1991. A 16-kilometre embankment along the Brahmaputra gave way, and 13,000 homes and rice and sugar cane crops were destroyed.

Cameroon During June 1991 thousands of people in Cameroon were in danger of death either by drowning or by asphyxiation because a giant natural dam on Lake Nyos in northern Cameroon was on the verge of collapse. Water levels were reduced to prevent this. However, lower water levels could have released clouds of suffocating gas dissolved in the lake. (In 1986 1,700 people were killed by the release of carbon dioxide, which had reached saturation level, the carbon dioxide being produced from volcanic vents.) The natural dam was formed out of a volcanic vent some several hundred years ago, but its thickness has been reduced from 300 to 30 metres by erosion.

Chile In June 1991 rescuers were continuing to search for more victims of a mudslide in Antofagasta. The mudslide was due to exceptionally heavy rains and the death toll was put at 72. As many as 750 people were injured and 20,000 people left homeless.

China An estimated 198 people died in heavy flooding which struck Hunan province in June 1990. It was one of the worst floods in the region in 40 years.

In July 1990 it was reported that floods had killed over 300 people in the Yangtze, Hunang and Zhijiang river valleys.

In August 1990, 98 people died in the Chinese coastal provinces in the wake of Typhoon Yancy.

Heavy rains caused a mudslide which killed more than 30 people in a village in south-west China in July 1991. A forecast of further heavy rain raised fears that the Yangtze could burst its banks for the first time in ten years.

In July 1991 it was reported that water levels in all the larger rivers of eastern China were continuing to rise. Eighteen of China's 30 provinces were affected with the worst flooding for many years, and over 1,000 people killed. Many thousands were trapped by floods, while 380,000 houses and 11,000 factories had been submerged.

Colombia On 22 September 1990 a mudslide caused extensive damage to the Calderas powerhouse and substation in the state of Antioquia, approximately 120 kilometres from Medellin. Landslides from the Betulia Mountain reached the Arenosa, damming it and causing an avalanche of mud which covered the powerplant and substation. Eight villages, eight road bridges, 20 kilometres of roads and the small town of San Carlos sustained serious damage. Twenty people were killed and 600 made homeless.

The Galeras volcano began spewing ash and rocks in August 1991. Experts called for the evacuation of people living in the area, about 500 kilometres south-west of Bogota.

Costa Rica At least 74 people were killed and over 800 injured in an earthquake which shook the Caribbean coasts of Costa Rica and Panama on the evening of 22 April 1991. It was thought that banana production, which is the main activity of the region, would be badly disrupted.

Ethiopia Famine worsened in northern Ethiopia in November 1989, threatening the lives of an estimated 4 million people. The drought situation caused a failed harvest which was accentuated by civil war. The worst affected areas were Eritrea and Tigre.

Europe In August 1991 torrential rains caused devastating floods in central and eastern Europe. The Danube flooded Passau; and in Romania at least 100 people died and 13,000 were left homeless. At least five people died in Bavaria and four in Austria in the worst floods for 30 years. Salzburg was declared a disaster area. In the USSR 16 people were drowned in the Black Sea coast area.

India The worst cyclone in more than a decade struck India's south-east coast in May 1990. As many as 962 people were known to have lost their lives in Andhra Pradesh state, but it was feared that the figure could actually be above 1,100. About 8.8 million people were affected by the cyclone. The storms also caused £400 million damage to crops and killed 100,000 farm animals.

The annual monsoon in India in July 1990 left over 100 people dead in northern and western areas. The floodwaters receded leaving 38 people dead in Gujarat and 46 in Rajasthan. Arunachal Pradesh was almost cut off from the rest of the country and in Himachal Pradesh 12 were reported dead. In Bangladesh, 26 people were known to have died with thousands more left homeless.

At least 86 people died in cold weather that swept through India during January 1991. Rivers and lakes were frozen and roads blocked by snow.

Hundreds of people were feared drowned after the Wardha River, in central India, burst its banks in July 1991. Up to 500 people were missing and more than 10,000 were made homeless.

Indonesia Two earthquakes occurred in Irian Jaya, one in April and one in May 1991. The second measured 5.1 on the Richter Scale.

Iran An earthquake measuring 7.3 on the Richter Scale with an epicentre in the Caspian Sea devastated a wide area of north-western Iran on 21 June 1990. During the week that followed the quake, the death toll fluctuated considerably and a precise figure was hard to establish, but conservative estimates put the figure at around 50,000. In some areas of Gilan province, there was 90% damage, including the town of Gilan itself. In the neighbouring towns of Manjil and Loushan, there was reported to be

70% destruction. The earthquake was the worst in Iran since September 1978 when a quake struck the same area.

As attempts were under way to relieve the devastation, a second earthquake of 5.7 Richter struck Rasht, the capital of Gilan. In the following 24 hours, no less than 22 aftershocks, some measuring as high as 5.5 Richter, hit the area. President Rafsanjani later revised the death toll in the June 1990 earthquake in Gilan province to between 35,000 and 36,000.

An earthquake measuring 7.0 on the Richter Scale hit a mountainous area of southern Iran in November 1990. There were no reports of casualties.

A rise in the level of the Caspian Sea during 1991 is threatening Iran's caviar exports. Floods are lapping the peninsula where over half the caviar is processed.

Japan An earthquake in the Pacific measuring 7.1 on the Richter Scale shook the north-east of Honshu Island on 1 November 1989. No casualties were reported.

New Zealand New Zealand suffered its coldest winter for 16 years in 1991, freezing sheep to the ground and killing thousands of birds. The cause seems to have been a shift in the El Nino weather pattern in the Pacific.

Pakistan The death toll from the earthquake which devastated a large part of Pakistan's North West Frontier province on 4 February 1991, measuring 6.7 on the Richter Scale, is estimated at 1,200. Thousands of houses were either destroyed or badly damaged in Malakand, Chitral, Majur and Hazara districts.

Peru At least 200 people were reported missing after a huge mudslide covered the village of San Miguel de Rio Mayo in February 1990. Recent heavy rains were blamed for destabilizing slopes.

An earthquake measuring 5.9 on the Richter Scale killed at least 89 people and injured hundreds more in or around Moyobamba in San Martin department, northern Peru, on 29 May 1990. Electricity supplies and telephones to the region were cut and urgent appeals went out for doctors and medicine.

Philippines In July 1990 an earthquake measuring 7.7 on the Richter Scale struck the city of Cabanatuan in Nueva Ecija province, 100 kilometres north of Manila, claiming at least 90 lives. Thirty-three children were killed when their school collapsed.

Typhoon Mike swept across the Philippines in November 1990 killing at least 72 people, destroying crops, sinking ships and leaving hundreds of thousands homeless. Extensive damage was reported in Cebu, the country's second industrial centre.

Mount Pinatubo erupted on 12 June 1991, shooting ash more than 25 kilometres into the air and raining down a cascade of molten rock. Thousands of people fled in panic and the United States Clark Air Base was evacuated. White-grey volcanic ash shrouded Manila and cut off air links as the fall-out was blown across a wide area by a typhoon. At the same time, earthquakes shook buildings in Manila. Thousands of people trekked through grey mud, using every possible means of transport and carrying their precious possessions, as they fled from the affected area. Rainstorms in turn set off mud-flows, and there were warnings that homes as far as 60 kilometres from the volcano were in danger of being buried. The worst affected provinces were Bataan, Pampanga, Tarlac and Zambales, and damage to crops and livestock has been put at $15.3 million.

Romania On 30 May 1990, one of the most powerful earthquakes ever recorded caused destruction across Eastern Europe. The epicentre was in the Carpathian Mountains of eastern Romania. The intensity was at first calculated as 7.5 on the Richter Scale, but was later reduced to 7.0. The official death toll was put at nine, although Tass claimed there had been more deaths in Moldavia but could give no details. The ostensibly low death toll was due to the fact that the epicentre was between 80 and 160 kilometres below the surface. The effects of the tremor were felt over a large area. In Moscow, the US embassy had to be evacuated.

At least 65 people were drowned, and many more missing, after flash floods devastated 17 villages in eastern Romania on 29 July 1991. Floodwaters burst a dam and left 13,000 people homeless near the city of Bacau.

Taiwan The death toll from Typhoon Ofelia in June 1991, the worst storm to hit eastern Taiwan in 30 years, rose to 17, with 23 missing.

Thailand Typhoon Gay struck southern Thailand on 3 November 1989, leaving over 250 people dead or missing. The majority of casualties were from a capsized ship in the Gulf of Thailand. At least 1,200 houses were destroyed in what was the most violent storm to hit the country for 35 years.

UK On 25 January 1990, 38 people died as 180 kph winds devastated the south of England. All of London's mainline train stations were closed because of fears for the safety of their roofs. Many motorways and main roads were closed due to lorries overturning.

In February 1990, 15 people were killed as 160 kph winds once more devastated the country. Among the incidents, the entire population of Towyn in North Wales (2,000 people) were evacuated after huge waves broke through the sea wall.

Thirty people were feared dead as emergency services continued their search for victims of the storms that swept the British Isles during the first week of 1991.

Underground water supplies dropped to a record low in parts of south-east England in early 1991. Three dry winters have seriously depleted the chalk aquifers on which the South-east depends for supplies.

Snow fell in parts of Sussex on 16 June 1991 and freak weather brought flooding to many parts of southern England. Up to 70 millimetres of snow fell in West Sussex, and parts of the A27 between Portsmouth and Chichester were flooded up to 1 metre deep. Three people were struck by lightning at Hungerford, Berkshire, and a ten-car pile-up occurred on the M3 during a hailstorm.

One of the dying rivers of southern England may be revived by recycling water, it was announced in July 1991. To overcome persistent flow failures of the Bourne in Hampshire, the National Rivers Authority proposes to collect water near Hurstbourne Viaduct and pump it 2 kilometres back upstream to St Mary Bourne, where the river bed is dry.

USA Tornadoes swept through Huntsville, Alabama, on 16 November 1989, killing 14 people.

As a result of estimations and assumptions, the previous casualty count for the October 1989 San Francisco earthquake was set too high. The new figure is put at 67 with several other deaths caused by related heart attacks. Most error was caused by assuming a high number of deaths in the collapse of the Bay Bridge access freeway.

On 19 June 1990 flash floods in Shadyside, Ohio, killed 21 people. Police said there was little hope of finding the remaining 16 missing people alive.

On 28 August 1990, 25 people were killed by a tornado in Illinois, south-west of Chicago.

A week of freezing temperatures in late December 1990 marked the coldest Christmas on record for most cities in the United States.

A tornado devastated parts of Oklahoma and Kansas in April 1991. Twenty-two people were killed when a caravan park was demolished near Wichita.

USSR In July 1990 at least 40 climbers were known to have died in an avalanche in the Pamir Mountains of Soviet Central Asia. The landslide, triggered by an earthquake, destroyed the camp of the expedition to Lenin Peak, the third highest mountain in the Soviet Union.

A powerful earthquake, 7.2 on the Richter Scale, rocked Georgia on 29 April 1991. It was more powerful than the one in December 1988 which killed over 25,000 people in Armenia.

A hurricane swept the Republic of Moldavia during the first week of July in 1991. Fifteen people were killed, 1,300 houses were underwater, and thousands of hectares of crops were ruined.

UK Population

Density Persons/ha., thousands and percentage change between censuses

■ Greater London

	Area km²	Population 1961 thousands	61 to 71 % change	1971 thousands	71 to 81 % change	1981 thousands	81 to 91 % change	1991 thousands	Density Persons/ha.
Greater London	**1,579**	**7,992.3**	**-6.76**	**7,452.3**	**-10.15**	**6,696.2**	**-4.75**	**6,377.9**	**40.4**
Inner London	321	3,492.8	-13.20	3,031.9	-17.65	2,496.8	-5.88	2,349.9	73.2
City of London	3	4.8	-12.5	4.2	38.99	5.9	-32.20	4.0	13.3
Camden LB	22	245.7	-15.87	206.7	-16.98	171.6	-0.64	170.5	77.5
Hackney LB	19	257.5	-14.45	220.3	-18.20	180.2	-8.88	164.2	86.4
Hammersmith and Fulham LB	16	222.1	-15.71	187.2	-20.89	148.1	-7.83	136.5	85.3
Haringey LB	30	259.2	-7.37	240.1	-15.37	203.2	-7.82	187.3	62.4
Islington LB	15	261.2	-22.70	201.9	-20.85	159.8	-2.88	155.2	103.5
Kensington & Chelsea LB	12	218.5	-13.87	188.2	-26.25	138.8	-8.07	127.6	106.3
Lambeth LB	27	341.6	-9.98	307.5	-20.10	245.7	-10.42	220.1	81.5
Lewisham LB	35	290.6	-7.60	268.5	-13.15	233.2	-7.68	215.3	61.5
Newham LB	36	265.4	-10.55	237.4	-11.84	209.3	-4.35	200.2	55.6
Southwark LB	29	313.4	-16.37	262.1	-19.23	211.7	-7.18	196.5	67.8
Tower Hamlets LB	20	205.7	-19.40	165.8	-13.75	143.0	7.34	153.5	76.8
Wandsworth LB	35	335.4	-9.87	302.3	-15.42	255.7	-7.12	237.5	67.9
Westminster, City of	22	271.7	-11.78	239.7	-20.44	190.7	-4.82	181.5	82.5
Outer London	1,259	4,499.5	-1.76	4,420.4	-5.00	4,199.4	-4.08	4,028.0	32.0
Barking & Dagenham LB	34	177.1	-9.20	160.8	-6.59	150.2	-6.86	139.9	41.1
Barnet LB	90	318.4	-3.71	306.6	-4.66	292.3	-3.18	283.0	31.4
Bexley LB	61	209.9	3.43	217.1	-1.06	214.8	-1.68	211.2	34.6
Brent LB	44	295.9	-5.17	280.6	-10.44	251.3	-10.03	226.1	51.4
Bromley LB	152	293.4	4.09	305.4	-3.57	294.5	-4.35	281.7	18.5
Croydon LB	87	323.9	3.09	333.9	-5.18	316.6	-5.37	299.6	34.4
Ealing LB	55	301.6	-0.17	301.1	-7.01	280.0	-5.86	263.6	47.9
Enfield LB	81	273.8	-2.12	268.0	-3.43	258.8	-3.83	248.9	30.7
Greenwich LB	47	229.8	-5.27	217.7	-2.71	211.8	-5.19	200.8	42.7
Harrow LB	51	209.1	-2.82	203.2	-3.54	196.0	-0.87	194.3	38.1
Havering LB	118	254.6	-2.71	247.7	-2.99	240.3	-6.62	224.4	19.0
Hillingdon LB	110	228.4	2.85	234.9	-2.34	229.4	-1.57	225.8	20.5
Hounslow LB	59	208.9	-0.96	206.9	-3.43	199.8	-3.20	193.4	32.8
Kingston upon Thames LB	38	146.0	-3.77	140.5	-5.77	132.4	-1.36	130.6	34.4
Merton LB	38	189.0	-6.19	177.3	-6.99	164.9	-1.88	161.8	42.6
Redbridge LB	56	250.1	-4.08	239.9	-6.21	225.0	-1.96	220.6	39.4
Richmond upon Thames LB	55	180.9	-3.48	174.6	-9.56	157.9	-2.09	154.6	28.1
Sutton LB	43	169.1	0.24	169.5	-0.65	168.4	-2.43	164.3	38.2
Waltham Forest LB	40	248.6	-5.59	234.7	-8.35	215.1	-5.44	203.4	50.9

■ Metropolitan Counties

	Area km²	Population 1961 thousands	61 to 71 % change	1971 thousands	71 to 81 % change	1981 thousands	81 to 91 % change	1991 thousands	Density Persons/ha.
Greater Manchester	**1,287**	**2,719.8**	**0.33**	**2,728.8**	**-4.91**	**2,594.7**	**-5.39**	**2,454.8**	**19.1**
Bolton	140	250.3	3.68	259.5	0.50	260.8	-2.88	253.3	18.1
Bury	99	151.8	15.02	174.6	1.15	176.6	-2.49	172.2	17.4
Manchester	116	662.0	-17.85	543.8	-17.40	449.2	-9.42	406.9	35.1
Oldham	141	215.7	3.85	224.0	-1.87	219.8	-3.82	211.4	15.0
Rochdale	160	189.8	7.01	203.1	2.07	207.3	-5.02	196.9	12.3
Salford	97	294.4	-4.93	279.9	-12.93	243.7	-10.59	217.9	22.5
Stockport	126	256.0	14.14	292.2	-0.86	289.7	-4.45	276.8	22.0
Tameside	103	204.2	8.13	220.8	-1.59	217.3	-2.58	211.7	20.6
Trafford	106	223.2	2.15	228.0	-2.89	221.4	-7.09	205.7	19.4
Wigan	199	272.4	11.20	302.9	1.98	308.9	-2.27	301.9	15.2
Merseyside	**652**	**1,718.3**	**-3.60**	**1,656.5**	**-8.66**	**1,513.1**	**-9.01**	**1,376.8**	**21.1**
Knowsley	97	151.4	28.20	194.1	-10.66	173.4	-14.01	149.1	15.4
Liverpool	113	745.8	-18.20	610.1	-16.36	510.3	-12.15	448.3	39.7
St. Helens	133	174.6	8.25	189.0	0.48	189.9	-7.69	175.3	13.2
Sefton	151	294.5	4.41	307.5	-2.44	300.0	-6.00	282.0	18.7
Wirral	158	352.0	1.08	355.8	-4.58	339.5	-5.13	322.1	20.4
South Yorkshire	**1,560**	**1,303.3**	**1.47**	**1,322.5**	**-1.57**	**1,301.8**	**-4.09**	**1,248.5**	**8.0**
Barnsley	329	223.8	0.80	225.6	-0.31	224.9	-3.38	217.3	6.6
Doncaster	582	268.1	4.85	281.1	2.74	288.8	-1.56	284.3	4.9
Rotherham	283	226.6	7.28	243.1	3.37	251.3	-1.67	247.1	8.7
Sheffield	368	584.8	-2.07	572.7	-6.27	536.8	-6.91	499.7	13.6
Tyne and Wear	**540**	**1,243.8**	**-2.58**	**1,211.7**	**-5.65**	**1,143.2**	**-4.92**	**1,087.0**	**20.1**
Gateshead	143	223.6	0.67	225.1	-5.95	211.7	-7.18	196.5	13.7
Newcastle upon Tyne	112	336.4	-8.35	308.3	-9.93	277.7	-5.29	263.0	23.5
North Tyneside	84	210.5	-1.24	207.9	-4.62	198.3	-4.79	188.8	22.5
South Tyneside	64	184.8	-4.17	177.1	-9.32	160.6	-5.42	151.9	23.7
Sunderland	138	288.5	1.66	293.3	0.61	295.1	-2.81	286.8	20.8

	Area				Population				Density
	km²	1961 thousands	61 to 71 % change	1971 thousands	71 to 81 % change	1981 thousands	81 to 91 % change	1991 thousands	Persons/ha.
West Midlands	**899**	**2,371.8**	**17.77**	**2,793.3**	**-5.27**	**2,646.1**	**-5.55**	**2,499.3**	**27.8**
Birmingham	264	1,183.2	-7.20	1,098.0	-8.31	1,006.8	-7.14	934.9	35.4
Coventry	97	318.3	5.78	336.7	-6.71	314.1	-6.88	292.5	30.2
Dudley	98	254.2	15.62	293.9	1.87	299.4	0.33	300.4	30.7
Sandwell	86	339.5	-2.74	330.2	-6.90	307.4	-8.26	282.0	32.8
Solihull	180	128.2	49.84	192.1	3.28	198.4	-2.17	194.1	10.8
Walsall	106	246.8	10.74	273.3	-2.63	266.1	-3.95	255.6	24.1
Wolverhampton	69	261.6	2.87	269.1	-5.65	253.9	-5.55	239.8	34.8
West Yorkshire	**2,039**	**2,005.4**	**3.10**	**2,067.6**	**-1.46**	**2,037.4**	**-2.59**	**1,984.7**	**9.7**
Bradford	370	452.8	1.99	461.8	-0.93	457.5	-1.84	449.1	12.1
Calderdale	364	201.2	-2.98	195.2	-2.00	191.3	-2.09	187.3	5.1
Kirklees	410	352.9	4.65	369.3	0.68	371.8	-1.13	367.6	9.0
Leeds	562	713.0	3.63	738.9	-4.59	705.0	-4.34	674.4	12.0
Wakefield	333	285.5	5.92	302.4	3.11	311.8	-1.76	306.3	9.2

■ Non-metropolitan Counties

	Area				Population				Density
Avon	**1,346**	**828.9**	**9.29**	**905.9**	**0.39**	**909.4**	**1.14**	**919.8**	**6.8**
Bath	29	84.1	0.71	84.7	-5.55	80.0	-0.12	79.9	27.6
Bristol	110	438.0	-2.60	426.6	-9.05	388.0	-4.56	370.3	33.7
Kingswood	48	67.6	15.09	77.8	7.97	84.0	3.69	87.1	18.1
Northavon	462	69.9	50.36	105.1	13.04	118.8	9.09	129.6	2.8
Wansdyke	323	60.8	16.61	70.9	7.62	76.3	3.15	78.7	2.4
Woodspring	375	108.7	29.53	140.8	15.20	162.2	7.46	174.3	4.6
Bedfordshire	**1,235**	**382.7**	**21.30**	**464.2**	**8.66**	**504.4**	**1.94**	**514.2**	**4.2**
Luton	43	140.1	15.20	161.4	1.61	164.0	2.01	167.3	38.9
Mid-Bedfordshire	504	69.1	29.96	89.8	13.70	102.1	5.78	108.0	2.1
North Bedfordshire	476	103.8	19.94	124.5	6.18	132.2	-0.08	132.1	2.8
South Bedfordshire	212	69.7	26.97	88.5	19.89	106.1	0.66	106.8	5.0
Berkshire	**1,259**	**516.8**	**22.17**	**631.4**	**7.00**	**675.6**	**6.05**	**716.5**	**5.7**
Bracknell Forest	109	43.2	48.38	64.1	27.77	81.9	14.53	93.8	8.6
Newbury	705	82.5	26.55	104.4	15.13	120.2	13.48	136.4	1.9
Reading	40	126.8	10.25	139.8	-5.22	132.5	-7.47	122.6	30.7
Slough	28	93.5	6.42	99.5	-2.51	97.0	1.65	98.6	35.2
Windsor and Maidenhead	198	108.1	14.62	123.9	5.00	130.1	-1.08	128.7	6.5
Wokingham	179	62.7	59.01	99.7	14.24	113.9	19.67	136.3	7.6
Buckinghamshire	**1,883**	**377.9**	**25.99**	**476.1**	**18.84**	**565.8**	**9.49**	**619.5**	**3.3**
Aylesbury Vale	904	90.9	25.85	114.4	16.00	132.7	8.21	143.6	1.6
Chiltern	201	72.3	22.96	88.9	2.81	91.4	-2.95	88.7	4.4
Milton Keynes	310	49.3	35.50	66.8	85.33	123.8	39.18	172.3	5.6
South Buckinghamshire	144	55.5	14.59	63.6	-2.20	62.2	-3.05	60.3	4.2
Wycombe	324	109.9	29.57	142.4	9.41	155.8	-0.83	154.5	4.8
Cambridgeshire	**3,409**	**436.5**	**15.83**	**505.6**	**13.35**	**573.1**	**11.80**	**640.7**	**1.9**
Cambridge	41	95.5	3.46	98.8	-8.50	90.4	11.73	101.0	24.6
East Cambridgeshire	655	45.1	8.87	49.1	9.16	53.6	10.63	59.3	0.9
Fenland	552	62.8	3.03	64.7	2.16	66.1	10.29	72.9	1.3
Huntingdonshire	924	69.0	40.58	97.0	27.32	123.5	13.93	140.7	1.5
Peterborough	334	90.2	17.07	105.6	25.28	132.3	12.47	148.8	4.5
South Cambridgeshire	903	73.9	22.33	90.4	18.58	107.2	10.17	118.1	1.3
Cheshire	**2,328**	**729.7**	**18.75**	**866.5**	**6.90**	**926.3**	**1.19**	**937.3**	**4.0**
Chester	448	102.0	13.43	115.7	0.78	116.6	-1.37	115.0	2.6
Congleton	211	55.3	28.39	71.0	11.27	79.0	4.94	82.9	3.9
Crewe and Nantwich	431	91.3	6.46	97.2	1.03	98.2	3.67	101.8	2.4
Ellesmere Port and Neston	82	56.6	38.69	78.5	3.82	81.5	-3.31	78.8	9.6
Halton	74	82.1	16.81	95.9	27.22	122.0	-0.49	121.4	16.4
Macclesfield	523	113.3	23.48	139.9	6.58	149.1	-1.41	147.0	2.8
Vale Royal	384	87.0	21.61	105.8	4.91	111.0	0.09	111.1	2.9
Warrington	176	142.1	14.36	162.5	3.88	168.8	6.34	179.5	10.2
Cleveland	**583**	**526.5**	**7.83**	**567.7**	**-0.32**	**565.9**	**-4.38**	**541.1**	**9.3**
Hartlepool	94	97.3	2.16	99.4	-5.03	94.4	-6.57	88.2	9.4
Langbaurgh-on-Tees	240	118.5	24.64	147.7	1.22	149.5	-5.22	141.7	5.9
Middlesbrough	54	164.8	-4.55	157.3	-4.77	149.8	-5.81	141.1	26.1
Stockton-on-Tees	195	145.9	11.93	163.3	5.45	172.2	-1.16	170.2	8.7
Cornwall and Isles of Scilly	**3,564**	**343.2**	**11.19**	**381.6**	**12.81**	**430.5**	**9.01**	**469.3**	**1.3**
Caradon	664	48.5	11.55	54.1	25.51	67.9	11.63	75.8	1.1
Carrick	461	63.4	10.57	70.1	8.70	76.2	8.53	82.7	1.8
Kerrier	473	64.7	15.61	74.8	10.96	83.0	4.10	86.4	1.8
North Cornwall	1,195	52.4	7.82	56.5	17.17	66.2	11.33	73.7	0.6
Penwith	303	49.8	3.01	51.3	7.99	55.4	7.22	59.4	2.0
Restormel	452	62.1	16.59	72.4	9.39	79.2	11.49	88.3	2.0
Isles of Scilly	16	2.3	4.35	2.4	8.33	2.6	11.54	2.9	1.8

	Area km²	Population								Density Persons/ha.
		1961 thousands	61 to 71 % change	1971 thousands	71 to 81 % change	1981 thousands	81 to 91 % change	1991 thousands		
Cumbria	**6,810**	**470.1**	**1.28**	**476.1**	**1.74**	**484.4**	**0.52**	**486.9**		**0.7**
Allerdale	1,257	95.4	-0.52	94.9	1.90	96.7	-0.41	96.3		0.8
Barrow-in-Furness	77	75.2	0.13	75.3	-3.59	72.6	-0.96	71.9		9.3
Carlisle	1,030	100.7	0.10	100.8	-0.10	100.7	-0.89	99.8		1.0
Copeland	737	73.5	-2.31	71.8	1.39	72.8	-2.88	70.7		1.0
Eden	2,158	42.5	-1.41	41.9	5.01	44.0	5.23	46.3		0.2
South Lakeland	1,551	82.8	10.39	91.4	6.89	97.7	4.30	101.9		0.7
Derbyshire	**2,631**	**846.8**	**4.70**	**886.6**	**2.29**	**906.9**	**0.85**	**914.6**		**3.5**
Amber Valley	265	102.8	2.43	105.3	3.51	109.0	0.64	109.7		4.1
Bolsover	160	75.4	-4.11	72.3	-2.63	70.4	-1.99	69.0		4.3
Chesterfield	66	94.1	2.23	96.2	1.77	97.9	1.84	99.7		15.1
Derby	78	212.7	3.24	219.6	-1.78	215.7	-0.79	214.0		27.4
Derbyshire Dales	795	62.6	3.99	65.1	2.15	66.5	1.80	67.7		0.9
Erewash	109	92.3	8.13	99.8	2.40	102.2	1.76	104.0		9.5
High Peak	541	70.3	12.66	79.2	3.66	82.1	2.07	83.8		1.5
North East Derbyshire	277	81.1	8.88	88.3	8.04	95.4	0.21	95.6		3.5
South Derbyshire	339	55.5	9.55	60.8	11.35	67.7	5.02	71.1		2.1
Devon	**6,711**	**822.7**	**9.20**	**898.4**	**5.97**	**952.0**	**4.85**	**998.2**		**1.5**
East Devon	817	83.6	16.27	97.2	9.16	106.1	0.09	106.2		1.3
Exeter	44	88.6	8.01	95.7	0.42	96.1	5.20	101.1		23.0
Mid Devon	916	47.1	11.04	52.3	11.09	58.1	9.47	63.6		0.7
North Devon	1,086	62.8	12.26	70.5	11.63	78.7	8.13	85.1		0.8
Plymouth	79	230.4	3.95	239.5	2.00	244.3	-2.25	238.8		30.2
South Hams	887	55.5	7.93	59.9	12.35	67.3	14.86	77.3		0.9
Teignbridge	675	80.4	11.94	90.0	6.22	95.6	12.03	107.1		1.6
Torbay	63	96.3	13.50	109.3	5.76	115.6	5.97	122.5		19.4
Torridge	985	40.5	10.37	44.7	5.82	47.3	10.15	52.1		0.5
West Devon	1,160	37.5	4.80	39.3	9.41	43.0	3.26	44.4		0.4
Dorset	**2,654**	**499.7**	**10.95**	**554.4**	**6.84**	**592.3**	**8.93**	**645.2**		**2.4**
Bournemouth	46	154.3	-0.26	153.9	-5.91	144.8	6.63	154.4		33.6
Christchurch	50	28.5	19.30	34.0	10.88	37.7	7.43	40.5		8.1
East Dorset	355	35.6	44.66	51.5	32.43	68.2	13.20	77.2		2.2
North Dorset	609	36.1	17.17	42.3	9.93	46.5	12.26	52.2		0.9
Poole	64	92.1	16.40	107.2	10.91	118.9	10.09	130.9		20.5
Purbeck	405	32.1	14.33	36.7	10.90	40.7	4.67	42.6		1.1
West Dorset	1,083	68.4	8.33	74.1	5.67	78.3	10.22	86.3		0.8
Weymouth and Portland	42	52.6	3.99	54.7	4.57	57.2	6.64	61.0		14.5
Durham	**2,436**	**605.3**	**0.30**	**607.1**	**-0.41**	**604.6**	**-2.45**	**589.8**		**2.4**
Chester-le-Street	66	42.7	13.11	48.3	8.07	52.2	-1.15	51.6		7.8
Darlington	198	95.1	2.94	97.9	-0.10	97.8	-1.12	96.7		4.9
Derwentside	271	99.9	-7.81	92.1	-4.34	88.1	-3.75	84.8		3.1
Durham	190	75.4	8.22	81.6	3.68	84.6	0.47	85.0		4.5
Easington	143	110.9	-1.80	108.9	-7.44	100.8	-4.46	96.3		6.7
Sedgefield	220	82.9	6.63	88.4	4.98	92.8	-3.88	89.2		4.1
Teesdale	843	26.4	-6.82	24.6	-0.81	24.4	-0.82	24.2		0.3
Wear Valley	505	72.0	-9.31	65.3	-2.14	63.9	-2.82	62.1		1.2
East Sussex	**1,795**	**586.2**	**10.44**	**647.4**	**0.80**	**652.6**	**2.76**	**670.6**		**3.7**
Brighton	58	163.1	-1.04	161.4	-9.48	146.1	-8.69	133.4		23.0
Eastbourne	44	60.9	16.42	70.9	9.45	77.6	7.22	83.2		18.9
Hastings	30	66.5	8.87	72.4	3.31	74.8	4.41	78.1		26.0
Hove	24	88.7	2.82	91.2	-7.13	84.7	-2.60	82.5		34.4
Lewes	292	56.6	27.56	72.2	7.34	77.5	10.19	85.4		2.9
Rother	511	64.0	10.94	71.0	6.06	75.3	6.51	80.2		1.6
Wealden	837	86.4	25.35	108.3	7.57	116.5	9.61	127.7		1.5
Essex	**3,672**	**1,103.6**	**23.06**	**1,358.1**	**8.16**	**1,468.9**	**1.82**	**1,495.6**		**4.1**
Basildon	111	88.7	45.89	129.4	17.08	151.5	3.96	157.5		14.2
Braintree	612	74.5	25.37	93.4	19.59	111.7	3.58	115.7		1.9
Brentwood	149	59.2	23.99	73.4	-1.91	72.0	-4.72	68.6		4.6
Castle Point	44	48.0	55.63	74.7	15.39	86.2	-2.32	84.2		19.1
Chelmsford	342	93.0	32.26	123.0	12.44	138.3	8.46	150.0		4.4
Colchester	334	93.8	26.01	118.2	13.11	133.7	5.53	141.1		4.2
Epping Forest	345	107.0	6.73	114.2	1.66	116.1	-2.58	113.1		3.3
Harlow	26	53.7	45.44	78.1	1.66	79.4	-7.43	73.5		28.3
Maldon	358	31.0	30.65	40.5	18.02	47.8	6.28	50.8		1.4
Rochford	169	49.4	38.66	68.5	7.30	73.5	0.68	74.0		4.4
Southend-on-Sea	42	165.1	-1.39	162.8	-3.75	156.7	-1.91	153.7		36.6
Tendring	337	80.4	27.36	102.4	11.13	113.8	9.93	125.1		3.7
Thurrock	163	114.1	9.55	125.0	1.52	126.9	-2.05	124.3		7.6
Uttlesford	642	45.7	19.26	54.5	12.48	61.3	4.24	63.9		1.0

	Area			Population					Density
	km²	1961 thousands	61 to 71 % change	1971 thousands	71 to 81 % change	1981 thousands	81 to 91 % change	1991 thousands	Persons/ha.
Gloucestershire	**2,643**	**426.3**	**9.57**	**467.1**	**6.92**	**499.4**	**4.25**	**520.6**	**2.0**
Cheltenham	35	79.9	5.63	84.4	-0.47	84.0	2.26	85.9	24.5
Cotswold	1,142	61.0	3.11	62.9	8.74	68.4	6.73	73.0	0.6
Forest of Dean	528	62.7	5.10	65.9	10.32	72.7	2.06	74.2	1.4
Gloucester	33	82.6	9.20	90.2	2.11	92.1	-0.33	91.8	27.8
Stroud	454	82.3	9.11	89.8	12.92	101.4	6.80	108.3	2.4
Tewkesbury	450	57.8	27.85	73.9	9.34	80.8	8.17	87.4	1.9
Hampshire	**3,777**	**1,150.6**	**19.33**	**1,373.0**	**6.21**	**1,458.2**	**3.68**	**1,511.9**	**4.0**
Basingstoke & Dean	637	68.3	51.39	103.4	25.63	129.9	8.08	140.4	2.2
East Hampshire	515	63.0	26.19	79.5	12.96	89.8	12.58	101.1	2.0
Eastleigh	80	61.0	28.52	78.4	17.98	92.5	11.57	103.2	12.9
Fareham	74	58.3	37.91	80.4	9.83	88.3	10.19	97.3	13.1
Gosport	25	62.4	21.96	76.1	1.58	77.3	-5.82	72.8	29.1
Hart	218	37.4	64.97	61.7	17.02	72.2	9.00	78.7	3.6
Havant	56	74.5	46.71	109.3	6.77	116.7	0.60	117.4	21.0
New Forest	753	107.7	21.63	131.0	10.76	145.1	8.20	157.0	2.1
Portsmouth	37	215.1	-8.23	197.4	-9.12	179.4	-2.62	174.7	47.2
Rushmoor	36	63.1	18.70	74.9	8.81	81.5	-1.35	80.4	22.3
Southampton	49	205.0	4.93	215.1	-4.97	204.4	-4.89	194.4	39.7
Test Valley	637	61.5	29.92	79.9	13.77	90.9	8.91	99.0	1.6
Winchester	659	73.3	17.19	85.9	5.01	90.2	6.10	95.7	1.5
Hereford and Worcester	**3,926**	**491.9**	**13.84**	**560.0**	**12.54**	**630.2**	**5.97**	**667.8**	**1.7**
Bromsgrove	220	64.6	19.35	77.1	14.14	88.0	2.05	89.8	4.1
Hereford	20	40.4	15.10	46.5	2.58	47.7	4.40	49.8	24.9
Leominster	932	33.2	0.00	33.2	12.05	37.2	4.84	39.0	0.4
Malvern Hills	902	70.5	9.36	77.1	6.10	81.8	6.36	87.0	1.0
Redditch	54	34.3	18.95	40.8	63.24	66.6	15.47	76.9	14.2
South Herefordshire	905	42.6	4.23	44.4	6.98	47.5	7.79	51.2	0.6
Worcester	32	66.1	11.35	73.6	3.26	76.0	6.58	81.0	25.3
Wychavon	666	69.8	17.77	82.2	14.23	93.9	6.28	99.8	1.5
Wyre Forest	196	70.4	20.88	85.1	7.52	91.5	2.08	93.4	4.8
Hertfordshire	**1,634**	**787.8**	**17.38**	**924.7**	**3.31**	**955.3**	**-0.40**	**951.5**	**5.8**
Broxbourne	52	53.3	33.40	71.1	11.95	79.6	-0.13	79.5	15.3
Dacorum	210	95.5	24.50	118.9	8.49	129.0	0.16	129.2	6.2
East Hertfordshire	477	80.0	27.88	102.3	4.69	107.1	6.63	114.2	2.4
Hertsmere	98	87.0	4.71	91.1	-3.62	87.8	-1.94	86.1	8.8
North Hertfordshire	374	85.5	16.37	99.5	8.24	107.7	0.84	108.6	2.9
St. Albans	161	107.5	12.93	121.4	2.88	124.9	-2.00	122.4	7.6
Stevenage	25	43.0	56.05	67.1	10.73	74.3	-0.81	73.7	29.5
Three Rivers	88	78.6	2.80	80.8	-3.84	77.7	-4.63	74.1	8.4
Watford	21	75.6	3.84	78.5	-5.22	74.4	-3.09	72.1	34.3
Welwyn Hatfield	128	81.8	14.91	94.0	-1.06	93.0	-1.51	91.6	7.2
Humberside	**3,512**	**797.1**	**5.22**	**838.7**	**1.07**	**847.7**	**-1.47**	**835.2**	**2.4**
East Yorks Borough of Beverley	404	81.4	24.45	101.3	4.34	105.7	3.60	109.5	2.7
Boothferry	647	54.0	2.22	55.2	9.24	60.3	4.64	63.1	1.0
Cleethorpes	164	49.2	35.77	66.8	2.10	68.2	-1.03	67.5	4.1
East Yorkshire	1,044	63.7	2.83	65.5	14.50	75.0	11.60	83.7	0.8
Glanford	580	46.3	24.19	57.5	14.43	65.8	6.38	70.0	1.2
Great Grimsby	28	98.0	-2.55	95.5	-3.56	92.1	-3.47	88.9	31.8
Holderness	540	33.2	20.48	40.0	14.75	45.9	8.71	49.9	0.9
Kingston-upon-Hull	71	304.0	-5.92	286.0	-6.19	268.3	-9.73	242.2	34.1
Scunthorpe	34	67.3	5.35	70.9	-6.35	66.4	-8.89	60.5	17.8
Isle of Wight	**381**	**95.7**	**14.42**	**109.5**	**7.95**	**118.2**	**7.11**	**126.6**	**3.3**
Medina	117	56.3	14.39	64.4	4.97	67.6	3.70	70.1	6.0
South Wight	264	39.4	14.47	45.1	12.20	50.6	11.46	56.4	2.1
Kent	**3,731**	**1,198.5**	**16.76**	**1,399.4**	**4.57**	**1,463.3**	**1.49**	**1,485.1**	**4.0**
Ashford	581	61.9	27.79	79.1	8.47	85.8	5.94	90.9	1.6
Canterbury	311	91.2	20.72	110.1	6.09	116.8	8.82	127.1	4.1
Dartford	70	79.4	4.91	83.3	-2.52	81.2	-3.45	78.4	11.2
Dover	312	94.4	4.87	99.0	1.82	100.8	1.79	102.6	3.3
Gillingham	32	72.9	19.20	86.9	7.94	93.8	-0.53	93.3	29.2
Gravesham	100	83.5	15.69	96.6	-1.66	95.0	-5.26	90.0	9.0
Maidstone	394	97.1	24.82	121.2	7.26	130.0	2.46	133.2	3.4
Rochester upon Medway	160	117.6	18.37	139.2	3.02	143.4	-0.98	142.0	8.9
Sevenoaks	371	87.2	14.11	99.5	7.74	107.2	-1.03	106.1	2.9
Shepway	357	73.8	11.11	82.0	5.00	86.1	3.60	89.2	2.5
Swale	369	84.3	19.69	100.9	8.52	109.5	3.84	113.7	3.1
Thanet	103	103.7	10.70	114.8	5.57	121.2	0.08	121.3	11.8
Tonbridge and Malling	240	70.2	33.19	93.5	3.21	96.5	2.69	99.1	4.1
Tunbridge Wells	331	81.3	14.76	93.3	3.00	96.1	2.29	98.3	3.0

	Area km²	**Population** 1961 thousands	61 to 71 % change	1971 thousands	71 to 81 % change	1981 thousands	81 to 91 % change	1991 thousands	**Density** Persons/ha.
Lancashire	**3,063**	**1,261.1**	**6.63**	**1,344.7**	**2.05**	**1,372.3**	**-0.52**	**1,365.1**	**4.5**
Blackburn	137	143.4	-1.60	141.1	0.50	141.8	-6.35	132.8	9.7
Blackpool	35	153.2	-0.85	151.9	-2.63	147.9	-2.30	144.5	41.3
Burnley	118	99.4	-2.82	96.6	-4.24	92.5	-3.78	89.0	7.5
Chorley	205	67.0	16.12	77.8	17.22	91.2	5.81	96.5	4.7
Fylde	165	58.4	14.73	67.0	2.09	68.4	2.49	70.1	4.2
Hyndburn	73	80.0	0.75	80.6	-2.11	78.9	-3.04	76.5	10.5
Lancaster	577	114.8	7.67	123.6	-2.18	120.9	3.89	125.6	2.2
Pendle	168	85.8	-0.47	85.4	0.23	85.6	-3.39	82.7	4.9
Preston	142	143.1	-5.87	134.7	-6.53	125.9	0.24	126.2	8.9
Ribble Valley	579	44.5	15.73	51.5	3.50	53.3	-4.32	51.0	0.9
Rossendale	138	65.7	-5.94	61.8	4.37	64.5	-0.78	64.0	4.6
South Ribble	111	66.5	29.77	86.3	12.28	96.9	2.99	99.8	9.0
West Lancashire	332	61.5	48.94	91.6	16.70	106.9	-0.28	106.6	3.2
Wyre	283	77.8	21.85	94.8	3.06	97.7	2.05	99.7	3.5
Leicestershire	**2,553**	**706.1**	**13.24**	**799.6**	**5.38**	**842.6**	**2.12**	**860.5**	**3.4**
Blaby	130	50.5	46.93	74.2	4.04	77.2	6.09	81.9	6.3
Charnwood	279	104.0	20.77	125.6	6.85	134.2	4.69	140.5	5.0
Harborough	593	42.5	24.71	53.0	14.72	60.8	8.88	66.2	1.1
Hinckley and Bosworth	297	64.2	17.60	75.5	15.89	87.5	6.97	93.6	3.2
Leicester	73	288.1	-1.35	284.2	-1.55	279.8	-3.29	270.6	37.1
Melton	482	34.4	13.08	38.9	11.31	43.3	2.77	44.5	0.9
North West Leicestershire	280	65.3	8.58	70.9	10.86	78.6	1.02	79.4	2.8
Oadby and Wigston	24	33.6	48.21	49.8	1.61	50.6	1.78	51.5	21.5
Rutland	394	23.5	17.02	27.5	11.64	30.7	5.54	32.4	0.8
Lincolnshire	**5,915**	**468.6**	**7.45**	**503.5**	**8.80**	**547.8**	**4.76**	**573.9**	**1.0**
Boston	360	47.3	3.17	48.8	7.79	52.6	0.00	52.6	1.5
East Lindsey	1,762	90.6	4.53	94.7	10.35	104.5	10.62	115.6	0.7
Lincoln	36	77.7	-4.38	74.3	3.23	76.7	6.78	81.9	22.8
North Kesteven	923	59.7	21.78	72.7	7.98	78.5	-0.13	78.4	0.8
South Holland	737	56.0	1.61	56.9	8.44	61.7	6.97	66.0	0.9
South Kesteven	943	75.0	14.00	85.5	14.39	97.8	9.61	107.2	1.1
West Lindsey	1,154	62.3	13.32	70.6	7.65	76.0	-5.00	72.2	0.6
Norfolk	**5,368**	**565.9**	**10.57**	**625.7**	**11.01**	**694.6**	**6.02**	**736.4**	**1.4**
Breckland	1,305	62.0	23.23	76.4	26.18	96.4	9.13	105.2	0.8
Broadland	552	67.7	28.06	86.7	13.38	98.3	6.31	104.5	1.9
Great Yarmouth	173	70.2	7.69	75.6	6.88	80.8	6.31	85.9	5.0
Kings Lynn and West Norfolk	1,426	106.1	3.68	110.0	10.82	121.9	5.33	128.4	0.9
North Norfolk	965	74.0	0.27	74.2	10.51	82.0	10.24	90.4	0.9
Norwich	39	121.1	0.83	122.1	0.16	122.3	-1.31	120.7	30.9
South Norfolk	907	64.3	25.51	80.7	14.99	92.8	9.27	101.4	1.1
Northamptonshire	**2,367**	**398.0**	**17.76**	**468.7**	**12.55**	**527.5**	**8.61**	**572.9**	**2.4**
Corby	80	40.0	31.50	52.6	0.19	52.7	-0.76	52.3	6.5
Daventry	666	37.3	28.95	48.1	19.96	57.7	6.76	61.6	0.9
East Northamptonshire	510	51.6	9.69	56.6	7.42	60.8	8.06	65.7	1.3
Kettering	234	59.9	9.85	65.8	8.36	71.3	5.47	75.2	3.2
Northampton	81	124.1	7.74	133.7	17.28	156.8	13.65	178.2	22.0
South Northamptonshire	634	40.9	36.43	55.8	14.87	64.1	7.33	68.8	1.1
Wellingborough	163	44.2	26.92	56.1	14.26	64.1	10.92	71.1	4.4
Northumberland	**5,032**	**274.3**	**1.90**	**279.5**	**7.30**	**299.9**	**0.23**	**300.6**	**0.6**
Alnwick	1,080	30.0	-7.00	27.9	2.87	28.7	4.53	30.0	0.3
Berwick-upon-Tweed	975	28.1	-8.19	25.8	1.55	26.2	0.76	26.4	0.3
Blyth Valley	70	54.5	11.38	60.7	26.52	76.8	1.56	78.0	11.1
Castle Morpeth	619	42.6	11.50	47.5	6.53	50.6	-1.78	49.7	0.8
Tynedale	2,221	52.3	1.53	53.1	3.77	55.1	2.36	56.4	0.3
Wansbeck	66	66.8	-3.44	64.5	-3.10	62.5	-3.84	60.1	9.1
North Yorkshire	**8,309**	**575.2**	**9.04**	**627.2**	**6.31**	**666.8**	**4.78**	**698.7**	**0.8**
Craven	1,176	45.1	2.88	46.4	3.02	47.8	3.97	49.7	0.4
Hambleton	1,312	59.0	14.07	67.3	10.25	74.2	4.58	77.6	0.6
Harrogate	1,334	116.3	9.72	127.6	9.48	139.7	0.93	141.0	1.1
Richmondshire	1,317	39.4	7.87	42.5	0.00	42.5	3.06	43.8	0.3
Ryedale	1,598	65.0	11.69	72.6	15.84	84.1	7.02	90.0	0.6
Scarborough	817	88.3	10.19	97.3	4.21	101.4	6.31	107.8	1.3
Selby	725	54.0	27.22	68.7	12.37	77.2	14.25	88.2	1.2
York	29	108.1	-3.05	104.8	-4.77	99.8	0.80	100.6	34.7
Nottinghamshire	**2,164**	**901.0**	**8.16**	**974.5**	**0.86**	**982.9**	**-0.23**	**980.6**	**4.5**
Ashfield	110	96.3	5.61	101.7	4.72	106.5	0.28	106.8	9.7
Bassetlaw	637	90.0	8.33	97.5	4.62	102.0	0.98	103.0	1.6
Broxtowe	81	86.9	13.00	98.2	4.68	102.8	1.75	104.6	12.9
Gedling	112	80.8	19.06	96.2	12.16	107.9	-0.28	107.6	9.6
Mansfield	77	85.0	12.35	95.5	3.98	99.3	-0.50	98.8	12.8
Newark and Sherwood	662	84.1	17.60	98.9	1.42	100.3	3.09	103.4	1.6
Nottingham	74	311.9	-3.62	300.6	-9.71	271.4	-3.65	261.5	35.3
Rushcliffe	410	66.0	30.15	85.9	7.80	92.6	2.48	94.9	2.3

	Area km²	1961 thousands	61 to 71 % change	1971 thousands	Population 71 to 81 % change	1981 thousands	81 to 91 % change	1991 thousands	Density Persons/ha.
Oxfordshire	**2,608**	**403.0**	**23.62**	**498.2**	**3.39**	**515.1**	**7.51**	**553.8**	**2.1**
Cherwell	590	70.1	34.66	94.4	13.24	106.9	8.42	115.9	2.0
Oxford	36	106.3	2.35	108.8	-9.47	98.5	10.66	109.0	30.3
South Oxfordshire	687	94.3	34.36	126.7	1.50	128.6	1.79	130.9	1.9
Vale of White Horse	581	76.4	22.25	93.4	7.82	100.7	8.44	109.2	1.9
West Oxfordshire	715	55.9	34.17	75.0	7.07	80.3	10.46	88.7	1.2
Shropshire	**3,490**	**297.7**	**13.23**	**337.1**	**11.42**	**375.6**	**6.92**	**401.6**	**1.2**
Bridgnorth	634	42.7	11.94	47.8	5.23	50.3	-1.19	49.7	0.8
North Shropshire	680	45.2	4.20	47.1	6.37	50.1	4.59	52.4	0.8
Oswestry	256	29.7	2.02	30.3	1.32	30.7	9.45	33.6	1.3
Shrewsbury	603	72.9	13.03	82.4	5.83	87.2	4.24	90.9	1.5
South Shropshire	1,028	32.6	-0.61	32.4	4.32	33.8	11.83	37.8	0.4
The Wrekin	291	74.6	30.16	97.1	27.19	123.5	11.01	137.1	4.7
Somerset	**3,451**	**345.4**	**11.87**	**386.4**	**9.99**	**425.0**	**8.02**	**459.1**	**1.3**
Mendip	739	69.5	13.81	79.1	9.99	87.0	9.54	95.3	1.3
Sedgemoor	567	69.9	15.88	81.0	9.88	89.0	8.99	97.0	1.7
South Somerset (Yeovil)	959	104.3	9.68	114.4	14.16	130.6	6.74	139.4	1.5
Taunton Deane	458	73.0	12.47	82.1	4.87	86.1	8.36	93.3	2.0
West Somerset	727	28.6	4.20	29.8	8.39	32.3	5.57	34.1	0.5
Staffordshire	**2,716**	**849.3**	**13.48**	**963.8**	**5.03**	**1,012.3**	**0.79**	**1,020.3**	**3.8**
Cannock Chase	79	60.7	30.15	79.0	6.96	84.5	3.43	87.4	11.1
East Staffordshire	388	86.9	8.63	94.4	0.53	94.9	1.37	96.2	2.5
Lichfield	330	51.4	56.81	80.6	9.80	88.5	2.49	90.7	2.7
Newcastle-under-Lyme	211	114.6	4.71	120.0	-1.75	117.9	-0.42	117.4	5.6
South Staffordshire	409	62.1	33.33	82.8	16.55	96.5	7.67	103.9	2.5
Stafford	599	92.9	19.70	111.2	5.76	117.6	-0.51	117.0	2.0
Staffordshire Moorlands	579	78.0	15.64	90.2	6.32	95.9	-1.98	94.0	1.6
Stoke-on-Trent	93	277.3	-4.33	265.3	-4.86	252.4	-3.01	244.8	26.3
Tamworth	31	25.4	58.66	40.3	59.55	64.3	7.15	68.9	22.2
Suffolk	**3,797**	**467.1**	**15.18**	**538.0**	**10.86**	**596.4**	**5.62**	**629.9**	**1.7**
Babergh	595	47.0	34.68	63.3	16.43	73.7	6.51	78.5	1.3
Forest Heath	374	31.7	24.92	39.6	31.06	51.9	10.21	57.2	1.5
Ipswich	40	117.4	5.03	123.3	-2.60	120.1	-3.83	115.5	28.9
Mid Suffolk	871	54.1	12.20	60.7	14.99	69.8	10.46	77.1	0.9
St. Edmundsbury	657	55.5	28.65	71.4	20.59	86.1	3.48	89.1	1.4
Suffolk Coastal	889	82.8	7.61	89.1	7.18	95.5	11.83	106.8	1.2
Waveney	370	78.6	15.27	90.6	9.49	99.2	6.35	105.5	2.9
Surrey	**1,679**	**905.6**	**10.67**	**1,002.2**	**-0.28**	**999.4**	**-0.14**	**998.0**	**5.9**
Elmbridge	97	106.2	8.76	115.5	-4.16	110.7	-0.72	109.9	11.3
Epsom and Ewell	34	71.2	1.54	72.3	-4.29	69.2	-3.18	67.0	19.7
Guildford	271	108.9	9.00	118.7	1.18	120.1	1.17	121.5	4.5
Mole Valley	259	73.2	5.87	77.5	-1.16	76.6	1.04	77.4	3.0
Reigate and Banstead	129	114.6	4.45	119.7	-2.92	116.2	-1.12	114.9	8.9
Runnymede	78	71.0	6.62	75.7	-6.08	71.1	0.56	71.5	9.2
Spelthorne	56	83.3	16.33	96.9	-4.13	92.9	-6.24	87.1	15.6
Surrey Heath	97	44.7	47.65	66.0	15.91	76.5	2.35	78.3	8.1
Tandridge	250	75.2	6.12	79.8	-5.01	75.8	-1.06	75.0	3.0
Waverley	345	89.8	15.92	104.1	4.61	108.9	2.39	111.5	3.2
Woking	64	67.5	12.59	76.0	7.11	81.4	3.19	84.0	13.1
Warwickshire	**1,981**	**386.8**	**17.76**	**455.5**	**3.97**	**473.6**	**0.72**	**477.0**	**2.4**
North Warwickshire	286	49.2	18.29	58.2	2.75	59.8	0.00	59.8	2.1
Nuneaton and Bedworth	79	88.5	21.58	107.6	5.48	113.5	1.59	115.3	14.6
Rugby	356	73.5	13.74	83.6	2.99	86.1	-3.14	83.4	2.3
Stratford-on-Avon	977	80.0	18.13	94.5	6.24	100.4	3.19	103.6	1.1
Warwick	283	95.6	16.74	111.6	1.88	113.7	1.06	114.9	4.1
West Sussex	**1,989**	**492.1**	**20.63**	**593.6**	**10.97**	**658.7**	**5.18**	**692.8**	**3.5**
Adur	42	48.6	13.58	55.2	5.25	58.1	-1.20	57.4	13.7
Arun	221	81.4	28.26	104.4	14.18	119.2	7.13	127.7	5.8
Chichester	787	80.6	13.15	91.2	7.02	97.6	2.77	100.3	1.3
Crawley	36	54.6	25.09	68.3	18.45	80.9	7.66	87.1	24.2
Horsham	533	67.0	27.76	85.6	14.02	97.6	9.94	107.3	2.0
Mid-Sussex	338	79.6	26.13	100.4	13.05	113.5	4.67	118.8	3.5
Worthing	33	80.3	10.21	88.5	3.73	91.8	2.51	94.1	28.5
Wiltshire	**3,481**	**423.0**	**15.13**	**487.0**	**6.41**	**518.2**	**6.77**	**553.3**	**1.6**
Kennet	958	54.8	17.52	64.4	-1.71	63.3	6.64	67.5	0.7
North Wiltshire	770	85.9	10.24	94.7	8.24	102.5	6.93	109.6	1.4
Salisbury	1,005	91.6	10.37	101.1	-0.20	100.9	2.28	103.2	1.0
Thamesdown	230	119.7	16.96	140.0	8.64	152.1	9.93	167.2	7.3
West Wiltshire	517	71.0	22.25	86.8	14.40	99.3	6.65	105.9	2.0
ENGLAND	**130,439**	**43,461**	**-0.95**	**46,019**	**0.45**	**46,226**	**-0.14**	**46,161**	**3.5**

■ *Wales*

	Area km²	1961 thousands	61 to 71 % change	1971 thousands	71 to 81 % change	1981 thousands	81 to 91 % change	1991 thousands	Density Persons/ha.
Clywd	**2,426**	**322.3**	**11.26**	**358.6**	**8.81**	**390.2**	**3.00**	**401.9**	**1.7**
Alyn and Deeside	154	51.9	26.20	65.5	9.47	71.7	0.00	71.7	4.7
Colwyn	553	38.9	16.20	45.2	7.52	48.6	12.96	54.9	1.0
Delyn	278	50.7	12.23	56.9	15.29	65.6	0.91	66.2	2.4
Glyndwr	966	38.4	-1.82	37.7	6.90	40.3	2.98	41.5	0.4
Rhuddlan	109	42.0	13.10	47.5	10.32	52.4	3.05	54.0	5.0
Wrexham Maelor	366	100.4	5.38	105.8	5.58	111.7	1.70	113.6	3.1
Dyfed	**5,768**	**315.7**	**0.22**	**316.4**	**4.30**	**330.0**	**3.52**	**341.6**	**0.6**
Carmarthen	1,182	50.3	-1.39	49.6	4.23	51.7	6.00	54.8	0.5
Ceredigion	1,793	53.6	2.43	54.9	4.55	57.4	10.80	63.6	0.4
Dinefwr	971	38.8	-6.70	36.2	1.93	36.9	2.98	38.0	0.4
Llanelli	234	79.0	-2.66	76.9	-1.95	75.4	-2.52	73.5	3.1
Preseli Pembrokeshire	1,151	61.3	2.45	62.8	10.35	69.3	0.43	69.6	0.6
South Pembrokeshire	438	32.7	10.09	36.0	9.17	39.3	7.12	42.1	1.0
Gwent	**1,376**	**424.0**	**4.10**	**441.4**	**-0.39**	**439.7**	**-1.68**	**432.3**	**3.1**
Blaenau Gwent	127	94.5	-9.42	85.6	-7.48	79.2	-6.06	74.4	5.9
Islwyn	99	65.4	1.38	66.3	0.15	66.4	-2.26	64.9	6.6
Monmouth	824	54.2	18.82	64.4	12.58	72.5	3.45	75.0	0.9
Newport	201	128.6	6.38	136.8	-3.87	131.5	-1.22	129.9	6.5
Torfaen	126	81.3	8.61	88.3	2.04	90.1	-2.11	88.2	7.0
Gwynedd	**3,869**	**213.7**	**3.23**	**220.6**	**4.49**	**230.5**	**3.51**	**238.6**	**0.6**
Aberconwy	606	48.3	5.38	50.9	3.14	52.5	3.05	54.1	0.9
Arfon	410	51.7	1.74	52.6	-0.57	52.3	4.40	54.6	1.3
Dwyfor	620	27.6	-6.52	25.8	1.94	26.3	8.75	28.6	0.5
Meirionnydd	1,518	34.4	-8.43	31.5	1.90	32.1	4.05	33.4	0.2
Ynys Mon-Isle of Anglesey	715	51.7	15.67	59.8	12.54	67.3	0.74	67.8	0.9
Mid Glamorgan	**1,018**	**519.5**	**2.33**	**531.6**	**1.15**	**537.7**	**-2.08**	**526.5**	**5.2**
Cynon Valley	181	72.6	-4.13	69.6	-3.30	67.3	-5.50	63.6	3.5
Merthyr Tydfil	112	66.9	-5.53	63.2	-4.59	60.3	-1.66	59.3	5.3
Ogwr	285	111.0	11.35	123.6	5.02	129.8	0.54	130.5	4.6
Rhondda	97	100.3	-11.27	89.0	-8.43	81.5	-6.38	76.3	7.9
Rhymney Valley	176	94.4	7.73	101.7	3.74	105.5	-3.89	101.4	5.8
Taff Ely	168	74.3	13.73	84.5	10.41	93.3	2.25	95.4	5.7
Powys	**5,077**	**102.3**	**-3.03**	**99.2**	**11.39**	**110.5**	**5.43**	**116.5**	**0.2**
Brecknock	1,794	39.6	-4.55	37.8	7.67	40.7	1.47	41.3	0.2
Montgomeryshire	2,064	44.2	-2.49	43.1	11.83	48.2	7.88	52.0	0.3
Radnorshire	1,219	18.5	-1.08	18.3	18.03	21.6	7.41	23.2	0.2
South Glamorgan	**416**	**380.3**	**2.63**	**390.3**	**-1.43**	**384.7**	**-0.36**	**383.3**	**9.2**
Cardiff	120	289.9	-0.79	287.6	-4.76	273.9	-0.47	272.6	22.7
Vale of Glamorgan	296	90.4	13.61	102.7	7.89	110.8	-0.09	110.7	3.7
West Glamorgan	**817**	**366.3**	**1.88**	**373.2**	**-1.58**	**367.3**	**-2.59**	**357.8**	**4.4**
Port Talbot	151	60.7	-2.14	59.4	-7.91	54.7	-8.78	49.9	3.3
Lliw Valley	214	55.7	1.08	56.3	5.51	59.4	3.87	61.7	2.9
Neath	206	69.9	-3.15	67.7	-1.33	66.8	-4.04	64.1	3.1
Swansea	245	180.0	5.44	189.8	-1.74	186.5	-2.36	182.1	7.4
WALES	**20,768**	**2,644.0**	**3.30**	**2,731.2**	**2.17**	**2,790.5**	**0.29**	**2,798.5**	**1.3**

■ Scotland	Area km²	1961 thousands	61 to 71 % change	1971 thousands	71 to 81 % change	1981 thousands	81 to 91 % change	1991 thousands	Density Persons/ha.
Borders	**4,672**	**102.2**	**-3.62**	**98.5**	**1.42**	**99.9**	**2.70**	**102.6**	**0.2**
Berwickshire	876	18.4	-7.61	17.0	7.65	18.3	2.73	18.8	0.2
Ettrick and Lauderdale	1,356	33.4	-2.99	32.4	-2.16	31.7	6.94	33.9	0.3
Roxburgh	1,540	36.3	-2.48	35.4	-0.56	35.2	-1.70	34.6	0.2
Tweeddale	899	14.1	-2.84	13.7	5.84	14.5	5.52	15.3	0.2
Central	**2,631**	**244.6**	**7.52**	**263.0**	**3.99**	**273.5**	**-2.01**	**268.0**	**1.0**
Clackmannan	161	41.9	10.02	46.1	3.69	47.8	-1.26	47.2	2.9
Falkirk	301	132.3	6.20	140.5	2.78	144.4	-3.74	139.0	4.6
Stirling	2,170	70.4	7.95	76.0	7.11	81.4	0.37	81.7	0.4
Dumfries and Galloway	**6,370**	**146.4**	**-2.19**	**143.2**	**1.40**	**145.2**	**1.31**	**147.1**	**0.2**
Annandale and Eskdale	1,553	35.6	-2.81	34.6	2.60	35.5	3.66	36.8	0.2
Nithsdale	1,433	55.8	1.25	56.5	-0.53	56.2	0.71	56.6	0.4
Stewartry	1,671	23.3	-4.72	22.2	4.50	23.2	1.72	23.6	0.1
Wigtown	1,713	31.7	-5.68	29.9	1.34	30.3	-0.99	30.0	0.2
Fife	**1,307**	**320.7**	**2.00**	**327.1**	**0.09**	**327.4**	**3.63**	**339.3**	**2.6**
Dunfermline	301	119.5	0.84	120.5	1.99	122.9	2.12	125.5	4.2
Kirkcaldy	248	140.1	3.50	145.0	-1.86	142.3	1.62	144.6	5.8
North East Fife	758	61.1	0.82	61.6	0.81	62.1	11.43	69.2	0.9
Grampian	**8,704**	**440.3**	**-0.39**	**438.6**	**7.59**	**471.9**	**4.51**	**493.2**	**0.6**
Aberdeen City	184	206.3	2.67	211.8	-3.73	203.9	-1.37	201.1	10.9
Banff and Buchan	1,526	75.4	-3.45	72.8	12.23	81.7	1.47	82.9	0.5
Gordon	2,214	49.2	-8.54	45.0	38.44	62.3	18.78	74.0	0.3
Kincardine and Deeside	2,548	34.1	-2.35	33.3	27.63	42.5	23.76	52.6	0.2
Moray	2,231	75.3	0.53	75.7	7.66	81.5	1.23	82.5	0.4
Highland	**25,391**	**163.9**	**7.14**	**175.6**	**14.01**	**200.2**	**4.60**	**209.4**	**0.1**
Badenoch and Strathspey	2,319	9.1	2.20	9.3	33.33	12.4	4.03	12.9	0.1
Caithness	1,776	27.4	1.46	27.8	-1.44	27.4	-4.74	26.1	0.1
Inverness	2,800	45.8	8.73	49.8	14.06	56.8	10.21	62.6	0.2
Lochaber	4,507	15.9	20.13	19.1	6.81	20.4	1.96	20.8	0.0
Nairn	422	8.4	32.14	11.1	-9.01	10.1	4.95	10.6	0.3
Ross and Cromarty	5,000	32.8	6.10	34.8	36.49	47.5	5.26	50.0	0.1
Skye and Lochalsh	2,701	10.3	-3.88	9.9	14.14	11.3	10.62	12.5	0.0
Sutherland	5,865	14.2	-2.82	13.8	2.90	14.2	-3.52	13.7	0.0
Lothian	**1,755**	**710.2**	**4.98**	**745.6**	**-0.95**	**738.5**	**-2.00**	**723.7**	**4.1**
East Lothian	713	74.8	3.48	77.4	4.13	80.6	2.98	83.0	1.2
Edinburgh City	261	483.9	-1.53	476.5	-8.31	436.9	-3.59	421.2	16.1
Midlothian	358	65.3	22.05	79.7	3.39	82.4	-5.34	78.0	2.2
West Lothian	423	86.2	29.93	112.0	23.66	138.5	2.17	141.5	3.3
Strathclyde	**13,537**	**2,584.0**	**-0.33**	**2,575.4**	**-6.65**	**2,404.1**	**-7.73**	**2,218.2**	**1.6**
Argyll and Bute	6,497	65.7	-1.07	65.0	5.85	68.8	-2.62	67.0	0.1
Bearsden and Milngavie	36	26.2	37.02	35.9	9.75	39.4	0.25	39.5	11.0
Clydebank	36	57.6	2.08	58.8	-11.39	52.1	-14.20	44.7	12.4
Clydesdale	1,325	54.6	-1.83	53.6	7.46	57.6	-1.04	57.0	0.4
Cumbernauld and Kilsyth	95	18.7	143.85	45.6	35.96	62.0	-1.61	61.0	6.4
Cumnock and Doon Valley	801	53.2	-8.46	48.7	-6.98	45.3	-7.28	42.0	0.5
Cunninghame	879	111.1	13.32	125.9	10.17	138.7	-2.88	134.7	1.5
Dumbarton	477	70.7	11.32	78.7	-0.38	78.4	-3.06	76.0	1.6
East Kilbride	285	40.8	81.86	74.2	11.86	83.0	-1.93	81.4	2.9
Eastwood	116	43.8	13.93	49.9	7.62	53.7	8.57	58.3	5.0
Glasgow City	198	1,140.1	-13.84	982.3	-22.06	765.6	-14.51	654.5	33.1
Hamilton	131	95.2	9.56	104.3	3.93	108.4	-4.89	103.1	7.9
Inverclyde	158	112.5	-2.76	109.4	-8.23	100.4	-12.25	88.1	5.6
Kilmarnock and Loudoun	373	79.0	2.53	81.0	1.36	82.1	-4.26	78.6	2.1
Kyle and Carrick	1,322	104.9	5.43	110.6	3.53	114.5	-0.79	113.6	0.9
Monklands	164	111.5	1.88	113.6	-2.55	110.7	-8.58	101.2	6.2
Motherwell	173	157.1	0.32	157.6	-4.89	149.9	-6.40	140.3	8.1
Renfrew	309	182.5	11.18	202.9	1.63	206.2	-6.11	193.6	6.3
Strathkelvin	164	58.8	31.63	77.4	12.79	87.3	-4.24	83.6	5.1
Tayside	**7,493**	**397.8**	**-0.05**	**397.6**	**-1.43**	**391.9**	**-1.68**	**385.3**	**0.5**
Angus	2,023	85.1	-1.06	84.2	10.45	93.0	-0.11	92.9	0.5
Dundee City	235	195.3	1.08	197.4	-8.97	179.7	-7.90	165.5	7.0
Perth and Kinross	5,235	117.4	-1.19	116.0	2.76	119.2	6.38	126.8	0.2
Orkney	**976**	**18.7**	**-8.56**	**17.1**	**11.70**	**19.1**	**2.09**	**19.5**	**0.2**
Shetland	**1,433**	**17.8**	**-2.81**	**17.3**	**57.80**	**27.3**	**-19.41**	**22.0**	**0.2**
Western Isles	**2,898**	**32.6**	**-8.28**	**29.9**	**6.69**	**31.9**	**-8.78**	**29.1**	**0.1**
SCOTLAND	**77,167**	**5,179.2**	**0.10**	**5,228.9**	**-1.88**	**5,130.7**	**-3.38**	**4,957.3**	**0.6**

■ Northern Ireland

	Area km²	1961 thousands	61 to 71 % change	1971 thousands	Population 71 to 81 % change	1981 thousands	81 to 91 % change	1991 thousands	Density Persons/ha.
Antrim	405	-	-	34.0	33.24	45.3	-2.43	44.2	1.1
Ards	368	-	-	46.8	23.08	57.6	11.11	64.0	1.7
Armagh	667	-	-	46.4	3.66	48.1	6.65	51.3	0.8
Ballymena	634	-	-	49.0	11.63	54.7	2.19	55.9	0.9
Ballymoney	417	-	-	21.9	4.57	22.9	4.80	24.0	0.6
Banbridge	441	-	-	28.7	4.18	29.9	10.70	33.1	0.8
Belfast	130	-	-	416.7	-28.51	297.9	-5.67	281.0	21.6
Carrickfergus	85	-	-	27.0	5.19	28.4	13.73	32.3	3.8
Castlereagh	84	-	-	64.4	-5.12	61.1	-0.65	60.7	7.2
Coleraine	478	-	-	44.6	6.50	47.5	7.37	51.0	1.1
Cookstown	512	-	-	26.1	0.77	26.3	17.49	30.9	0.6
Craigavon	280	-	-	67.7	4.87	71.0	4.79	74.4	2.7
Derry	373	-	-	84.9	82.9	14.23	94.7		
Down	638	-	-	46.9	13.01	53.0	8.49	57.5	0.9
Dungannon	763	-	-	42.6	-3.52	41.1	10.46	45.4	0.6
Fermanagh	1,669	-	-	51.0	1.96	52.0	4.40	54.3	0.3
Larne	337	-	-	29.9	-1.34	29.5	-0.68	29.3	0.9
Limavady	585	-	-	23.8	10.92	26.4	10.23	29.1	0.5
Lisburn	436	-	-	70.7	17.68	83.2	-41.35	48.8	1.1
Magherafelt	562	-	-	31.5	-2.22	30.8	16.56	35.9	0.6
Moyle	494	-	-	14.0	2.14	14.3	2.10	14.6	0.3
Newry and Mourne	886	-	-	72.4	0.28	72.6	13.22	82.2	0.9
Newtownabbey	151	-	-	66.9	7.62	72.0	2.36	73.7	4.9
North Down	72	-	-	52.6	24.90	65.7	6.54	70.0	9.7
Omagh	1,124	-	-	41.2	-0.24	41.1	10.95	45.6	0.4
Strabane	861	-	-	34.4	1.74	35.0	2.00	35.7	0.4
NORTHERN IRELAND	**13,483**	**-**	**-**	**1,536.1**	**-2.99**	**1,490.2**	**5.35**	**1,570.0**	**1.2**

■ Isle of Man

	Area km²	1961 thousands	61 to 71 % change	1971 thousands	71 to 81 % change	1981 thousands	81 to 91 % change	1991 thousands	Density Persons/ha.
Isle of Man	572	48.1	-34.3	56.3	14.92	64.7	-4.17	62.0	1.1

■ Channel Islands

	Area km²	1961 thousands	61 to 71 % change	1971 thousands	71 to 81 % change	1981 thousands	81 to 91 % change	1991 thousands	Density Persons/ha.
Jersey	116	59.5	21.0	72.0	5.69	76.1	10.38	84.0	7.2
Guernsey	63	44.9	14.5	51.4	3.70	53.3	9.76	58.5	9.3
Alderney	8	1.5	13.3	1.7	23.53	2.1	9.52	2.3	2.9
Sark	5	0.5	20.0	0.6	0.00	0.6	0.00	0.6	1.2

Totals are the figures on the night of the census in April of the year concerned.

UK Trade 1990

IMPORTS £ million
from (countries in order of total trade)

	Total trade	Percentage of world total	0 Food and live animals	01 Meat and meat preparations	04 Cereals and cereal preps	05 Fruit and vegetables	07 Coffee, tea, cocoa and spices	1 Beverages and tobacco	2 Crude materials (inedible)	24 Cork and wood	28 Metalliferous ores and scrap	3 Mineral fuels, lubricants etc	33 Petrol and petrol products	4 Animal and veg. oils and fats	5 Chemicals	51 Organic chemicals	52 Inorganic chemicals	53 Dyeing and tanning materials
West Germany	19,907.1	15.8	586.7	69.7	61.5	75.1	97.8	240.5	242.8	20.2	63.3	66.7	37	24.4	2,220	436.4	169.7	174.3
USA	14,357.5	11.4	371.2	6	24.3	151.1	2.9	69.6	615.6	72.6	226.5	253.8	76.2	4.9	934.6	202.8	82	56.4
France	11,758.5	9.3	1,038.1	159.5	261.7	255.7	34.3	695.6	176.4	11	17.9	292.8	172.4	20.6	1,682.9	371.7	264.3	85.2
Netherlands	10,483.6	8.3	1,547.1	439.9	80.3	535	69.3	112.3	430.7	22.2	44.2	871.8	716.4	101.9	1,609	467.3	168.1	71.3
Japan	6,761.6	5.4	20.7	0	0.6	0.9	0.2	2.8	27.3	0	3.8	2.8	0.3	6.2	249.5	90.3	52.7	14.4
Italy	6,735.5	5.3	422.8	14.5	75.9	268.8	9.8	172.7	82.7	5.3	13.9	30.7	29.2	10.7	423.5	78.2	13.7	14.2
Belgium-Lux.	5,732.4	4.5	390.7	55.2	53.4	113.2	26.1	38.3	156.6	7.5	21.9	192.2	153.5	34.1	1,053.1	205.8	52.1	56.3
Irish Rep.	4,498.6	3.6	1,192.9	273.5	61.9	58	131.1	120.5	203.2	27.1	51.9	56.7	34.5	8.5	490.4	222.2	15.7	9.4
Switzerland	4,252.8	3.4	52.2	0.1	2.3	20.1	12.3	3.2	44.2	0.1	7.8	10.6	10.5	0.8	480.5	159.5	9.3	69.4
Norway	4,235.3	3.4	86.7	0	0.7	0.2	0.1	0.3	133.9	20.3	46.7	2,876.9	2,212.9	7.6	165.8	30	34.6	5.2
Sweden	3,594.5	2.8	19.4	1	7.9	1	5.1	0.6	369.4	228.5	18.5	112.6	111.7	4.1	269.5	44.4	11.3	7.4
Spain	2,884.7	2.3	398.1	7	32.3	329.2	12.1	88.1	115.8	3.5	11.2	83.5	77.9	6.6	152.8	25.3	17.6	13.7
Denmark	2,278.6	1.8	814.7	458.3	27.6	14	3.3	8.5	95.5	5.3	5.4	53.2	51.7	22.9	165.9	24.5	1.1	11.7
Canada	2,259.1	1.8	161.4	0.9	34.8	37.5	0.2	20.1	806.6	352.6	312.3	40.8	3.2	0.6	64.8	7.6	19.1	1.7
Hong Kong	1,972.2	1.6	15	0.1	1.1	5.4	1.1	0.5	18.4	0.5	4.6	0	0	0.4	13.9	2.1	3.6	0.6
Finland	1,775.8	1.4	7.5	0.1	2	0.2	2.8	0.7	304.2	170	4.4	54.4	24.5	1.2	100.3	38.7	4.5	13.7
Taiwan	1,212	1	3.5	0	0.3	1.3	0.3	1.5	12.6	0.2	2.3	0.1	0	0	19.8	4.8	0.5	2
Portugal	1,176.2	0.9	35.1	0.1	0.8	15.1	0	39.5	13.8	66.6	2.5	27.5	26.5	0.1	36.4	14	0.9	1.3
South Africa	1,078.5	0.9	161.3	0.2	8.1	140.7	9.3	11.1	273.4	4.8	168.3	9.5	2.2	0.3	22.6	3	13.6	1.1
Australia	1,039.1	0.8	80.4	41.8	2	27.2	0.2	22.6	212.8	3.9	167.4	128.2	0	0.7	18.5	0.8	1.5	0.6
Singapore	1,021.1	0.8	21.3	0.2	1.2	2.4	4.1	1.1	24.7	12.4	2.1	0	0	0.3	39.8	35.5	0.1	0.3
South Korea	963.8	0.8	6	0	0	0.2	0	9.8	3.7	0	1.4	0	0	0	20.7	9.4	0.9	1.6
Austria	957.8	0.8	11	0.1	0.9	0.8	4.3	1.3	14.4	3.3	5	0.1	0.1	0	83.8	15	2.8	0.7
USSR	917.7	0.7	17	0	0	0.2	2.2	1	188.9	169.5	2.1	514.3	489.9	0.2	25.5	12.6	8.8	0.5
World Total	126,165.8	100	10,409.2	1,887.8	785.1	2,964.9	904.4	1,907	5,721.2	1,410	1,479.2	7,840	6,254.9	377.4	10,834.6	2,593.1	1,000.2	651.3
Percentage of world total	100		8.3	1.5	0.6	2.3	0.7	1.5	4.5	1.1	1.2	6.2	5	0.3	8.6	2.1	0.8	0.5

EXPORTS £ million
to (countries in order of total trade)

	Total trade	Percentage of world total	0 Food and live animals	01 Meat and meat preparations	04 Cereals and cereal preps	05 Fruit and vegetables	07 Coffee, tea, cocoa and spices	1 Beverages and tobacco	2 Crude materials (inedible)	24 Cork and wood	28 Metalliferous ores and scrap	3 Mineral fuels, lubricants etc	33 Petrol and petrol products	4 Animal and veg. oils and fats	5 Chemicals	51 Organic chemicals	52 Inorganic chemicals	53 Dyeing and tanning materials
West Germany	13,169.4	12.7	354.2	85.7	50.5	46	39.1	83.2	283.8	2.4	81.2	1,325.6	1,307.4	11.9	1,519.2	417.1	126.3	141.5
USA	12,998.5	12.5	155.5	0.4	15.3	7.6	21.5	397	85.2	0.7	18.1	1,632.2	1,621.1	1.9	1,076.2	339	137.9	71.3
France	10,885.8	10.5	749	285.6	53.7	14.9	35.3	224.7	159.8	1.5	29.5	938.1	865.9	5.9	1,411.5	394.7	73.5	107.1
Netherlands	7,516.6	7.2	383.7	36.4	90	26.8	11.9	188	137	1.3	46.5	1,392.6	1,353.8	11	1,105.1	445.6	46.4	60.1
Belgium-Lux.	5,648.6	5.4	254.9	35	95	7.8	6.6	102.4	240.1	1.5	100.4	218.7	199.3	3.5	854.5	375.7	67.3	37.4
Italy	5,612.8	5.4	227	33.8	92.4	5.1	11.6	85	260.3	1	65.4	250.8	250.4	2.1	789.2	215.1	34.2	95.4
Irish Rep.	5,311.5	5.1	599.7	50.1	127.9	69	66.1	71.6	88.9	9.4	17.2	460.3	431.2	22.4	664.6	54	41.2	44.3
Spain	3,750.1	3.6	224.1	28.8	57.7	12.7	12.7	149.2	193.2	1.6	113.8	161.9	150.6	2.2	429.7	124.3	57.2	37.4
Sweden	2,712.8	2.6	53.8	0.4	10	7	12.9	21.5	55.8	2.5	13.3	281.5	263.4	1.4	346.1	34.5	45.2	33.9
Japan	2,631.3	2.5	76	0.9	25.1	4.4	8.5	201.4	45.2	0	13.1	2.9	2.8	0.8	372.3	94.9	30.1	36.3
Switzerland	2,358.5	2.3	33.8	4.9	2.5	2	5.1	21.3	23.7	0.3	1.1	2.3	2.3	0.8	326.9	92.3	21.7	74.2
Saudi Arabia	2,012.6	1.9	86.7	1	20.7	2.4	29.2	51.9	8.2	0.1	4.1	2.3	1.7	1.6	225.6	7.4	3.8	18.7
Canada	1,901.9	1.8	71.2	0.1	9.3	4.1	43.9	45.8	24.7	0.1	15.2	512.1	511.9	3.7	179.3	46.3	10.7	11.3
Australia	1,645.6	1.6	31.3	0.1	4.3	2.7	13.1	50	17.8	0.1	2.4	1.9	1.7	0.5	243.5	34.8	23	12.8
Denmark	1,413.7	1.4	54.2	6.7	12.7	3.3	5.1	16.7	18.1	0.2	0.5	83.4	70.8	1.5	196.3	11.8	14.5	18.5
Norway	1,289.8	1.2	47.6	0.4	12.8	2.3	6.3	6.2	18.2	0.3	1	126.7	94.1	0.4	128.5	17.5	8.4	18.1
India	1,264.2	1.2	0.9	0.1	0	0	0.1	7.5	49	0.3	34.6	21.6	21.2	0.2	73.8	27.1	3.8	11.5
Hong Kong	1,238	1.2	33.9	4	4.4	2.2	9	121.7	8.2	0	0.4	2.1	1.3	0.1	165.1	12.1	27.3	44.3
South Africa	1,113.4	1.1	19.7	2.4	10.7	0.9	0.7	50.1	22	0	1.1	4.9	2.4	0.5	204.4	56	13.8	15.9
Finland	1,041.7	1	20.3	0.3	3.1	1.3	5.2	9.3	64.7	0.6	4.8	105.2	91.5	0.7	152.2	13.2	22.2	10.6
Singapore	1,040.2	1	15.7	1.9	2.1	1.3	2.9	77.1	6.8	0	3.6	7.7	7.6	0.4	146.5	33.1	12.1	17.8
Portugal	1,033.3	1	28.5	2.2	4.6	4.5	4.6	36.7	28.7	1.6	3.6	69.4	39.7	0.4	134.9	25.1	5	21.5
Austria	705.9	0.7	8.4	0.8	1.4	0.5	1.8	3.9	11.8	0.2	1.3	0.6	0.5	0.1	97.2	12.7	9.8	9.4
Greece	682.9	0.7	34.5	1.5	15.5	1.8	3.5	87.3	13.5	0.1	0.6	4.4	4.2	0.3	103.6	11.5	2.7	21.4
World Total	103,911	100	4,324.6	609.2	1,045.5	263.7	438.7	2,770.2	2,162.6	27.7	633.5	7,801.4	7,477.6	87.7	13,182.5	3,351.1	952.2	1,193.5
Percentage of world total	100		4.2	0.6	1	0.3	0.4	2.7	2.1	0	0.6	7.5	7.2	0.1	12.7	3.2	0.9	1.1

0.0 means zero or less than £50,000.

IMPORTS £ million
from (countries in order of total trade)

	54 Medicinal and pharm. prods	57 Plastics - primary	58 Plastics - non-primary	6 Manufactured goods	64 Paper, board and pulp	65 Textile yarns and fabrics	67 Iron and steel	68 Non-ferrous metals	7 Machinery and transport equip.	71 Power generating mach.	75 Office mach. and computers	76 Telecomm. and sound equip.	77 Electrical mach. and equip.	78 Road vehicles	8 Miscellaneous manufactures	82 Furniture	84 Clothing	85 Footwear
West Germany	235.4	488	285.6	**3469**	670.4	582	669.3	445.5	**10,784.2**	725.6	815.9	328.4	1,144.2	5,104.5	**2,217.8**	214.3	360.3	43.8
USA	1.7	152.5	79.7	**1,171.6**	153.4	193.1	55.9	169.9	**8,000.7**	900.1	1,871.2	396.8	1,185.9	221.7	**2,564.1**	43.4	58.8	17.4
France	146.2	318.5	83.9	**1,791.4**	339.2	313.6	280.6	222.9	**4832**	314.3	473.5	189.1	645.5	1,973.3	**1,197.5**	99.4	163.7	37
Netherlands	105	444.1	91.2	**1,470.2**	288.5	250.7	291.7	263.9	**3,304.3**	118.6	1,042.7	100.5	414.6	980.9	**985.7**	68.9	119.9	16.2
Japan	17.4	25.9	17.3	**442.5**	15.7	118	80.1	28	**5,126.5**	244	939.6	948.1	876.5	1,467.2	**789**	2.6	11.3	0.4
Italy	41.5	89	95.6	**1,269.1**	78.5	423.8	153.8	56.1	**2,744.7**	175.5	538.3	51.2	533.4	506.9	**1,566.1**	218.7	340.3	308.5
Belgium-Lux.	67.6	356.6	132.4	**1,632.9**	115.8	430.1	307.3	178.7	**1724**	72.2	133.4	139.8	159.8	759	**501.2**	38	90.4	4.7
Irish Rep.	86.5	17.3	28.4	**538.6**	64.3	171.6	41.8	20.9	**1,231.5**	35.8	672.2	42.7	235.5	66.3	**646.6**	44.3	162.7	17.1
Switzerland	119.6	26.3	27.2	**2,066.1**	36.4	118.1	13.8	49.7	**684.8**	47.7	34.3	27.3	149.8	9.2	**905.5**	8	20.9	4.9
Norway	8.1	46.8	10.7	**589.1**	178.8	6.9	92.8	253.5	**221.1**	24.6	19	16.8	16.7	10.3	**96.3**	4.7	2.6	0.1
Sweden	74.5	60.4	29.2	**1,264.9**	784.6	22.5	194	92.4	**1,273.6**	100.1	94.1	124.6	113.1	452.6	**276.8**	51.5	17.4	0.7
Spain	14.6	36.7	6.8	**521.3**	36.1	73.4	137	29.2	**1,192.1**	122.9	68.3	24.4	78.9	598.3	**322.5**	21.9	12.1	128.1
Denmark	51.4	3.8	40.1	**263.9**	37.4	63	28.7	14	**478.6**	26.7	38.5	51.6	67.9	40.7	**362.4**	88.1	46.7	7.8
Canada	12	8	2.6	**431.2**	173.1	25.9	17.9	128.3	**609.4**	143	60.5	52.3	42.4	12.7	**113**	9.2	2.9	1.1
Hong Kong	1.4	0.4	0.7	**148.3**	10	57.4	0.8	2	**464.5**	10.9	79.1	189.4	143.3	18.7	**1,305.3**	4.8	693.3	16.9
Finland	6.3	16.1	13.2	**1,017.4**	787.8	9.6	53	81.5	**221.6**	12.4	46.8	38.6	32.9	4.6	**67.4**	4.9	16.4	0.4
Taiwan	0.1	5.4	3.6	**177.4**	4.6	40.4	0.9	0.5	**564.5**	4.7	232.9	85.8	106.3	59.4	**431.1**	28.7	70.4	71.9
Portugal	1.5	10.7	2.1	**338.9**	12.9	171.9	6.5	1.1	**192.1**	22.3	4.3	7.1	48.6	68.9	**370.2**	3.8	233.1	101.8
South Africa	0.5	1.2	0.1	**492.4**	14.6	20.2	22.6	399.2	**57.1**	17.6	7.7	1.4	6.3	5.9	**48.6**	11.4	11.7	0.7
Australia	9.3	0.1	3.2	**187**	0.2	6	1	165.8	**289.2**	43.3	19.8	14	14.1	8.9	**77.7**	3.1	3.1	0.1
Singapore	0.2	0.9	0.7	**47.1**	1.9	6.4	0.6	0.6	**639.9**	17.7	275.2	174.6	130.8	7.1	**238.2**	10.5	64	2.8
South Korea	0.2	4.4	2.8	**111.3**	2.8	47.7	7.7	1	**368.9**	5.2	99	117.6	96.1	16.5	**442.4**	0.7	127.8	134.5
Austria	5.9	22.4	30.2	**385.4**	110.6	85.6	51.9	23.8	**306.9**	53.1	6.7	48.4	88.3	15	**149.4**	7.2	45	2.1
USSR	0	0.1	0.1	**85.4**	2.4	0.6	9.7	26.6	**54.9**	1.2	1.1	1.5	6.5	36.5	**30.1**	1.7	0.1	0
World Total	1,157.8	2,212.5	1,014.9	**21,899**	4,016.6	3,936.2	2,676.7	3,003.6	**47,289.6**	3,518.4	7,714.6	3,486.4	6,924.2	12,586.8	**18,252.1**	1,112	3,905.9	1,169.1
% of world total	0.9	1.8	0.8	**17.4**	3.2	3.1	2.1	2.4	**37.5**	2.8	6.1	2.8	5.5	10	**14.5**	0.9	3.1	0.9

EXPORTS £ million
to (countries in order of total trade)

	54 Medicinal and pharm. prods	57 Plastics - primary	58 Plastics - non-primary	6 Manufactured goods	64 Paper, board and pulp	65 Textile yarns and fabrics	67 Iron and steel	68 Non-ferrous metals	7 Machinery and transport equip.	71 Power generating mach.	75 Office mach. and computers	76 Telecomm. and sound equip.	77 Electrical mach. and equip.	78 Road vehicles	8 Miscellaneous manufactures	82 Furniture	84 Clothing	85 Footwear
West Germany	175	205.6	100.2	**2,084.2**	208.6	300.6	512.2	440.1	**5,940**	618.2	1,262	522.8	903.1	1,118.7	**1,462.1**	79.1	222.1	22.9
USA	207.2	75.4	54.3	**1,460.6**	92.8	180.6	251.3	239.2	**6,110.9**	1,435	634.7	191.9	677.2	863.9	**1,975.1**	89.7	127.9	25.1
France	236.4	147.5	103.8	**1,420.4**	221	260.1	264	208.6	**4,622.5**	313.3	862.9	290.9	585	1,150.5	**1,266.6**	56.7	165.4	26.2
Netherlands	175.7	129.3	64.4	**912.1**	116.6	192.2	171.2	119.9	**2,507.5**	239.6	874.7	158.6	260.5	298.5	**822.9**	53.4	96.5	14.3
Belgium-Lux.	65.3	111.2	35	**1,690.9**	108.8	146.6	165.7	107.8	**1,892**	320.2	170.8	66.8	171.8	744.5	**373.5**	26.2	67.6	4.9
Italy	136.7	108.6	44.8	**842.2**	51.3	170.5	210.9	154.5	**2,504.3**	165.7	564.4	225.2	406.2	481.4	**627.6**	20.2	84.8	14.1
Irish Rep.	154.2	67.7	53.5	**981.2**	198.9	175.1	146.1	95	**1,401.2**	61.7	262.3	87.1	318.4	271.6	**929.3**	54.8	276.3	67.5
Spain	43.1	38.5	25.2	**448.1**	52.8	59.6	146.4	32.4	**1,747.3**	157.7	245.8	134.1	163.1	476.9	**380.6**	11.6	65.3	7.4
Sweden	65.4	46.3	36.6	**467.4**	38.8	68.4	96.2	72.3	**1,060**	83.2	162.1	74.5	194.5	145.2	**407.8**	21.5	104	4.9
Japan	112.8	17.9	9.9	**404.6**	9.3	105.7	9.1	162	**727.1**	59.4	51.6	34.7	139.7	237.2	**773.1**	14.8	86.3	9.2
Switzerland	40.3	22.4	20.9	**452.7**	21.1	40.7	73.8	159.2	**720.4**	61.1	199.3	55.2	105.5	97.6	**748.2**	10.5	44.2	3
Saudi Arabia	74.4	11.6	5.8	**101.9**	7	22.6	17.7	16.1	**1,099.5**	80.5	15	47.3	45	26.5	**126.2**	4.8	17.9	1.9
Canada	54.8	10.7	13.1	**235.3**	27	36.7	51.3	18	**588.4**	187.1	35.6	34.3	59.5	46.3	**215.5**	6.9	21.8	6.2
Australia	63.8	26.4	20	**202.1**	32.2	44.7	20.5	14.6	**763**	108.6	73.8	42.7	79.2	61.3	**293.2**	10.5	7.6	3.2
Denmark	29	20.7	22.6	**254.7**	26.6	57.2	80.3	19.2	**573**	35.4	156.3	56.7	72.3	38.3	**195.4**	8.9	36.1	2.3
Norway	16.2	15.4	10.1	**197.7**	15.3	33.1	60.9	21.5	**490.3**	51.8	59.4	26.4	61.1	22.4	**184.6**	6.2	38.3	3.4
India	6.2	4.1	0.9	**486.7**	11.1	3.8	42	12.1	**503.3**	76.9	12.1	14.4	44	9	**85.6**	0.1	0.7	0.4
Hong Kong	30.7	11.5	7	**222.7**	22.3	42.2	60.7	14.4	**474.3**	84.5	27.5	34.1	141.5	53.6	**192.4**	7.2	25.1	7.6
South Africa	25.3	29.5	10.8	**154.7**	25.2	20.7	23.7	6.4	**520.4**	68.7	68.1	24.6	85.3	71.8	**123**	0.9	2.9	1.5
Finland	23.8	17.7	11.2	**183.7**	26.4	35.9	40.7	16.2	**377.9**	24.8	74.2	24.9	59.6	49.8	**120.5**	5.5	16.1	1
Singapore	24.8	11.5	4.5	**123.6**	14.8	17.3	28.4	14.9	**490.8**	62.6	51.8	29.6	125.2	56.7	**154.9**	1.8	5.9	1.2
Portugal	26.2	13.1	6.7	**162.1**	9.3	61.6	36.9	9.4	**480.3**	46.5	44.9	42	39.8	172.6	**86.7**	2.1	16.5	2.2
Austria	23.6	11	6.4	**122.8**	11.6	29	14.2	28.3	**348.4**	12.8	76.2	35.5	77.5	49	**108.3**	2.9	19.2	2.6
Greece	23.8	8.3	7.2	**147.9**	7.3	21.8	71.5	6.5	**198.8**	16.8	18.3	19.8	20.5	53.5	**88.7**	1.5	15.8	1.4
World Total	2,258.1	1,342.4	781.7	**15,821**	1,539.4	2,447	3,036	2,193.2	**42,155.4**	5,251.4	6,341.2	2,687.5	5,648.5	7,300.6	**13,347**	533.2	1,699.4	274.4
% of world total	2.2	1.3	0.8	**15.2**	1.5	2.4	2.9	2.1	**40.6**	5.1	6.1	2.6	5.4	7	**12.8**	0.5	1.6	0.3

United Kingdom balance of payments

Credits
£ million

	1970	1975	1980	1985	1988	1989	1990
Services	**3,444**	**7,857**	**15,787**	**24,217**	**27,488**	**30,402**	**32,096**
Government	51	139	315	483	551	449	432
Sea transport	1,357	2,651	3,789	3,211	3,526	3,870	3,847
Civil aviation	316	780	2,210	3,078	3,192	3,758	4,358
UK airlines	234	560	1,255	2,048	2,307	2,742	3,124
Overseas airlines	82	220	955	1,030	885	1,016	1,234
Travel	432	1,218	2,961	5,442	6,184	6,945	7,784
Holiday	340	951	2,226	4,149	4,332	4,913	5,592
Business	92	267	735	1,293	1,852	2,032	2,192
Financial services	439	1,025	6,192	12,083	14,035	15,380	15,649
Interest, profits & divs.	1,452	2,841	23,681	52,269	56,723	74,170	81,287
Total invisibles	**5,126**	**11,457**	**41,041**	**80,022**	**88,041**	**108,465**	**117,350**
Goods							
(Visible Exports f.o.b.)	8,063	19,330	47,149	77,991	80,346	92,389	102,038
Food and beverages	514	1,388	3,233	4,937	5,450	6,466	6,990
Basic materials	271	556	1,495	2,199	2,124	2,349	2,248
Fuels	206	827	6,414	16,777	6,213	6,129	7,775
Semi-manufactures	2,776	5,851	13,867	19,921	23,984	26,763	28,875
Manufactures	4,032	9,987	21,010	32,319	40,620	48,378	53,879
Other goods	261	721	1,130	1,838	1,955	2,304	2,271
Total Credits	**13,276**	**30,787**	**88,190**	**158,013**	**168,387**	**200,854**	**219,388**

Debits
£ million

	1970	1975	1980	1985	1988	1989	1990
Services	**2,963**	**6,342**	**11,520**	**17,530**	**22,914**	**25,717**	**26,869**
Government	360	709	1,165	1,781	2,351	2,698	2,753
Sea transport	1,437	2,562	3,739	3,508	3,566	3,737	3,591
Civil aviation	270	675	1,863	2,877	4,097	4,298	4,674
UK airlines	146	343	905	1,399	1,500	1,679	1,885
Overseas airlines	124	332	958	1,478	2,597	2,619	2,789
Travel	382	917	2,738	4,871	8,216	9,357	9,916
Holiday	315	752	2,217	4,952	6,768	7,741	8,061
Business	67	165	521	1,131	1,448	1,616	1,855
Interest, profits & divs.	898	2,068	23,865	49,624	51,676	70,083	77,258
Total invisibles	**4,269**	**9,637**	**39,555**	**73,799**	**81,937**	**104,271**	**113,055**
Goods							
(Visible Imports f.o.b.)	9,051	22,663	45,792	81,336	101,970	116,987	120,713
Food and beverages	2,052	4,089	5,515	8,660	9,992	10,769	11,610
Basic materials	1,364	1,967	3,505	5,041	5,470	5,936	5,528
Fuels	946	3,912	6,552	10,271	4,691	6,098	7,388
Semi-manufactures	2,510	5,355	12,532	20,338	28,043	31,050	31,565
Manufactures	2,070	6,746	16,917	35,667	51,998	61,284	62,629
Other goods	110	594	771	1,359	1,776	1,850	1,993
Total Debits	**12,453**	**32,300**	**85,347**	**155,135**	**183,907**	**221,258**	**233,768**

Balances
£ million

	1970	1975	1980	1985	1988	1989	1990
Services	**481**	**1,515**	**3,653**	**6,687**	**4,574**	**4,685**	**5,201**
Government	-309	-570	-850	-1,298	-1,800	-2,249	-2,321
Sea transport	-80	89	50	-297	-40	133	256
Civil aviation	46	105	347	201	-905	-540	-316
Travel	50	301	223	571	-2,032	-2,412	-2,132
Interest, profits & divs.	554	773	-182	2,646	5,047	4,088	4,029
Invisible balance	**+857**	**+1820**	**1,487**	**6,222**	**6,103**	**4,195**	**4,295**
Goods							
(Visible balance)	-988	-3333	1,357	-3,345	-21,624	-24,598	-18,675
Food and beverages	-1538	2701	-2,282	-3,723	-4,542	-4,303	-4,620
Basic materials	-1093	-1411	-2,010	-2,842	-3,346	-3,587	-3,280
Fuels	-740	-3,085	-138	6,506	1,522	31	387
Semi-manufactures	266	-3085	1,335	-417	-4,059	-4,287	-2,690
Manufactures	1,962	496	4,093	-3,348	-11,378	-12,906	-8,750
Other goods	151	127	359	479	179	454	278
Current balance	**-131**	**-1513**	**2,843**	**2,878**	**-15,520**	**-20,404**	**-14,380**

Printed and Bound in Great Britain by
Hartnolls Limited, Bodmin, Cornwall.